*Evolutionary History of the Marsupials and an Analysis of Osteological Characters* has a dual aim. Its overriding concern is the critical assessment of the extant and fossil morphological evidence against which phylogenetic hypotheses of marsupial phylogeny and taxa should be tested. In addition to a reexamination of published character complexes and the proposed taxa, the author presents detailed analyses of new fossils and osteological information previously not considered in phylogenetic studies of marsupials. All the higher taxa of metatherians are diagnosed and discussed in detail in the book, and a final chapter presents an analysis of the evolution and paleobiogeography of the marsupials.

In addition to the primary aim of the book, which is to arrive at an interim, but most probable rather than parsimonious, phylogeny and biogeography of the marsupials, emphases are placed both on the methods of phylogeny reconstruction and the special significance of the skeleton. It is argued that the use of osteological properties, considering all the available information that these contain, such as their functional–adaptive aspects, is critical for plausible phylogenetics. The author argues that phylogenetic analyses of groups require more than cladistics, but that any phylogeny must be based on noncircular analyses of character transformations to reveal their contingent and unique attributes. The temporal data and the morphology are provided by the fossils, and there should also be a proper consideration of ancestry, in addition to sister-group relationships.

This book is the only single-authored, comprehensive volume on the history of marsupials that deals with both living and extinct groups. As such, it will interest all paleontologists and biologists concerned with mammalian evolution.

T0192538

# Evolutionary history
# of the marsupials and an analysis
# of osteological characters

*Heavily burdened didelphid mother (Didelphis virginiana) with weaned and variously attached young (Courtesy of the American Association for the Advancement of Science; with permission)*

# Evolutionary history
# of the marsupials and an analysis
# of osteological characters

FREDERICK S. SZALAY

*Hunter College, City University of New York*

CAMBRIDGE
UNIVERSITY PRESS

CAMBRIDGE UNIVERSITY PRESS
Cambridge, New York, Melbourne, Madrid, Cape Town, Singapore, São Paulo

Cambridge University Press
The Edinburgh Building, Cambridge CB2 2RU, UK

Published in the United States of America by Cambridge University Press, New York

www.cambridge.org
Information on this title: www.cambridge.org/9780521441698

First published 1994
This digitally printed first paperback version 2006

*A catalogue record for this publication is available from the British Library*

*Library of Congress Cataloguing in Publication data*
Szalay, Frederick S.
    Evolutionary history of the marsupials and an analysis of
osteological characters / Frederick S. Szalay.
        p.    cm.
    Includes bibliographical references (p.   ) and index.
    ISBN 0-521-44169-2 (hc)
    1. Marsupialia – Phylogeny.   2. Marsupialia – Morphology.
    3. Marsupialia – Classification.   4. Marsupialia, Fossil.   I. Title.
    QL737.M3S93   1994
    599.2 – dc20                              94-109
                                               CIP

ISBN-13  978-0-521-44169-8 hardback
ISBN-10  0-521-44169-2 hardback

ISBN-13  978-0-521-02592-8 paperback
ISBN-10  0-521-02592-3 paperback

To **W. D. Matthew** and **G. G. Simpson** *for inspiring so many,*
*and*
To **W. K. Gregory** and **W. J. Bock** *for their Darwinian pursuit of connections among form, function, and adaptation in order to discover and understand diversity and descent.*

# Contents

# Preface

The roots of this book reach back to my graduate student years when I was quite proud of myself for the discovery of what I mistakenly believed was the first Asiatic marsupial among the undescribed treasures of the American Museum of Natural History. That it turned out to be a didymoconid placental is immaterial, because after that all fossil marsupials had a special meaning for me for a complex of reasons. Simply put, their morphological diversity, constrained so differently from what one was accustomed to in placentals, coupled with the many unknown aspects of their history, had a great appeal. It seemed that fossil marsupials could provide one with additional perspectives on evolutionary morphology beyond what placentals could offer. It was also the subsequent realization in the 1970s that postcranial remains in the fossil collections of major museums were waiting, so to speak, to be incorporated into a synthesis of metatherian evolution.

More than any other support, a Fellowship from the John Simon Guggenheim Foundation enabled my family and me to spend time in South America where I could study fossil marsupials, and a sabbatical year in 1980 perusing the extinct and extant marsupials in Australia. A number of grants from the City University of New York have provided support for the countless tasks that seem to be without end when a larger project is brought to fruition. James Warren, at that time the chair of the Department of Zoology of Monash University, provided me with excellent facilities during my studies in Australia. Many curators in many museums have been especially helpful by providing access to their collections. The paucity of fully skeletonized specimens of recent mammals often necessitated the extrication of pedal skeletons from study skins, which was not only a chore of labor, but often I had the difficult task to convince some rather conservative curators of mammals of its importance. Besides the museum specimens, a large (and initially smelly) collection of Australian species as well as New Jersey *Didelphis* road kills are to be honored for their corporeal contribution to this work. Of my colleagues and mentors who were especially helpful in obtaining either important fossil specimens or who provided their trust and help, I especially single out from many W. J. Bock, D. de A. Campos, R. C. Fox, C. F. Howell, W. P. Luckett, D. E. Savage, and G. G. Simpson. Without their help, much of this work would not have been possible. For their warm hospitality, stimulating discussions, their contributions to marsupial evolution, and sharing their knowledge of Australian natural history, I am especially grateful to Michael Archer, Tim Flannery, Peter Murray, Pat and Tom Rich, and Rod Wells. Most of the drawn figures were prepared by Anita Cleary and Patricia Van Tassel. I am grateful for the critical reading of parts of this book by W. J. Bock and R. L. Cifelli.

Finally I thank my colleagues in the Department of Anthropology at Hunter College of the City University in New York, students of humans, who have tolerated among them a student of marsupials and other mammals.

Most of all I thank my wife Jeanne for her love and understanding which continue to sustain me.

# 1

# *Introduction*

*An Opossom hath a head like a Swine, & a taile like a Rat, and is of the Bignes of a Cat. Under her belly shee hath a bagge, wherein shee lodgeth, carrieth, and sucketh her young.*

Smith (1612, p. 14)

*What is history all about if not the exquisite delight of knowing the details and not only the abstract patterns?*

Gould (1990, p. 17)

This book provides a historical–narrative explanation for a number of osteological features of all groups of marsupials, tests a number of group hypotheses against these, and based on the tests, offers what at present appears to be the most plausible phylogeny, an evolutionary history of these animals, albeit in many ways an inadaquate one. Why my aim is not a "genealogy" but rather a phylogeny, or evolutionary history, is explained more fully below. Any truncated abstractions of phylogeny, as are the cladogram summaries of taxa, should be based on, and in fact readily fall out of, the analysis and testing of numerous interdependent paleontological, historical, and functional–adaptive problems. The scientific validity of phylogenetic reconstruction, and of course any level of confidence in it, depends on such studies (Szalay & Bock, 1991).

There are a number of outstanding general accounts of marsupial biology and evolution (Archer, 1981, 1984a,b; Tyndale-Biscoe, 1973, and others), and books especially dedicated to development, organ systems, and molecules (Tyndale-Biscoe & Renfree, 1987; Tyndale-Biscoe & Janssens, 1988), as well as important scholarly compilations (Cabrera, 1919; Clemens & Marshall, 1976; Marshall, 1981c). The volume edited by Archer and Clayton (1984) is a unique and extremely valuable tome, a veritable cornucopia, in which the evolutionary history of vertebrates in Australia, including of course the marsupials, is treated encyclopedically. The primary journal literature and other edited volumes (see especially Archer, 1987; Collins, 1973; Hunsaker, 1977a; Smith & Hume, 1984; Stonehouse & Gilmore, 1977) present and review the growing number of theories and provide new empirical information about marsupials. There is no single volume, however, in which one student addresses problems of evolutionary (phylogenetic) relationships of character systems of the hard anatomy and of the taxa.

Hard tissues are of particularly special significance in taxonomy as they ensure a continuity between the living and the extinct forms. Studies on the taxonomic diversity and the phylogenetic relationships of marsupials of the taxon Metatheria, like the analysis of parallel problems for other fossil mammals, has been, and continues to be, based primarily on hard parts, particularly on dental and cranial criteria, and, increasingly, on postcranial characters as well. Much of the evidence of the mammalian fossil record is dental, and therefore the judgments regarding taxonomic diversity and relationships are often based on morphological differences that are dental. This is an understandable and often necessary constraint on taxonomy. Yet there are no a priori reasons why an animal's dental or cranial components should be more (or less) likely to reflect their evolutionary divergence or phylogenetic relationships, or do so more accurately, than any other area of the phenotype. I believe, therefore, that it is important to examine and amplify theories on phylogenetic relationships that are based primarily on dental, cranial, and biomolecular taxonomic properties. This can be done by thoroughly examining the already-cited rich sources of data, as well as other systems that are well represented in the fossil record, and that have been *understood*, at least to some degree, in living forms.

While I aim to present a new synthesis of marsupial evolutionary history, I shall also try to reinforce what should be obvious: the relevence of biologically and paleontologically oriented analyses of characters to phylogenetic analysis. The latter is expressed, but never fully, in a classification of higher taxa. In addition to introducing a useful series of characters – those from the lower leg and the foot (Table 1.1) – I shall present an analytical survey of the still poorly studied [from a fully functional-adaptive and historical, (i.e., evolutionary) perspective] cranioskeletal system. While a great deal of information on the soft anatomy and molecular characteristics is highly relevant, data on the soft anatomy are used sparingly as fossils obviously lack such tissues. These other areas need to be reviewed by those better qualified dealing with the complex assumptions that lie at the base of such data accumulation (see papers in Szalay, Novacek & McKenna, 1993a, b). The analysis of character complexes, the great diversity of genetic and molecular evidence included, should be always carried out on their own merit. One should omit the frequent intrusions into evolutionary decision making the influence from the transformational (either ontogenetic or phyletic) understanding of other character clines.

For the past twenty five years I have been surveying and gathering information on mammalian osteological diversity in several

Table 1.1. *Abbreviations used in text and on figures for anatomical and functional patterns of selected areas of postcranium.*

| | |
|---|---|
| | Bones, joints, and joint facets |
| ACJ | astragalocuboid joint |
| ACu | astragalocuboid |
| AFi | astragalofibular |
| AN | astragalonavicular |
| ANJ | astragalonavicular joint |
| ANl | lateral astraglonavicular |
| As | astragalus |
| ATi | astragalotibial |
| ATia | anterior astragalotibial |
| ATid | distal astragalotibial |
| ATil | lateral astragalotibial |
| ATim | medial astragalotibial |
| ATa | astragalotibiale |
| ATip | posterior astragalotibial |
| Ca | calcaneus |
| CaA | calcaneoastragalar |
| CaAa | auxilliary calcaneoastragalar |
| CaAd | distal calcaneoastragalar |
| CaAs | secondary (macropodid) calcaneoastragalar |
| CaCu | calcaneocuboid |
| CaCua | auxilliary (australidelphian) calcaneocuboid |
| CaCud | distal calcaneocuboid |
| CaCul | lateral calcaneocuboid |
| CaCum | medial calcaneocuboid |
| CaCup | proximal calcaneocuboid |
| CaFi | calcaneofibular |
| CaMt5l | calcaneus-fifth metatarsal ligamentous connection. |
| CCJ | calcaneocuboid joint |
| CLAJP | continuous lower ankle joint pattern |
| CNJ | calcaneonavicular joint |
| Cu | cuboid |
| CuEc | cuboidoectocuneiform |
| CuMt4 | cuboid-fourth metatarsal |
| CuMt5 | cuboid-fifth metatarsal |
| Ec | ectocuneiform |
| EcMc | ectocuneiform–mesocuneiform |
| EcMt3 | ectocuneiform–third metatarsal |
| EMJ | entocuneiform-mesocuneiform joint |
| EMt1 | entocuneiform-first metatarsal |
| EMt1l | lateral **EMt1** |
| EMt1m | medial **EMt1** |
| EMt1J | entocuneiform–first metatarsal joint |
| EMt2J | entocuneiform–second metatarsal joint |
| En | entocuneiform |
| EnMc | entocuneiform-mesocuneiform |
| EMt2 | entocuneiform-second matatarsal |
| EnPh | entocuneiform–prehallux |
| EPJ | entocuneiform–prehallux joint |
| Fe | femur |
| FFJ | femorofibular joint |
| Fi | fibula |
| FTJ | femorotibial joint |
| LAJ | lower ankle joint |

Table 1.1. (cont.)

| | Bones, joints, and joint facets |
|---|---|
| Mc | mesocuneiform |
| Mcp | metacarpal |
| MMt2 | mesocuneiform–second metatarsal |
| Mt | metatarsal |
| Na | navicular |
| NaCu | naviculocuboid |
| NCJ | naviculocuboid joint |
| NECJ | naviculoectocuneiform joint |
| NEc | naviculoectocuneiform |
| NEJ | naviculoentocuneiform joint |
| NEn | naviculoentocuneiform |
| NMc | naviculomesocuneiform |
| NMJ | naviculomesocuneiform joint |
| Ph | prehallux |
| SLAJP | separate lower ankle joint pattern |
| Su | sustentacular |
| Sua | accessory sustentacular |
| Suad | distal accessory sustentacular |
| Sus | superior sustentacular |
| Ta | tibiale |
| TF | tibiofibular |
| TFJ | tibiofibular joint, distal |
| TFpJ | tibiofibular joint, proximal |
| Ti | tibia |
| TiFi | distal tibiofibular |
| TMTJ | tarsometatarsal joint |
| TTJ | transverse tarsal joint |
| UAJ | upper ankle joint |
| UnMcp5J | unciform-Mcp5 joint |

| | Topographical (nonarticular) bony details, ligaments, tendons, and anatomical directions |
|---|---|
| aacl | anterior (or dorsal) astragalocalcaneal ligament |
| ah | astragalar head |
| ac | astragalar canal |
| adt | astragalar distal tuber |
| afl | anterior fibular ligament |
| ampt | astragalar medial plantar tuberosity |
| an | astragalar neck |
| anl | astragalonavicular ligament |
| at | anterior plantar tubercle |
| atl | astragalotibial ligaments |
| ccl | calcaneocuboid ligament |
| cfl | calcaneofibular ligaments |
| cfla | calcaneofibular ligament attachment |
| cflf | calcaneofibular ligament facet |
| cnl | calcaneonavicular (spring) ligament |
| clp | calcaneal lateral process |
| cump | cuboidal medial process |
| cupp | cuboidal proximal process |
| d | distal |
| do | dorsal |
| edbt | extensor digitorum brevis tendon |
| ehlt | extensor hallucis longus tendon |

Table 1.1. *(cont.)*

| Topographical (nonarticular) bony details, ligaments, tendons, and anatomical directions | |
| --- | --- |
| fdbt | common tendon of flexor digitorum brevis |
| fft | flexor fibularis (= flexor hallucis longus) tendon |
| ftt | flexor tibialis (= flexor digitorum longus) tendon |
| ge | groove for extensors |
| gf | groove for flexors |
| gtpl | groove for tendon of peroneus longus |
| jca | joint capsule attachment |
| l | lateral |
| lca | ligamentous contact on astragalus |
| lcc | ligamentous contact on calcaneus |
| lu | lunula |
| m | medial |
| mc | meniscus |
| mtm | medial tibial malleolus |
| nel | naviculoentocuneiform ligament |
| p | proximal |
| pacl | posterior (or plantar) astragalocalcaneal ligament |
| pbt | peroneus brevis tendon |
| pfp | parafibular process |
| plt | peroneus longus tendon |
| pp | peroneal process of calcaneus |
| ppl | process for peroneus longus on **Mt 1** |
| ps | posterior shelf of distal tibia |
| ptt | peroneus tertius tendon |
| ret | retinaculum |
| se | sesamoid |
| sa | sulcus astragali |
| sc | sulcus calcanei |
| syn | syndesmosis |
| tat | tibialis anterior tendon |
| tc | tuber of calcaneus |
| tmtls | tarsometatarsal ligaments (they form the peroneal canal) |
| tnl | tibionavicular ligament |
| tpt | tibialis posterior tendon |
| tst | triceps surae tendon |

*Notes:* The first letter of the abbreviations of specific bones and articular facets are capitalized. Specifically, both broadly and restrictively defined, joints are abbreviated using a combination of the first letters, in capitals, of the names of those units that contribute to the joint and the letter J for joint. Abbreviations entirely in lower case designate landmarks on specific bones, anatomical directions, or ligaments, tendons, or muscles. These abbreviations are listed under two separate headings in order to facilitate retrieval of information. Most of the characters in this table denote given areas of anatomy in terms of specific *homology designations* rather than only topographical descriptive terms.

large taxonomic groups. The perspective of study has been always primarily focused on phylogenetic relationships. Yet efforts at understanding were particularly directed towards becoming increasingly aware of the importance of evolutionary morphology of such areas as the crurotarsal or carpal complexes, or dental and cranial structures in extant species, and the scattered and unidentified fossils of these anatomical areas in fossil collections. The conviction has grown steadily that all areas of hard anatomy (or any other morphological complex) traditionally studied by taxonomists must be appraised along similar lines for causal evolutionary understanding (see particularly Bock, 1981). The overwhelming amount of new and significant information on bones, ligaments, muscles, and particularly joint form-function in the hand and foot which is readily available for biological analysis rivals other systems in its potentially great phylogenetic content. The understanding of this data is unquestionably some of the most important for assessing phylogeny because of its potential connection with the fossil record. In spite of some tired, and I dare say obviously unwarranted, pessimism expressed about morphology by some taxonomists, it cannot be sufficiently emphasized how likely it is that future studies concentrating on the functional-adaptive morphology of the full record of the extant and fossil skeletal systems are likely to be. These undertakings, based on the urgent need to understand the ecomorphology of living species, along with the recovery of new fossils and a far fuller understanding of them along with the known ones, will increasingly shape our view of mammalian evolution.

The continued and far more refined use of the fossil record is dependent not only on many new and needed developmental studies but also on the ecological morphology of living species. Yet this book sorely lacks a more detailed developmental, functional, and adaptational perspective. This current synthesis undoubtedly suffers because of this, as future reassessments will surely reveal. Such studies, relating to the several assumptions which are made and discussed below, will be needed on a number of problems which I touch on. Their execution is beyond the powers of this book. The only excuse I offer is that this study stems from an eagerness and fascination with mammalian evolution in general, with marsupial morphological and ecological diversity and phylogeny in particular, and the deep conviction that the often instructive time constraints of the fossil record and the need for evolutionary morphology in character analysis have been unwisely discarded by many current (and past) pundits of taxonomic theory. If at least I have sown a seed for a few more functional-adaptive studies in the service of phylogenetic understanding, then my efforts have not been wasted.

The analysis in this volume attempts to account for the diversity of cranial, dental, and postcranial traits. The emphasis is on the analysis of cruropedal evidence in living forms and in as many of the fossil American and Australian marsupials as I had the opportunity to study. I also attempted to identify numerous fossil crurotarsal remains from known collections and allocate them to taxa known primarily by teeth, analyzed them briefly in some functionally and adaptationally perhaps significant ways rather than along exhaustive "desriptive" avenues. I have tried to remedy temporarily the lack of modern morphological accounts by many illustrations, discussed a number of transformation hypotheses, and ventured to integrate the most probable ones to provide an interim phylogeny of marsupials in addition to the character phylogenies.

I wish to reemphasize here that a lasting change and increased confidence in any one specific hypothesis about phylogeny in the future will depend equally on new understanding of how the phenotype is coded and unfolds, i.e., developmental process, on newly perceived information derived from new and known fossils, as well as on the numerous, sorely needed, comparative developmental, functional–adaptive, and ecomorphological studies. I believe that the frameworks offered by formalist taxon focused approaches which stress the primacy of distribution analysis of often a large and vastly heterogeneous collection of characters and declare their sole validity, in reality sidestep the theoretical foundations provided by evolutionary theory for taxonomic method. Evolution of organisms which reproduce sexually is the result of both phyletic (patristic) and cladistic (branching) events of lineages of organisms. While strictly cladistic analyses will continue to generate new classifications isomorphic with cladograms, these cannot reflect the actual history of all samples of life, and they are not likely to help a causal understanding, and therefore the testing, of competing phylogenetic hypotheses. General enthusiasm for the current cladistic analysis, and the macroevolutionary theory derived from it, appears to be a consequence of the disillusionment with a messy but real world of probabilities. The latter are part of causal analysis and empirical testing within the framework of historical-narrative explanations, so necessary for phylogenetics. But for any taxonomic practice to be evolutionary, its conceptual methodology must be derived from the full understanding of evolutionary theory based on the known and tested mechanisms of evolutionary change and diversification.

# 2

# Phylogenetics of characters and groups, and the classification of taxa

> But I must explain my meaning more fully. I believe that the arrangement of the groups within each class, in due subordination and relation to the other groups, must be strictly genealogical in order to be natural; but that the amount of differences in the several branches or groups, though allied in the same degree in blood to their common progenitor, may differ greatly, being due to the different degrees of modification which they have undergone; and this is expressed by the forms being ranked under different genera, families, sections, and orders.
>
> Darwin (1859, p. 420; [italics are Darwin's])

> The great objection against the account . . . is that he has not tried to distinguish more clearly between more or less primitive characters and therefore he has been unable to use the characters for tracing phylogeny.
>
> Winge (1941, p. 298; written between 1887–1917)

> Advocates of pure morphology continue to maintain that nothing – absolutely nothing whatsoever – occurs as a legitimate premise in an inference about the history of life on earth, other than shared attributes of individual specimens. Of course they are not consistent nor are they a monolithic faction. But to be consistent they have to make some very strong claims. They must exclude attributes that are predicated, not of organisms but of populations . . . . They must rule out considerations of chorology (distribution in time and space), of function in the sense of how things work, and of such contextual information as niche and habitat. More importantly, the laws of nature are strictly excluded from the evidence . . . . A legitimate historical narrative places individual bodies and events in a sequence such that all the individuals obey the laws that apply to them. In a hypothetico-deductive approach to history, hypothesized events must be consistent with all true laws. But since more than one hypothesized event may be consistent with a given law, the valid form of inference has to be refutational – that is to say, to reject those hypotheses that contradict laws of nature.
>
> Ghiselin (1991, p. 290)

> The wars in systematics have concerned not the substance of science but its expression: having whatever information and methods one wants, what criteria should be used to incorporate the results into a classification? This question lacks an objective answer . . . . Only power can arbitrate genuinely basic questions of taste, and that is why there has been war in systematics.
>
> Van Valen (1989, p. 100)

9

The practice of taxonomy based on a Darwinian adaptational perspective is difficult; in fact it requires the handling of more complex issues than any other approach. It requires a broad and thorough grounding in evolutionary theory, in diverse aspects of natural history and biology of organisms, and in taxonomic theory *based* on the best current understanding of evolutionary and functional biology. Its effective practice demands the stretching of limits of a researcher's ability to apply the many constraints required for effective scientific thinking, but without restraining the necessary search for novel hypotheses that all scientific research eventually requires. Such perspectives bring up formidable, yet welcome difficulties. Those related to the understanding of organisms and their various attributes are particularly welcome, as they pose the legitimate challenges to phylogenetics. Rarely has this activity – the pursuit of phylogenetic understanding – been better expressed (inadvertently, related to other issues of carefully educated human perception) than by the quote that it "is seeing something noticeable which makes you see something you weren't noticing which makes you see something that isn't even visible" (Maclean, 1976, p. 98, quoted by Lombard, 1991, p. 754).

Another kind of perspective, however, is embodied in some of the currently professed versions of phylogenetics that are called various forms of cladistics. Perhaps the most significant issue is the vehemence and depth of anti-Darwinian attitudes, recently reiterated by Nelson (1992), with which such views have been stated. Such reactions appear to be in response to an ever present implicit or explicit challenge of paleontologically and biologically founded neo-Darwinian research program, one that relies heavily on functional–adaptive approaches in addition to the cognizance of distributional evidence. This Darwinian view of taxonomy confronts that of an often abiologically conceived approach to a conflated micro- and macrotaxonomic practice cum theory which axiomatically wants to rid itself of the putative pollution of evolutionary biology of organisms and of neo-Darwinian theory itself without offering any genuinely scientific refutation. Thus, G. Nelson's (1992, p. 146) interesting remarks place some of these big issues of phylogenetics in an appropriate and revealing perspective. He states: "Indeed, one of . . . [cladistics's] greatest merits in my eyes is the fact that it occupies a position of complete and irreconcilable antagonism to *that vigorous and consistent enemy of the highest intellectual, moral and social life of mankind –Darwinism*" [italics supplied]. Commenting on Darwinian theory based phylogenetics, Nelson adds (p. 146): "their attempt will meet with the fate which the Scripture prophesies for all such. . . ."

There is a great and often ignored gap between the logically packaged rules of taxonomic practice based on taxonomic (as opposed to evolutionary) theory and the choice, understanding, and integration of characters of an organism to form reliable taxonomic properties. Often characters of vastly varying perceptions are chosen by taxonomists to serve the needs of axiomatic procedures to cluster rather than to help recognize and solve the critical problems related to the riddles of descent. In light of the extensive and often fiercely partisan literature dealing with methodologies of phylogenetic analysis, therefore, I consider it important that I should attempt to summarize my particular theoretical approach for analyzing evidence on which all views of relationships (of taxa and phyla) rest. I shall also discuss, briefly, the construction of classifications that are based on phylogenetic analysis itself, and examine all those axioms and categorical statements in the literature that are antithetical to the approach advocated here. The evolutionary or phylogenetic method (and to me the two are the same) is an eclectic combination of tracking historical contingencies and their constraints within the framework of lawlike (nomological–deductive) explanations (N-D Es), considering them as adaptive imperatives in a specific context, thus closely appraising the causal constraints of not only developmental dictates but also of all relevant processes. Each level of understanding must be explicated so it may be tested against objective data, which is simply information that others agree with. This methodology is both syncretic and synthetic.

It is fair to say that the aim of all science is the explanations of causality, which in geology and the life sciences reside in the process of understanding mechanisms and their products and their history. Yet one aspect of the more recent general debates in phylogenetics is often focused on the question whether the "chronicle" of phylogeny can be attained without cause and process based explanations. This issue was recently discussed by Szalay and Bock (1991), and it will not be repeated in detail. It is clear however, that "genealogy," a phylogenetic diagram of any sort, is not an explanation. It *may* become one when it is tested against evidence of all sorts, and the corroborations increase confidence in the probability that the pattern-connection of the empirical information that it claims to depict reflects the most probable evolutionary change. Evolutionary change, the base component of phylogeny, occurs in a specific geographical context, and it is reflected in recognizable differences among populations of organisms (Fig. 2.1). These differences are the results of the evolutionary process (Bock, 1979; Stuessy, 1987) that manifests itself in (1) phyletic (patristic, anagenetic) relationships, (2) cladistic (branching) relationships, (3) phenetic difference or distance, and (4) chronistic (temporal) differences (Szalay, 1993c). Phyletic relationships are tested by vertical comparison of characters and taxa, whereas cladistic ties are established by horizontal comparisons (Bock, 1977a). While differences between different organisms can be obviously atemporal, it is one of the seemingly banal but foundational axioms of evolution that these differences developed through time.

*Figure 2.1.* Diagram depicting the theoretical context and empirically ascertainable aspects of a phylogeny that yield data, against which hypotheses of characters, taxonomic properties, and taxa may be tested. A to G represent lineages; T1 and T2 are two time planes; dagger denotes extinction. From such tested hypotheses the most probable theory of descent and branching of organisms are constructed. No confidence is warranted in any phylogenetic hypothesis if the taxonomic properties are weak or untested (see text for discussion).

In a geographic context, the history of lineages (not taxa) should be conceptualized as the phyletic and branching evolution of these through time, usually assessed by the contextually specified similarities and differences, the taxonomic properties based on them (such as transformation series, i.e., polarized characterclines, apomorphies, or synapomorphies). All of the latter are theory laden testable hypotheses, which need to be analyzed using developmental, functional, and adaptive perspectives. These tested hypotheses of taxonomic properties are the evidence against which lineages and the pragmatically delineated species taxa are tested in a temporal and spatial context. Qualitative discrimination (in principle

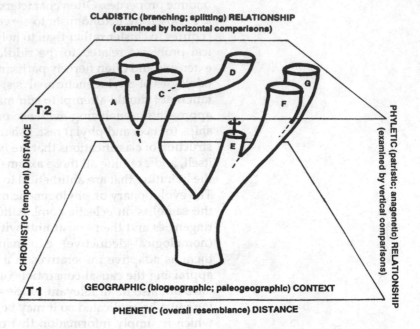

all differences are quantifiable) of the time–space expression of traits is obtained through both horizontal and vertical comparisons of samples during character analysis in a functional–adaptive context. Such procedures,

using the techniques of phenetics (for delineating species taxa), phyletics, and cladistics, establish the relative time of appearance of these traits in organisms. (Modified and redrawn after Stuessy, 1987.)

Phylogenies of organisms become increasingly convincing through a series of complex and interdependently corroborated explanations of the characters. Such explanations, if correct, account for not only the distribution of all the characters but also for the meaning of their biological differences. The practice of science is in the tests and corroborations, in the explanations of the sundry and all relevant evidence that bears on that postulated evolutionary history. As methods properly fall out of theory, the valid taxonomic methods are those that are evolutionary (and not based on "taxonomic theory") (Bock, 1981; Szalay & Bock, 1991). The goals of any such scientific effort are the establishment of the tested reality of the lineages and groups and the phylogenetic ties of these. It is through the study of the traits that a taxonomist delineates a part of the diversity of organisms, and thus formally recognizes groups, the taxa. One does not only delineate diversity of organisms by their characters, but it is also through characters of the organisms, and the *comparative biology* of the proximal mechanisms responsible for them, that one attempts to understand their evolutionary history. This history is the result of causes acting on individual organisms. Bock (1989a, p. 54) has put it succinctly that

"Basically, when considering evolutionary causes, the proper units of evolution are individual organisms. But for evolutionary processes, the units of evolution are populations (for sexually reproducing organisms only)." The causes and processes involved in the unfolding of phylogeny, therefore, are not the splitting of lineages only. It is in fact the dynamics of selection of individuals with specific attributes and the resulting lineage transformation, phyletic evolution, that are at the heart of the formation of any and all new lineages, whether coupled with speciation or not. This view does not in any way deny the fact that diversity *at any one moment in time* in biological history is the result of speciation.

As a phylogeneticist, my efforts to understand evolutionary change of characters and taxa are based on, and attempt to be consistent with, all the constraints that an expanded evolutionary theory (an ongoing modern synthesis of evolution that is a fully Darwinian enterprise accommodating all tempos and modes of change) permits. This framework certainly integrates without conflict, contrary to what is often claimed, a large part of re-emphasized or newly developed evolutionary theory (see particularly accounts in Mayr & Ashlock, 1991), *if* the evidence warrants it in any specific instance. This approach is evolutionary, Darwinian, and of course phylogenetic. It is based on a conviction that phylogenetic change must be understood in all groups, and rigorously tested against objective evidence of all sorts (see also Bock, 1977a, 1981; Cartmill, 1981; Szalay, 1977a; Szalay & Bock, 1991; and references therein). It is very different from such views as those of transformed cladism. Yet contrary to many studied efforts of other cladists to distance themselves from transformed cladistics, the axioms of the latter crisply formulate the quintessential messege of such evolutionary cladists that students find in the well-known works, for example Eldredge and Cracraft (1980), regarding a view about the nature of connection of pattern with process. C. Patterson (1988, p. 65), who along with N. I. Platnick and G. J. Nelson, argues for pattern cladism, admirably explicates in detail his additional assumptions regarding the lack of need for any Darwinian framework, namely that "sampling of genotypes results in an unambiguous concept of relationship, since inferred genealogical relationship and similarity coincide at the level of the genome. Neutralism provides theoretical justification for differences in rate at both phenotypic and genotypic levels." This is an argument for pure phenetics.

The politically driven aspects of the field that have reached the very marrow of taxonomy (national funding, publication access, tenure, etc.) has marred, in my view, the very fabric of scientific discourse in systematics. It has created tremendous additional and unnecessary pressures on the younger generation of scientists who must live in a real world, whatever their theoretical or methodological perspective may be. As Ghiselin (1991, p. 293) stated: "Therefore it is common practice to decide what is true on one set of criteria, while arguing on the basis of quite different criteria in their publications. The reasons for this are intelligible. . . . The reward system is set up so as to reward scientists for being right in the long run; but short term success depends upon getting one's views presented to the scientific community."

The complexity of the connections between the phenotype and ecology of organisms is particularly vexing for some taxonomists, and the appeal of assumptions of wholesale neutralism and a rejection of selectional theory might even be understandable in some instances. Yet uninterpreted lists of characters of numerous recent efforts, based exclusively on parsimony-driven decision-making algorithms, without the integrative input of developmental and functional-adaptive analysis (i.e., bioscientific thinking) within the context of the interpreted data of an existing fossil record, in my view, and those of others, "leads into a maze of cladistic paradoxes for which there is no apparent escape" (Murray, Wells & Plane, 1987, p. 461). All sorts of data, the genetic code, molecules, distance measures, or supramolecular characters are the evidence. All of these have to be compared both vertically and horizontally, and they all have to be interpreted within the context of tested theory as well as within the frame of evolutionary biology (the various nomological-deductive explanations) to be useful for testing phylogenies.

To base the understanding of evolutionary history (phylogeny), as well as evolutionary process, on a framework derived from the entities delineated in taxonomy or taxonomic practice (as this is widely and axiomatically assumed), and not on the numerous causes and processes affecting populations, organisms, and character transformations, is, simply, putting the cart before the horse. While one searches for the antecedent historical context, this activity is constrained and guided by one's understanding of the numerous nomological–deductive explanations of biology. Taxonomy has been profoundly transformed through the influence of the Modern Synthesis (detracting and revisionist analyses, e.g., that of de Queiroz, 1988, notwithstanding), and the accretionary change in evolutionary theory, exemplified by today's amplified and expanded Modern Synthesis, continues to influence the practice of taxonomy. Methods should be the consequences of those evolutionary theories that have been thoroughly tested, and taxonomic practice itself is fundamentally dependent on the advances made in various branches of biology.

It should be noted here that some researchers (e.g., Eldredge & Salthe, 1984, Vrba & Eldredge, 1984; Eldredge & Novacek 1985; and Rieppel 1988, p. 59), while discussing theory, constantly shift between a taxic (holophyletic) and an evolutionary realm, thus making a methodological derivation of taxonomic theory from the potentially testable evolutionary one virtually impossible. While a holistic taxonomic endeavor should deal with the ordering of clustered samples of life through time, such necessary taxonomic practice cannot serve as the theoretical foundation for evolutionary theory. That theory is built on the causal understanding of proximal and distal mechanisms involved in survival and reproductive strategies of individual organisms, and lineage transformation and multiplication. Species taxa are arbitrary (but not unnatural or unreal) segments of seamlessly continuous lineages from the beginnings of "sexually reproducing" organisms (Bock, 1986). This is only slightly different from Van Valen's (1988c, p. 59) usage, in which all taxa are "adaptively modified parts of phylogeny," a designation that often means "modified" only in practice. Because a classification can never reflect the full dimensions of phyletics, cladistics, phenetics, and chronistics (the facets of the evolutionary realities), pragmatic considerations must sober its practice. Van Valen (1988c, p. 59) captures the relationship between phylogeny and taxonomy when he states that: "reality has no relation to discreteness." Monophyletic taxa (i.e., both paraphyla and holophyla) thus are both real and arbitrary at their point of origin and in the case of paraphyla at points where other such taxa are delineated from them. How discrete they may be, however, depends on the *nature of evidence on the lack of continuity* with other taxa, and the scientific acumen of the taxonomist.

Many accounts of the conflicting taxonomic approaches miss the most fundamental aspects of post-Darwinian taxonomy. This practice is forged not only by the reality of evolution, or the existence of highly corroborated processes, but by the recognition that parallelisms and homoplasies exist which, when not explicated, result in "incongruent cladograms," and which cannot be "resolved" without causal research and subsequent a priori weighting. It should become obvious below that neither a classically Simpsonian or Mayrian approach is advocated here for phylogenetics (see especially Szalay & Bock, 1991). Nevertheless evolutionary theory establishes (or should establish), as an absolute given for any taxonomic method derived from it, the arbitrariness of discontinuous samples (taxa) in separate time slices of the same lineage. This is simply a consequence of descent, rather than the key evidence for the essentialism that was pervasive both before and well after Darwin. The current body of corroborated evolutionary theory also prohibits the dependence of phylogenetic history reconstruction on theory founded on taxon-based assumptions. It is not realistic to state that the same or similar histories may be generated by the either "evolutionary theory" or the "pattern"-based approaches to the understanding of diversity [but see Kemp (1985), a functionally oriented student of synapsid evolution, for a different view on the potentials of various taxonomic methods]. In cladistics the concept of "phylogeny" (nested groupings based on horizontal, but not vertical, comparisons) often takes on the appearance (with many outstanding exceptions, however; e. g., MacFadden 1984) of a pre-evolutionary pursuit. In spite of the frequently stated and implied rhetoric, the sublimation of unattempted understanding of characters, their biological history, and the taxa into cladograms "first," as in the advocacy of the primacy of a cladistic "genealogy," will not render these cladistic shortcuts into spare, strong, and true statements. Such procedures will not become mistake proof by attempts to strip the "distribution of synapomorphies" from the required probabilities of process based judgments.

Modern scholarly interest in the evolution of groups, without an equally deep commitment to the understanding of the evolution of their character complexes is a contradictory, and potentially sterile, perspective. If commitments to character evolutionary studies, i.e., character analysis, are made, however, then the "resolution" of cladograms or phenograms (without descent) becomes completely irrelevant to the actual process of understanding character evolution. I strongly suspect that taxon-focused individual efforts (of any one school of taxonomy) without the primacy of (evolutionary) character analysis result in much less enduring contributions than those that test taxon phylogeny hypotheses against the results of in-depth character research.

Bock (1991) and Szalay and Bock (1991) have concluded that beyond an inductive hypothesis of phylogenetic patterns, the confidence in any phylogenetic hypothesis is directly proportional to its testing against objective evidence. These authors propose the following *nonconsecutive* steps for phylogenetic reconstruction, a methodological procedure which is followed here.

1. Formulation of **group hypotheses** (phylogenetic ones about phyla, i.e., relationships of lineages, and classificatory ones about recognized taxa). While pheneticists stated that phyla are unknowable, the tenet of cladistics today is that taxa and phyla must be identical. This is rejected below on the grounds that phyla are nonarbitrary, whereas taxa must be arbitrarily delineated at some point in time (see Figs. 2.1 and 2.2).

2. Testing of group hypotheses against the appropriate **taxonomic properties** of features (see Fig. 2.2A). The conflation of all the interrelated but distinct taxonomic properties into "synapomorphies" is an artificially limiting formalism, which does not help testing in phylogenetics in a complete sense. It only suits the construction of classifications that are claimed to be (but are not) isomorphic with real phylogeny.

3. Formulation of hypotheses about **taxonomic properties** of features. These properties are **homologies, apomorphies, synapomorphies, and transformation series,** etc. (see Fig. 2.2A).

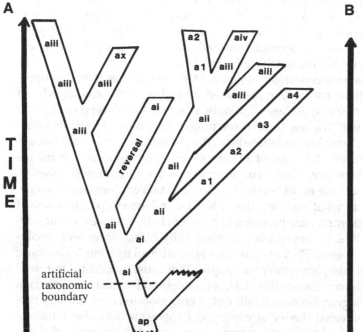

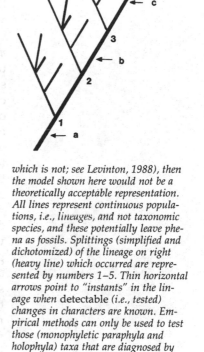

**Figure 2.2.** *Simple models for evolution of character states (A) and lineages (B). A. Theoretical model of character transformation of a homologous trait derived from tested evolutionary, as opposed to* taxonomic, *theory. The various taxonomic properties (homology, transformation series, apomorphy, and synapomorphy) shown with examples are derivable from an understanding of evolutionary transformations and splittings of the ancestral character state "a."*

*Examples of potential interpretations of the character states in the study of this hypothetical organism through time with a single homologous character, a; a with various designated modifiers = homology; ai–aiv, a1–a4, ai–a1, and ai–*

*ax = transformation series; ai = diagnostic apomorphy of ancestor of group delineated by broken line; ap = condition antecedent (primitive) to ancestry of diagnosed group; ai = reversal ("plesiomorphy"); aii, aiii = synapomorphy and parallelism; a2 = parallelism ("synapomorphy").*

*B. Theoretically tested evolutionary model, and not a cladogram, of transformation through time of lineages. Cladograms always depict discontinuous "units" (taxa) of phylogeny, and therefore never, strictly and theoretically speaking, hypotheses of actual phylogeny, which is a continuous process. If punctuated equilibrium as an expression of the underlying mechanisms related to speciation was corroborated in living organisms (but*

*which is not; see Levinton, 1988), then the model shown here would not be a theoretically acceptable representation. All lines represent continuous populations, i.e., lineages, and not taxonomic species, and these potentially leave phena as fossils. Splittings (simplified and dichotomized) of the lineage on right (heavy line) which occurred are represented by numbers 1–5. Thin horizontal arrows point to "instants" in the lineage when detectable (i.e., tested) changes in characters are known. Empirical methods can only be used to test those (monophyletic paraphyla and holophyla) taxa that are diagnosed by the apomorphies of the last common ancestor at these points and not at the furcations that in this model show no character change.*

4. Testing of the hypotheses about taxonomic properties within the context of the relevant nomological–deductive explanations of evolutionary theory against **objective empirical observations.** Neither the strictly parsimony-based algorithms nor distributions of characters on which these are based can substitute for the latter. Ontogenetic character transformation is not a proper model for evolutionary transformation.

Group hypotheses cannot be meaningfully tested until confidence is attained about taxonomic properties. It is obvious, therefore, that steps 3 and 4 (character analysis) must precede step 2. Needless to say, character analysis should be rigorously independent of both the formulation and testing of group hypotheses; otherwise the undertaking is circular.

It has been repeatedly stated within the confines of highly operationalized theory (e.g., repeated by Ridley, 1986, among others) that *patristic* hypotheses are only a "specialized" type of *cladistic* statement! The use of such seemingly (monotonic) logic is unfortunate. Branching relationships, the cladistic ties of organisms, are recognizable because of the necessary assumptions one makes about common *descent* and inheritance from a common ancestry. Descent is the quintessential idea behind all evolutionary change, and it is what patristic (i.e., phyletic) relationships are about. I want to emphasize here that I consider cladistics an excellent *technique*! It is, however, not a method fully in harmony with evolutionary theory. Phylogenetic analysis, with all its complexities and several complementary techniques that must be faithful to tested evolutionary theory (Fig. 1.1), is the method for the reconstruction of phylogeny because it falls out of corroborated evolutionary theory. As tested theory of evolution evolves, so will the details of methodological considerations. Exegeses of the various cladistic techniqes that attempt to recast and remold all aspects of evolutionary theory create a grotesque reflection in a mirror which is far from the best that we have available.

Given this general introduction, the critical issues that are discussed below have to do with the analysis of characters and with the testing of both character, phylogenetic, and taxonomic group hypotheses. These fundamental issues are simple. One may ask: (1) What is primitive, and what is advanced in a contextual spectrum of homologous characters? (2) How do character patterns test and corroborate (or do not) statements of phylogeny and monophyletic classifications (see Figs. 2.1 and 2.2)? Under "**Darwinian Phylogenetic Analysis,**" I will discuss the questions raised in 1, and under "**Trees and Classification**" I will consider question 2.

## Darwinian phylogenetic analysis

In human and vertebrate anatomy, the disciplines that provide the usual foundations for training of vertebrate taxonomists, accounts and explanations are heavily oriented toward a developmental and mechanical (i.e., form–function) explanation of the organism. This is an important and valuable heritage from the main thrust of the anatomical sciences. The morphology (and even functional explanations seemingly independent of taxonomy) utilized in phylogenetics has been greatly influenced by this tradition. Bock (1973, 1988, 1990, p. 262), however, has defined evolutionary morphology as including "all analyses of adaptation of morphological features, how they modify (evolve) with respect to time (excluding, of course, ontogenetic changes of features within the life span of individuals), and the clarification of the past evolutionary history of these features. Evolutionary morphology is a far more inclusive subject than phylogenetic analyses using morphological features (including phylogenetic reconstruction in all of its diverse meanings), and the two should not be confused. . . . Any complete explanation in morphology must be done within the context of evolutionary morphology . . . ."

It is clear to me that a convincing phylogenetic understanding of complex morphological features (i.e., adaptations; hence Darwinian phylogenetics), a connected analysis of **form–function** (mechanics) and their **biorole** in the total external environment (Bock & von Wahlert, 1965) is (or will be) highly relevant for understanding character evolution. The nascent and difficult field of ecomorphology (Goldschmid & Kotrschal, 1989; Bock, 1990) is an area where critical perspectives and information are likely to arise, in addition to the absolutely fundamental paleontological data (for groups which have one) for the increasingly refined character analyses and consequently better tested phylogenies.

The principle of multiple evolutionary pathways of adaptations (Bock, 1959) is a key concept which should be routinely applied. Different, **paradaptive** (Bock, 1963) solutions to fulfill specific feeding and locomotor bioroles are the strongest tests for hypotheses of origins for various functionally tied character complexes, and therefore for taxonomic properties. The fossil record (see especially Cartmill, 1981; Godinot, 1985; Laws & Festovsky, 1987), but without the necessity of a utopian stratophenetic paradigm (e.g., Gingerich, 1979), form–functional correlations, and the external environment related biorole research supply the necessary contexts and boundary conditions for both *generating* and *phylogenetically ordering* new and known evidence. Such an explicit research program, as opposed to mere "character recognition," should be the foundation of modern taxonomic aspirations, the goal of which is the formulation of taxonomic properties and taxa in a phylogenetic context (see also Szalay & Bock, 1991).

The basic data of taxonomy are not only the distribution of features – the latter is the very same data with which all taxonomists must deal. The nature of the characters and the mechanisms responsible for them are the most critical aspect of these data distributed in space and time. Different researchers will recognize, consider, or use different characters. Some taxonomists will consider a trait to be important while others will give it extremely low weight; in all cases functional–adaptive tests (analyses of all sorts and testable arguments) should be offered for a particular choice of character. The manner of character evolution, as tested by all sorts of evidence, should be common in the considerations of all taxonomists.

A simple and general model of evolutionary character transformations is shown in Figure 2.2A. This model shows the character states of a single homology of a rooted group. It is from such an array of well tested homologous characters, shared among many lower rank taxa, that one tries to derive those other taxonomic properties against which the concepts of taxa should be tested (see Bock, 1981; Szalay & Bock, 1991). A brief examination of this figure shows that the identification of such taxonomic properties as apomorphies (whether unique or not), transformation series, or synapomorphies must be sorted out from character reversals, convergences, or parallelisms (="apomorphic tendencies," "inside parallelism," "underlying synapomorphy," and "non-universal derived character state"). While their undoubted homology ensures the establishment of a tested *monophylum* at an arbitrary point where "a" was a nascent feature, any further refinement depends on understanding the transformation series. This testing can be accomplished only with methods that derive from an underlying theory that incorporates diverse developmental and evolutionary causalities, in light of what most probably happened in evolution of the character states.

Sober (1989) in his recent book on phylogenetic evidence all but admits that everything in phylogenetics depends on the understanding of character evolution and the causality and processes that affect them. In spite of this brief vision he nevertheless goes on to concentrate on "logical strength," which he admits has nothing to do with plausability or evidential support. Based on the roots of such scholasticism so well unearthed and explicated by Sober, many taxonomists today dismiss research on character transformation based on ecomorphology as if such research had little to do with the understanding of phylogeny. It is disappointing to read in Sober (1989, p. 11) that "The less we need to know about evolutionary processes to make an inference about pattern, the more confidence we can have in our conclusions . . . ." What does produce character patterns if not the numerous proximal causes and processes of heredity, development, tissue biology (i.e., the causal laws of lineage specific biology), mechanical constraints, or the behavioral and ecological causes dependent on time, context, or environmental change? When can the taxonomist say that all of these operate equally in all of the characters and therefore they are ignorable? The less we know about probable causes of shared traits the less confidence we should have in the phylogenetic usefulness of their subsequently poorly assessed similarities or differences.

It is important to note here that in the character-coding related analysis of Mickevich and Weller (1990, p. 161) these authors state that: "Cladogram construction requires transmodal characters. An initial set of character state trees is needed to construct the first cladogram." They define transmodal characters on p. 168, as: "Character state trees whose transformations are produced by the application of a model of character evolution." While the terminology is questionably complex in that contribution, their stand is a clear recognition that without processual input there are no (meaningful) phylogenetic patterns to be obtained. Transmodal characters, by any other name, are those whose transformational history (historical–narrative explanation, H-N E) has been determined in the specific context of constraints derived from N-D Es from all aspects of evolutionary biology. So Sober's statement above, to paraphrase him, that the less we should know in order to be more confident about a phylogeny is not exactly the proper rallying cry for the future exploration of the frontier of the *specific evolutionary biology of organisms*. The sphere of evolutionary biology, of course, subsumes the phylogenetic history of organisms.

Many claim that cladistic analysis (alone) is thoroughly concordant with the Darwinian "descent with modification." It is also quite widely recognized that while cladistics, like phenetics, is rigorous about clustering groups, it does consider "phylogeny" of both characters and taxa to be revealed by a hierarchically nested system of classification of taxa, determined by synapomorphies that are ordered by the taxonomic, cladistic outgroup. All else is "scenario." In fact the issues of *cladistic outgroups* and "scenarios" have come to figure very prominently in the literature. Many theoretical studies continue to attempt to get around the probabilistic nature of scientific judgments in order to accommodate the cladistic paradigm. What Eldredge and Cracraft (1980, p. 63) have called a "rather simple methodological principle" is, according to these authors, applicable as follows:

> if we are attempting to resolve the scheme of synapomorphy within competing three-taxon statement cladograms, those character states occurring in other taxa within a larger *hypothesized* monophyletic group that includes the three taxon statement as a subset can be *hypothesized* to be primitive . . . and those character-states restricted to the three taxon statement itself can be *hypothesized* to be derived. . . . In short, the procedure is one of mapping the distribution of character-states within and without the groups under consideration. [italics supplied] It would seem that it is not possible to decipher the level of synapomorphy of one similarity without tentative acceptance of a higher level of synapomorphy defined by a different similarity. [p. 65] Synapomorphies are tests of cladistic hypotheses (cladograms), but reciprocally these cladograms, with their expected pattern of congruent synapomorphies are also tests of the synapomorphies.

This procedural rule of taxonomy in general rests on the assumption that the *taxic* outgroup had most likely preserved the primitive character states. While this is a genuinely phylogenetic notion, but which is open to some serious debate, the references themselves in the definition to *hypothesized*, but never tested, cladograms make for a classically circular argument from cladogram to synapomorphies to cladogram.

As exemplified by the statements of Eldredge and Cracraft (1980), in such a paradigm an idealogy tied to a parsimony based evolutionary belief system drives polarity determinations. It is not based on the evaluation of probabilities. The cosmology in which such an approach is nestled may or may not be distantly related to one that was probably most clearly enunciated by Haeckel in his various accounts of "phylogenesis" as a process, and particularly of human evolution. For Haeckel, of course, a consumate and brilliant taxonomist with an unconstrained belief in the specifics of evolutionary mechanisms, all phena were closed groups. Similar to the views of Haeckel, perhaps because of a historical connection, the structure of cladistics has never come to grips with the difference between what a phenon and a corroborated lineage are (see Szalay, 1993c). Haeckel also implicitly maintained that evolutionary splitting of groups (i.e., phena cum taxa) meant the "progress" of one and the "retention of the primitive" by the other. (Haeckel and others certainly believed in the consequently progressionally hierarchic and fundamentally racist nature of the "mental" and "spiritual" (i.e., cultural) "attributes" of specific human groups. These were subsequently "rigorously" and taxically delineated by many practitioners of this particular belief system of phylogeny and taxonomy using these absurd and far reaching notions of phenon–phylogeny.)

Objectivity in science resides in testing against public (generally agreed on) observations (usually involving qualified experts). This objectivity of phylogenetic contributions lies not in their cladogram based methodology, a widespread mistaken belief, or in the more recent additions of consistency schemes generated by parsimony based algorithms of an extreme range of properties, all declared to be taxonomic characters. Objectivity cannot be built on *declarations* of synapomorphies that are based on maximum congruence of an outgroup driven cladogram of taxa. Objectivity should reside in the clear rendering, delineation, explanation, and ordering of character distribution based on the method of **null–group comparison** (N-G C, see below, and Szalay & Bock, 1991), and functional–adaptive analysis (F-A A) outlined by Bock (1981). In science the objectivity of a system certainly does not mean that it is so unrevealingly axiomatic that it is shielded from (contextually evolutionary and biological) probability judgment.

As hinted above, one of the most persistent, unfounded, yet widely accepted notions among both cladists and many evolutionary biologists peripheral to taxonomy is that there is a science dedicated to the uncovering of *pure genealogy* (see particularly Patterson, 1988), independent of the rest of evolutionary biology. The equation of "genealogy" in the literature with phylogeny has been unfortunate, and virtually all taxonomists (myself included, in the past) have used these terms synonymously. Yet the concept of genealogy is a poor model, or even an analogy, for the nexus of relationship of lineages (they do not have two parents) or methods through which one constructs phylogeny. To belabor the obvious, humans and domestic stock have a *recorded* genealogy of individual organisms *pairing* with another and thus producing other individual organisms. One of course may push this precise concept in taxonomy and say that *taxa* (without pairing) gave rise to other taxa. But obviously, nothing is recorded for the evolution of lineages and their splits – we must determine that ourselves from the interpretation of the samples of life at our disposal. The "phylogenetic pattern" is not exactly given by any one rule (see below). In addition, the notion of various taxa (species included) being "individuals" or "holophyletic" is not a "fact," but merely a perspective, and sometimes a flawed one at that (see below; also Szalay & Bock, 1991; Szalay, 1993c). Yet cladistics posits that such a taxon genealogy, produced solely by the cladistic methods that order objective evidence, is to be the *bases* against which other hypotheses must be tested, and that it is something which is to serve as a further framework for the addition of a fossil record, and, finally for the understanding of the "scenarios."

This concept of "scenario" was defined by Tattersall and Eldredge (1977, p. 204) as "essentially a phylogenetic tree fleshed out with further types of information, most commonly having to do with adaptation and ecology." This description, particularly as amplified below, was a poorly masked attempt to buttress the foundational nature of cladogram centered research protocol of cladistics. This scheme produced a patronizing caricature of H-N Es. These authors added (p. 207): "But as vague as the methodology for constructing trees is, there is no methodology at all for the formulation of a scenario, with all its various aspects of evolutionary relationship, time, adaptation, ecology, and so forth. In devising a scenario one is limited only by the bounds of one's imagination and by the credulity of one's audience. . . ." In their explicit view, only a cladogram on which a tree and scenarios must be based is scientifically defensible. So, cladistics has and continues to decree that how, when, and why something happened during evolutionary history is really not relevant to the unraveling of that history. This view, of course, is the exegesis of the very heart of what has become known as evolutionary, or neo-cladistics, practiced in many varieties of rigor. This is a slight but far reaching modification of Hennig's (1950, 1965, 1966) approach, a view which struggles to maintain its theoretically clear identity from the even more rigorous, explicitly nonevolutionary and abiological formal logic of pattern cladistics. Neo-cladistics (as articulately proselytized in texts by Eldredge and Cracraft, 1980, and Wiley, 1981, their differences notwithstanding) holds in addition to a strict adherence to taxic outgroup analysis, that yes, evolution is reality, but that factors of biological causality and process affecting organisms, and the available spatiotemporal frame, cannot be relevant because they cannot be rendered "rigorous" for taxonomic procedures.

Bock (1977a, 1981), Szalay (1977a, b), and Szalay and Bock (1991), as well as others, have outlined in some detail why character analysis based on a cladistic outgroup methodology (as, for example, detailed by Eldredge & Cracraft, 1980) is circular. Character transformations for a *group of organisms* should not be determined by the choice of a taxic cladistic outgroup. In this vicious circle the characters that are supposed to test a sister group relationship of taxa are "rooted" by the very choice of the cladistic outgroup. Clearly, however, comparison of groups of traits is the essence of comparative evolutionary biology. The concept of a well established procedure of comparative biology, reaching back into the nineteenth century, a null-group comparison (N-G C; see Szalay & Bock, 1991) is a procedure concerned with the testing of taxonomic properties independent of the testing of group hypotheses of taxa. The etymology of the term is related to the notion that there are no specific phylogenetic relationships assumed for the *taxa* (or the organisms) whose characters are being assessed. One starts taxonomic analysis by first establishing homologies. Beyond that level of accepted certitude, taxonomists should try to ascertain successive stages of postulated transformations, or their parallel occurrence, or their convergence. The simpler the characters are that we examine, the less successful or convincing our efforts are likely to be. Furthermore, the great numerical preponderance of simple or very low weight traits cannot substitute for the greater probabilities (for or against) offered by a few complex traits. This is so because probabilities of individual trials (tests) are not additive. So a null–group is merely an initial collection of specific homologues of species or higher taxa, which are most similar to the ones in the group being analyzed. The comparisons are for the purposes of understanding the likelihood, and specifics, of a particular character transformation. The null–group does not represent a taxon in any preconceived phylogenetic scheme as in a cladistic outgroup; it does not assume taxic relationships. Because of the assumed reality of varying degrees of mosaic evolution (resulting in the many different kaleidoscopes of character patterns of taxa made up of "heritage" and "habitus" traits, early, in 1910, recognized by W. K. Gregory) and because of an imperfect fossil record, the null–groups for different homologues of the group studied (i.e., the ingroup) can be often expected to be from different taxa. If the null–groups for the various features occur in the same taxon, then this is certainly telling us something. After the various null–group comparisons of diverse characters in the study of any group, the consideration of time factors supplied by the fossils and a priori weighting, the latter dependent on biological criteria, probability based decisions are made and supported by the necessary evidence and its interpretation. Horizontal and vertical comparisons result in the assessment of sister group and ancestor-descendant relationships of both traits and groups, independent of one another.

In contrast to a null–group based approach of phylogenetics, a rigorous cladistic outgroup comparison dictates a posteriori weighting of the characters. Cladistically nested taxa reveal little or nothing of transformations of characters, phyla, and taxa (i.e., the missing components necessary for the construction of a real phylogeny. These cladograms have no empirical content that concern the most important single aspect of phylogenetic studies, namely the tested transformation of known traits. Groupings of taxa do not simply yield transformation–series. These are derived from the explicit rendering of evidence and the research protocols applied in making the relevant comparisons of characters and decisions.

So an inseparable corollary of the "cladogram first, tree next, and scenario last" approach to scientific legitimacy for the study of evolutionary history – a "genealogist's creed" – is the proposed recognition and testing of "phylogenetic patterns" with the rigorous *exclusion* of microevolutionary theory, evolutionary causes and process, from the testing procedure, as forcefully echoed by Eldredge and Cracraft (1980). As many have discussed, and as has been aptly expressed recently by Friday (1987, p. 70): "Probability statements about evolutionary change are usually regarded [by cladists] as drenched in theory and inadmissible. In the following sections I shall briefly review some models of evolutionary change and an approach to estimation of evolutionary trees in which probability statements are of the essence." Such a methodological perspective on the use of probability as expressed by Friday and others fairly reflects my own approach to the understanding of the evolution of characters, the decision about choosing of taxonomic properties, and the delineation and diagnosis of a supraspecific taxon by the attributes of the most recent common ancestor included in that taxon.

It is obvious by now that variously understood concepts about the nature of biological patterns are at the heart of many taxonomic debates. These concepts involve both evolutionary causality (acting on individual organisms), and molecular, organismal, and population based processes, and the resulting patchwork of organic diversity. While one observes many patterns of nature inductively, only some are of critical importance for understanding both processes and the resulting patterns. Biological patterns of theoretical significance should be consistent with nomological–deductive explanations (Szalay & Bock, 1991), a perspective which is in direct contrast to the allegedly process-free use of "patterns" (e.g., phylogeny) to get at evolutionary processes (e. g., as advocated by Eldredge & Cracraft, 1980). An issue that should be addressed may be stated with a question. Namely, how is it possible during the inductive phase of taxonomic practice to recognize biologically and subsequently historically meaningful patterns without considering the constraints imposed by the "internal" and "external" causes of the evolutionary (phylogenetic) process?

The **form–pattern** of living and extinct organisms, pertaining to their ontogeny, established obligate activities, physiology, and morphology, is the available empirical evidence of systematic biology. These form–patterns are either the individual organisms or their parts, manifestations, or aspects. They are the bases for inductions as well as the evidence against which one tests deductions.

But beyond form–patterns in taxonomy there are the cause- and process laden **character–patterns, population–patterns** and **species–patterns,** the latter expressed in the form of various **taxonomic–patterns.** It is the **phylogenetic–pattern** (of taxa) to which Eldredge and Cracraft (1980) applied the notion of "pattern" juxtaposed to process, and stated their view regarding the separation of evolutionary processes from the recovery of phylogenetic–patterns. As detailed in Bock (1981) and Szalay and Bock (1991), phylogenetic–patterns are H-N Es, the reality or validity of which (and that means high probability and the attendant confidence) is directly proportional to the amount and nature of testing. Much of this testing procedure, the character analysis, is directly related to the functional–adaptive analysis.

The significance of functional research is not perceived by all taxonomists in the same way. For example, Lauder (1990), a functional anatomist, advocates an approach that weds functional analysis to cladistics, and as a consequence it illustrates a functional perspective which, rather than being an independent testing ground for branching and phyletic relationships, follows the prescription of the "cladogram to scenario" approach. Lauder accepts taxonomic outgroup driven cladistics as the first step that should direct "causal" analysis. Additional recent polemic support for such a cladistic approach to evolutionary history has come from O'Hara (1988) who considered the "chronicle" of human history to be the equivalent of the cladogram (genealogy), and "history," an analytical account, the correct analogy for "scenario." These statements were made in spite of the fact that it should be evident that in human history a chronicle of records is kept by human witnesses to events, whereas in biological history we only have the preserved extant and extinct samples of life. There is no "chronicle of life," a disembodied genealogy, without the analytical, process laden causal analysis (examined), assumption laden interpretation of the evidence, and the testing of various H-N Es with process-drenched taxonomic properties. I conclude here that the attainment of a phylogenetic–pattern of any significant probability is exactly the reverse to that proposed by Eldredge and Cracraft (1980), O'Hara (1988), and Lauder (1990). This activity of phylogenetic reconstruction becomes science of consequence only, and not a circular exercise, as a result of the understanding and application of evolutionary biology to the evidence in a rigorous temporal framework.

The original tests of taxonomic properties involve a heavy investment of process-oriented procedures before confidence can be placed in them. The simple assumption, therefore, that runs through this book, is that the understanding of the historical relationships of groups is dependent on the rigorous testing of homologies, apomorphies, synapomorphies, and elucidations of transformation series (Fig. 2.2A). Functional–adaptive analysis of characters supplies the context, perspective, and the information necessary for gaining the most probable character transformations (polarity) that are essential for the establishment of common descent, as well as for branching. Disagreement with a purely cladistic, a cladogram-cum-taxon-based, methodology is not founded only on the rejection of the axiomatic idea of that school that all sample differences are "speciation related." This disagreement is also a consequence of the methodological issue professed here; namely that the analysis of character transformations is of foundational importance in phylogenetics.

The form and function (i. e., mechanics) of a character (Bock & von Wahlert, 1965) allows clear and traditionally laboratory oriented assessment in various taxa, but it is the combination of this information with assessments of the biological roles (bioroles) of specific aspects of organisms in the total environment (umwelt), however, which has become known as ecological morphology (see review in Bock, 1990). The establishment of causal relationships between the proximal mechanisms of morphogenesis of form–function complexes and their adaptive significance, or bioroles ("functions"), in as many species as possible helps to lay the foundation for a repertoire of empirical relationships which can be used and further tested. Ecological morphology, like the fossil record, is extremely important in testing and guiding transformational analysis. This perspective has played a critical role in my attempted understanding of marsupial evolution, and subsequently it has influenced the construction of the phylogenetic tree in Figure 2.3.

Tested homologies are the taxonomic properties that are the bases for the establishment of synapomorphies either for sister lineages resulting from horizontal comparisons, or for the source of transformation series one obtains from vertical comparisons. The tests are process based analyses of similarities, and these are dependent on functional–adaptive assessments or tested assumptions. It is a simple theoretical consequence of *descent from earlier to later* time that synapomorphies (the remnant, and therefore shared, identical apomorphies of an ancestor; Fig. 2.2) of lineages

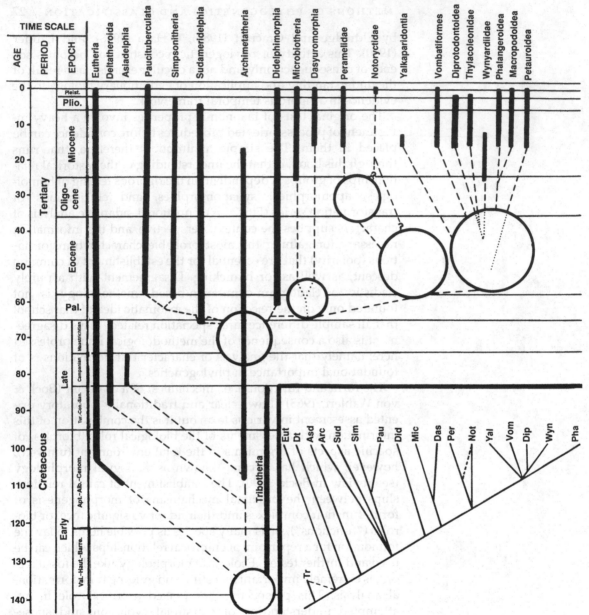

**Figure 2.3.** *Phylogeny of metatherian relationships. Phylogenetic tree, in which all named groups are monophyletic (i.e., paraphyletic or holophyletic). Circles represent the major temporal, phyletic, and/or cladistic uncertainties. Broken line leading to the Oligocene-Miocene and younger microbiotheriids suggests the uncertainty of recognizing these forms based on the incomplete dental record alone.*

***Below in inset in lower right corner:*** *Cladogram, derived from the phylogenetic hypothesis (the phylogenetic tree) shown, with considerable loss*

*of corroborated phylogenetic pattern. Functional-adaptive analysis of homologies, within the context of the fossil record, provided tests for taxonomic properties against which the taxon relationships shown in the tree above were tested. The cladogram is therefore derivative and not foundational in a fully Darwinian phylogenetic analysis which I consider theoretically, and consequently methodologically, sound. The methodological use of cladograms of taxa (not merely character complexes) with taxic outgroups to polarize characterclines in a methodologically*

*consistent cladistic paradigm, one that axiomatically demands causality and process free "pattern" establishment, and reflection on this with the aid of the cladogram, has the analytical consequence of being impervious to much critical information and explanatory theory, in addition to being circular. The cladistic "scenario" reconstruction from the "pattern" of cladograms represents a misunderstanding of the theoretical primacy of non-circular functional-adaptive analysis of homologies.*

can be present in sister groups and in samples of an undivided lineage. The attributes (apomorphies), therefore, which may be shown to change (rather than stagnate) through time, and result in transformation series are the tests, in addition to stratigraphic provenance, of lineage evolution or descent. "Transformational synapomorphy" is a misnomer since the successive conditions differ. Synapomorphies are silent on the direction of change, and therefore nonsynchronous samples tested by them may represent either undivided vertical sequences or split lineages (see Figs. 2.1 & 2.2). The resolving power of synapomorphies alone without the fossil record, the antecedent primitive condition (plesiomorphies), therefore, is limited for the analysis of evolutionary events.

A complex and controversial literature exists on the various approaches to weighting (Hecht & Edwards, 1976; Neff, 1986; Szalay & Bock, 1991). I basically practice the procedure of a priori weighting of those features that are not only most widely available (the latter to a degree are an artifact of the perceived record by me and by previous students) but are also complex both in their form and development as well as in their functional–adaptive biology. Causal studies tie many previously unintelligible minor characters into fewer but functionally integrated, and proportionally more significant, complex ones. Functionally highly integrated features of relatively independent character complexes are preferred over size- or growth-dependent attributes, or simplified or lost traits. Although these may not seem often available in fossils, the analytical understanding of fossils and the evolutionary morphology of the living can, and does, result in the establishment of such attributes. The functional understanding of integrated features that are the targets of directional selection allow one to use fewer, but more heavily weighted, attributes in contrast to a proliferation of often confusingly distributed minutiae frequently employed as taxonomic properties. Relevant comparisons and functional–adaptive analysis allow, by the process of "factoring out," the recognition of historical residues of the most recent evolutionary heritage of the groups studied. Focused and causality-oriented comparative analysis yields the important "noise" in phylogenetic analysis, and, therefore, the significant shared apomorphies. These comparative studies help estimate the common underlying genetic–epigenetic constraints of monophyletic groups. Such "factoring out" without inquiry into the functional and ecological aspects of traits is rarely possible. The literature is replete with impressive and long lists of characters (marsupial studies included). These lists "favor" or "resolve" this or that cladogram, but many such long lists leave little room for confidence because of the either unanalyzed significance, or extremely low weight of the traits.

One of the few open and explicitly reasoned criticisms of the use of functional and adaptive criteria for phylogenetic reconstruction is that of Cracraft (1981). However, he evades a discussion of the critical role of ecological morphology, and of character integration and its subsequent significance, in character evaluation before taxonomic properties are formulated. He clearly states (p. 30), like other taxonomists before him who restrict themselves to cladistic analysis and a posteriori weighting (the latter not restricted to cladistics), that the issue of evolutionary history of characters ("imagining intermediate stages," according to him) is simply not of primary importance for the reconstruction of "nature's hierarchy." Cracraft rejected probability based decisions in taxonomy, and, as did Eldredge and Cracraft, (1980), considered taxic outgroup comparison and development as the methods of importance.

## Trees and classification

Van Valen (1989, p. 99), as cited in the quote at the opening of this chapter, has perceptively quipped that "Only power can arbitrate genuinely basic questions of taste, and that is why there has been war in systematics." Power in taxonomy resides with those who control collections and funds, the access conduits for taxonomic research. As this power base becomes more extensive, its keepers become more blatantly influential and coercive. This power extends not only into national funding agencies but also into the editorial policies of journals. In addition to the unfortunate sensitivity of taxonomic "consensus" to the access of resources, the specific issues of what is a "natural," "real," "nonarbitrary," or monophyletic taxonomic delineation of organisms has fueled the often vicious internecine feud within taxonomy. From the outset I reject the axiomatic, cladogram-dictated ranking demanded by the Hennigian paradigm of conflated phylogenetic analysis and classification. It logically leads to n-1 taxonomic categories (or the rungs in the recent practice of "ranking" without categories), n representing the number of named species taxa (Szalay, 1977b). The utter impracticality of such a stated theoretical aim for classification and its recurrent attempted expressions in practice needs no further comment.

The significant issue under this heading is the concept of **monophyly** in taxonomy, and its expression in a Linnaean classification as either **holophyletic monophylum** or **paraphyletic monophylum** (for definitions see Ashlock, 1971, 1972; Mayr & Ashlock, 1991). It may be asked: What choices are there for expressing our tested understanding of evolutionary history in taxonomic terms? While phyla of organisms are continuous, sequential, and branching, taxa are usually well-delineated entities and are diagnosed at their base. Nevertheless, taxonomy (phylogenetic analysis and classification) deals with the discovery and formalization of these two kinds of monophyletic entities in classification (see Bock, 1977b). Mayr (1981, p. 511) stated that a classification consisting of monophyletic groups is merely an ordered index to the plethora of relevant information known about groups. It is not and cannot be a system that "stores phylogeny," as many cladists would have it. Before I discuss the still-controversial concept of monophyly, I will briefly digress to the topic of trees and other pictorial representations of the history and/or clusters of organisms, and their sundry kinds of relationships expressed by these visual (and for cladistics actually methodological) devices.

While the methodological tenets of cladistic analysis were comprehensively discussed by Eldredge and Cracraft (1980) and others, they left, apparently unrecognized by them, a major unresolved inconsistency. For example, Eldredge and Cracraft (p. 10) admitted that, as repeatedly pointed out in the literature, phylogenetic trees (e.g., Fig. 2.3, upper section) are more detailed and precise hypotheses than are cladograms (e.g., Fig. 2.3, lower section). Paradoxically, in spite of the alleged adherence to a purely hypothetico–deductive approach which axiomatically demands hypothesis precision to facilitate refutation, they chose the notion of phylogeny restricted to splittings and referred to cladograms as the "less extreme view" (p. 11), "tempering" the difficulties. Consequently, the studied and intentional imprecison of cladistics (i.e., when more precise expressions of phylogenetic hypotheses can be tested and therefore do exist) and an accompanying distortion of phylogenetic topology has come to be considered and practiced as a virtue. In that perspective whatever cladistics avoids doing or seeking (such as a priori weighting, or postulating ancestral conditions or groups) is considered a desirable procedure, specifically that of avoiding a "confusing issue" (p. 12). Science is interesting and is practiced because of the challenge of clarifying confusing issues.

Some taxonomic positions and assumptions are often viewed by taxonomists as revealed truths. For example, the notion of one group *including* (as opposed to *originating* from) other groups has its roots in pre-evolutionary days. It is one of the sacrosanct perspectives of pre-Darwinian taxonomy – in fact the system of biological classification initiated by Linnaeus is based on it. Groups of course do not include other groups, only their classification (but not their evolutionary topology) is so rendered through one's taxic perceptions of **inclusive hierarchies**. Groups can be defined by shared attributes, or diagnosed (in phylogenetics) by their specially shared traits or their clearly understood traces. These are their *taxonomic properties which are tested (through corroborations) against evidence and that are shown to have originated at a particular point in their history*. It is my belief that taxonomic efforts that continue to aim at the production of "genealogies," and their isomorphic expressions in classifications (or their latest version of unranked and parenthetically subordinated nomina) with an explicitly stated independence from causal biology, will continue to reap for taxonomists, in general, the loss of the needed (for pragmatic purposes) respect from the remaining biological community. The alternative is very simple, and certainly more challenging and interesting: Why not concentrate on producing analytical accounts of the evolutionary change of character systems from which we can generate both (at least theoretically) complete and robust phylogenies and useful monophyletic classifications about organisms?

Species and lineages are real, but our grasp of their phylogeny is often problematic. All taxa are delineated by the taxonomist, and therefore the distinction between the history of the diversity of life and formally recognized groups should be always kept in mind. While one can strive to make classifications increasingly consistent with the actual history of organisms (Darwin, 1859; Simpson, 1961a; Hennig, 1950, 1966), this theoretical goal cannot be completely fulfilled – because of the nature of taxa. If we had a perfect record of evolution, then this problem would be even more obvious, as blatantly artificial boundaries would have to be drawn far more frequently than we must do now, as often and eloquently pointed out by Simpson (1961a). So what is the seething and seemingly never-ending conflict in the taxonomic literature about "arbitrariness," "naturalness," and "artificiality" of various kinds of taxa and classifications? Perhaps it is important to note that even such committed cladists as Ax (1987) or Tassy (1988) have a view equally skeptical to that of a phylogeneticist of my own ilk of "the" classification. Tassy (1988, p. 52), somewhat enigmatically, states: "I have tried to show, using the examples of Proboscidea, that for a variety of reasons it is impossible to accept definitive rules of classification for every systematic problem. The method of formal classification in phylogenetic systematics must remain pluralistic. The arbitrary decisions I have taken follow previous conventions or rules, and naturally, other choices could have been taken. . . . A final quotation from Hennig . . . would support this statement: 'a hierarchically arranged list of names of monophyletic groups' is only . . . the phylogenetic system in the narrower sense'. Still, this narrow sense weighs heavily on systematic practices and results." In fact the concept "monophyly," and what has become of it in the alleys of the jungle where taxonomic warfare at all levels has been conducted, is the key to understanding of how classifications should proceed and what they are.

Charig's (1982) critical analysis of various "clado-evolutionary" and "natural order" (pattern cladistics) systematics has vividly brought to focus some of the evolutionary or biological bases (or their lack) of the conflicting approaches to classifications. Charig has dealt a devastating blow in his deconstructional analysis to the procedures and aims of pattern cladistics (which generates the most "natural" of classifications in a pre-Darwinian sense). He has not, however, gone to the heart of the methodological issues that determine taxonomic practice and that still divide approaches to character analysis, as I have noted above. Charig, I believe, has not fully appreciated that many pattern cladists have striven to represent rigorously the most logical exegesis (in a formal scholastic sense devoid of nonmonotonic reasoning: e. g., Nelson and Platnick, 1984) of some of Hennig's own fundamental assumptions. Pattern cladists have realized the need for a particularly rigorous theoretical perspective and have striven for it. But in that quest, I believe, they have come to equate, mistakenly, the mirage of a logical exercise based on untested, untestable, or patently uncorroborated assumptions with that of a rigorous theory tested against known and well corroborated mechanisms of evolution, and against the foundational nature of the fossil record both for the theory of evolution and the practice of phylogenetics itself.

It is important to recognize that just as Simpson (1961a) has obfuscated the *theoretical evolutionary concept* of monophyly with *taxic* compromises, Hennig's confusingly presented taxonomic-cum-evolutionary concept of paraphyly has been disastrously emended and confused even more by Nelson (1971), Platnick (1977), and Farris (1974). It was Ashlock (1971, 1972, 1979) and Charig (1982) who clarified these critical concepts related to group ancestry in *taxonomy*. It is well to remember that for the ontology of a theoretical concept it does not matter what Haeckel did or did not mean by the term "monophyly." All modern taxonomists reject **polyphyletic groups** (taxa that do not include the last common ancestor of their lineage members). All phylogeneticists ("cladists" or "evolutionists") agree that **monophyletic groups** (taxa that include the last common ancestor, a taxonomic species, of their members) are necessary for a phylogenetic classification. Almost all of the debate surrounding classifications today, however, centers on either the desirability or total rejection of paraphyletic groups. The clear concepts of **holophyletic groups** (monophyletic taxa that include all the descendants of the last common ancestor) and **paraphyletic groups** (monophyletic taxa that do not include all the descendants of the last common ancestor) continue to be mishandled, sometimes on purpose, in order to win advocates for one or another classification or "taxonomic philosophy." Such political term-switchings are particularly unattractive aspects of cladistics as they continue to inhibit the scholarly exchange of ideas.

Paraphyletic groups are not grades. They are groups derived from a single common ancestor (one lineage) which can be diagnosed by *apomorphies in contrast to its origins.* **Grades,** however, retain their original meaning in which polyphyly was significant (e.g., previous concepts of Mammalia as a "grade"). In light of the fact that all ancestral groups (taxonomic species) are paraphyletic when they give rise to descendants, their designation as "artificial," versus the "natural" holophyla, is bizarre and biologically meaningless. While ancestors do represent a **stage** of evolution, this concept is anything but "unnatural." The studied conceptual use of "natural" versus "not real" or "artificial" is again an ill considered taxonomic practice, which is based on the concepts of "crown" versus "stem" groups, concepts that do not fall out of sound evolutionary theory, beyond the arbitrary expediency of a taxonomic practice. This is a practice based on a taxically conceived, and therefore in many ways arbitrarily delineated (but not unreal), world, one that is necessary for communication. A single paraphyletic (taxonomic) species (i.e., one that gave rise to descendant lineages or taxa), however, does represent a stage in evolutionary history. Therefore, attributes of such paraphyla may be considered stages in character evolution which are "real" and "natural."

Classification of a vertical continuum (i.e., the history of a lineage of life) is the making of a series of *arbitrary* cuts of a continuum – even holophyletic taxa, of course, are the result of such "chopping" at their bases. The arbitrariness is gradational. Paraphyla are additionally arbitrary compared to holophyla, but both of these monophyletic groups are **diagnosed,** as opposed to **defined** (see Rowe, 1987, for holophyla only; and Szalay, Rosenberger & Dagosto, 1987), *based on the derived attributes of the last common ancestor* of the group members. Both types of groups are therefore *phylogenetically diagnosed.* For *defining* (as opposed to phylogenetically diagnosing) groups need to have shared attributes of all the included subgroups within them. This is a most unsatisfactory procedure as attributes transform in phylogeny, and even direct ancestor–descendant samples may have different, albeit homologous, attributes. The apomorphies of the group's ancestor often change through time (Fig. 2.2A). Characters do not stop evolving, and therefore the various transformation series that lead from them are the tests of group membership rather than the definition oriented possession of unvarying synapomorphies by all members. At any rate identifiable groups derived from a paraphylon have their own diagnosing apomorphies, those of their included last common ancestors. The Reptilia is a perfectly natural taxonomic group with its diagnostic attributes retained from the protoreptile, as no member of it has the feathers and the dominant right aortic arch of the protobird. Definitions of higher taxa are more suitable for the pragmatic taxonomic keys, rather than for phylogenetic analysis or increasingly subtle expression of its results in a classification. While both holophyla and paraphyla are abitrary, but natural in a phylogenetic sense, only polyphyla are *artificial*. Designation of paraphyla as "artificial" suggests a crisis of creative scientific thinking in the minds of those who attempt to forge "new macroevolutionary realities" from the trivia of partisanly loaded relabeling.

A phylogenetic diagnosis of a monophyletic group, then, is the total of the known apomorphies (and this includes process based interpretation) of the last common ancestor of that group in contrast to its predecessor. The diagnosis is tested against all taxonomic properties of the descendant part of the lineage involved. The whole lineage is the one that should be arbitrarily divided into an antecedent and descendant taxon.

The merits of the unavoidable paraphyletic groupings of assemblages of closely clustered forms may lie not only in the pragmatic usefulness of such taxonomic practice but also in what they reveal about evolutionary variation (not quite the regularity, and hence lawlike implication of "pulses") at the long recognized level of fundamental structural organizations ("bauplans"), which persist for great lengths of time because of the complexities of epistatic genetic integration. In fact here lies the closest correlation of evolutionary change in evolving lineages of organisms with the reality of monophyletic *taxa*. It has long been obvious, of course, that biological processes in evolution have close bearing on phyla and taxa. Erwin (1990 p. 99) has recently commented on the correlated progression model of evolution of Thomson (1988) for the origin of higher taxa as follows: "Under the correlated progression model, change is so rapid that synapomorphies are lacking between major groups and phylogenetic analysis is difficult; character complexes will appear suddenly and apparently already well integrated." To repeat and amplify, this has obvious implications for taxonomy. If groups or adaptive radiations, facilitated by rapidly changing habitat expansion related to latitudonal and geographical change, appear suddenly with well integrated characters or numerous sibling species like lineages, and if we pay any attention to the various probabilities of the nature of integrated change in character complexes, then such groups will defy attempts at rendering all taxa holophyletic. Forced theoretical perspectives, even if clearly articulated, can do serious damage to understanding of actual history (e.g., that of Wyss & de Queiroz, 1984). Furthermore, unlike some taxonomists whose interests are pristinely (and often narrowly) claimed to be limited to "resolved holophyla," most other taxonomists are interested in the actual evolutionary biology and history of these organisms. As paraphyla, these taxa, however, will greatly aid biological understanding and phylogeny, as well as the admittedly contestable goals of classification. All of this should be particularly applicable to many stem groups of the more recent extremely successful higher vertebrate taxa, such as the teleosts, birds, mammals, and their various subsets. It is slowly becoming obvious even to those taxonomists who are the most ideologically motivated by a taxonomic (as opposed to an evolutionary) theory that considering a stem group (and which higher group did not have a stem group?) or a "plesion" as holophyletic entities, in even a purely cladistic analysis, is truly pushing the application of the concept of holophyly well beyond its epistemological limits. Recalling Hennig's (1966) views, recently Ax (1987) has reaffirmed the paraphyletic nature of non-crown groups. It is obvious that *all* monophyletic groups can be diagnosed by the apomorphies of their last common ancestor, and that stem groups ("metaspecies" – a sanitized cladistic term for ancestor) do abound (even among living species and genera). But does it have to follow that in a growing literature, in order to adhere to restrictive cladistic tenets, taxonomic concepts of groups of organisms be placed, increasingly, in quotation marks to depict their paraphyletic nature? Not infrequently, of course, one taxonomist's untested holophylon is another's tested paraphylon.

Figure 2.2B is a model of evolutionary transformation, through time, of lineages consisting of organisms; note however that this is a schematic rendering of continuous *populations or species, and not taxa,* and therefore this is *not* a cladogram. All the lines depicted are continuous populations (i.e., lineages and not taxonomic species) with a potential to leave phena as fossils. While cladistic theory demands the recognition of taxa at each postulated node of splitting in evolution, in the real world the practice of this perspective is entirely dependent for its corroboration on evidence. Phenon differences of allopatric or allochronous samples is not such evidence. If the appropriate sample differences are not available, then not only is the split impossible to substantiate, but the taxon so constituted cannot be diagnosed by the apomorphies at that node, as depicted in Figure 2.2B. But even in such a framework of a taxonomic, rather than evolutionary, ontology for which there is no known epistemology (see Szalay & Bock, 1991), the group below the one first diagnosed (its stem group) is paraphyletic based on the available character evidence, and so are the others below that, and so on, even if atemporal cladograms on which such histories may be shown were given temporal significance along their common axis. For example, while it is helpful to know that diagnoses of Life, Amniota, Aves, and Mammalia are those of holophyletic taxa, many groups within these must be paraphyletically accommodated units if they are to be testable and tested groups.

Figure 2.2B shows the splittings (simplified) of the lineage on the right (heavy line), represented by numbers 1–5. The thin horizontal arrows point to those moments in time in this model lineage when *detectable* (i.e., tested) changes in characters had occurred. These are the first manifestations of new characters. These traits are the bases for any inference in taxonomy, unless we assume that somebody was there watching and recording the choreography of events. The latter would be a real *chronicle* (in opposition to the usage of that concept advocated by O'Hara, 1988). Empirical methods can only test those (monophyletic) taxa based on such a model which are diagnosed by the apomorphies of the last common ancestor *at these points,* and not necessarily at the furcations. The cladistic, compared to a phylogenetic, model of (taxic) evolution, bereft of the components of the evolutionary process except for splitting, lacks both a tested ontology and has no epistemology for the discovery of those kinds of taxa its theoretical foundation subscribes to.

Note that a holophyletic "crown group" could be delimited at points **a** to **d** on Figure 2.2B. The diagnosable (and therefore monophyletic) clusters *between* **a** and **d**, however, could not be delineated holophyletically. If such groups are diagnosed, no matter how the cut was made, then they are *real* and *natural* monophyletic groups, i.e., paraphyla. The cladistic system, however, never grapples with the phylogenetic reality of paraphyla. *The product of the cladistic paradigm, the isomorphic taxa–cladogram, the taxogram, is not a real abstraction, but a subtle and fundamental distortion of what the tandem usage of the two types of monophyletic taxa are capable of summarizing about phylogenetic research.* Notice that in Figure 2.2B all the diagnosable groups at points **a, b, c,** and **d** depend on the recognition of new characters, and therefore character evolution. All are natural, tested taxic segments of the phyletic continuum and splittings.

Construction and tests of monophyletic groups, the aim of phylogenetics, is independent of formally naming or ranking taxa, at least in principle. No theory exists that can deliver in practice classifications that are not arbitrary to some degree (see above). My preference is usually, but not always, for phylogenetically well-diagnosed, preferably holophyletic taxa at the increasingly higher category levels when there is clear evidence of structural–functional reorganization of at least one well understood character complex – when the evidence is unequivocal. Holophyla, however, often cannot be delineated more precisely than a postulated, well corroborated origin from some member of a well diagnosed monophyletic radiation, a paraphylon. Paraphyla, therefore, as shown in Figure 2.2B, are unavoidable in any realistic and useful classification that attempts to reflect the transformational biology of often sparsely sampled phyletic events. A balanced and biologically sound discussion of the subject of classification is given by Bock (1977b) (see also Ashlock, 1979; Mayr & Ashlock, 1991; Warburton, 1967). A good classification, in addition to Mayr's (1981) concept of an "ordered index," also mirrors the most secure knowledge of monophyletic groups diagnosed by some traits of adaptive significance (as far as this can be ascertained) and their most probable ties. For that reason the *formalization* of group hypotheses should attempt to be commensurate with the confidence in well corroborated groups.

A note should be added in relation to the terminology. Because, as mentioned above, I diagnose monophyletic groups by the derived traits (*apomorphies at the designated level*) of their last common ancestor of the included members in contrast to their *antecedents*, it logically follows that such diagnostic, apomorphous features of a group are primitive (*plesiomorphous or ancestral at the designated level*) for the subsequently *derived* lineages and groups (Szalay et al., 1987). For example, while ovovivipary might have been an *apomorphous* feature of a therian (or some lineage prior to the Theria as diagnosed here) and therefore for the *group* derived from it, this attribute is a *primitive* trait for all other *descendant* therians. Of course one cannot be sure that it was the first mammal with a diagnostic tribosphenic tooth, the first tribotherian, that was ovoviviparous. This trait obviously can also be considered as a *synapomorphy* that corroborates the monophyly of the therians. This is somewhat in contrast to the customary expression that a synapomorphy "demonstrates" or "defines" a group. I chose my usage because evolutionary change and its conceptually preferable analysis (i.e., *from the bottom up*), necessitates the recognition of groups that may have no more obvious synapomorphies than the sharing of the ancestral apomorphous constraints. Lack of "obvious" synapomorphies in a monophyletic group, therefore, may be due to both significant divergence and subsequent change, ascertained as a result of character analysis, derivative from the diagnostic condition of the group.

## Metatherian classification

In classifications of mammals, as in other groups, there is an increasing pressure exerted toward a rococo and pedantic conformity, that all groups in a formal system of classification must be holophyletic, even though they cannot be. This single Hennigian generality, perhaps taken beyond its original meaning, is creating an attitude in taxonomy which is comparable only to the notion that the medium is the message. All taxonomists should (ideally) be careful that as "precision" in understanding group ties increases, confusion can spread among those who want to use classifications of others as a guide to evidential support for their understanding of evolutionary biology, history included, but cannot afford full-time tracking of the efforts of the taxic experts. All biologists, I believe, would like to be able to see evidence and analysis (the level of interest in this is highly variable, of course) that some conditions, systems, or groups originated from another one, based on some readily available information, even if the exact species or genus cannot be recognized! In fact that is a (most) significant form of evolutionary understanding for the paleontological and biological community in general, and if a tree and accompanying classification cannot reflect this, then such efforts are not in touch with phylogenetic reality and the needs of fellow travelers. It did not take the rise of cladistics to show, in spite of published claims to the contrary, that a statement of origin of one

supraspecific taxon from another merely meant an approxima-
tion, or an available level of resolution, for the understanding of a
particular case of descent.

I have abandoned the taxonomic term "Marsupialia" in favor of
the concept of Metatheria as a taxon. My reasons for this, I believe,
transcend taxonomic practice and are rooted in the use of the
descriptive term "marsupial" in evolutionary biology. There appears to be a general agreement among researchers on early mammals that many therians (and perhaps even some pretherians)
have had a reproductive and developmental system more similar
to opossums than to any eutherians. The term for such a system
has been widely associated with the concept of "marsupial," a
useful vernacular describing a mode of reproduction and
development (lacking an inner cell mass) more primitive than that
of the earliest eutherian. Having a taxon called "Marsupialia" and
a vernacular derived from it which is applicable to the most
characteristic attributes of animals that are not included in it, is, I
believe, confusing. Therefore, it is a practice that should be abandoned. Some researchers may choose to unite the Ameridelphia
and Australidelphia in the Marsupialia (e.g., Kielan-Jaworowska
& Nessov, 1990). I have not done that. If distinctions are to be
made in text between the Ameridelphia and the other groups not
derived from it, then it should suffice to refer to the Ameridelphia
and such groups (e. g., the Deltatheroida or Asiadelphia) within
the Metatheria.

In accordance with the dictates of the phylogenetic and
classificatory approach adopted here, the classification of the Metatheria that I propose is based on a combination of (1) the most
probable transformational understanding of the characters analyzed, (2) the usage of the fossil record for establishing some polarities, (3) judgments on character evolution derived from a
(rudimentary) functional anatomy and ecological morphology of
the characters, (4) a synthetic appraisal of highly corroborated
monophyletic groupings, (5) the hypothesized phylogeny of taxa
and phyla (see Fig. 2.3), and (6) the extent of phenetic divergence
of the taxa from one another compared with generally accepted
ordinal groupings within the Eutheria.

The occasional stylish absence of ranks in several recent contributions of taxonomy is a final exegesis of the idea that classification and expressions of diversity and multiplicity should not be
connected in any way. I do maintain that ranking serves the purpose, imperfectly, of reflecting both the cladistic and phyletic aspects of taxa, particularly when coupled with an intelligent
explanation of the phylogenetic trees accompanying such classifications. To minimize the explosive growth of categories that require proper nouns I prefer to use the prefix *semi*, by itself or
together with other prefixes commonly used, attached to widely
accepted Linnaean categories. Addition of a prefix like *semi*, to a

category means a subdivision of that rank, and thus the prefix and the root word designate a rank below that of the root word (see Szalay et al., 1987). For example, the semiorders of the order Syndactyla allow the usage of subordinal designations in the semiorders. Further subdivision of suborders, for example, into semisuborders, as in the case of the subdivisions of the suborder Vombatiformes, permits the retention of infraordinal groupings, as refinement of phylogenetic understanding proceeds without inflating the traditional structure of the basic Linnaean categories. Prefixes with clearly attached meaning of ranking have the advantage of immediately signaling super- or subordination in classification without the necessity for new names for the widely recognized and used suprafamilial categories.

Based on the phylogenetic tree in Figure 2.3, partly on the nature of the evidence and interpretations cited above and detailed below, the classification at the end of this chapter is consistent with both the tested tree and the cladogram in that figure. It is my conviction that when compared with the Eutheria, the classification of marsupials is highly inflated on both the ordinal and even family levels. The reasons for this are complex and cannot be detailed here. This practice, however, is partly driven by the formalism of taxonomy in general that often equates node recognition with new ranking. It is further clearly exacerbated by cladistic ideology, which strives for taxograms, even when some cladograms are of the most tenuous validity. Marsupials do not appear to have been more diverse, and probably considerably less so, reproductively, chromosomally, and morphologically, than the generously combined phenetic range of the Insectivora, Carnivora, Primates, and Rodentia. In addition to my accepted and supported allocation of the orders Deltatheroida and Asiadelphia (Trofimov & Szalay, in press) to the Metatheria, three additional orders are sufficient to reflect the extent of known diversity of marsupials in comparison to the eutherians. Fossil discoveries are likely to change this current assessment of diversity.

## Classification of Theria and Metatheria adopted in this study

Subclass **Theria** Parker and Haswell, 1897
Infraclass **Tribotheria** Butler, 1978
Order **Aegialodontia** Butler, 1978
Family **Aegialodontidae** Kermack, Lees, and Mussett, 1965
Order **Pappotherida** Butler, 1978
Family **Pappotheriidae** Slaughter, 1965
Infraclass **Eutheria** Gill, 1872
Infraclass **Metatheria** Huxley, 1880
Cohort **Holarctidelphia,** new
Order **Asiadelphia** (incl. *Asiatherium* Trofimov and Szalay, nomen nudum)

Order **Deltatheroida** Kielan-Jaworowska, 1982
  Family **Deltatheroididae** Kielan-Jaworowska, 1982
  Family **Deltatheridiidae** Gregory and Simpson, 1926
    Subfamily **Deltatheridiinae** (Gregory and Simpson, 1926)
    Subfamily **Sulestinae** Nessov, 1985
Cohort **Ameridelphia** Szalay, 1982
Order **Didelphida** Szalay, 1982
  Suborder **Archimetatheria,** new
    Family **Stagodontidae** Marsh, 1889
    Family **Pediomyidae** (Simpson, 1927)
      Subfamily **Alphadontinae** Marshall, Case, and Woodburne, 1990
      Subfamily **Pediomyinae** Simpson, 1927
      Subfamily **Glasbiinae** Clemens, 1966
      Subfamily **Peradectinae** Crochet, 1979
  Suborder **Sudameridelphia,** new
    Infraorder **Itaboraiformes,** new
      Family **Caroloameghiniidae** Ameghino, 1901
        Subfamily **Eobrasiliinae** Marshall, 1987
          Tribe **Eobrasiliini** (Marshall, 1987)
          Tribe **Derorhynchini** (Marshall, 1987)
          Tribe **Protodidelphini** (Marshall, 1987)
        Subfamily **Caroloameghiniinae** (Ameghino, 1901)
        Subfamily **Mirandatheriinae,** new
        Subfamily **Monodelphopsinae,** new
    Infraorder **Polydolopimorphia** Ameghino, 1897
      Family **Prepidolopidae** Pascual, 1981
      Family **Polydolopidae** Ameghino, 1897
      Family **Bonapartheriidae** Pascual, 1981
    Infraorder **Sparassodonta** Ameghino, 1894
      Family **Borhyaenidae** Ameghino, 1894
        Subfamily **Hathlyacyninae** (Ameghino,1894)
        Subfamily **Hondadelphinae** Marshall et al., 1990
        Subfamily **Prothylacyninae** (Ameghino, 1894)
        Subfamily **Proborhyaeninae** (Ameghino, 1897)
        Subfamily **Borhyaeninae** (Ameghino, 1894)
      Family **Thylacosmilidae** (Riggs, 1933)
  Suborder **Glirimetatheria,** new

Infraorder **Paucituberculata** (Ameghino, 1894)
  Family **Caenolestidae** (Trouessart, 1898)
    Subfamily **Sternbergiinae,** new
    Subfamily **Caenolestinae** (Trouessart, 1898)
    Subfamily **Abderitinae** (Ameghino, 1889)
    Subfamily **Palaeothentinae** Sinclair, 1906
Infraorder **Simpsonitheria,** new
  Family **Gashterniidae** Marshall, 1987
  Family **Groberiidae** Patterson, 1952
  Family **Argyrolagidae** Ameghino, 1904
  Family **Patagoniidae** Pascual and Carlini, 1987
Suborder **Didelphimorphia** (Gill, 1872)
  Family **Didelphidae** Gray, 1821
    Subfamily **Caluromyinae** Kirsch and Reig, 1977
    Subfamily **Didelphinae** (Gray, 1821)
      Tribe **Didelphini** (Gray, 1821)
      Tribe **Chironectini** (Haeckel, 1866)
    Subfamily **Herpetotheriinae** Trouessart, 1879
  Family **Sparassocynidae** (Reig, 1958)
Cohort **Australidelphia** Szalay, 1982
Order **Gondwanadelphia,** new
  Suborder **Microbiotheria** Ameghino, 1889
    Family **Microbiotheriidae** (Ameghino, 1887)
  Suborder **Dasyuromorphia** (Gill, 1872)
    Family **Dasyuridae** (Goldfuss, 1820)
    Family **Myrmecobiidae** Waterhouse, 1838
    Family **Thylacinidae** Bonaparte, 1838
Order **Syndactyla** Wood-Jones, 1923–25
  Semiorder **Peramelina** (Gray, 1825)
    Family **Notoryctidae** Ogilby, 1892
    Family **Peramelidae** (Gray, 1825)
      Subfamily **Peramelinae** (Gray, 1825)
        Tribe **Peramelini** (Gray, 1825)
        Tribe **Chaeropini** (Gill, 1872)
      Subfamily **Thylacomyinae** (Bensley, 1903)
  Semiorder **Diprotodontia** (Owen, 1866)
    Suborder **Phalangeriformes** (Szalay, 1982)
    Superfamily **Phalangeroidea** (Thomas, 1888)
    Family **Phalangeridae** Thomas, 1888
      Subfamily **Phalangerinae** (Thomas, 1888)
        Tribe **Ailuropini** (Flannery, 1987)
        Tribe **Phalangerini** (Thomas, 1888)

Subfamily **Ektopodontinae** (Stirton, Woodbourne, and Plane, 1967)
Subfamily **Pilkipildrinae** Archer, Tedford, and Rich, 1987
Subfamily **Miralininae** Woodburne et al., 1987
Superfamily **Petauroidea** (Gill, 1872)
Family **Petauridae** (Gill, 1872)
Subfamily **Petaurinae** Gill, 1872
Tribe **Petaurini** (Gill, 1872)
Tribe **Dactylopsilini** (Kirsch, 1977)
Tribe **Pseudocheirini** Winge, 1893
Subfamily **Burramyinae** Broom, 1898
Tribe **Burramyini** (Broom, 1898)
Tribe **Acrobatini** (Aplin, 1887)
Family **Tarsipedidae** Gervais and Verreaux, 1842
Superfamily **Macropodoidea** (Gray, 1821)
Family **Hypsiprymnodontidae** (Collett, 1887)
Subfamily **Hypsiprymnodontinae** Collett, 1887
Subfamily **Propleopinae** Archer and Flannery, 1985
Family **Macropodidae** Gray, 1821
Subfamily **Potoroinea** (Gray, 1821)
Tribe **Potoroini** (Gray, 1821)
Subtribe **Potoroina** (Gray, 1821)
Subtribe **Bettongina** (Flannery, 1990)
Tribe **Bulungamayini** (Flannery, Archer, and Plane, 1983)
Subfamily **Macropodinae** (Gray, 1821)
Tribe **Balbarini** (Flannery, Archer, and Plane, 1983)
Tribe **Macropodini** (Gray, 1821)
Tribe **Sthenurini** (Glauert, 1926)
Tribe **Dendrolagini** (Bonaparte, 1850)
Suborder **Vombatiformes** Woodburne, 1984
Semisuborder **Vombatomorphia** (Aplin and Archer, 1987)
Superfamily **Phascolarctoidea** (Owen, 1839b)
Family **Phascolarctidae** Owen, 1839b
Superfamily **Vombatoidea** (Burnett, 1830)
Family **Vombatidae** Burnett, 1830
Family **Ilariidae** Tedford and Woodburne, 1987
Semisuborder **Diprotodontiformes,** new

Superfamily **Wynyardioidea** (Osgood, 1921)
    Family **Wynyardiidae** Osgood, 1921
Superfamily **Thylacoleonioidea** (Gill, 1872)
    Family **Thylacoleonidae** Gill, 1872
        Subfamily **Thylacoleoninae** (Gill, 1872)
        Subfamily **Wakeleoninae** Murray et al.,
        1987
Superfamily **Diprotodontoidea** (Gill, 1872)
    Family **Diprotodontidae** Gill, 1872
        Subfamily **Diprotodontinae** (Gill, 1872)
        Subfamily **Nototheriinae** (Lydekker, 1887)
        Subfamily **Zygomaturinae** Stirton et al.,
        1967
    Family **Palorchestidae** (Tate, 1948)
Suborder **Yalkaparidontia** (Archer, Hand, and God-
thelp, 1988)
    Family **Yalkaparidontidae** Archer et al.,
    1988

# 3

# Problems in understanding metatherian evolution

*. . . the present writer has pointed out the probability that the aplacental condition of most Marsupials is actually primitive, and that the placental connection in* Perameles, *like a multitude of other characters in which Marsupials resemble Placentals, has been independently acquired; in other words, that it represents a convergent or homoplastic development.*

*However this may be, we have the more definite fact that the marsupials and Placentals are collateral and, in a certain sense, equivalent groups of common parentage; and this conception may be welcomed as clearing the way for a better perception of the details of their secondary evolution or adaptive radiation.*

Bensley (1903, p. 85)

Importance of determining the order of appearance of diagnostic characters. – *The relative age of different characters should in all cases be a primary object of research. The historical method (although open to many pitfalls) when judiciously applied seems more likely to lead to lasting phylogenetic results than the time honored method of setting down resemblances and differences between two animals. . . ."*

Gregory (1910, p. 112)

The simple well corroborated theme in this section that underlies the scrutiny of a few areas of marsupial biology is that answers without history are very incomplete. Yet, this history cannot be deciphered without considerable input from a most current understanding of causality (proximal as opposed to more distal) of sundry biological processes and mechanisms. Accounts of historical relationships are not particularly confidence generating unless they are tested against the paleontology and functional–adaptive biology of as many features as feasible (Bock, 1981; Szalay & Bock, 1991). Separation of functional and historical biology is perhaps pragmatically, but only temporarily, justifiable for the study of many ongoing problems, but conceptually, and therefore theoretically, it is not. If the latter is true, however, methods based on theory should play an important role in the empirical work itself.

An only seemingly trivial issue concerning the use of language in marsupial taxonomy has long plagued the literature, and it is briefly discussed here to stem, hopefully, this unfortunate confusion. The problem is the use and perceived connotation of the concept primitive (ancestral) versus advanced (derived) in the comparative literature dealing with therian evolution with regard to development, reproduction, metabolism, encephalization, and all other aspects of natural history causally related to these. In any phylogenetically meaningful discussion, the taxonomic properties of groups have some relevance to the first members of that group, for example either the first therians, or the protometatherians, or the protoplacentals. If a certain diagnostic attribute is retained in the descendants of these, then of course we may speak of these as characteristically "therian," "metatherian," or "eutherian." It is most unfortunate, however, that the concepts of "inferiority" and "superiority" have become admixed in the accounts and polemics of outstanding students of marsupial biology. An assumed corollary of the nineteenth-century view of marsupial reproduction was the label "inferior." The reaction of some modern students of marsupials to such nineteenth-century assumptions, but also to the continued designation of the reproductive system as representing the primitive therian condition, has spawned some unnecessary generalizations in the literature and complex arguments for marsupials being "alternate" mammals.

First of all, it is still often routinely and wrongly assumed in the literature dealing with various traits that all characters found in (living) marsupial anatomy must be plesiomorphic among therians compared to the eutherian conditions (e.g., Lewis, 1963, 1989). This is an imprecise and taxic "outgroup" biased assumption dating to the late nineteenth century, probably based on aspects of the highly corroborated transformation hypothesis of the reproductive system in the Mammalia (see Renfree, 1993, for a recent synthesis of that system, but slightly differently viewed than the one summarized below). But there were more general consequences of this thinking. Marsupials (the whole animals, and not only selected character complexes) have come to be referred to as "lower," "basal," and "primitive" mammals in general, due to a Haeckelian, "split-cum-progress in the branch most similar to humans"-oriented, hierarchic perspective of evolution. The (debatable) notion of a "living fossil" when extended to all attributes of some organisms often influences phylogenetics. The counter reaction of some students of marsupials to such generalizations (pertaining to the variously stated putative competitive inferiority of marsupials versus placentals) resulted in the somewhat paraphrased and often repeated self evident statement that the two groups are the "results of different selection pressures." What was needed instead, I believe, was to address the issue that the reproductive and developmental systems of the (inferred) protometatherians are either phylogenetically antecedent (plesiomorphic) in a therian context, or they are an alternate and unique apomorphy (an autapomorphy) of marsupials among therians.

Tyndale-Biscoe and Renfree (1987) and Renfree (1993) have argued that the protomarsupial manner of reproduction (of the taxon Metatheria) is not a primitive therian attribute, but it is highly derived on its own right from a condition ancestral to both Metatheria and Eutheria. These authors contend that the major therian reproductive differences in the two groups have evolved due to differing metabolic needs related to increasing body size. They have not specifically addressed the issue of the origins of the protometatherian and protoeutherian reproductive attributes, an examination which must be done along with any such stand. Renfree (1993) maintains that the condition of the oviducts and ureters are central to this debate. In marsupials the oviducts are separated by the ureters which pass in between them. Renfree believes that because the oviducts cannot meet in the middle to form a single median uterus and vagina, this limiting feature is one of the reasons for the small size of marsupial neonates. The implications of these arrangements for the evolution of ontogeny in marsupials, however, has not been fully explored.

The statements pertaining to the across-the-board "superiority" of eutherians are as obviously imprecise and without useful evolutionary meaning as are the lack of rigorous accounts of what it is to be an "alternate" mammal. Character complexes are phylogenetically either primitive or advanced, or merely divergent, in relation to one another. The Darwinian meaning of improvement only allows for "better" versus "less good" assessment of phyletically successive populations that improve their survival and reproductive strategies in a given context. "Inferiority" versus "superiority" of biological "design," or a bauplan in general, is entirely context dependent, but in a given similar or identical ecological context such judgments are usefully made by students who study metabolism, energetics, rates of increase, and other phenomena within the field of behavioral ecology (e.g., Nicoll, 1987). These issues may or may not be causally related to the evolutionary history of particular traits (but see below). In fact, Thompson (1987, p. 206) in his study in which he estimated reproductive potential, has recently stated that:

> While the long [conception to weaning time] and low [intrinsic rate of natural increase] of marsupials may reflect important evolutionary constraints, the rigidity of these constraints is questionable given the observation that at least some marsupials have managed to compensate for these constraints through a diverse set of reproductive tactics and strategies. Thus considering the extent to which evolutionary outcomes of competitive interactions are influenced by reproduction per se, and reproductive rate in particular, marsupial reproduction does not generally appear to be competetively inferior to that of eutherians. Reproductive differences between marsupials and eutherians reflect compromises of phy-

logeny and ecological tactics not reproductive inferiority or superiority on the part of marsupials.

Nicoll (1987) and Thompson (1987) in their studies on basal metabolic rates and energetics suggest that the highest initial and total energy investment in reproduction is in marsupials. Nicoll (1987) states (p. 8):

> Thus, our comparison of marsupials with eutherians having low basal metabolic rates does not support claims that the short gestation and prolonged lactation of marsupials is energetically advantageous compared to the prolonged gestation and short lactation of eutherians. Indeed, we suggest a continuum of reproductive energetic strategies among therian mammals based not on phylogeny but on basal metabolic rates, which is overlain by a diverse array of developmental patterns.

Similar conclusions are noted in Lillegraven et al. (1987).

Preexisting conditions nevertheless do represent the powerful constraints for adaptive solutions for all lineages. The "effect hypothesis" is a restatement of this old adage. Obviously, the matter of a metatherian versus eutherian reproductive system is fundamental in this regard in ways that lie beyond the comparative evaluation of these systems alone. This does not make the resulting new lineages, after the split of the ancestral one, into an "inferior" versus "superior" subdivision, and it does not alter the critical specific evolutionary roles of the reproductive–developmental systems. In fact, the question that arises is the one that pertains to the nature of the ancestral reproductive and developmental patterns. As Lillegraven (1979) argued repeatedly (see details of this below), there is no meaningful evidence that contradicts the hypothesis that the primitive metatherian mode of reproduction (implying that this was the condition of the therian ancestry common to these two groups) was the stage from which the primitive etherian reproductive system evolved. Furthermore, the obligate commitment of the marsupial arm and shoulder to the task of propelling the neonate to the pouch has fundamentally constrained the possible range of development of adult responses to environmental opportunities and competetive interactions, in spite of arguments by Kirsch (1977) or Parker (1977). It is not the reproductive system's ability to respond to pressures of adaptation or competition that is in question, but rather its evolutionary potential in limiting, or in some cases facilitating development of areas of the organism (and therefore ultimately the bioroles of its sundry features). Such a view offers a significant perspective on marsupial evolution.

What attributes may be objectively judged as either primitive or derived have nothing whatsoever to do with the status and *competetive outcome* related concepts of "inferiority." In light of avian competitive success among vertebrates, the stem–amniote mode of reproduction is clearly a highly successful system, albeit intricately and contextually dependent on other survival aspects of avian biology – flight, endothermy, and parental behavior included. My own conviction is that the development and reproductive strategy of the *primitive* living marsupials is a near (but not complete, see Renfree, 1993) faithful preservation of the (living) therian ancestral condition. Furthermore, the eutherian blastocyst and trophoblast (the absence of a form of the latter in marsupials is not suggested) is a phylogenetic advance derivable from such a condition. Stating this in such terms is merely the rigorous phylogenetic usage of concepts (and language), and as such it is totally devoid of the inuendo of "superiority." Without the analytical and terminological conventions of evolutionary theory employed in taxonomy, there is no objective study of evolutionary history. Phylogenetic designations, in a Haeckelian framework of "phylogenesis," which would link hierarchy with optimality related "progress," can, of course, clearly reflect preconceived biases of biological "superiority." For a long time such designations did. The anxieties expressed by various students of marsupials about the concepts of primitive and advanced, however, may be related to the vernacular, but *taxonomically invalid*, meaning attached to these terms. Most taxonomists, who usually use these concepts with appropriately narrow rigor, have long decoupled the vernacular connotation. Whether one system of characters serves organisms better or worse is judged by behavioral ecologists differently than the phylogeny related concepts used by evolutionists who reconstruct historical–narrative explanations. I suspect that the earliest tribosphenic (therian) mammals, were we able to study them in toto, would be considered "marsupials," based on their reproductive biology and development. While it is only tangentially relevant, the recent account of Cockburn (1989) on adaptive aspects of marsupial reproduction must be noted. His arguments that large, as opposed to small, young is primitive in marsupials are unconvincing to me. The polytocous (multiple young per brood) and tiny young seen in most didelphids are still the most convincing hypothesis for the last common ancestor of living marsupials.

Given the foregoing, would not the retention of the earliest therian traits of the reproductive pattern and early development make living marsupials, along with monotremes, the source for some of the most fundamental questions and answers about the biology of "mammalness," and a key to the understanding of mammalian character transformations? There is no "mainstream" in the study of mammalian evolution, only anthropomorphic perceptions of it. The most fundamental aspects of mammalian biology are those that are among the most ancient and those that were present in the last common ancestor. Such attributes had the potential of channeling all subsequent changes that followed.

**Developmental biology**

The study of marsupial and eutherian development is almost as ancient as the interest in embryology, although the scientific discovery of marsupials by European naturalists obviously post-dates that of eutherians. Developmental attributes, therefore, have long played a significant role in the understanding of the major outlines of mammalian evolution. Perhaps more than just a historical curiosity, but rather an instance which gives some insight into his grapplings with taxonomy, Darwin in a letter to Lyell (September 23, 1860) has outlined his ideas of therian evolution. He indicated in a sketch that a highly evolved marsupial condition was preceded by "true marsupials lowly developed" and that the placentals diverged from the "highly developed true marsupials" (F. Darwin, 1887). His usage of the term "marsupial" was certainly in a paraphyletic sense. Even today, developmental studies supply some of the most important criteria, at least in terms of the tested developmental adaptations and constraints (Alberch, 1980, 1982; Maynard-Smith et al., 1985), within which one should contextually evaluate the evolution of all characters and groups.

In contrast to the eutherian manner of inner cell mass formation, early embryogenesis of marsupials follows the pattern seen in other amniotes. Marsupials retain an eggshell that disintegrates at about two thirds of the pregnancy, thus allowing the intrauterine "hatching" of embryos in these ovoviviparous mammals. This is in contrast to the eutherian condition, which lacks an eggshell, and very early in development forms first, similarly to marsupials, a yolk sac placenta, and subsequently a chorioallantoic placenta. Renfree (1993), on the other hand, suggests that marsupial origins involved the degeneration of the allantoic placenta, the curtailment of intrauterine life (implying a preexisting longer state), in addition to a neonate more immature than in the protoeutherian.

The presence of a eutherian trophoblast (whether or not this designation should be used for homologous tissues in marsupials), which prevents immune rejection of the young by the mother, is considered by Lillegraven (1975, p. 720) as "the single most important evolutionary event" in the origin of eutherians, a view that is favored here in light of the morphological similarity of the early marsupials and placentals (see particularly the discussions in Lillegraven, 1979 and Maier, 1993). Nevertheless, Renfree's (1993) recent analysis raises some still unanswered questions. The issue of the significance and homologies of the trophoblast is a contentious one, as expected, considering its potential evolutionary significance (see Lillegraven, 1979, 1985, and for alternate interpretations see Padykula & Taylor, 1977, and Tyndale-Biscoe & Renfree, 1987). The trophoblast and its derivatives in eutherians allow continued intimate, placental association of the intrauterine young with the mother. The lack of this comparable innovation in marsupials, in light of the reality of an immunological attack after the inert eggshell disintegrates, "forces" (in terms of survival related selective imperatives) the neonate to be born.

While Lillegraven (1969, 1974, 1975, 1979) has been a champion of the cause of evolutionary understanding of both the developmental and reproductive evidence in Metatheria and Eutheria, he has been sharply criticized by Tyndale-Biscoe (1987) and Renfree (1993). In addition to the integration of the large embryological literature (much of it in German), Lillegraven has contributed balanced analytical evolutionary interpretations concerning the developmental–reproductive consequences of the marsupial–placental dichotomy. He has reviewed and synthesized significant recent contributions of Müller (1967, 1968a, 1968b, 1969a, 1969b, 1969c, 1972–1973) on metatherian ontogeny, and other closely relevant literature. An extended and detailed discussion of much of this literature can be found in Lillegraven (1979). Lillegraven (1975, p. 719) summarized some of the more basic aspects of marsupial development in contrast to the more familiar eutherian ones as follows:

> Living marsupials are born following a short period of internal gestation at a remarkably immature stage of organogenesis. As described by Dr. Fabiola Müller, they have a series of contemporary closures for the mouth, eyes, and ears that develop shortly before birth and are functionally significant after birth in attachment to the nipple, development of the secondary jaw joint, and in protecting the eyes and ears against desiccation in the new atmospheric environment. Living eutherians generally have longer periods of internal gestation than do marsupials, and some have greatly lengthened intrauterine development times. Even eutherians that may be born in precocial states, however, intrauterinely form transitory mouth, eye, and ear closures at early stages of development that may be lost well before birth. These closures are today without function in eutherian nidifuges and seem only to be conservative features retained from eutherian ancestors that were born, as are marsupials today, in an extremely nidiculous state following a short period of internal gestation.

In addition, I also quote here a supplemental summary by Hill and Hill (1955) concerning some other attributes which they described in dasyurids:

> Notable at birth are 1) the relatively advanced state of the forelimbs, 2) the presence of deciduous claws upon the manual digits and their later replacement by definitive claws, 3) the presence of the oral shield as a specialization upon the muzzle, 4) the unique [i.e., to dasyurids] cervical swelling, 5) special features of the tongue and larynx, with the intranarial epiglottis, 6) the reptilian state of lungs, 7) the advanced condition of the stomach and duodenum compared with the rest of the gut, 8) the state of development of brain and sense organs, the olfactory parts being especially forward in development, together with those parts of the nervous system necessary for controlling the movements of the

forelimb, sucking and respiratory movements, 9) the indifferent state of the gonads.

In light of what is known of the developmental evidence, it appears nearly certain that the primitive marsupial condition of development and reproduction, birth, and post-neonate nipple-attachment and growth of an "embryo" was closely similar to that which was *antecedent* to the eutherian common ancestor. There is no known evidence that would contradict this hypothesis. It appears that the ovoviviparous (probably "marsupial-like") lineage that led to the common ancestor of the *living therian* mammals evolved from an oviparous and altricial (nidiculous, i.e., remaining in the nest for a period after hatching – a term borrowed from ornithology) ancestor, a mammal (atribosphenic or tribosphenic) in which the hatchling (rather than a neonate) was capable of initial movements that took it to the nipples without the mother aiding in this process. The diagnostic (apomorphous, but ancestral for subsequent lineages) *therian* condition (of the living groups) is probably retained by the primitive metatherian strategy of reproduction and development. The *mammalian* condition ancestral to the therian one, however, may be closely represented by the reproductive–developmental system of the advanced oviparous monotremes (see Renfree, 1993). The relatively (developmentally) precocious strategy of the first placentals was built on the therian–marsupial strategy, emphasized by the previous references. The term "nidifugous" is mistakenly applied to the first placentals. The word and the concept it stands for is used in ornithology, and describes a condition of rapid development of *hatchlings* that leave the nest shortly after *hatching* (one may substitute birth here). Used for eutherians in contrast to marsupials, it implies, mistakenly I believe, a hypothesis of transformation from a nidiculous *marsupial* condition. While the two terms correctly contrast degrees of development before birth, neither marsupials nor eutherians hatch, nor is the assumption of quick nest departure necessarily warranted for the earliest eutherians (see below).

In his developmental studies, Klima (1987) usefully standardized the various stages of marsupial development as **intrauterine embryo, neonate,** and **pouch young.** His studies deal with the development of the pectoral girdle, which is adaptively critical and developmentally one of the most precocious functional areas in the life of marsupials. These works (see references in Klima, 1987) have again drawn attention to the fact that selective forces and subsequent adaptations in marsupials must be evaluated in light of the entire life history strategies, particularly the critical constraints imposed by the developmental mechanics. I believe that this fundamentally functional, historical, and adaptive perspective should be taken into account in evaluating all aspects of marsupial biology.

Although a vestigial "egg-tooth" persists in marsupials (to perform lactation related, and presumably new therian, bioroles, according to Renfree, 1993), there can be no doubt about the adaptive significance of the "larval claws" in tearing through the embryonic membranes during parturition, and subsequently their biorole in clinging to the mother's fur during the passage to the nipples (Hill & Hill, 1955). Similarly, the changes in the development of the pectoral girdle in marsupials, particularly the shoulder–breast arch, are one of the most interesting vertebrate examples. The adaptive requirements that bridge the neonate to pouch young stage lay the foundations for continued development. This pattern of development, while specifically adaptive for the neonate, at the same time must, to some degree, constrain the development of at least this part of the post-neonate animals. Furthermore, as long as the pattern of development requires the climbing-crawling locomotion of the neonate, the conservativeness of the early development of the pectoral girdle must be preserved, irrespective of the various potential adult-related adaptive demands. Yet, the adaptive modifications of the the breast–shoulder apparatus, combined with a developmental perspective related to post-pouch young feeding and postural (locomotor and positional) habits, have not been studied. This could be profitably undertaken in a comparative framework. A hypothesis clearly articulated by Lillegraven (1975, p. 715–16) may be thus tested. He stated that:

> Marsupial development is restricted in that all pathways leading to the newborn must result in a creature that can, on its own, reach the nipple and survive there breathing air. . . . Many [eutherian] alterations in ontogenesis result from profoundly different adaptations to the environment, and some of these kinds of adaptations would be simply impossible in marsupials because of the necessity of that arduous independent "Australian crawl" from down under to the nipple at birth.

This view is supported, even if indirectly, by Hall and Hughes (1987, p. 269), who state that: "The birth mode has had severe limitations on the evolutionary diversification within the Marsupialia, particularly in regard to the adaptation of the forelimbs and possibly brain size."

Klima's (1987) studies on the ontogeny of amniote and mammalian breast–shoulder apparatus have produced a wealth of data and some very perceptive analytical interpretations of this area. His work illustrates how the establishment of homologies and testing of evolutionary transformation hypotheses is intimately related to functional–adaptive considerations within a wide comparative perspective of the entire ontogeny. The first significant fact apparent to me is that the therian diagnostic pattern of the breast–shoulder arch was probably the nearly uniform neonate as well as adult patterns seen in marsupials. This suggests a marsupial-type embryonic transformation, serving the specific biological roles noted, in the last common ancestor of living therians. This conclusion is based on the similarities (with significant functional differences) of the embryonic patterns of the anlage between marsupials and monotremes, and on the near uniformity of the adult therian pattern.

The analysis of the functional and adaptive aspects of the developing breast–shoulder apparatus in marsupials and monotremes was clearly the determining factor in Klima's conclusions concerning the evolution of this area. Although in both groups the postulated proximal-pectoral area consists of the same individual parts, the evolutionary transformation, and therefore the homologies of the adult bones, are critically different in the two groups. The monotreme and marsupial embryonic conditions are topologically, but not functionally, similar. In monotremes, significantly, there is not the strong bracing of the embryonic elements from the shoulder to the breast plate to the other shoulder as we see in marsupials. Yet, the actual minor topological differences between marsupials and monotremes relate to the presence of the paired dermal elements of the interclavicle in the latter. The unpaired chondral element of the interclavicle in the embryo of monotremes, which forms part of the interclavicle in the adult, is the same as the unpaired central element in therians, which in the latter becomes part of the manubrium sterni. Monotreme adults have not only an ossified interclavicle of dual composition, but also a procoracoid (praecoracoid or epicoracoid), as well as a metacoracoid (the posterior element). Therians, lacking an interclavicle (although they incorporate part of its anlage into the manubrium), have only small ossified remnants of the procoracoid anlage (praeclavium in marsupials, and possibly the suprasternal ossicles in placentals), and incorporate the metacoracoid portion of the shoulder anlage into the coracoid process of the scapula.

## Evolution, locomotion, reproductive biology, and physiology

Reproductive biology and development are obviously related to one another, yet mention of some issues, while relevant, can only be tangential to the emphasis in this book on the skeletal system in character analysis. Energetics of reproduction and their physiological and ecological implications have played an important central role in recent discussions of marsupial evolution, as noted at the beginning of this chapter. Many studies dealt with the fascinating physiological, ecological, and behavioral consequences of marsupial and placental developments, and I will only briefly focus on the possible differences in evolutionary causality that may follow from the two distinct patterns. Pond's (1977) and Lillegraven's (1979) reviews of lactation and reproduction in Mesozoic mammals point in a direction where the fossil record will be increasingly more fully interpreted, and the degree of confidence in the various hypotheses tested against osteological observations will grow (see also Guilette & Hotton, 1986). Lillegraven's (1979) essay stands out among discussions of the developmental, reproductive, and evolutionary factors in the origins and differentiation of mammals. This essay is particularly important as a background for any causally oriented detailed study of marsupial evolution. A recent thorough examination of the origin of lactation and the evolution of milk by Blackburn, Hayssen and Murphy (1989) is an equally important contribution in which the authors parenthetically, and significantly, state (p. 20) that: "Ideally, evolutionary scenarios are an early stage, rather than a final step, in reconstructions of evolutionary history."

Müller (1972–73), based on developmental data, has argued that marsupials are capable only of "phylogenetic" *expansion*, that is, a developmental prolongation of intrauterine time, but not what she called "phylogenetic" *extension*, that is, a developmental complexification, which is, as she defined it, the simultaneous increase of intrauterine time as well as an increase in organizational complexity of the embryo. The latter phenomenon is characteristic of eutherian lineages but not of metatherian pre-neonate ontogeny. While the near conflation of the concepts "phylogenetic" and "ontogenetic" is not ideal, Müller's observations are significant. The tightly argued papers of Lillegraven (1975, 1979) and Lillegraven et al. (1987) for the thesis that eutherians have a development- and reproduction-related, fundamental, evolutionary advantage over marsupials are well known, as are the alternate perspectives of Sharman (1970), Kirsch (1977), Parker (1977), Tyndale-Biscoe and Renfree (1987), and others. That debate had been fruitful and enlightening, but it is not directly relevant to the historical–narrative account of marsupial diversity. Ideally, as I noted previously, the debate should be a comparison of the ancestral species of the two groups of living therians. It can, however, serve as necessary background information for understanding the perceptions and approaches of previous students to topics dealing with marsupial phylogeny.

I propose, after a relevant discussion of the epipubic bone, an adaptively and phyletically more complex perspective on the adaptive differences of the two kinds of therian reproductive and developmental systems than what has been considered before. Lillegraven (1979, p. 272–3) argued for the adaptive advantages of both the young and mother in the placental system of carrying young securely in the uterus while locomoting for a living, in contrast to the more vulnerably exposed marsupial mother and her attached young.

There is one area of the metatherian postcranial anatomy that has long been tied to living marsupials as being one of their characteristic attributes. We do know today, however, that the so-called "marsupial" or properly named epipubic bone is not an originally metatherian, but a rather more ancient cynodont acquisition, although clearly part of the adaptive strategy of living marsupial mammals. Thus, while a primitive trait, the epipubic bone is also obviously an adaptation serving a number of bioroles (White, 1989). Nevertheless, particularly for the reason of its antiquity and the light it may shed on the understanding of the function, biological role, and origins of critical adaptive solutions in synapsid evolution, functional–adaptive analyses of the epipubic bone are of special significance for mammalian phylogenetics. For a long time, ever since marsupials were known to zoology, the presence of epipubic bones has been associated with the marsupium and its loaded-down condition when carrying a litter of neonates. While this causal association has never been fully accepted, it had some general support particularly in light of the sexual dimorphism of this bone. White's (1989) recent analysis, addressing exactly this problem, using scaling theory, bone measurements, and reproductive strategy information, is both unique and penetrating in its contribution to the understanding of the evolution of the epipubic bone. White (1989) has taken linear dimensions of the epipubic bone in 61 species, one of which was the platypus, and scaled them to mass. He has carefully detailed both the "marsupium support hypothesis" and the "locomotion hypothesis" as these may be related to allometry and dimorphism of the epipubis. His detailed functional outlines of both hypotheses were tested by least-squares regression analyses. White presented regressions of log epipubic length/log mass, log epipubic width/log mass, and log epipubic length/log epipubic width (both for the groups, and for individual specimens of both sexes of *Didelphis* and *Trichosurus*).

In stating the first of the two hypotheses he discussed, the "marsupium support hypothesis," White noted (p. 344) that "Two characteristics of epipubic bone architecture are suggested by this hypothesis: epipubic bones may be sexually dimorphic and epipubic bones in females should scale to the maximum mass of the young which must be carried in the pouch." White confirmed the long recognized sexual dimorphism in the expression of this bone. Yet, his empirical findings were that in those species that have marsupia the epipubic bones did not scale as well to expectations as in those species without a marsupium. White noted that the fact (p. 351–2)

> That there is sexual dimorphism in taxa without marsupia indicates a functional difference for epipubic bones independent of support of a marsupium. . . . Scaling theory suggests that the lower slope observed for females could be explained by the loading of the epipubic bones associated with litter mass; they are observed to become broader with increasing litter mass. . . . This suggests an alternate hypothesis that epipubes of pouchless taxa

do serve to support litter mass, perhaps directly supporting the mammae and attached young. How stress related to litter mass might be associated with the epipubic bones in pouchless taxa is unknown. As a similar relationship between epipubic width and mass does not exist in taxa possessing marsupia, a different regime of epipubic loading must be assumed. This implies that the presence of a marsupium is sufficient for supporting litter mass and, as a result, epipubic bones may be allowed to assume configurations suited to perform other functions.

So White's data suggest the puzzling trend that of the pouched species, for a given mass, the females had, as a rule, longer epipubes than the males. That, he suggests, may be the result of a (putative) relaxed constraint on the epipubis as a result of the presence of the pouch. Yet, he added, scaling theory would predict that these females, rather than the males should have the shorter and relatively wider, and therefore stouter, bones for pouch support. This observation led White to another complementary, rather than necessarily alternate, hypothesis relating the epipubis to locomotor adaptations.

White (1989, pp. 345 & 347) stated that

> The basis of this [locomotor] hypothesis is my observation that the epipubic bones form a kinetic link between the trunk and the pelvic limbs. The epipubic bones articulate on the pubic rami of the innominate bones with hinge-like joints. The epipubic bones serve as a site of insertion for the hypaxial muscles of the trunk and the site of origin for the pectineus of marsupials and the gracilis of monotremes. As a result, the shortening of the hypaxial muscles (e.g., rectus abdominis, pyramidalis, external oblique) whose action is bending of the trunk is also effective in adducting the epipubic bones. In turn, this action lengthens the pectineus or gracilis and serves to assist protraction of the pelvic limb especially if the pectineus or gracilis actively resists lengthening . . . [p. 347]. The length of the epipubic bones approximates the effective lever arm and mechanical advantage of the hypaxial muscles. Mechanical advantage is determined by the ratio of the lever arms of the hypaxial muscles and the pectineus. The lever arm of the pectineus is always small since it normally takes its origin near the base of the epipubic bone. Because the hypaxial muscles attach along the entire length . . . their effective lever arm is determined by this length. If epipubic length is assumed to scale to maintain a length (mechanical advantage) sufficient to allow the hypaxial muscles to overcome limb inertia and assist in limb protraction, then a geometrically similar mechanical advantage can be determined.

White's data and conclusions emphasize that the similar scaling of the epipubes and effective limb length suggest similar loading regimes. His data also support his locomotion hypothesis as just stated. In general White's findings suggest that the epipubic bone in pouchless taxa has relative dimensions that are much less likely to fail, given equivalent loads, than their counterparts in the pouched species. But, as he notes (p. 354): "This observation, however, cannot be used to falsify the marsupium support hypothesis." As stressed by White (1989; see also particularly Chapter 4), the biological role of the epipubes in marsupials is not a single one, but it probably represents a compromise solution for different roles. He astutely remarked (p. 344): "This study does not presume to suggest how epipubic bones function in individual taxa. It is an attempt to elucidate epipubic form across all available taxa in order to infer their generalized function."

I consider White's "generalized function" to mean the roles performed by the phylogenetic and adaptive compromises of that complex in members of a lineage which came to carry their young on their teats, from an ancestry that developed the epipubis for lineage specific locomotor strategy. The epipubis and the attendant musculature and mechanics probably first arose as a character complex related to locomotion. The egg-laying and nesting habits of the relic monotremes appear to support that. Marsupium and/or litter support is an added adaptive modification built on the preadapted locomotor-related arrangements in the cynodonts and the earliest mammals. The eutherians, almost certainly tied to their divergent development and reproductive strategy have lost this trait. But it is likely that the earliest cladistic eutherians may have retained this conservative feature in a number of lineages (see Kielan-Jaworowska, 1975b). The question remains intriguing: Were these cladistic eutherians still reproductively "marsupial"?

Marsupial females are burdened by the attached and lactating young, and the construction of the *therian* nipples themselves (with an expanded tip) might have been an adaptation for *attaching* the altricial young and to feed them continuously, as required by their extremely altricial condition. These characteristically constructed nipples may be called *postareolate* inasmuch as the monotreme ones, while have areolae (hence *areolate*) lack the expanded tip of therians. It is also well to remember that monotreme and marsupial neonates are at a comparable stage of development (see Tyndale-Biscoe and Renfree, 1987). A highly stressed terrestrial lineage of marsupials subsequently may have "invented" (as suggested by the role of the trophoblast) the protection of such embryos and carried them in the uterus beyond the usual three weeks in marsupials. To use Müller's (1972–3) perceptive designations, the first developmentally and reproductively eutherian lineage phylogenetically extended and complexified intrauterine development of its marsupial ancestry. This strategy permitted by the prevention of immunodiffusion would allow taking advantage of the opportunities offered by more precocial neonates compared to those of the (on the average) three week old marsupial antecedents.

These protoplacental neonates would be called altricial by placental standards. Modern precocial young of eutherian, as seen in the highly cursorial ungulates, represent highly derived modifications among placentals. I suggest here that the locomotor-related demands on an early terrestrial form with a marsupial mode of development (a marsupial) was the initial and boundary condition, the setting for the selection of a eutherian-mode of young retention and development. The initial advantages for the protoplacentals with the novel reproductive strategy of extended development, beyond that attained by a marsupial ancestor, was permitted by the trophoblast (see Lillegraven, 1985, and references therein). But the slightly extended development of the protoplacental young and the subsequent relative locomotor independence of the mother from the young may be causally the most significant. By having neonates that could be temporarily left independent of the lactating mother, the advantages for the mother were probably survival-related.

The longer the young may have been retained in the earliest placental which overcame the immunodiffusion problem, the more vulnerable the foraging pregnant mother became because of the burdensome load of the developing young in utero. Selection at that stage may have favored a condition that allowed a neonate to be "precocial-enough" to be left unattended for periods of time, compared to the antecedent and altricial marsupial standards, but which in fact was still very altricial by phyletically subsequent eutherian ones. What, then, was the adaptively significant reason for this particular balance between a developmental condition in utero more advanced beyond the marsupial one, but by no means precocial by any living eutherian standard? I suggest that it required a certain level of ontogenetic development for these placental neonates to be able to remain in a (hidden) nest, but not developed enough to wander, without a constant source of nourishment. So at the particular point of development when the placental neonates were developmentally advanced enough to be left behind in a nest, the time of vulnerability for the pregnant mother became shorter, compared to an antecedent marsupial mode of rearing young. Thus, the unburdened placental mother, while lactating, may have enjoyed greater survival (and consequently reproductive) advantages than the preexisting marsupial-stage strategy allowed. Slight ontogenetic extension coupled with complexification of intrauterine development and nesting, then, might have been the critical factors in the development of the protoplacental strategy from a reproductively marsupial one.

## Genetic and molecular evidence

The successful analysis of the phylogenies of various molecules and a wide variety of conceptual approaches related to genetic and molecular phylogenetics are well known, and the general expectations based on the results of molecular techniques and methods have been widely circulated to the extent that entire journals are devoted to this area of research. In numerous groups, particularly those with no fossil record, any and all molecular information, albeit some very expensively obtained, should be welcome and integrated judiciously with other data for a most corroborated evolutionary history of groups. Different molecules will offer evidence against which different evolutionary problems may be tested, perhaps taxic ones at different levels, but no molecular evidence will help determine the history of all groups (as so clearly stated by K. A. Joysey, personal communication). This is, of course, exactly what one should say about any other aspect of the phenotype, even the time bridging hard anatomy.

Genetic and molecular evidence, however, has received the kind of proselytizing, and not even from its practitioners, that should be pointed out here. Some students of the mammalian fossil record, with access to rich repertories of morphological data and its potential garnering into taxonomic properties, have often turned to molecular phylogenies as the real hope for "pure genealogies." But none have articulated this enthusiasm in quite the same way as has the evolutionary biologist and eminent literary figure, S. J. Gould (1985, 1986). This is of particular significance because Gould is both a morphologist and a taxonomist whose brilliant and deeply insightful writings enriched the field of evolutionary biology. His peerless (and not infrequently agenda laden) natural history essays are widely read, and through these he has had enormous potential influence on the thinking of students who have tasted enthusiasm, but often without the sobering joy of some understanding. Nonhistorical biologists and anthropologists also tend to look upon Gould's essays on evolutionary analysis as the most widely accepted distillation of the wisdom of historical biology.

Referring to the technique of DNA–DNA hybridization, Gould (1985, p. 18–25) proselytized that:

> We should all rejoice in the success of molecular phylogeny because its techniques have probably solved *the* biological problem of the ages, so recognized ever since Aristotle (and surely, in unrecorded form, well before – whenever people puzzled about their place in nature): namely, how, and why are organisms interrelated. . . . Sibley's work, and other efforts of molecular phylogeny, have received a fair slug of press lately, but I do not fully understand why we are not proclaiming the message from the housetops [italics supplied].

Elsewhere Gould (1986, p. 68) in a resoundingly positive paper championing historical biology and in a commemorative accolade to Darwin whom he perceptively hailed as a historical methodologist, stated his own view of character analysis: "Once we map homologies properly, we can finally begin to ask interesting biological questions about function and development – that is, we can use morphology for its intrinsic sources of enlightenment, and not as *an inherently flawed measure* of genealogical relationships" [italics supplied]. It is somewhat surprising that Gould, the morphologist who has in the past called for the rebirth of the science of form, appears to imply the "proper" mapping of homologies to be devoid of the complex and precise level of resolution that such "flawed" morphology permits. In such hasty comparisons of morphology with molecular evidence it should matter that at least DNA–DNA hybridization, for example, one of the several promising quantitative molecular approaches, is proving to be all but ineffective for higher level mammalian relationships, at least those beyond 15 to 20 million years. Similarly, the promise of mtDNA studies is mired in the yet uninvented complexities of the necessary statistical analyses. There is little hope, in my view, for rendering most of the precious molecular evidence relevant to phylogenetics unless the functional–adaptive attributes of specific molecules are carefully taken into account as the context of qualitative analyses of gene products. And it matters, and Gould the paleontologist must know that the understanding of innumerable intricate homologies of the phenotype, as well as detailed convergences, is a triumph of the post-Darwinian morphological approach. One should also add the fact that molecular phylogenies almost invariably fall back on some fundamental assumptions in their time scales based on the fossil record, as in fact all morphological studies ultimately must (see especially Cartmill, 1981). Furthermore, groups of the fossil record can be integrated into a *complete* history of organisms only through the study of their morphology. This is so in spite of the fact that Darwin, Haeckel, and their many contemporaries made some astonishingly accurate assessments of phylogeny based on morphology of living species, without a good fossil record.

More specifically related to one particular molecular approach, there is a curious contradiction that exists in the literature concerning the use of serology, be it estimates of antigen–antibody reactions in a liquid medium (sometimes measured nephalometrically), or micro-complement fixation and immunoprecipitation in gels. The serological distances, which are extremely valuable phenetic measures, are nevertheless thought, by advocates of "molecular clock" models, "usually" to reflect cladistic ties. This has served as the base for using serology in considerations of phylogenetic hypotheses, particularly for marsupials. The avowedly, and usefully, phenetic nature of the data, however, stands in direct contradistinction to one particular way of employing serology in a cladistic analysis of a group. Reig, Kirsch and Marshall (1987, p. 52) in fact state that an assumption of rate constancy and the fact that "serological distances approximate cladistic affinities among opossums . . . provide the basis for choosing among other hypotheses of relationships." I must point out that in that particular study of marsupial relationships (see also Kirsch, 1977, 1982) the assumption of rate-constancy underlies not only biochemical systematics, but, as the authors clearly imply, the weighting of other characters they use. These authors consider a serology based tree as an "initial cladistic hypothesis" (p. 54). This is, of course, a methodologically circular use of morphological data.

An integration of much valuable and robust molecular research into the understanding of phylogenetics will undoubtedly happen, independently of the morphological data, as it should. As stated by Morris and Cobabe (1991, p. 307): "molecules can provide significant phylogenetic information when accompanied by a careful analysis of their structrure, synthesis, genetics, and function." The examination of the proximal mechanisms and bioroles responsible for the structure of morphological as well as molecular attributes is the key for making both optimally revealing about phylogeny. To paraphrase, in homage, V. Sarich's quips in his sundry pioneering efforts in molecular taxonomy, the chips must fall where they may, but without, a priori, considering any line of evidence as primary. The lasting contributions of the various subfields of historical research will undoubtedly reinforce one another in a noncircular manner.

## Overview of the cranioskeletal system

One of the most critical areas of research for understanding significant aspects of hard-tissue morphology, the relative absence of which is often lamented in this book, is the study of marsupial behavioral and ecological morphology, particularly positional (postural and locomotor) and feeding behaviors. Without continued elucidation of all the particular evidence from the living, research concerning the evolution of teeth and bones, with its potential to bridge time in phylogenetic studies, is likely to stagnate, or merely continue to focus on taxic schemes. A critical aspect of character analysis is the functional–adaptive analysis of interrelated attributes of organisms, because the naturalistic behavior of the animals is partly causal in the modification of their anatomy. Yet this empirical area of study, potentially capable of yielding significant and robust taxonomic properties of the skeletal system, has been lagging behind the appeal of manipulating raw morphological data via theoretical solutions derived from cladistic techniques built around various algorithms. Perhaps this is correlated with the increase in the faith of many investigators that molecular techniques have a kind of "objectivity" and "primacy" because they are somewhat more proximal to the genome (of living organisms). Kirsch et al. (1991), for example, consider one of the taxonomic properties from the tarsals of marsupials as an "insight" rather than what it is – a taxonomic property which, if accepted by others, should be viewed as a potential and boundary condition for a number of hypotheses. Given the fact that these authors (1) dismiss the explanatory power of a hypothesis that deals with the shared apomorphies of the tarsus in all of the australidelphian marsupials, (2) dismiss syndactyly, and (3) consider *Dromiciops* as the sister group of diprotodontians, suggests to me a veritable chasm in the understanding of the significance of taxonomic properties based on morphological research or the nature of what taxonomic properties are. This is perhaps not surprising in the light of the numerous contributions where the total number, and not the nature or significance of characters, that "supports" a cladogram and the built-in a posteriori approach to "homology" (i.e., synapomorphy) recognition have come to be synonymous with "proper" phylogenetics. Furthermore, "standard" taxonomic accounts with their rejection of the theoretical constraints of evolutionary morphology can misread and miscast morphological evidence because they eschew implicit or explicit use of functional–adaptive understanding of a whole complex of related traits (e. g., Hershkovitz, 1992). The significance of morphology and the technique of vertical comparison has been undoubtedly diluted by both the practitioners of the "old" as well as the "new" taxonomic approach. But most taxonomists using cladistic techniques exclusively consider the foundational role of evolutionary morphology, with all the necessary "nontaxonomic" (extra) training and research programs it requires, as "scenario" and therefore nonessential in the training of taxonomists. This kind of misunderstanding reveals that even complex and robustly tested morphological taxonomic characters are looked upon either by students of molecular biology or by the practitioners of a more conventional taxonomy as merely more of the countless disposable attempts that follow the vagaries and circularity of cookbook procedures based on the often chimeric and nonbiological assumptions that can form the bases of some algorithms. In light of how glibly many taxonomists using morphology often construct character matrices for algorithm-driven cladistic analyses (and the attendant assumptions), or even most of the traditional atemporal endeavors, perhaps students of molecular taxonomy cannot be held accountable for their perspectives on morphological attributes.

As may be obvious from the foregoing discussion, the evaluation of the evidence, or rather the recovery or observation of the features from the specimens themselves, is critically dependent on the observer's perspective and methodology, as well as a host of assumptions about biology, mechanics, and evolutionary processes (Szalay, 1981). It is not only the characters, but the taxonomist's largely training based ability to link features, and thus facilitate the testing of transformation hypotheses, that determines the success of phylogenetic endeavors in taxonomy. Very often students who claim the independence of their cladistic efforts using morphological data from any functional–adaptive foundations make extensive comparative use of characters and transformational tests that are the results of exhaustive non-taxic and character oriented evolutionary research of others, and thus they soundly base their comparisons on significant character complexes. Phylogenetic analysis of morphology is critically dependent on the understanding of constraints which channel change. Unfortunately, most of current understanding of these constraints is still restricted to a semantic and theoretical realm. Research methodologies with testable ontology and accessible epistemology are sorely needed. Meanwhile the perspective that certain aspects of a given area of morphology play a role in directing change, be they phylogenetic, morphogenetic, or adaptational, is a critical element in the decision-making process of character analysis. Recognition of group specific phylogenetic (heritage) factors are the objects of such research, precious epiphenomena of sundry biological and evolutionary studies.

Not rarely the descriptive literature of osteology, and of fossils also, is not as useful as it should be, often merely signaling that some interesting taxa or specimens exist. There are several reasons for this shortcoming. Perhaps the first and most important one is that relatively few adaquate or well-ordered modern comparative studies have been made on the musculoskeletal morphology of most living species. Equally rare are modern works dealing with the evolutionary, particularly ecological, morphology of selected areas of the skeleton. Given this relative paucity in modern zoological morphology, there are relatively few standards of comparison. There is also a relative scarcity of contributions on theoretical and empirical guidance that would direct the process of integration and interpretation of fossils.

Finally, very few institutions or departments in the world focus on the integrated interpretation of evolutionary morphology, compared to those in anatomy, molecular, cell, and genetic biology. When in unique national institutions, founded and designed for the collection and housing of specimens of organisms, departments of comparative anatomy have been abolished and those devoted to molecular phylogenetics are established, a clear message is sent to the young new talent entering taxonomy. Graduate training tends to be haphazard, and often the very theory or the empirical training that is relevant is not carefully instructed when morphology is studied. The results of such a background for morphological studies manifest themselves in many published analyses. Such analyses are often driven by a paradigm that promises that it is not what one puts into the algorithm but the algorithm itself that is important, often employing "characters" described in a biologically vacuous framework, akin to static, process-purged descriptions of land forms. While such a depiction of many morphological studies may appear exaggerated to some, it is a fact that morphology is an area of "phylogenetic endeavor" which is used routinely even by those with a minimal training in anatomy itself. In fact anatomy, mistakenly, is the most often encountered synonym of (evolutionary) morphological analysis. It is my bias that the study of the osseous morphology of the living and fossil forms, taken together, cannot fulfill its great promise for phylogenetics apart from the complete and integrated approach of evolutionary morphology (see particularly Bock, 1990, 1991).

## Dental morphology

A good fossil record of mammalian teeth is usually an excellent guide to the analysis of form–function of a critical part of the feeding adaptations, and consequently, to the nature of the divergence and shared ancestral constraints of particular groups related to the feeding mechanism. In fact, groups in mammalian paleontology, beyond microtaxonomy, are often justifiably erected based on dental morphology clusters. Few outside the practice of paleomammalogy appreciate the pleasures of acquiring a functional and evolutionary understanding of dental diversity. But the temptation of mammalian paleontologists to regard teeth as primary evidence always, or at least most of the time, is based on many sources. Most important is the fact that in the fossil record teeth are more abundantly represented than skulls and postcrania, and they appear to evolve more rapidly. This appearance is also reinforced by the greater focus necessarily placed on the fossil dental record, compared to the equally rich storehouse of other characters in the extant forms. Fossil morphology cannot be fully appreciated unless the living homologous systems have been subjected to evolutionary analysis.

The proverbial rapidity of phyletic change of the form–function in mammal teeth, so evident from their enormous diversity, is causally driven by the selective forces of new food combinations, physical attributes of foods such as hardness, and proportions of different foods in the diet, and it often confounds paleontologists. Selection produces convergent and parallel cheek tooth solutions which, with an absence of intermediate conditions, can make dental character analysis difficult or sometimes impossible. Dental evidence can be particularly difficult when one must use it by itself to test higher taxon evolutionary relationships. This problem, however, is often balanced by the diversity and temporal position of similar samples, and therefore reliable judgments may be made about homologies and transformations. A dental record of highly diversified groups gives an excellent opportunity to study the evolutionary transformation of an important aspect of the feeding mechanism and to track the adaptive change in the diet of the taxa. But teeth, like anything else, do not evolve by themselves, and when they evolve little or not at all in many branches of a larger group, the tendency is to squeeze unwarranted phylogenetic information out of such evidence.

When the fossil record is poor and the evolutionary change in a character complex is relatively limited, particularly when such a record is spread over enormous periods of time, the problems that can surround dental evolutionary analysis are much more difficult to surmount than when the change is rapid. This is exactly what faces students of early therians, the tribotherians, metatherians, and eutherians. They must rely largely on dental evidence when attempting to determine the early ancestral-descendant and branching sequences. The consequences of the early dental record of metatherians is that we have a good understanding of taxonomic *diversity*, perhaps even feeding diversity, but necessarily a very incomplete understanding of phylogenetic relationships among several major groups or within relatively dentally stereotyped ones. In diagnosing metatherian taxa I make liberal use of the dental evidence along with other hard parts. I often discuss the nature of usually critical dentition based data and contributions. Yet I restrict illustrations of teeth in this book to a few synthetic diagrams and one plate of selected Cretaceous molars. The reason for this is simply that most of the recent relevant paleontological literature is very well described and profusely illustrated. (For a full bibliography see especially the recent critical evidence and its analysis in Cifelli, 1990a–f, 1993, and a discussion of some of this evidence by Clemens & Lillegraven, 1986.)

The nature of dental differences and similarities of most Cretaceous tribotherians, metatherians, and Paleogene metatherians is difficult to assess, and their confusing influence on attempts to understand the early evolution of these groups is obvious from various titles, or designations such as "opossum-like." While whole dentitions suggest relatively clear identifications of species as either metatherian or eutherian, or some of the early tribotherian patterns, our general ignorance is also evident from the very difficult and confusing array of taxonomic allocations by many highly competent students of early marsupials of the early ameridelphian genera to either the Pediomyidae, Peradectidae, Didelphidae, Microbiotheriidae, or even the Borhyaenidae. For example, the concepts of *Alphadon* and *Pediomys* (e.g., Fox, 1987; Archer & Aplin, 1987; Cifelli, 1990a, b; Marshall et al., 1990; etc.) are still in disarray because of the extreme difficulties one encounters in evaluating similar phenetic clusters. Clearly, the difficulty of separating ancestral from derived patterns has reached a zenith with the dentition of the tribotherians, and the various Cretaceous Asian and American metatherians are the tip of a probably holarctidelphian diversity, in addition to the "opossum-like" ameridelphians. What did emerge from my survey of dental and tarsal characters of the ameridelphian "opossum-like" taxa is that, while grouping of dental attributes is possible, particularly with an increasingly good fossil record (see Eaton, 1990; Cifelli, 1990a–f, 1993, and references therein), it is more than likely, contra Reig et al. (1987), that a few simple functional changes have repeatedly evolved among the amazingly bradytelic molar patterns of early and some living ameridelphians. This view appears to be supported by Cifelli (1990a, b; 1993) as well. A glance at Figure 8.7 will reveal, even to relatively uninitiated students of dental morphology, how narrow the limits of form are that characterize some selected Cretaceous genera. In light of the variety of *incertae sedis* taxa based on molar morphology in the Campanian Late Cretaceous of North America, Cifelli (1993) makes the convincing point that the diagnostic features of the Metatheria (i.e., the concept Ameridelphia, as used here) may be restricted to a combination of protocone enlargement and the corresponding enlargement of the talonid. As discussed in chapter 8, the concept of the Metatheria and the various monophyla comprising it may be diagnosed somewhat differently based both on dentition as well as on some additional criteria.

This gratifyingly growing, expertly analyzed, but potentially greatly confusing evidence that has bearing on the critical, early (i.e., Mesozoic and Paleogene) relationships of ameridelphian marsupials was recently extensively studied by Reig et al. (1987). The living Didelphidae, discussed in some detail below, present the most important reservoir of information that may be used to discipline one's thinking concerning the Mesozoic and Paleogene ameridelphian dental evidence. Reig et al. (1987) recognized this, yet made what I would consider somewhat less than fully effective use of the information from the dentition. This was largely due, I believe, to their failure to apply available expertise on dental development and functional–adaptive criteria to the distributional evidence. Most of the numerous characters chosen in their analysis of ameridelphians were of no supraspecific or suprageneric significance. In fact, the range of variation that has been used to group genera with families, and the latter with one or another order, was largely based on a few simple traits. These were: (1) the conformation of the crest between the two main buccal cusps of the upper teeth, the paracone and metacone; (2) the nature of the stylar cusps; (3) the size and development of the sylar shelf; (4) relative size of the paracone and metacone; (5) the proportions of the upper molars in general; and (6) the alleged emphasis of the metaconule in "later didelphoids." Shearing teeth or bunodonty, the degree to which conules develop, the relative size of the trigonids and talonids, and a large number of important minutiae that properly differentiate species and genera are often not well suited for use in the analysis of higher taxon relationships when the appropriate paleontological data is entirely lacking, and particularly when many of these groups have retained an all but primitive ameridelphian molar crown morphology. Work with dental evidence in many other groups in addition to metatherians convinces me of the evolutionary lability of such mechanically important features, and therefore of the rampant fluctuations and parallelism that were likely to occur in molars of mammals of relatively similar genetic background.

The most critical dental morphocline, and its significance in the interpretation of the dental evidence in the living Didelphidae, has eluded the analysis of Reig et al. (1987). In many ways promisingly cogent, their study was turned into an essentially phenetic, morphometric analysis. They failed to come to terms with the cause and effect aspects of functional factors in molar row evolution in the living opossums, and therefore almost certainly in all primitive ameridelphians. For example, the buccal crest of didelphids, including the mesial preparacrista, the centrocrista (the crest between the paracone and metacone), and the postmetacrista, varies between the rectilinear (*pediomyimorph*, straight and essentially running apically and basally on the crown, not W-shaped in an occlusal view, but rather U-shaped) or dilambdodont (*peratheriumorph*, in which the centrocrista sweeps buccolingually in addition to apically and basally on the crown, W-shaped in an occlusal view) conditions. In the didelphid Caluromyinae (the woolly, black-shouldered, and bushy-tailed opossums) the buccal crest (the ectoloph) is "pediomyimorph," whereas in the other didelphids it is "peratheriumorph." Furthermore, the relative size of the paracone and metacone is subequal in the caluromyines, or more precisely, the relative metacone size seems to covary with the height of the trigonid shearing distal to it. The combination of paracone stunting and/or metacone "hypertrophy" appears to be causally correlated with the enlargement of the mesial shearing wall of the occluding lower trigonid. If this is the case, then we are observing the consequences of a constantly fluctuating aspect of primitive ameridelphian dentitions: How much or how little trigonid shear is optimal?

In the Caluromyinae (and in the Microbiotheriidae) we see the similar size of the paracone and metacone correlated with a trigonid row which does not increase posteriorly, a dentition which is not committed to the distal ectoloph shearing to the extent that it is in other didelphids. So at least in the Didelphidae the ancestral pattern of molars may or may not have been rectilinear or dilambdodont. The reduction of the paracone and the consequent "enlargement" of metacone seems to vary with the relative importance of bite and shear in the tooth row, given a certain basic adaptation of the cheek dentition. Nevertheless, the disparity between the two major buccal upper cusps is ancient and possibly diagnostic of the Ameridelphia (a reverse of the ancestral therian condition). If that were so, then the recently described and diagnosed taxa, based on molars (but not complete toothrows which are unknown as yet) by Cifelli (1990a, b), *Iqualadelphis*, *Protalphadon*, *Turgidodon*, and *Anchistodelphys* from the late Cretaceous (see Fig. 8.7) may not, by such a rigid and perhaps unrealistic criteria, fully qualify to be ameridelphians. But of course even a subequal paracone and metacone on the upper molars may signal to us the diagnostic *beginnings* of paracone reduction (or metacone hypertrophy) within the Metatheria or even Ameridelphia. The genus *Iugomortiferum* from the same formation (see Fig. 8.7) described by Cifelli, who properly doubted its affinities with metatherians (ameridelphians, as I interpret it), however, really does not qualify as an ameridelphian by this criterion as it clearly preserves the primitively larger paracone compared to the metacone. Cladistically, of course, it may well have been an ameridelphian, and certainly it might well have been a metatherian.

When allocating isolated upper molars, it should be kept in mind, however, that the size disparity between the buccal cusps in Metatheria can vary with the position of the tooth. In the fossil microbiotherian *Microbiotherium*, as in caluromyines, the metacone of the first upper molar occludes against the second strongest trigonid, that of the **M/2**. As predicted, the first upper molars have slightly larger metacones than the teeth behind them. No doubt these attributes are heritable, but sharing of such apomorphies are not convincing signs of synapomorphy, of common descent, because of the high likelihood, due to the (assumed) relative simplicity of such a progressive change, of parallelism. While usually the utmost care is taken for sorting the rare and precious dental phena into taxa, variation of intertaxonally important characters within the same row of molars in better known animals can hold obviously disturbing caveats for metatherian molar taxonomy.

What, then, does the causal correlation of paracone reduction, and metacone "enlargement"(?) mean concerning ectoloph shape? Some of the Cretaceous *Alphadon* have a very well developed W-shaped ectoloph, but no significant metacone "enlargement"; the latter is perhaps an expression of relatively minor paracone reduction. *Monodelphopsis* from the Itaboraí fissures of Brazil, however, has a U-shaped ectoloph with a rectilinear centrocrista, yet has an extremely developed disparity in size between the smaller paracone and much larger metacone! This is dissimilar to the conformation of the Cretaceous *Pediomys*, with which Marshall (1987) and Marshall et al. (1990) grouped *Monodelphopsis* within the Microbiotheria. So it would appear that these two characters can vary independently from one another, perhaps related to nothing else than the degree of paracone reduction. But more importantly, they are "minor" changes of crown form, not appearing to affect morphology of the teeth very much. Their presence or absence in various taxa might not represent a character of importance equal to such features as *occluding* accessory cusps in some taxa that we know to manifest themselves even when evolutionary alterations of great magnitude have occurred in a lineage. Yet centrocrista shape as well as fluctuations in paracone/metacone proportions are exploited far beyond their usefulness in the three recent attempts to understand early ameridelphian evolution (Reig et al., 1987; Marshall, 1987; Marshall et al., 1990). Paracone reduction, because it is a trait that involves progressive simplification of crown morphology, should not be weighted beyond the fact that it *may be* uniquely present at the base of *some* of the early lineages of the Metatheria. Functional–adaptive analysis of this character complex, imaginatively conceived, is essential before such data can be used reliably.

It should be seriously considered that (1) if the complex tarsal adaptations of didelphids are synapomorphies uniting all living genera that share it, and (2) if the non-carnassiform dentition of the caluromyines are a primitive retention from the first didelphids, then (3) the carnassiform molar row of the Didelphinae evolved independently from those of other North American Cretaceous and Paleogene taxa that show this character. Clearly, therefore, the shared similarities of the cheek teeth and of the combined molar row of the Caluromyinae, Microbiotheriidae, Peradectinae, and Pediomyinae are probably ancient Cretaceous retentions, and therefore not proper bases for the evaluation of one of the early radiations of the marsupials, the arguments of Reig et al. (1987) and Marshall (1987) notwithstanding.

Something else may be stated about molar data in the Ameridelphia that is strongly supported by the tarsal evidence presented and analyzed below. The emphasis of the shearing dentition of borhyaenids, an emphasis on the metacone and paraconid with a de-emphasis on the talonid, must have been independent from the stem of the Didelphidae as we must infer from the living didelphids. The tarsal pattern of undoubted borhyaenids are not likely to be derived from the highly *modified* patterns of the ancestral Didelphidae, as this is discussed in detail below. The detailed accounts of the fossil dentitions by Clemens (1966, 1979), Clemens and Lillegraven (1986), Lillegraven (1969), Fox (1971, 1979a–c, 1981, 1983, 1987), Crochet (1977a, b, 1979a, b, 1980, 1984), de Paula Couto (1952a–c, 1962, 1970), Marshall (1987), Marshall and de Muizon (1988), Cifelli (1987a, b; 1990a–c; 1993), Cifelli and Eaton (1987), and of others should be consulted by those interested in examining the relevant evidence in specific details. Reig et al. (1987) have published some excellent illustrations of whole dentitions of didelphids and *Dromiciops*.

## Cranial morphology

Recently Maier (1993) has proposed a major hypothesis which may account for the cranial bauplan of not only Metatheria, but for the constraining original form–function of the skull in the therian ancestor of living marsupials and placentals. He suggested that the cranial construction of didelphid neonates, which, after being born, must immediately engage in the activity of full-time lactation, is not only fundamentally adapted to this activity, as expected, but that this pattern has far reaching consequences for adult cranial form as well. Because these animals are practically "embryonic" at this stage from a placental perspective, the very strong chondrocranium determines structural construction of the later ontogenetic pattern, long before the appearance of the dermal elements of the skull. The cranial attributes of eutherians appear to be linked to the extended intrauterine development afforded to them, according to Maier. Given the strong early similarities of cranial ontogeny of therians and the great variety displayed by eutherians that appear to have a heterochrony mediated diversity of later ontogenetic patterns, Maier suggested that the basic didelphid cranial construction is not only primitive for the metatherians, but also for the ancestral therians. Maier proposes that the cranial adaptations of the neonate marsupials to pump-sucking suggest a very likely ancient therian constraint for both adult marsupial skull morphology and for the cranial construction of the last common therian ancestor; this is a critically important functional–adaptive hypothesis for studies that deal with therian cranial evidence.

The inflection of the angle located below the ear region of the skull in living and fossil marsupials is a well-known diagnostic feature of the last common ancestor of the group. Its explanation, however, remained elusive for a long time. Mellett (1980) offered the following functional hypothesis, but not tests, for this phenomenon:

> In typical eutherians (placental) mammals, the angular process of the mandible is directed downward and to the rear of the jaw ramus. In marsupials, however, the angular process is reflected medially, and forms a distinct shelf at the rear inner jaw margin. The inner wall of the shelf serves as an insertion area for the medial pterygoid musculature; the outer part of the shelf provides insertion area for the superficial branch of the masseter musculature. This arrangement allows medial and lateral rotation of the mandibular ramus about its long axis. The inflected angle arose sometime in the Mesozoic, after the development of the protocone, to maintain occlusal relations between the buccally displaced stylar cusps, and the trigonid. Eutherians later reduced their stylar cusps and enhanced the development of conules between the protocone and paracone-metacone. With stylar cusps becoming less important in occlusion the inflected angle was subsequently lost in the eutherians. One consequence of mandibular rotation about its long axis is the crowding of lower incisors against the upper incisors as the jaws achieve centric occlusion. Marsupials responded to this by reducing the number of lower incisors to provide more room for them to maneuver against the uppers. Continuation of this trend toward lower incisor reduction led to the development of diprotodonty in a number of lineages.

I see three serious objections to this hypothesis. (1) Early eutherians like *Prokennalestes* or *Paranyctoides* not only have the same number of incisors as didelphids, but they also have large stylar shelves and some well developed stylar cusps, yet they only have a small and less inflected angle. (2) Marsupials that lose their stylar shelves and stylar cusps reduce the number of their incisors and switch to an entirely different dental strategy (e.g., kangaroos, among many others) but retain a robustly inflected angle. (3) Several lineages of eutherians become diprotodont without the imperative of an inflected ancestral angle.

More recently Maier (1987b) has undertaken a detailed developmental study in the didelphid *Monodelphis* in order to gain an insight into the role of the angular process in marsupial evolution. Unlike eutherians, marsupials retain a close connection between the cartilaginous anlage of the dentary, the tympanic (angular), and the prearticular (gonial). In neonates both the dentary and tympanic are vertical in orientation but in the following two weeks these become more oblique due to braincase growth. What is significant is that the tip of the angular process, free of muscular insertions, retains contact with a rectangular fenestra on the medioventral floor of the bulla, specifically with a membrane formed of loose connective tissue that covers this fenestra. Maier (1987b, p. 123) states that

> During juvenile and adult life stages, the process becomes somewhat removed from the fenestra for obvious reasons, but at a gape of about 40–50 degrees it inevitably must touch the 'inferior tympanic membrane' and possibly also the tympanic ring. It is speculated that the relationship between the angular process and the tympanic [i.e., auditory] bulla represents a specific form–function complex for sound transmission, which may be a modified retention from archaic mammalian conditions.

The above points are emphasized, since the taxonomically oriented study of group-specific differences of ear regions of not only marsupials but all other mammals has acquired a special mystique approaching that which surrounds dental studies. Given the potentially complex changes that must occur at this crossroad of the skull, this faith may not be unjustified most of the time. Perhaps because of its co-adapted complexity, this area is more conservative morphologically and functionally, but it is adaptively less well understood than the dentition. Unlike Maier's papers cited above, however, many basicranial studies still lack a developmentally and functionally based causal and analytical framework. Such holistic endeavors would obviously facilitate explanations for either similarity or differences of patterns. Like all other character complexes, the ear region may be a good indicator of certain monophyletic clusters of lineages, and may show unique specialization defining a group very well but without indicating any clues about the affinities of the unique group. Equally plausible, although not too many cases are known, convergent adaptive modification has resulted in confusing assessments that were corrected subsequently. The pertinent literature on the evolution of the mammalian middle ear is enormous, and the basicranial literature in the Theria itself is daunting (see Wible, 1990, 1991; Presley, 1993; and numerous other refrences in Szalay et al., 1993a, b).

The monographic study of Fleischer (1973, 1978) is particularly important for understanding some of the functional principles and boundary conditions that are likely to channel change in mammalian middle ear evolution. But until more is known of the probable path of evolutionary transformation of the ear region in marsupials based on the mechanical factors of feeding and hefting of the face on the neurocranium, all of which play an important causal role in changing cranial morphology, the exact description of cranial morphoclines by itself, given the subsequent uncertainties regarding transformations, will not provide reliable tests of taxon phylogeny without additional functional–adaptive analysis. Significant recent contributions to marsupial cranial and basicranial development and analysis of evidence have been made by Maier (1987a, b; 1989a, b), who presented invaluable evidence for developmental dynamics; by Wible (1990), who extensively analyzed the available Cretaceous evidence; and by Murray et al. (1987), who provided a view and a general model of analysis of the process-guided phylogeny of the syndactylan ear regions.

In Fleischer's (1978, pp. 8–10) description of the ancestral therian middle ear, which may also be considered the primitive metatherian condition, the tympanic is held in place by connective tissue. The tympanic is connected to the malleus by the gonial process (prearticular), which is fused to both. The incus is attached to the petrosal by its short arm, whereas it articulates with the stapes with its long arm. As Fleischer noted, "Since the tympanic membrane, as well as its peripheral connective tissue, is mass loaded by soft tissues, only the thin tympanic membrane is free to vibrate easily due to the air of the outer acoustic meatus on one side and the air of the middle ear cavity on the other" (Fleischer, 1978, p. 8). In spite of the relative stability of this ancestral mammalian and therian middle ear, the vulnerability of such an arrangement to deformation, hence potential for pressure change, supplied the selective forces for increased refinement of low frequency hearing in small mammals. One result is the variety of independent solutions for ossifying the ventral walls of the middle ear cavity. Much of what may be understood regarding the change in the middle ear in phylogeny, then, is that the regulation of air pressure by means of cavity stability, cavity size change, pneumatization (rendering the bones cancellous and thus capable of becoming air-chamber extensions), or encasing of the vessels traversing them into bony tubes, all relate to the physics of improving hearing at a certain frequency. As animal size, feeding habits, and locomotion change, both the protection from and alteration of acoustic ability by various types of interferences that can potentially deform the cavity takes place as a result of forces such as those generated from the harvesting of foods, and from the masticatory, respiratory, and circulatory systems, and the alterations that usually keep pace with these changes. The expansion of other areas, such as the brain, can undoubtedly also influence the necessarily altered form–function solutions. As noted, in small animals low-frequency hearing is of particular significance, and the morphology of the external auditory meatus can be shaped by species-specific mastication related selectional forces (Packer & Sarmiento, 1984, and references therein).

In the formation of the auditory bulla, the bones which are employed in different lineages to surround the middle ear cavity can have a variety of origins in eutherians (see especially Novacek, 1977). In the various marsupials, however, the predominant element that becomes involved in bulla formation, repeatedly, is the tympanic wing of the alisphenoid (see Fig. 3.1). Presley (1984) noted that the development of that region in therians (present concept) "suggests that the early stages of evolution of the bulla involved the second arch cartilage forming a bridge over a posterior part of the tympanic cavity, the fissura craniotemporalis." Whatever the merits of the various views on the beginnings of an ossified bulla, there is obviously great need for caution in interpreting at least the bullar-floor evidence. The persistence of the alisphenoid wing in various shapes and sizes coupled with the added floor element produced by the rostral (petrosal wing) or caudal (petromastoid wing) extensions of the petrosal bone, or both (the tympanic processes of the petrosal; see Fig. 3.1) must temper any phylogenetic account based on the ear region, particularly the bulla, alone. There is no evidence for the *consistent* presence of so-called entotympanics in marsupials (Szalay, 1982, p. 636; Maier, 1989a), or for bullae formed exclusively from the ectotympanic or the petrosal.

Webster and Webster (1980, and references therein) have conclusively demonstrated the importance of the increase in middle ear air cavity (bulla inflation) and low-frequency hearing, and connected the biorole of this form–function complex to predator avoidance in rodents. This solution would be, of course, equally useful for prey detection in other animals. Because middle-ear cavities must have a certain volume for efficient low-frequency discrimination, some of the small mammals have enormous bullae. So there is clearly strong selection for the independent evolution of enlarged middle ear cavities in therian mammals – size alone being an exceptionally poor indicator of phylogenetic relationship.

Another significant phenomenon of mammalian basicranial construction that adds to the diversity so important for analyses of

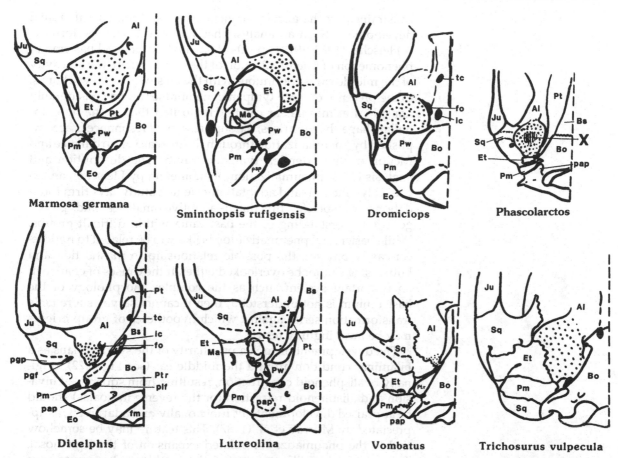

**Figure 3.1.** Selected basicranial patterns in metatherians. Stippled area: tympanic wing of the alisphenoid. Abbreviations: **Al**, alisphenoid; **Bo**, basioccipital; **Bs**, basisphenoid; **Eo**, exoccipital; **Et**, ectotympanic; **fo**, foramen ovale; **fm**, foramen magnum; **ic**, entry for promontory branch of internal carotid; **Ju**, jugal; **Ma**, malleus; **pap**, paroccipital process (of exoccipital); **pgf**, postglenoid foramen; **pgp**, postglenoid process; **plf**, posterior lacerate foramen; **Pm**, petromastoid; **Pw**, petrosal wing; **Pt**, pterygoid; **Ptr**, petrosal (promon-torium); **Sq**, squamosal; **tc**, transverse canal; **X**, ligamentous attachment of hypertrophied adult bulla to mandibular angle (in Phascolarctos). (Marmosa, Sminthopsis, Lutreolina, and Trichosurus were redrawn after Maier, 1989a.)

evolutionary change is the development of a bony meshwork, the pneumatization noted above, in this region of the skull. Taxon specific pneumatization of the bones surrounding the inner ear bearing petrosal, the squamosal, alisphenoid, or the wings of the petrosal itself is common in both eutherians as well as metatherians. The adjustment of basicranial morphology becomes necessary whenever the bulla itself needs to be reshaped for reasons other than hearing. The forces of mastication transmitted by the realignment of tendons and ligaments, or space needed for the hyoid are only a few of the several selective factors that must be mediated by the anatomy. Among the marsupials, the koala offers an amazing example of bulla modification related to chewing yet constrained by the ancient phyletic aspects of marsupial cranial development. Such an analysis has not as yet been performed.

Fleischer (1973; 1978, pp. 20–21) related the general basicranial phenomenon of pneumatization of the bones, where the air cavity of the middle ear is continuous with the meshwork of these bones, to the presence of large tympanic membranes and a considerably nonspherical middle ear cavity. He posited that as the need for bullar shape change arises, the suppression of resonance is accomplished by pneumatization through the increase of turbulance and friction in this extended tympanic cavity. The description and analysis of the pneumatization phenomenon by Fleischer, an exclusively function- and adaptation-oriented study, lays firm foundations for specific functional–adaptive analyses and phylogenetic understanding of the basicrania within difficult groups. While basicranial pneumatization is likely to be related to middle-ear-cavity physics, the possible relationship to mastication and buttressing cannot be overlooked either. If the causes of change in an important system such as the complex morphology of the basicranium is not understood, then it cannot serve as a reliable transformation series against which hypotheses of group origins may be meaningfully tested.

The development of such a peculiarity of the syndactylans as a bilaminar condition around the middle ear (see Fig. 8.22), originally of alisphenoid construction, resulting from squamosal invasion and alisphenoid retraction (or the reverse in koalas) should be appraised developmentally, functionally, and adaptively, as appreciated by Murray et al. (1987). This feature may be somehow tied to the pneumatization related expansion of the squamosal. This relationship, like numerous other problems, has never been investigated. The repeated use of the same components by diverse lineages can make it nearly impossible to detect the phylogenetic constraints that may be diagnostic of a particular cluster of lineages when the appropriate causal inquiry into morphogenesis and cranial mechanics is not performed. The phylogenetic understanding of this area depends on the developmental and functional–adaptive information that underpins character choice, and not only on the sorting of gleaned characters.

A recent study by Norris (1993) on the changes in the composition of the bulla in *Phalanger orientalis* is of particular importance not only for the Diprotodontia, but also for the study of basicranial morphology of therians in general. In examining a large sample (well over 50) of this species, he found that the bulla remained membranous even when the third molar erupted. At this stage he recorded three areas of incipient ossification: one lateral to the foramen ovale, another medial to the glenoid fossa, and the third anterior and lateral to the posterior lacerate foramen (**plf,** see Fig. 3.1). According to Norris, although the two cranially located ossifications were periosteal extensions of the alisphenoid, the caudal one appears to be independent within the membrane from any of the surrounding bones and therefore (according to him) qualifies to be called an entotympanic. Norris reports that by the time the fourth molar erupts, the rapid ossification of these centers obliterated any sign of independent ossifications in the floor of the bulla. In addition to the complex early stages of bulla formation, Norris also reports variability in the composition of the bulla, namely variation extended to the ontogenetic replacement of the "normal" alisphenoid bulla being occasionally replaced by a squamosal expansion, resulting in a squamosal bulla. He reports a variety of intermediate conditions, even individuals that have a squamosal bulla on one side, and an alisphenoid bulla on the other! This variation was closely linked to the geographical distribution of the populations in the southern Solomon Islands, to the point that some islands consistently have individuals with squamosal bullae. While Norris carefully points to one of the important conclusions concerning this range of variation in bulla construction within a species, namely that for the purposes of auditory roles the extent or origin of the bulla elements is not important, there is a far more significant lesson in that important study for metatherian and eutherian phylogenetics.

Although for reasons of the observed variability Norris has cautioned about using bulla composition as a taxonomic character, his analysis has added specifically significant methodological strength to the argument that functional–adaptive analysis is necessary for constructing robust taxonomic properties from patterns of bulla composition. As MacPhee (1981) noted, and was cited by Norris (1993), it is the primary fibrous membrane of the floor of the bulla that will be the arena closely influenced by the surrounding bones. In fact, it is the expansion of the various bones, and therefore their relative rate of growth compared to one another, that will be the final influence on the makeup of the bony bulla itself. If either of two bones (in the case of the phalanger example studied by Norris, the squamosal or the alisphenoid) invades either the anterolateral or anteromedial centers of ossifications (given the formative influence of the fibrous membrane itself; MacPhee, 1981), the result in subsequent lineages can be a bulla formed *primarily* by either of these bones. It is quite clear, however, that given the complex masticatory and subbasal cranial dynamics of mammals, selection will strongly influence the proportions, the extent, and therefore the relative growth of the bones surrounding the middle ear. Such selection-mediated aspects of skull form will have a profound subsequent influence on bulla composition which then can lead to fixation and relative (evolutionary) permanence of bullar composition. It is also clear that transformational understanding of basicranial homologies is not very likely without the appropriate functional–adaptive analysis – the taxic outgroup method with its mostly abiological (i.e., nonadaptational) assumptions is irrelevant here.

Functional and adaptive insight into the nature of convergence of such relatively complex features as the external auditory meatus, the bulla, and the postglenoid region are provided by the study of Packer and Sarmiento (1984). In light of the importance of this part of the anatomy for phylogenetics, a detailed treatment of the marsupial basicranium is certainly warranted in the future. The transformation into a half-tubelike structure of the ectotympanic, or any other bony element that borders the external auditory meatus on the skull, is an important variation of cranial structure – a source for characters for phylogenetic analysis. In order to make the most judicious use of such evidence, the evolutionary morphology of this area needs to be considered. Packer and Sarmiento (1984, p. 18), for example, noted the following in their investigation of this area in relation to some pernicious problems in primate phylogeny.

> In summary, the bony meatal tube is an adaptation designed to reduce physiological noise produced by motions of the mandible during mastication in species where the jaw joint borders directly on the external ear. The wall of bone is thin but relatively much more rigid than the easily movable cartilage. The latter would act as a radiating surface for disturbing sounds capable of masking important, relevant auditory signals. In other species, there is enough room in the infratemporal region for the imposition of a postglenoid process of the temporal bone between the jaw joint and the cartilaginous meatus. In primates, the structural relation between the jaw joint and the meatus may be dependent on body size, the degree of facial and/or basicranial flexion, the position of the foramen magnum, specializations of the masticatory apparatus . . . , auditory specialization (e.g., bulla expansion), or more likely a total morphological pattern of the skull influenced by several of these factors. Since none of these causal features is an indicator of phylogenetic affinities per se, the bony meatal tube must also be considered a weak indicator of affinities (i.e., it is subject to convergence).

This caveat has particular significance for the evaluation of the fascinating variation seen in the great diprotodontian radiation. It obviously does not mean that one can glibly entertain hypotheses of synapomorphy or notions of a meatal form easily evolving into a nonmeatal one, or the reverse.

Another potentially confusing issue related to meatal ossification and elongation known in eutherians (see Cartmill, 1975; Szalay, 1975a,b; Novacek, 1977) is the ossification of the anulus membrane (an ossified anular bridge) connecting the tympanic to the external meatus. The role of this phenomenon must be independent of the external tube. Packer and Sarmiento (1984, p. 19) speculate that

> The anulus membrane of tupaiids and lemuriforms may act as an admittance in parallel with the tympanic membrane admittance. In this respect, its shunting effect would be qualitatively similar to that portion of the tympanic membrane not coupled to malleus motion. . . . The stiffness and relatively small surface area of the anulus membrane in living tree shrews and lemurs indicates that the shunting effect of a broader anulus membrane could have been large enough to favor ossification. . . .

This in fact is the case in tarsiiforms (Szalay, 1976) and plesiadapids (Szalay et al., 1987), respectively, but almost certainly convergently. Yet rather than the shunting effect, the further stabilization of the middle-ear volume in these forms might assure a less effected low frequency hearing.

Awareness of the biomechanically sensitive nature of the otic region to the demands and forces imposed not only by the auditory system but also the masticatory one can have a profound influence on the interpretive frameworks of evolutionary studies. Yet in spite of the compelling evidence of the functional constraints on the change of this functionally and adaptively complex, multirole area of the skull so convincingly outlined by Fleischer, some workers continue to give indiscriminate weight to the cladistic outgroup method in rooting transformation of basicranial features.

One particular area of basicranial studies is the assessment of the circulatory patterns in both living and fossil forms. The information on the circulatory patterns of fossils is critically dependent on studies in the living, as Presley's (1979, 1993) and Wible's (1984, 1990) incisive studies so convincingly demonstrate. Several classic works on the basicranium of marsupials (Tandler, 1899; van Gelderen, 1924; Archer, 1976c) have laid the foundation for the analysis of transformation related to circulation, one area of which was recently reevaluated by Wible (1990). A number of attributes need to be kept in mind, as they appear to have been points of focus for the assessment of change. Following Presley (1979), it is clear that only a single internal carotid is present primitively in the immediate vicinity of the promontorium of the petrosal in therians; this conclusion is supported by Cretaceous evidence (Archibald, 1979; Wible, 1990, 1991; Rougier, Wible & Hopson, 1992). Arteries leave virtually no trace on the ventral surface of the petrosal of early metatherians or in didelphids, and the single vessel, the internal carotid, that bifurcates into stapedial and promontory arteries, is often impossible to trace on osseous remains alone (Wible, 1990; Trofimov & Szalay, in press; Figs. 8.1–8.6, courtesy of J. R. Wible). The entry into the bulla (whether it is fibrous-cartilaginous or osseous) is through the posterior carotid foramen; the entry of the promontory branch (not the internal carotid, as called by Wible, 1984) into the skull has been called by a number of names, but anterior carotid foramen (*fide* Wible, 1984) is preferable. Different positions in relation to the overlying promontorium, and emphases as well as deemphases of these two vessels in relation to the developing auditory bulla in the Theria, among others, represent the most often used characters in trying to understand transformation in this complex of features. In fact the early (i.e., Cretaceous) apparent loss of the stapedial artery in marsupials coupled with the medial position of the promontory artery (Archibald, 1979; Wible, 1984; 1990, 1991; Rougier et al., 1992) may be one of the diagnostic characters of Metatheria that we could cranially distinguish from a stem group of therians. Diagnostic alterations in the basicranial circulation of some Cretaceous and all living (holophyletic) Metatheria was corroborated by Rougier et al. (1992). They note that in addition to the reduction of the stapedial system, the arteria diploetica magna and the lateral head vein are reduced. To what degree, if any, this has occurred in the Holarctidelphia is unknown. The Gurlin Tsav deltatheroidean and the Udan Sayr Late Cretaceous Mongolian metatherian (Trofimov and Szalay, 1993; in press) show no signs of grooves for the promontory and stapedial arteries.

## Postcranial morphology

Feeding and locomotor adaptations are at the core of survival, and as such they are the coveted entry tickets into the lottery of differential reproduction. Trying to understand adaptive success only through a perspective of reproductive advantages without knowing the specifics of such survival equipment and their ties to all behaviors handicaps any undertaking aimed at fully deciphering evolutionary causality (cf. Szalay & Costello, 1991). While the teeth and skull must closely track feeding demands in evolution, the motion of the animals in the pursuit of eating and avoiding to be eaten is acutely dependent on the mechanics of their postcranial musculoskeletal system. Postcranial evidence of all sorts derived from the entire ontogeny (e.g., Klima, 1987) represents a critical portion of the skeletal framework that taxonomists employ to understand evolutionary history. While emphasis in the study of mammals has been often placed largely on cranial and dental evidence, the postcranium offers numerous important details for functional–adaptive analysis. The direct comparability of the postcranial evidence, along with that from the skull, permits a continuity of comparisons with the fossil record which alone should give the whole skeleton an overriding importance in phylogenetics.

There are a number of serious misconceptions prevalent in the mammalian literature which sometimes considers the out of context "plasticity" of bone as a reason for its neglect in phylogenetic studies. I noted this before (Szalay, 1977b, p. 320), and my views in this regard have remained largely the same. I reiterate here that studies of fossil mammals often ignore postcranial evidence on the basis of the notion that because bone is much more "plastic functionally" (i.e., ontogenetically) than teeth, this fact must be true phylogenetically, so that anything may be expected due to "convergent functions." Such logic assumes that phylogenetic inertia is more characteristic of teeth than of other components of the skeleton. This perspective (which also assumes that teeth are phylogenetically less responsive to functional demands than bone) overlooks the fact that evolutionary change is partly the result of changes in some gene frequencies. The implied lack of heritability of bone and joint form–function and its alleged unsuitability for detailed phylogenetic analysis in the fossil record as compared with teeth are arbitrary. Ontogenetic change in bony features (grossly overestimated for mammals, and usually based on quite irrelevant experimental manipulations or pathological readjustments in rare specimens) does not in any way negate or contradict taxon specific heritability (in spite of its relative ontogenetic plasticity) as high as for dental traits. It is an undisputed fact that fossil skeletons of the various Mesozoic and Paleogene groups that figure prominently in the understanding of evolutionary history of mammals are rare and fragmentary. It is equally true, however, that skeletal characters of the more abundantly represented Neogene forms, or extant mammals, have not been adaquately studied from the perspective of evolutionary morphology. This is particularly true for marsupials, with the exception of such pioneering studies as, for example, those of Barnett and Napier (1953), the various foundational papers by F. A. Jenkins, Jr. and associates (see references), and a few others.

While epipubic bones are not metatherian acquisitions, their analysis by White (1989) is relevant here. The functions and biological roles of the epipubics, because of their possible dual role in the evolution of locomotor change and reproductive strategies were discussed above, in this chapter.

As a general guideline for using musculoskeletal evidence in phylogenetic analysis, the critical issue is whether some areas or elements of the system will or will not show traces of phyletically precedent conditions. Testing homology hypotheses against features that have complexity of structural pattern in a functional–adaptive context, along with continuity from anlage of early ontogeny (if available), is the only way to avoid using homoplastic characters. The humerus and femur and their distal and proximal articulations have shown themselves to be so often so similarly affected by functional demands driven by nearly identical biological roles that their usefulness by themselves is somewhat limited. Demands on the elbow and knee joint on the interacting elements have also produced many detailed convergences which renders their usefulness questionable within the infraclasses of mammals. Nevertheless the study of whole articulating joint complexes, when possible, is far more useful than that of the individual elements alone. While I have examined a number of postcranial features and conducted a brief survey of the carpal patterns (Figs. 3.2 and 8.16), my efforts have been focused in the greatest detail on the lower crus (tibia and fibula) and the osseous foot in attempting to test taxonomic properties against features derived from these areas. The reason for this is simple. In depth understanding of both the adaptive and complete morphological diversity of any one functionally complex and interdependent area of the anatomy is extremely time consuming. The only other temporary excuse for neglecting the in depth assessment of the carpus or the rest of the skeleton lies in the fortuitous preponderance of tarsal bones in the fossil record. These tarsals can be, even with a relatively unsophisticated understanding of details, quickly placed in higher categories, and therefore their detailed analysis can properly focus on problems of adaptation, evolution, and diversification.

The understanding of change in skeletal elements is first and foremost a combination of developmental, adaptive, and phylogenetic *perspectives*. The methods for analysis are widely available in the literature, but without integration. An excellent and accessible small volume is *The Mechanical Adaptations of Bones* by J. Currey (1984), which is not only authoritative, but it is zoologically and adaptively focused. This stands in contrast to the often solely mechanics dominated presentation of bone and joint biology. In his book, Curry presents a most convincing argument against antiadaptationist perspectives as far as manifestations of the skeletal system is concerned. An earlier volume on human joint biology by Barnett, Davies, and McConaill (1961) is another outstanding, if less accessible, source on diarthrodial joints. In an excellent and relevant recent review, Biewener (1989) discussed the hypothesis of stress similarity. This view of bone stress maintainance across evolutionary change, in spite of great fluctuations in size in some monophyletic groups, powerfully supports the often conservative nature of many osteological attributes used in phylogenetics. In another brief review, Gordon (1989) succinctly discusses the ontogenetic and phylogenetic aspects of change in bone structure. The specific conceptual background for the evaluation of the pedal evidence is discussed below in Chapter 4, and that evidence is examined in detail in Chapters 5–7.

Although I have not, as noted above, analyzed the details of the carpus in marsupials, I have recorded some patterns of this complex (Figs. 3.2 and 8.16) which is of some use in this phylogenetic analysis. Unfortunately, this will have to suffice until more time and effort are spent on this area of the skeleton. One of the most interesting general observations is that relatively early (Paleocene) and many living groups of eutherians retain the centrale, but the ancestry of living marsupials has eliminated it, probably through incorporation into the scaphoid (i.e., the radiale). While I note the seemingly universal patterns in a number of taxa and

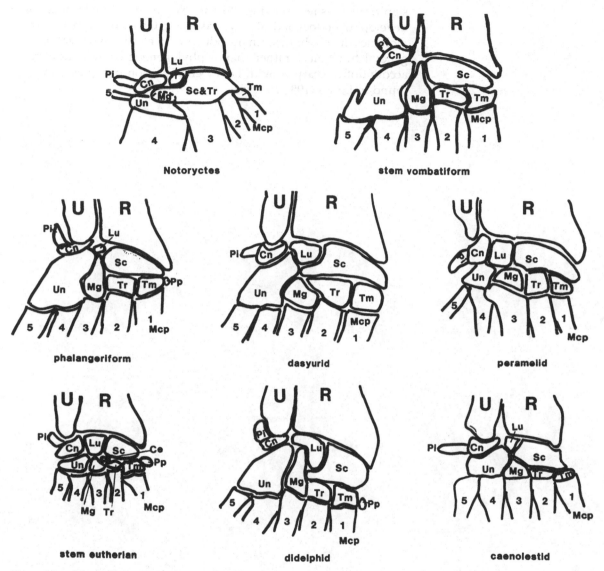

**Figure 3.2.** *Schematized patterns of carpal arrangements of the right hand in representative metatherians and the hypothetical stem eutherian (see also Fig. 8.16). Neither the pattern diversity of the whole carpus nor the detailed form–function of the individual carpals have been studied in any detail comparable to that of the crus and the tarsus presented in this study. Abbreviations: **Cn**, cuneiform (= triquetrum); **Lu**, lunate; **Mcp**, metacarpal; **Mg**, Magnum (= capitate); **Pi**, pisiform; **Pp**, prepollex; **R**, radius; **Sc**, scaphoid; **U**, ulna; **Un**, unciform (= hamate); **Tr**, trapezoid; **Tm**, trapezium. In referring to carpals in the text the quadrupedal mammalian stance is considered to be the "normal" anatomical position, and not the human one.*

suggest hypotheses of transformations, future functional–adaptive studies of carpal evolution in marsupials are potentially some of the most significant ones for an increasingly refined understanding of the phylogeny of this group. Shubin and Alberch (1986) is a general reference of importance, which, among other significant issues relating to developmental constraints of developing endochondral bone, takes a critical look at carpal homologies in vertebrates. An introductory overview and a valuable general functional (rather than a phylogenetic or evolutionary) account of the carpus, with heavy emphasis on primates, can be found in Lewis (1989).

# 4

# Form–function, and ecological and behavioral morphology in Metatheria

*It may be said that natural selection is daily and hourly scrutinising, throughout the world, every variation even the slightest; rejecting that which is bad, preserving and adding up all that is good; silently and insensibly working, whenever and wherever opportunity offers, at the improvement of each organic being in relation to its organic and inorganic conditions of life. We see nothing of these slow changes in progress, until the hand of time has marked the long lapse of ages, and then so imperfect is our view into long past geological ages, that we only see that the forms of life are now different from what they formerly were.*

Darwin (1859, p. 84)

*Morphologists have been slow in realizing that the basic principles of Darwin's theory of evolution by natural selection meant that understanding the evolution of anatomical features must be based on an analysis of how these structures interact with demands of the external environment on the organisms.*

Bock (1990, p. 254)

**Some assumptions**

Mammalian skeletal elements and coadapted areas within the body reflect the loading of the bones, the mechanics of locomotion, and habitual postures to varying degrees. The hand and foot are in particularly close physical contact with the details of the surface of the substrate. On the microscopic or tissue level the arrangement of different types of osteons in the Haversian system or the trabecular patterns further mirror the specific epigenetic responses of the tissue to stresses in addition to inherited patterns of bone form. The multiboned units, like the manus and pes, are the immediate arbitrators of many physical forces at the environment–organism interphace. Considerable adjustments of the hand and foot are brought about by minor movements of the individual elements in relation to one another. Major evolutionary changes, therefore, in the mechanical abilities of the carpus and tarsus can result from relatively minor form–function alterations of the individual bony elements. This is one reason why these areas of the skeleton are relatively conservative. The other reason is that there are a limited number of ways of locomoting on either terrestrial or arboreal substrates. The many constraining elements of the carpus and tarsus and the noted small adjustments needed combine together to offer skeletal composites in which changes from the preexisting conditions are well reflected. At the same time these areas show consistent macrotaxonomic variation and specificity at lower categorical levels as well.

In fact the interpretation of the form–function of an area of a bone as a structure and the interpretation of the pattern of configuration of the synovial joints (a subfield of morphological expertise) in a particular adaptive context are *often the most critical in the ordering of relevant skeletal information*. While studies in behavioral ecology are relevant to some extent, most such endeavors usually (unfortunately) omit morphological form–function considerations of the organisms as these relate to life-history adaptations. Bridges are needed to link behavioral and ecological morphology with behavioral ecology, and these connections will undoubtedly affect the interpretive perspective on characters used by taxonomists.

I have summarized below a list of assumptions that should be considered while comparatively (empirically) ordering skeletal, but particularly diarthrodial, joint evidence. This list, in numerical order, is obviously incomplete; its coherence may be justifiably questioned by colleagues with greater understanding of this field than mine; and some of these points are partial overlapping restatements of observations in some of the other points in this list. A perusal of such foundational accounts of bone biology and arthrology as those found in MacConaill (1946a–c, 1948, 1953a, b), Barnett et al. (1961), and Currey (1984) should be of particular additional help. The review by Rubin and Lanyon (1985) and the others suggested above are additional important sources for understanding patterns in skeletal structure.

1. Form–function (i.e., shape and mechanics) of skeletal details (both on the macro- and microlevels) reflect the compromise between conflicting selection pressures, the physical and developmental properties of the materials, and the historical constraints expressed in the genotype. A purely "genealogy" oriented appraisal of traits is likely to miss the functional–adaptive context within which these evolved, and therefore also some of the most important clues for the ordering of a transformation series. At the same time, however, structures and functions have often been maintained not for one but for several biological roles that may have evolved sequentially. The taxonomic properties based on skeletal features, then, should be constructed, when possible, in an adaptive context.

2. Articular kinetics and statics of joint surfaces (the study of loading, axes of motion, and the degrees of freedom allowed in

rotation and translation, etc.) closely reflect movements and posture of animals, in spite of the phylogenetic constraints often obvious in the construction of joints. Yet understanding the former as they relate to behavioral and ecological morphology is critical to one's ability to recognize the latter.

3. Patterns of bone shape in mammals, but particularly the specific diarthrodial solutions of many monophyla, are amazingly buffered from the predictions of allometric surface-to-volume relationships in successive members of lineages or between closely related lineages (in spite of some general textbook statements to the contrary). This appears to hold true within a great latitude of size change of organisms – from tiny animals to forms up to 300 kg. "The increase in peak stress predicted by the scaling of bone geometry . . . contrasts sharply with the experimentally determined values of stress. Indeed, the increase in peak stress predicted by long bone scaling suggests that, under strenuous conditions, terrestrial mammals greater than 10 kg body mass would likely fracture their bones. Instead, the empirical values of stress verify the stress similarity hypothesis" (Biewener, 1989, p. 778). Previously widely advanced explanations that allometry accounts for all sorts of morphological divergence in bone patterns are often invalid, and similarly such allometry-deduced accounts fail to grapple with persistent synapomorphies of diarthrodial patterns in morphologically relatively unmodified animals that widely differ in body mass.

4. Because bone may be loaded much more in compression than by shear, its ability to tolerate shear stresses is much more influenced by selection. Properties of bone, therefore, limit the possibilities of change. The direction of force passing along a bone can be altered only gradually, as weakness rapidly develops where the curvature is not appropriate.

5. "It appears . . . that the forces generated within the bones of the limb and those transmitted through them interact throughout the load-bearing period to provide a restricted load distribution. . . . This constancy of loading is maintained not only throughout each stance phase at one speed but also throughout the animal's entire speed range. This restricted manner of loading should allow for a unique (and optimal) structural solution at each skeletal location" (Rubin & Lanyon, 1985).

6. "Optimal" solutions should be always understood to mean "optimality" within the historical constraints created by specific genotypes and the epigenetic pathways available. Historical constraints often dictate the expression of "optimality." (see Item 1 for the notion of compromise designs for varying roles).

7. Unclear and confusing structural solutions in bone suggest a diversity of loadings, that is, behaviors that may include a variety of movements and postures.

8. A large bulbous enlargement of bones near the joint is a

strong indication that the joint is loaded with large forces in widely differing directions.

9. The paradox of joints is that while they allow movements, bones articulating with other bones are a potential source of weakness of the supporting skeleton.

10. The mechanism that allows mammalian joints to remain stationary, either during rest of the whole body or while allowing other parts of the body to function, is partly related to their surface anatomy and their ligamentous facilitation and constraints.

11. Any mated system of articulating bones has mechanically set limits of mobility and stability. Those joints that facilitate movement in one direction only are usually well stabilized against disarticulation. These limits are one of the more significant constraints to recognize and evaluate.

12. Often several articulating bones are closely coadapted functionally to accommodate species-specific bioroles for a given part of the organism, such as in the motions and supports required of the carpus and tarsus.

13. The greater the incongruity between the surfaces of the male and female components of a joint in a particular cross sectional plane (manifested by the disparity in the radius of curvature of these components), the more likely it is that the mechanics of these areas are related to mobility rather than to load-bearing stability (which is necessarily reduced at this area).

14. During close-packed position of a joint, two bones are in tight and congruent contact, although full congruence of joints does not necessarily mean close-packing. Habitual motions are usually away or toward close-packing, and involve swings combined with rotations. Ligaments tightened by the swing toward the close-packed position are twisted taut by rotation, making the bones function as one unit.

15. Usually the most dangerous loads generated on the skeleton require close-packing. Yet this position is also potentially the most dangerous because of the relatively poor resistance of bones to torsion compared to compression or tension.

16. Weight is usually carried near the close-packed position. Loose-packed positions of joints involve minimal contact between male and female surfaces – for practical purposes, sometimes a point.

17. During full congruence of a joint, the most extensive male area contacts the largest possible surface of the female conarticular area.

18. Knowledge of the articular relationships of even a single bone in the carpus or tarsus usually involves understanding three to five distinct articular constraints and interactions.

19. The clusters of bones that make up the wrist and the ankle allow great flexibility – the clusters, not so much the individual bones, are readily deformed – even if there is only a small amount

of motion permitted at any one of the articulations. In lineages not extremely committed to a limited form of positional or locomotor behavior, the ability resulting from the small additive motions of individual bones to adjust the carpus and tarsus can be enormous; arching and twisting movements allow a wide range of change. In such cases only potentially small evolutionary morphological changes are needed to modify substantially the adaptive traits of the lineage.

20. Compared to the joints of only two bones, changing the articular relationships of complex multi-bone systems, such as the carpus and tarsus, involves the potentially far greater probability of retaining more distant ancestral solutions (i.e., phylogenetic constraints, the apomorphies of a monophylum) in the complex system. *The more complex the articular relationships, the more they are going to be constrained by, and therefore reflect, phylogeny.*

## General accounts

The best general guide to descriptive aspects and broad comparisons of mammalian limbs and feet can be found in Brown and Yalden (1973), who note that "There are real differences between reptilian and mammalian limbs, but the mammalian limb cannot be characterized in a summary of a few words; the matter is too complex." What complicates this issue further are the sundry phylogenetic changes that occurred in the specific continua of synapsid evolution and the further diversification of the mammals themselves. Several authors [see particularly Kemp's (1982) comprehensive book] review some of the functional aspects of mammalian modifications in the limbs from a synapsid ancestry, and Brown and Yalden (1973) establish important criteria for judging evidence from the limbs. These authors carefully detail useful descriptive concepts related to locomotion and functional anatomy. A. B. Howell's (1944) book on speed in animals is a classic that continues to be relevant to these issues.

In therian mammals, retraction of the humerus and femur, along with knee and cruropedal, or upper ankle, joint (UAJ) extension and flexion, will move the animal, in contrast to the more reptilian axial rotation of the humerus and femur. It is because the joints of mammals so well reflect the motions of retraction, protraction, flexion, extension, adduction, abduction, supination, pronation, inversion, and eversion that even fossils can reveal intricate details of function and the dependent skeletal activity patterns of mammals that are long extinct.

Although the hands and feet reflect to varying degrees the habitual substrate preference and locomotion, it is an acknowledged fact that the forelimbs and the manus in particular are often critical for habitat exploration, food manipulation, and even social activities. It is for this reason that the independent analysis of carpal diversity and transformations have the potential to reveal complementary aspects of marsupial phylogeny in addition to that which is presently supported by the cranial, dental, pedal, and tarsal evidence. I consider this a difficult and time-consuming undertaking (judged from my own ongoing studies on the tarsus and my initial efforts on the carpus), but potentially one of the most revealing of the major comparative studies yet to begin to advance marsupial phylogenetics. Nevertheless, the fundamental constraints imposed by metatherian reproduction discussed above probably set as yet unspecified limits to the extent that hand (but not necessarily carpal) evolution occurred in these animals.

Unlike the forelimb and the manus, the hindlimb and the foot are relatively rigidly controlled by the forces responsible for adaptations to a particular substrate and by a posture and locomotion regime of a certain manner and at given limits of speed. It is also possible, but as yet uninvestigated, that the genetic control of limb differentiation is sufficiently uniform early in ontogeny such that constraints affecting the embryonic and neonate bioroles of the manus in marsupials also strongly constrain the pes and, therefore, tarsal development. While Klima (1987) has contributed an excellent developmental survey of the shoulder complex, the best general account of hindlimb adaptations in marsupials is the now quite dated work of Elftman (1929). The latter related morphological differences in the musculature and the pelvis to locomotor behavior for some groups of marsupials. Such an outstanding beginning as Elftman's has not been followed by detailed studies of ecological morphology in living marsupials (exceptions being studies on *fossil* marsupials), in contrast to some other mammalian groups where such endeavors are an actively pursued area of research.

I noted in 1984 (Szalay, 1984, p. 225) that "Due to the physical proximity of the tarsus to the substrate . . . the foot becomes a much more sensitive tracker of phylogenetic changes in substrate occupancy than the upper part of the hindleg, because the animal primarily accommodates itself through complex movements of the tarsus." Whether such a generalization will hold is yet to be determined from widely ranging mammalian studies on the ecological morphology of the entire skeleton.

**Morphological and functional framework of the pedal and tarsal evidence**

I will emphasize here some of the more general points as they relate to the specific traits discussed, those characters that are also noted under the section dealing with the taxa and phylogeny of marsupials. The interpretive framework for the evaluation of all the fossil tarsal evidence is mostly derived from the detailed assessment of the morphology of the pes in the living radiations. Yet, perhaps significantly, the impetus to study the living forms in their osteological details came primarily from evolutionary questions raised when I first gleaned puzzling information from the fossils. Diversity at any level of time has close bearing on the objective understanding of evolutionary history, and no caveats should prevail in emphasizing any one time level, such as the Recent (in spite of the clamor of some cladists who became transformed). It is, of course, clearly understood that the neontological record is special in providing complete systems to analyze from developmental, functional, and adaptive perspectives. The analyses in this book are based on the recognized acceptance of this tenet. On the other hand, fossil evidence is equally fundamental in providing *various levels of proximity, and therefore probability unless the morphological evidence is contrary,* to the actual – "bottom-up" – direction of evolutionary change. Fossils provide anchoring points through time (1) for the time of first appearance of morphs, (2) for diversity through different slices of geological time, and (3) for testing full phylogenies (descent and diversification) of characters, of organisms (taxa) as opposed to only the cladistic relationships of taxa. The technique-bound taxonomists who restrict themselves to a process of delineating and ordering taxa and their traits in a single time–plane perspective are not only depriving their endeavor of the full phylogenetic method, but also become immersed in a perception of diversity prevalent in the pre-Darwinian days of taxonomy.

The reference framework for the interpretation of the fossils is the relatively detailed understanding of the form–function as well as aspects of the ecological morphology of the living radiations. This is often emphasized below because it is regularly, and axiomatically, rejected by many taxonomists from all schools. What is "living marsupial," or what are other level metatherian characters are explicitly formulated hypotheses of morphology of the last common ancestors for members of the monophyletic groups in question. These diagnostic apomorphies of the ancestors of monophyletic groups are detailed in Chapter 8. As noted previously, such hypotheses comprise a set of analyzed and interpreted characteristics of the first representative of that taxon.

## Morphology and external surface of the foot (Fig. 4.1)

A general review of limbs and feet (both manus and pes) can be found in Wood-Jones (1923–1925) and Brown and Yalden (1973). Detailed illustrations and descriptions exist for the surface of the foot in a variety of marsupials, and Bensley (1903) and Wood-Jones (1923–1925), in particular, are repositories of information on this subject. Stein (1981, p. 133), in an account of primarily the sole of the foot as a representative didelphid condition, states that

The pes of *Didelphis* is smaller and more anthropoid in appearance than that of *Chironectes*. The skin is thick and coarse as in the manus

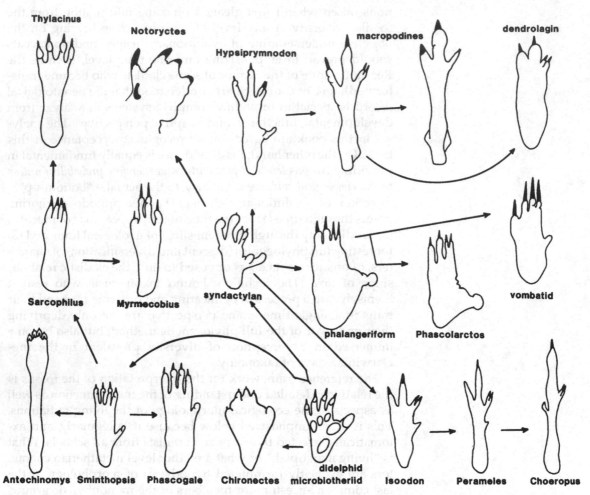

**Figure 4.1.** *Evolution of the pes in some metatherians, showing the plantar surface of the foot. The didelphid–* *microbiotheriid pattern (with the pads shown) is the most primitive living one, and derivations of outlines of the* *major types of feet are shown from this structural ancestry.*

with a regular pattern of ridges on the plantar surface. . . . Although the digits are independent, there is a web of thick skin that stretches between the digital pad on the hallux and Digit II. There are four interdigital pads at the base of the five toes, with a single pad centered between Digits III and IV. The tip of each toe is swollen, forming smaller digital pads which have a whorled pattern. A cartilaginous spur is found on the radial side of Metatarsal I.

The cartilaginous spur Stein refers to is underneath the skin within the foot and it is perhaps best called a **cartilaginous pad;** I have seen it in all taxa of didelphids and phalangeroids that I have dissected. Its position is roughly under the thenar (inner) metatarsal pad, and it may well be indicative of the fact that the last common ancestor of the *living* Metatheria was a committed arborealist like the majority (and the first member) of the Didelphidae and the Phalangeriformes. In the light of the hallucial reduction in the caenolestids, this is difficult to judge but I strongly suspect this structure to be primitive for the living marsupials.

A well known but important point to emphasize here is the virtually complete absence of any keratinous remnant of a falcula (or any other derivatives of claws) on the first digit of didelphids and other marsupials. The exception known to me is *Notoryctes* (almost certainly an evolutionary reversal), and the occasional scaly remnants I have noticed on a few specimens of caenolestids.

Figure 4.1 summarizes some of the major aspects of the literature on surface morphology of the pes in marsupials, while several figures throughout the chapters (Figs. 5.1, 7.1, 7.13, 7.24, 7.32, 7.43, 7.50, 7.62, & 7.74) show samples of the diversity of the whole osseous foot.

### Bone homologies and development

Some of the extensive early literature on mammalian tarsal bone homologies has been reviewed by Lewis (1964a). All the bones discussed in this work occur in all marsupials with the exception of some reductions in cheiridia (rays) in a number of groups. Although the presence of a tibiale articulating with the underside of the astragalus can be easily confirmed in monotremes (as first shown by Emery, 1901), there seems to be no evidence for the occurrance of this bone in any marsupial.

The degree of ossification of some bones can be a source of extreme variation in morphology in mammals. Yet in marsupials it appears that the articular patterns of adult animals are established early, even before the adult size is reached (see Fig. 4.2). The carpals and tarsals of marsupials, like those of eutherians, are each ossified, with the exception of the calcaneus, from a single center in a cartilaginous blastema (for a useful discussion of various developmental and functional factors see Chapter 3 in Lewis, 1989). It is only the calcaneus that has an epiphysis at the posterior end of the tuber. The metacarpals and metatarsals all have two centers of ossification: one for the body of the bone and one for the epiphysis. On metacarpals and metatarsals 2–5 one center of ossification is found proximally for the body, and this section includes the proximal end of the bone articulating with the carpals and tarsals. The second, epiphyseal, ossification is distal. The first metacarpal and metatarsal, however, like all the phalanges, have proximal (but not distal) epiphyses which articulate with the trapezium and entocuneiform, respectively, and one ossification for the body of the bone that includes the articulation for the distal end of these long bones. I am not aware of a detailed functional–adaptive explanation of this important pattern or of its exact point of acquisition in mammalian or synapsid phylogeny. The possibil-

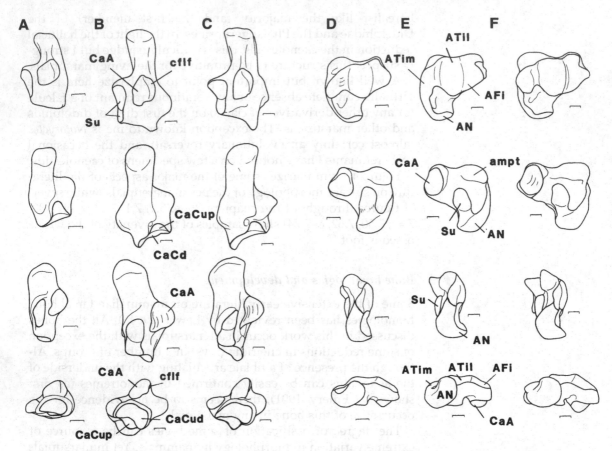

**Figure 4.2.** *Three growth stages of the calcaneus (A, B, and C) and of the astragalus (D, E, and F) of* Philander opossum. *From juvenile to adult (from left to right) the **LAJ** and **UAJ** facets are kept constant on the drawings to show the change in size and proportions in such a way as to maximally reflect ontogenetic variation. In spite of evident changes in some aspects of morphology, the articular facets remain specifically recognizable by genus. From top to bottom: dorsal, plantar, medial, and distal views. For abbreviations see Table 1.1. Scales = 1 mm.*

ity exists that the unique (phalanx-like) patterning of the development of the first metacarpal and metatarsal of therians (or mammals) may be specifically related either to a post-neonate or to a post-weaned stage adaptive solution of either the hand or foot of the ancestral condition. It is potentially an intriguing avenue of inquiry into synapsid or early mammalian adaptations. This pattern may have been related at its inception to either manual or pedal grasping, but when exactly in synapsid phylogeny is not even remotely understood.

The marsupial navicular, as seen in eutherians (with some interesting unresolved issues remaining with archaic eutherian arctocyonids and dinoceratans in the Paleogene, and with the extant hyraxes), contains the ossification formed from the blastema derived of the anlage of the embryonic tibiale. The prehallux, widely present in marsupials, is also developed from the blastema formed from the tibiale. The studies of Emery (1897, 1901) on the embryology of monotreme and metatherian hand and foot morphology are the only ones of their kind I am aware of. His work documents the independent chondrification of the tibiale in monotremes and therians, ossification in the former, and a post-chondrification fusion to the navicular in the latter.

The small bones, usually two, on the posterior part of the upper ankle joint in marsupials that have been called "intermedium tarsi" in the literature are the supporting lunulae of the highly characteristic meniscus between the astragalofibular facets of the fibula and astragalus. Lewis (1964a, 1980a) expressed strong conviction that this meniscus is a primitive mammalian character. While it may have been present in multituberculates (see discussion below), it does not occur either in known living reptiles or in monotremes, and there is no indication of its presence during development in the Eutheria (see Szalay, 1984). It all but disappears in the peramelids which are terrestrial. Its history is tied to the upper ankle joint (**UAJ**) – and that history is as yet not clearly understood.

The prehallux has been discussed by Lewis (1964a). Not only marsupials, but several groups of eutherians retain this bone. Tenrecoids, tupaiids, probably stem archontans, and some erinaceids, for example, have a prehallux, but in large mammalian taxa like the Rodentia the issue is complicated by the presence of multiple sesamoids and alleged tibiale (see discussion in Szalay, 1985). The prehallux is supposed to be present in some living primates (Lewis, 1972), but the homology of the small ossicle present in the articulation of the hallucial joint (**EMt1J**) in some anthropoids has been questioned by Szalay and Dagosto (1988).

### Muscles and ligaments

Analytical studies are needed on the comparative morphology of the musculature and the architecture of ligamentous constraints, variation, and any aspect of the anatomy of the foot in living marsupial taxa. The few excellent references that exist, almost exclusively on myology, will be noted in addition to my own observations when appropriate. The early studies of Glaesmer (1908, 1910), the more modern works of Lewis (1962a, b, 1983, 1989, etc.), and the detailed limb myology of two didelphids by Stein (1981) are the foremost and most reliable references. My dissections of *Didelphis virginiana* specimens have confirmed the path of major flexor, extensor, abductor, and adductor tendons that go dorsal and ventral to the bones of the foot and of the important retinaculae that anchor and shunt, and therefore localize forces on the bony foot. These basics, along with the major ligamentous bindings of the tarsus of *Didelphis*, are illustrated in Figure 4.3 (A and B are tendons; C and D are ligaments). An understanding of the elementary mechanics of a few of the important pedal muscles, realized through the path and position of their tendons, is essential for the appreciation of the osteological differences among the taxa discussed here. The muscular and skeletal (i.e., musculoskeletal) systems are inextricably interwoven in their mechanics, development, and evolution. Neither bones nor muscles can be understood by themselves from either a developmental or phylogenetic perspective. The information noted here was almost certainly present in the last common ancestor of living metatherians.

The large common tendon, made up of the tendons of the flexor fibularis (**fft** = flexor hallucis longus) and that of the flexor digitorum brevis (**fdbt**), the latter a small muscle that originates about 2–3 cm above the tuber of the calcaneus from the common tendon of the flexor fibularis in *Didelphis* (Stein, 1981), is in a tunnel formed by the plantar (flexad) side of the calcaneal sustentaculum and the calcaneal tuber and the astragalus (see the details in Fig. 4.3A,B). In a similarly typical therian fashion the plantar-flexing tendons of the triceps surae (**tst**) attach to the calcaneal tuber. The extensor groove (**ge**) on the distal fibula shunts the tendons for the peroneus longus (**plt**), the peroneus brevis (**pbt**), the peroneus tertius (**ptt**), and the extensor digitorum brevis (**edbt**). In many fossil marsupial calcanea the dorsal surface of the peroneal process bears a distinct groove (**gtpl**). It is almost certain that this is a structural solution for the stabilization of the tendon of the peroneus longus (**plt**) that makes a sharp turn distal to the peroneal process as it enters the canal formed by the tarsometatarsal ligaments (**tmtls**). The path of this ligament (**plt**) is a particularly variable feature, superficially, in fossil and living marsupials. In living didelphids a strong retinaculum straps the tendon of the peroneus longus onto the dorsal surface of the peroneal process, and this was probably the case for the archaic marsupials as well. In addition to being an adductor of the hallux, the peroneus longus, together with the other peroneal musculature (peroneus brevis and peroneus tertius), forms a powerful everter complex of the foot. Most importantly, the peroneus longus brings about the rotation of the distal tarsus on the distal articular surfaces of the calcaneus and astragalus. This is of particular importance when the extremes of eversion are required.

Held to the posterior surface of the medial malleolus of the tibia by a retinaculum on the surface of the astragalotibial ligament, the tendons of the flexor tibialis (**ftt** = flexor digitorum longus) and the tibialis posterior (**tpt**) pass distally. The tendon of the tibialis posterior inserts onto the dorsal surface of the navicular, whereas the tendon of the flexor tibialis goes past the prehallux and attaches to the medial slip of the flexor digitorum brevis (Stein, 1981). Lewis (1962b) has reported that the flexor tibialis attaches to the prehallux, a fact I could not confirm in several dissections,

**Figure 4.3.** *Some of the major tendons of the foot (A, dorsal view and B, plantar view), and some of the ligaments binding the tarsus and crus in* Didelphis *(C, dorsal view and D, planter view). For abbreviations see Table 1.1.*

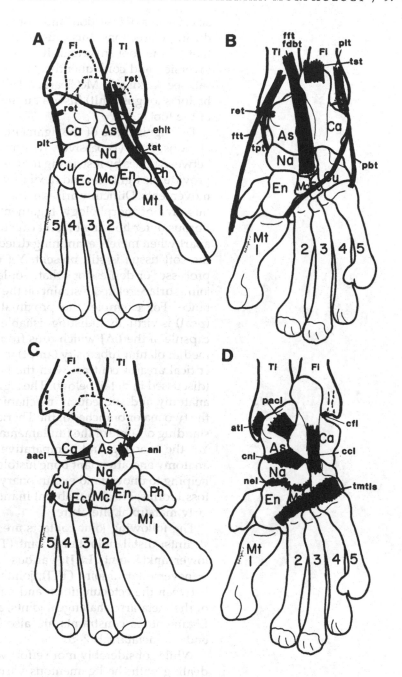

primarily, I suspect, because this is an exceptionally delicate, and perhaps variable, area, and my dissections have not been nearly as extensive as those of Lewis (1962a, b, 1983).

The tendon of the tibialis anterior (**tat**) attaches to the medial and dorsal (extensad) border of the entocuneiform, next to the entocuneiform–prehallux (**EnPh**) articulation. Judging from its size, its area of insertion, and from the manipulation of the tendon during dissections, this muscle is undoubtedly the most important of the foot inverters. It plays a particularly critical role in inversion and consequently also in "hyperinversion" (foot reversal, see Jenkins & McClearn, 1984). The tendon of the extensor hallucis longus (**ehlt**) inserts on the distal phalanx of the first toe of the foot.

The significance of the ligaments (see details in Fig. 4.3C,D) is obvious. They (1) passively determine the direction of movement between bones by restricting movement along certain axes and (2) provide the tightness and flexibility needed for small but essential movements. Difficulty arises in the absence of detailed accounts of this area of morphology. Ligamentous restraints and the tough fibrous outer housing of joint capsules are often impossible to tell apart when merely examining dried skeletal specimens on which the soft tissue is still present. Yet correct identification of bony processes or depressions that would signal either ligamentous or joint-surface related restraint on the fossils is of paramount importance. For example, the proximal calcaneoastragalar ligament (**pcal**) is virtually indistinguishable from the exceptionally tough capsule of the **LAJ** which runs from an area called the astragalar medial plantar tuberosity (**ampt**) in marsupials. This is probably a critical area of contrast with the homologous part of eutherians (discussed in detail below). The significance of the associated soft anatomy and subsequent mechanical differences of the joints in the two major branches of the Theria can, of course, lead to understanding of some of the fundamental differences in between them. Yet the exacting and comparative explorations of ligamentous anatomy and attendant bone histology, potentially far reaching in helping to understand evolutionary morphology of the living and fossil forms, are all but absent in much of the nonhuman, particularly marsupial, literature.

The following joint contacts are mediated by ligamentous restraints: distal tibiofibular joint (**TFJ**), upper ankle joint (**UAJ**), lower ankle joint (**LAJ**), various critical subcomponents of the transverse tarsal joint (**TTJ**), joints between the cuneiforms and between the ectocuneiform and cuboid, various subcomponents of the metatarsophalangeal joints, and the interphalangeal joints. Ligamentous constraints are also important between the distal ends of metatarsals 2–5.

While considerably more effort would yield valuable results in dealing with the ligamentous variation of marsupial feet, I will mention again Currey's (1984) book as a succint source on the mechanical adaptations of bones, and principles relating to bone, muscle, ligaments and joint mechanics; it is also a primer for the vast literature on functional osteology and kinesiology. Studies by Lewis (1980a–c, 1983, 1989, and references therein) supply useful descriptive and functional information on the muscles and tendons of some marsupial taxa. Good human anatomy texts and Gebo's (1985, 1986, 1987) recent detailed work on the ligamentous restraints of the foot in strepsirhine and tarsiiform primates are also valuable references, and these may be sometimes relevant to comparative aspects of marsupial pedal anatomy. There are unfortunately no sources available for the comparative details and mechanical significance of ligaments in the manus and pes of marsupials.

### Plantarflexion and dorsiflexion

In both eutherians and metatherians the entire foot is moved through the contact of the astragalus (and the calcaneus, when this applies in tricontact UAJs) and the crus.

### Inversion and eversion

This movement of the foot along its long axis in primitive living Metatheria is only partly the result of the movement of the calcaneus under the astragalus (and under the crus, when this applies in tricontact UAJs), and through the calcaneus the more distal tarsals are also affected. In eutherians motion occurs primarily in the LAJ and TTJ. In metatherians this pedal motion is accomplished in the UAJ (see discussion below), LAJ, ANJ, the joints between the navicular and cuneiforms, as well as the joints between the cuneiforms. While some inversion of the foot of didelphids and phalangeroids occurs in the LAJ, in general this compound movement is relatively much less significant than that seen in eutherian arborealists. In metatherians the movements that accomplish inversion–eversion appear to be combinations of adduction–abduction and plantarflexion–dorsiflexion of the whole foot. In most terrestrial marsupials (but see under the DASYUROMORPHIA, below in Chapter 7) the inversion–eversion component of UAJ mobility is largely eliminated.

### Abduction-adduction

In eutherians this motion of the whole foot occurs about the long axis of the tibia, whereas in metatherians, except in highly modified cursor–leapers, the motion of the whole foot probably occurs about the long axis of the crus (i.e., both the tibia and fibula). As noted above, this motion is coupled with inversion–eversion in marsupials, and it is greatly mediated by the meniscus and their ossified lunulae in the joint. In eutherians abduction–adduction occurs in the sundry varieties of adjustments of the foot to different substrate surfaces (Langdon, 1986). This is probably also

*Figure 4.4.* *Comparison of the right distal crus in three metatherians and one eutherian (anterior and distal views) to show corresponding, homologous areas. The distal crus, like the tarsal articulations, is the source of some of the most significant taxonomic properties of therian groups. Reduction of the **ATip** facets, enlargement of the medial tibial malleolus, and mortiselike function of the distal fibula (complementing the medially restricting tibia) are convergently evolved in Metachirus, the representative macropodid, and the eutherian. While Metachirus is the most terrestrial living didelphid, it resembles other arboreal didelphids in most morhplogical detail rather than the eutherians or kangaroos. For abbreviations see Table 1.1. Scales = 1 mm.*

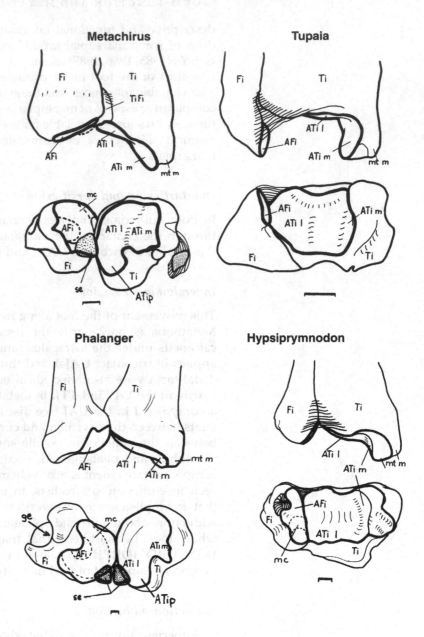

true for marsupials that have a more independently mobile foot from the crus than do the eutherians (Jenkins & McClearn, 1984; Szalay, 1984; see Fig. 4.4).

### Supination–pronation

In eutherians this complex motion involves the twisting of the foot about its long axis, and involves motion in the **LAJ** and the joints between the tarsals. According to Langdon (1986, p. 42) "The motions of supination and pronation are frequently produced by the action of external forces on the foot, particularly on an uneven substrate, or by ground resistence to movement of the lower limb. Inversion at the subtalar joint is likely to be accompanied by pronation of the forefoot if there is a resisting substrate." As far as I know there are no studies on marsupials that would shed light on this or other related motions.

**Joints, their facets, and form–function in marsupial hind feet**

Throughout this study the concept of a joint is used to designate articulation(s) between two or more bones, and it specifically refers to one or more pairs of conarticular male and female facets of two bones. This is a pragmatic departure from the standard treatment in human anatomy where the concept often is restricted to the contents of a single diarthrodial capsule. In light of the fragmentary nature of the fossil record that requires detailed analysis, I find an approach such as the one I employ here (see also Szalay, 1984, 1985, etc.) more conducive to precise and terse description and analytical comparisons. It does not mean that I abandon those joint designations, for example, that are anatomically, functionally, and therefore phylogenetically closely tied together, such as the upper and lower ankle or the transverse tarsal joints.

As in many eutherians, there are a total of eight tarsal bones, including the prehallux (**Ph**). The maximum number of joints (and probably the primitive therian condition in which the calcaneus has a double articulation with the astragalus and it also articulates with the crus) is a minimum of twenty separate sets of articulations, counting the various contacts with the metatarsals. As I noted, the term "joint" as I use it here with macrotaxonomic aims in mind (i.e., homology designations) means either the anatomically capsule-restricted contacts, or the *articular contacts between the individual tarsals, and between the proximal tarsals and the fibula and tibia*. The number of paired joint facets, however, is larger than twenty, as some bones (astragalus, calcaneus, entocuneiform, etc.) have double articulations with other bones. The important point to note here, as discussed above, is that the pattern formed by such a large number of complex articular contacts is not easily eradicated in evolution. One may judge this from the nature of behavioral–locomotor diversity across species, and by the virtual lack of any higher taxon pattern variation within species, genera, and often families. Deviation from these numbers is noted under the taxa that can be characterized by additions, modifications, or reductions. In the following section I briefly discuss these various joints, and their facets and mechanics, referring primarily to *Didelphis* (a representative, *at the general level discussed*, of the primitive condition for not only the didelphids but for all living marsupials), and the morphological characters with known or unknown mechanical consequences that are of adaptive importance. Following this section the abbreviations given in parentheses and also in Table 1.1 will be used instead of the full name of the joints and other morphological units, except for the names of the bones themselves.

The foot is in immediate contact with the substrate. Its ability to perform the necessary mechanics for a variety of bioroles is, presumably, severely and rapidly tested during the animal's life. An astonishing aspect even of the smallest details of cruropedal osteology (and the rest of the skeleton, for that matter) is the recognizable and understandable mix of the historical constraints (heritage, lineage specific constraint, or local constraint of Maynard-Smith et al., 1985) with lower-level lineage-specific details, often clearly associated with specific activities and consequently causally explicable by specific load bearing and motion management (i.e., facilitation/restriction). By causal explanation I mean selectional, evolutionary causality guided by developmental constraints.

The usual load bearing mechanics of the hindfoot can be associated with the specific biological roles of (1) propulsion (walking, running-jumping, hanging, climbing, swimming, digging, etc.) and (2) resting postures. A variety of miscellaneous bioroles such as grooming can also pose biomechanical demands on the foot in marsupials, which may or may not be contained in the inherited morphogenetic program of a taxon. Whatever mechanical demands the behavior (active or passive) of the animal places on the cruropedal complex, these movements will be performed along the movements and motions of the foot in relation to the crus. The general principles that facilitate the understanding of these movements are well described in anatomy and kinesiology texts (see particularly Hildebrand et al., 1985). Even these movements, of course, are directed by historical contingencies beginning with the earliest adaptations of the first tetrapod, a quadrupedal quasi-terrestrial animal. Several recent studies of joint mechanics in various therians and multituberculates have focused on the meaning of mechanics of pedal joints as they are causally related to specific bioroles of posture and locomotion on various substrates. The works and references and those cited by the following authors have contributed to the analytical understanding of joint morphologies and movements in mammals: Dagosto (1986), Decker and Szalay (1974), Fleagle (1977), Gebo (1986), Grand (1967), Hall-Craggs (1966), Jenkins (1974), Jenkins and Krause (1983) Jenkins and McClearn (1984), Krause and Jenkins (1983), Lewis (1989), Szalay (1984, 1985), Szalay and Decker (1974), Szalay and Drawhorn (1980), and edited volumes such as Strasser and Dagosto (1988).

In addition to dorsiflexion, plantarflexion, adduction, abduction, inversion, and eversion, morphological patterns and the movements which they allow within the foot itself are critical documents of mechanical and historical solutions against which the testing and understanding of homologies hypothesized for cruropedal characters can proceed. Yet the comparative kinesiology of marsupial cruropedal morphology is virtually nonexistent. The recent study of Jenkins and McLearn (1984), in which they examine movements and bony relationships in *Didelphis* along with a number of other mammals, is the best beginning in the literature for an empirical evaluation of some of the comparative kinesiology of the pedal joints in selected marsupials and placentals. The functional aspect of my account here, a necessary prerequisite for the analysis and interpretations of the evidence of cruropedal characters, is based on my manipulations of numerous freshly killed specimens (roadkills of *Didelphis virginiana*, and a whole range of Australian dasyurids, peramelids, phalangeriforms, and wombats), preserved specimens, skeletal preparations, and extensive comparisons with better understood, yet still barely studied, eutherians. I will at first briefly compare the shared motions, or at least the commonly applicable terminology, of the foot between eutherians and metatherians, and then examine in general terms the nature of motion between specific bones under the various joint designations. Jenkins and McClearn (1984) outstanding study is a primary source for correlating some of the osteological attributes with observed motion of the animals. My discussion below of didelphids, represented by *Didelphis*, relies heavily on their account. Yet, as I will note, I do occasionally disagree with answers to several questions raised in their study concerning evolution of characters and therefore implied phylogenetic transformation. These disagreements, along with other aspects of the evolutionary synthesis, are discussed below. For the motions of the foot, based on eutherians, I follow the definitions by Langdon (1986).

Under some of the subheadings below I review the relevant form–function attributes that probably characterized the ancestral conditions in the Didelphidae and Caenolestidae, that is, the living Ameridelphia. These form–function solutions, given their taxon specific ways (i.e., constrained by a particular heritage), are assumed to reflect, in addition to phylogeny, certain bioroles in the last common ancestor. This section does not attempt or pretend to be an adaquate substitute for the many required functional studies of living marsupials – their comparative, evolutionary musculoskeletal biology is in dire need of scrutiny. Additionally, studies of behavioral and ecological morphology are needed, beyond the summaries provided below, in order to proceed to the correlation of bioroles and form–functions for a causal appreciation of morphological patterns. Meanwhile, however, some interim hypotheses for the attributes of living form-function complexes are needed in order to begin an analysis and explanation of the diversity.

### Tibiofibular joints (TFpJ and TFJ)

In two broadly comparative papers Barnett (1955) and Barnett and Napier (1953) have discussed tibiofibular articulations and movements in both eutherians and metatherians and have attempted to relate the mobility of the crus to astragalar shape. Although their contribution is from a functional rather than a synthetic evolutionary perspective, these papers, in spite of a few errors, are some of the most seminal works on this anatomical area in the mammalian literature.

Figure 4.5 is a summary of the patterns of contacts between femur, tibia, and fibula, and the crus with the foot in representative groups of Mammalia, the form of presentation being patterned after that of Barnett and Napier. In the postulated primitive metatherian condition (altered in several lineages, and discussed below) the fibula articulates with the femur via a cartilaginous disk which was partly continuous with the lateral meniscus of the knee joint (Fig. 4.5C). This primitive metatherian femorofibular joint (**FFJ**) was continuous with the knee joint itself (the **FTJ**). The **FFJ** articulation, via a thick cartilaginous disk (a meniscus), is nearly vertical in orientation, a general condition probably primitive in cynodonts and even synapsids. Unlike primitive synapsids, however, but like cynodonts (Jenkins, 1971), there is proximal tibiofibular contact in all marsupials. The most likely primitive metatherian distal tibiofibular joint (**TFJ**) is closely approximated in several groups of living marsupials. This condition, discussed in detail below, involved a synovial articulation between the tibia and fibula, a considerable articular contact of the tibia with the astragalus and the fibula with both of the two proximal tarsals.

### Upper ankle joint (UAJ)

In marsupials, but not in eutherians and monotremes, this joint contains a prominent meniscus with a lunula (an ossification in the meniscus) between the fibula (**Fi**) and astragalus (**As**) (see the marsupials in Fig. 4.4). There is evidence for a lunula in that position in the multituberculates (Krause & Jenkins, 1983; see below), but not in other groups. The meniscus and its relationship to the bones has been described by Lewis (1964a–c). The meniscus is thick posteriorly, and in the space between the tibia and fibula it is further strengthened by a pyramidal ossification within it, the lunula. The meniscus itself has strong, triple distal calcaneal, fibular, and tibial ligamentous attachments. The primitive cynodont (Fig. 5.1; Szalay, 1993a), tritylodontid (Fig. 5.4), morganucodontid (Fig. 5.5), monotreme (Figs. 5.2 & 5.3), multituberculate (Figs. 5.8 & 5.9), and therian conditions (Szalay, 1993a) almost certainly had extensive calcaneofibular (**CaFi**) contact (contra arguments in

**Figure 4.5.** *Comparison of the right distal end of the femur and crus in selected mammals to show the nature of fibular contact with the femur and the **UAJ** articulation formed by the distal crus. A, Tachyglossus; B, borhyaenids; C, didelphids (and most other living marsupials); D, tupaiids (and most other eutherians); E, macropodids (except dendrolagines); F, dendrolagines; G, peramelids; H, Notoryctes. For abbreviations see Table 1.1.*

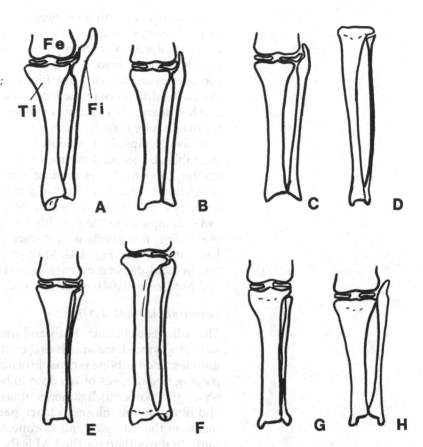

Lewis, 1983, 1989) in addition to the astragalofibular (**AFi**) contact seen in most living marsupials. In fact it is almost certain that the enlarged **AFi** contact in living taxa is an advanced modification, and a much expanded one, in contrast to the Cretaceous and prevalent Paleogene conditions (Szalay, 1984; 1993a, and below). The fibula, however, even in living forms that have lost the **CaFi** articulation, continues to be an important load bearing, bracing, and motion mediating element in marsupials.

In general, the differences in the following characters are of some importance in the analysis of mechanics, homology, and bioroles of this area: (1) presence or absence of **CaFi** contact and their extent and configuration; (2) extent and configuration of **AFi** contact; and (3) configuration and proportions of **ATi** contact, medial and lateral (**ATim** and **ATil**). The shape, curvature, wedging, and asymmetry of the astragalar trochlea are characterized by the descriptions and analyses of **AFi**, **ATim**, and **ATil** facets, and their relationship to one another. In contrast to the study of some eutherians, particularly primates, marsupial pedal mechanics, with the exception of kangaroos, has been a practically unstudied subject. Consequently, all phylogenetic assessments of the meaning of these characters related to osteology must often be considered only tentatively. For example, in eutherians asymmetry of the trochlea may allow for conjunct inversion with plantar flexion and conjunct eversion with dorsiflexion. In metatherians, however, due to the difference in the form-function of the **UAJ** (see Jenkins and McClearn, 1984; Szalay, 1984) such an assessment of asymmetry may not be correct.

In the didelphids, the neutral stance of the foot involves considerable abduction (Jenkins, 1971) in the **UAJ**. It would be misleading to state that eutherians completely lack the abduction–adduction component, but in general eutherian **UAJ**s are more severely constrained by their anatomy (see particularly Dagosto, 1985), compared to the possibly less-derived therian condition of the living metatherians, represented by didelphids or phalangeriforms (see Fig. 4.4). Most of plantarflexion and dorsiflection in didelphids takes place in the **UAJ**, but, as noted by Jenkins and McClearn (1984), some also occurs in the **LAJ**.

### Lower ankle joint (LAJ)

This joint, both in eutherians and metatherians, is almost universally ginglymoid, inasmuch as the astragalus is not moved by any muscles, and the bone is capable of a continuum of motion encompassing two degrees of freedom in both rotation and translation. Short and powerful ligaments attach the astragalus to the tibia and fibula, to the calcaneus (**aacl, pacl,** and the interosseous ligaments in the astraglar and calcaneal sulci), and to the navicular (**anl**). In the eutherians the **LAJ** is the most important area for the facilitation of movements on inclined surfaces (Szalay, 1984). In arboreal eutherians the motion around the longitudinal axis of the foot results in inversion–eversion and the concomitant movement in the **LAJ**. The movement of the calcaneus slackens the ligaments attached to it and to the navicular and cuboid, and consequently motion is allowed in the **TTJ**. In *extant* primitive metatherians, however, this motion is much less, given the mechanism of the **UAJ** and the morphology of the conarticular surfaces. In eutherians the close-packing of the sustentacular (**Su**) facets occurs during supination, and the close-packing of the calcaneoastragalar (**CaA**) facets is in pronations (i.e., congruency is the greatest in both cases). In fact inversion–eversion of the foot is increased in eutherians by the development of a helical calcaneoastragalar (**CaA**) facet that increases the range of movement in the **LAJ** (see Szalay & Decker, 1974; Decker & Szalay, 1974; Szalay & Drawhorn, 1980). There is no evidence for an equivalent or similar solution in marsupials – and as previously noted (Szalay, 1984), metatherians appear to accomplish inversion of the foot primarily in the **UAJ**.

Among the known tarsally most primitive marsupials (and these are not didelphids), the **Su** facets are remarkably helical, in contrast to the **CaA** facets, which are not. The significance of this is discussed below under the respective taxa. It is sufficient to note here that this condition may not be a modification related to inversion but rather it may be a **LAJ** solution for the parasagittal movement of the foot somewhat lateral to and under the astragalus. If close continuity of contact is maintained between the astragalar and calcaneal **LAJ** facets, then in effect the **LAJ** is restricted, an adaptive solution carried to spectacular extremes in the locked **LAJ** of kangaroos. The macropodids parallel the eutherian perissodactyls (and not the artiodactyls to which kangaroos are often compared ecologically) in the mechanics, but not in the historically contingent morphological pattern, of that joint.

## Transverse tarsal joint (TTJ)

This area of movement involves two individual pairs of bone contacts, the astragalonavicular joint (**ANJ**) and the calcaneocuboid joint (**CCJ**). These are principal areas of flexion–extension and also, but to a lesser degree, supination–pronation. The absence of one or the other movement, and the relative extent of each depends on the particular adaptive solutions of the sundry taxa. Contact between the navicular and cuboid can range from rigidity across a nearly flat surface to slight rotation on slightly arched contacts.

**TTJ** movements probably help increase the range of further adjustments on uneven surfaces. In the marsupial cursors the foot appears to maximize stability and mobility in flexion–extension, similar to the mechanics displayed by the **TTJ** of eutherian runners–leapers. This, as discussed below, may be a significant departure point for understanding evolution of the primitive eutherian pedal character complex.

Interestingly, the astragalonavicular (**AN**) facets in marsupials, even in the most arboreal species, do not become spherical as they do in eutherian arborealists. In metatherian arborealists, like most of the didelphids and many australidelphians, movement in the **TTJ** appears to be centered on supination and pronation, and this is closely coordinated with the movements of inversion–eversion executed in the **UAJ** and to a minor degree in the **LAJ**.

The highly reliable correlations (predictions) about the extent of habitual inversion or eversion based on the morphology of the head of the astragalus in eutherians (see Szalay & Decker, 1974; Szalay, 1984, 1985) is only slightly indicative of these mechanics, and therefore the associated behaviors, in marsupials. This is clearly due to the different underlying pattern of **UAJ** versus **LAJ** movements. In eutherians during inversion the calcaneus rotates on its long axis laterally and the sustentaculum of that bone is brought distal and more flexad to the medial border of the head of the astragalus. As navicular contact shifts laterally on the head of the astragalus (see Jenkins & McClearn, 1984), the spring ligament comes into articular contact with the medial side of the head. This movement is reflected in the astragalar head of habitual eutherian inverters (arborealists), as was shown for the earliest primates and archontans (Szalay & Decker, 1974; Szalay & Drawhorn, 1980). The medial extension of this "navicular" facet on the astragalar head is consequently the area where loads are taken up during inversion. Habitual inversion, therefore, is reflected in this charac-ter alone among placentals. During eversion (usually the push-off phase of the gait) the lateral side of the head is loaded, and in largely nonarboreal forms this side is usually the most robust among the placentals.

In marsupials, rather, the orientation of the "head" (a relatively "neckless" one) holds some predictive value. The head is trans-verse in habitual inverters with an extensive medially sweeping **AN** facet, and dorsoplantar (or extensad–flexad, in relation to the body of the astragalus) in terrestrial and primarily hopping–running forms.

The size of the astragalar head in eutherians is proportional to the loads mediated on it, as it has been shown for primates by Langdon (1986), and this is very likely a correct causal correlation for all mammals. In fact if the tritylodontid and morganucodontid condition is any indication (see below), presence of a "neck" may well be the primitive mammalian condition. Because most living marsupials are either powerful graspers or descendants of them, the virtually "neckless" condition of didelphid and most aus-tralidelphian astragali is indicative of this relationship. Because the astragali of most metatherians are "neckless," the deviation of the neck is not a helpful predictor of mechanics. This condition is probably either a therian, or metatherian, or Ameridelphian heri-tage feature, possibly a reflection of a powerful grasping pes. In fact, a narrowing of the neck occurs in such cursor–leapers as caenolestids.

In relation to the mechanics of the **TTJ**, the form of the navicular must be discussed in some detail. This bone articulates with the greatest number of other tarsals – with almost all of them except for the calcaneus. Its overall dimensions reflect adaptive signifi-cance, shared only to some degree by both eutherians and meta-therians. Relatively long dorsoplantarly deep naviculars are usually clear expressions in eutherians of running–leaping, whereas short and primarily transversely aligned ones are indica-tive of a highly mobile and easily adjustable tarsus important for climbers. The relative size of the naviculoentocuneiform (**NEn**) facet is probably indicative of the nature of loading from the hal-lux. In a robustly grasping form, the loading is much more exten-sive than in the foot of a cursor–leaper that places little or no load on a relatively small first ray.

The **CCJ,** either on the calcaneal female facet or on the cuboid male facet, is a clear predictor of the ability of the foot to rotate either on its long axis or around some midtarsal vertical axis. Taxon-specific, characteristic aspects of this joint diagnose and positively test the phyletic unity of entire radiations. In addition, the significance of this feature is that it, along with other tarsals it potentially delineates the mechanical foundations of locomotor adaptations of groups.

### Joints between the cuneiforms

Relatively little is recorded in this study about any of these joints. Their configuration appears to reflect none of the clearly adaptively and phyletically determined form-functions of the cuboid and the more proximal tarsals. But perhaps it would be more appropriate to state that much more work would be needed to determine the reasons for their minor differences.

### Tarsometatarsal joint (TMtJ)

Although this is an artificial conglomeration of contacts, it can be useful to designate the mechanical nature of this area. The **TMtJ** is discussed below in relation to the various potentially ancestral patterns of synapsid morphology of the entire foot skeleton. The distal articular facets of the tarsals that contact the cheridia (the rays) allow and maintain the bone relationships as forced by the more proximal bones and articular facet, but will not usually channel specific movements. In other words, this area will contain (*fide* Langdon's, 1986, concept of containment) a large variety of mechanical requirements that are reflected in the more proximal joints, without the highly diagnostic mechanical specializations often seen in these joints. It follows that the general "predictive" value of the distal facets of the distal tarsals (with the notable exception of the entocuneiform) does not appear to be good. The proximal facets on the metatarsals (with the notable exception of **Mt1**) allow slight translation (gliding) and some supination–pronation. Undoubtedly, however, intertaxonal differences that reflect the extent of flexion, extension, or wedging of the rays can be predictive of mechanical differences which often reflect bioroles. Further study will surely add to this general assessment.

### Entocuneiform–Metarsal 1 Joint (EMt1J)

This joint is highly indicative of both grasping ability (therefore usually climbing) as well as taxon specific solutions to the mechanical demands of abduction–adduction and the subsequent loading of the joint facets (**EMt1**). The joint equally well reflects the nongrasping or rudimentary nature of the medial area of the foot, yet it invariably retains, even in a highly reduced condition, the morphotypic pattern from which the **EMt1J** was transformed. Several striking examples are presented below, and many others will be described in eutherians elsewhere.

In evaluating the distinct patterns of the **EMt1J** in marsupials and other mammals it is important to consider the ability of **Mt1** to rotate while it flexes. In studying higher primates, Langdon (1986) found that there is a lateral shift in the axis of curvature on the entocuneiform from the superior to the inferior end of the facet. This of course results in a helical movement. In living marsupials, grasping mechanics are significantly distinctive between didelphid and australidelphian graspers. These phyletically successive stages of form-function solutions represent plesiomorph and apomorph states of a transformation series.

# 5

# *Background to the analysis of metatherian cruropedal evidence*

*More than a century has thus elapsed since the first Mesozoic mammal was made known. In that time, which includes almost the entire history of the science of vertebrate paleontology, most students of fossil mammals have been concerned in some way with the group, from Cuvier to our contemporaries. Yet the known mammalian faunas stand out like lights in the vast darkness of the Age of Reptiles – and very dim lights most of them are. This mammalian prehistory is two to four times as long as the "historical period" which followed it, and yet the materials for the latter are literally many thousand-fold those for the former. This however, only makes close scrutiny of the Mesozoic mammals which are known the more necessary, and the results which are to be obtained from them the more precious.*

Simpson (1928, p. 7)

*The bony skeletons in a museum give such good feeling for what the animals must have been like that we tend to forget what a skeleton is. We are looking only at the compression members of a structure whose tension members have rotted away. . . . Bones have no functional meaning without their muscles, tendons, and ligaments, which move them and hold them together. Contrarily, of course, muscles and tendons must act against something comparatively unyielding, such as bone, cuticle, or water.*

Currey (1984, p. 185)

The transformational hypotheses developed in this chapter are based on the fact that the therapsid–mammal character clines involve a series of conditions ranging from the proximal tarsals (the astragalus and calcaneus) being sagittally parallel to those that are superimposed on one another. So in a very real sense the fossil record is the final arbitrator of transformational hypotheses. These conditions of the tarsus make up the objective evidence against which hypotheses of pedal and tarsal evolution in the Mammalia must be tested (Jenkins, 1970a, b, 1971; Szalay, 1984; 1993a). I have previously argued that the development of the sustentacular contacts was originally unrelated to superposition, but rather was an ancient expression of the mobility and stability required of the bones arranged in parallel with one another. Mammalian superposition of the astragalus on the calcaneus bridges an enormously long period of geological time, ranging from an astragalar "lean" to the "complete" superposition in most living eutherian mammals. In fact, as suggested by Kemp (1982, p. 122), early in therapsid history the axial rotation of the tibia probably caused the distally closely anchored astragalus to rotate on the calcaneus with the resulting origins of the LAJ articulations. The concept of a "lower" ankle joint is, of course, extracted from a fully eutherian perspective. The sustentacular process of the calcaneus is a later development, postdating the LAJ, associated presumably with an improvement in the flexibility of the foot (Schaeffer, 1941). The latter, however, is not a satisfactory explanation as there is no evidence that reptilian tarsi, or carpal configurations without sustentaculum-like calcaneal processes, are any less flexible than mammalian feet. These interpretations and caveats are basic to the understandig of transformation series in the Theria.

An assessment of the transformation of the musculoskeletal system of synapsids into mammals and of the specifics of early mammalian differentiation is essential for an understanding of the origin of any of the monotreme, therian, metatherian, and eutherian characters (for reviews see Jenkins, 1970a, b, 1971, and Kemp, 1982). As a necessary background for the analyses to follow, I will briefly outline here my interpretation of some of the known clusters of Mesozoic tarsal complexes along with those seen in the monotremes. It should be emphasized again that both the carpus and tarsus represent complex character associations, and that these are not necessarily either numerous single characters or only one complex trait. Each of these two systems changes in such ways that while the elements making them up are coadapted, the components are capable of somewhat independent evolutionary change within the complex, depending on the paradaptive and adaptive imperatives.

## Relevant groups

I have restudied an African Middle Triassic right foot (BMNH TR.8), the Manda cynodont, reconstructed the articular connections of the foot, and made further comparisons (see Jenkins, 1971; Szalay, 1993a). The early cynodont stage (at least as represented here by the Manda specimens; compare Fig. 5.1 with Figs. 5.2 & 5.3) can be characterized as one in which (1) the tibial contact of the astragalus (ATi) is widely separated from the fibular articulation (AFi); (2) the UAJ is tricontact (i.e., the fibula is broadly in articulation with the calcaneus and only narrowly with the astragalus); (3) the astragalus and calcaneus (the latter with a small tuber and an enormous peroneal process) is essentially sagittally parallel, although less so than monotremes; and (4) the calcaneocuboid contact forms an angle of about 45 degrees with the long axis of the foot (Szalay, 1993a). The small ectocuneiform is correlated with the slight articulation of Mt3 with the cuboid in addition

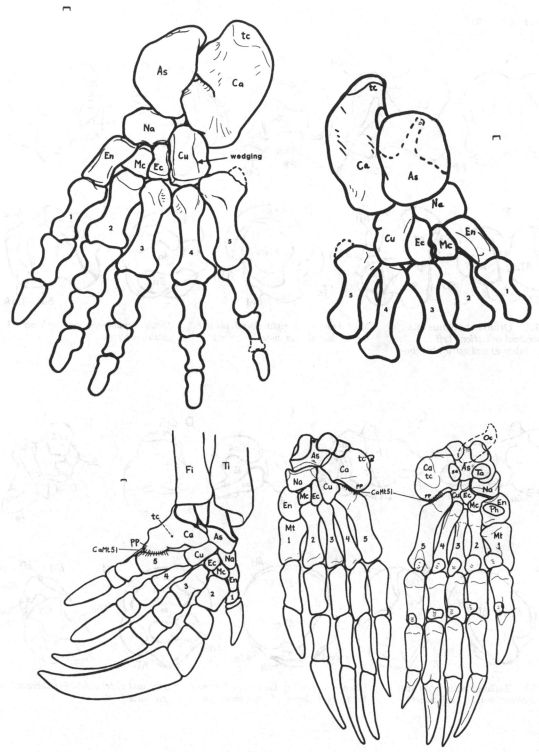

**Figure 5.1.** Above: *dorsal (right; phalanges not shown) and plantar (left) view of the right foot of the African Middle Triassic Manda cynodont* (BMNH Tr 8). Below left: *right foot and distal crus of* Tachyglossus *in dorsal view.* Below right: *left foot of* Ornithorhynchus *in dorsal (left) and plantar (right) view; dashed outline is that of the os calcis. For abbreviations see Table 1.1.*

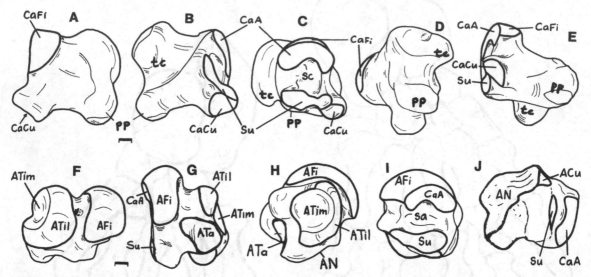

**Figure 5.2.** Ornithorhynchus anatinus *(personal collection) Left calcaneus* (above) *and astragalus (below). From left to right: dorsal, plantar, medial, lateral, and distal views. For abbreviations see Table 1.1. Scales = 1 mm.*

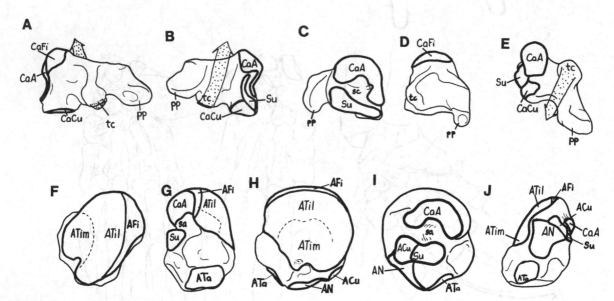

**Figure 5.3.** Tachyglossus aculeatus *(personal collection). Left calcaneus* (above) *and astragalus (below). From left to right: dorsal, plantar, medial, lateral, and distal views. For abbreviations see Table 1.1.*

to the ectocuneiform. The astragalar canal, although rudimentary, is present. A weight bearing or movement-mediating role of the fibula probably also existed; at least the contact of this bone with the astragalus, in addition to the calcaneus, is evident.

This pes shows a particularly interesting mosaic of traits. The tricontact **UAJ** (Szalay, 1984) or the phalangeal formula of 2,3,3,3,3 cannot be considered mammalian apomorphies as they are present in at least this cynodont. Similarly, the well defined calcaneal tuber, the articular contact between the calcaneus and astragalus (**CaA** and **Su** articulations) and between the calcaneus and cuboid (**CaCu** articulation), and the well defined peroneal process are also obviously well established before the last common ancestry of the Mammalia. The largely distal rather than distomedial orientation of the **CaCu** facet on the calcaneus is unusual in the context of these comparisons. The **LAJ** articulations of the calcaneus and astragalus show that the movements of the calcaneus involved a slightly helical rotation of the calcaneus not only on the **CaA** facets but primarily on the convex proximomedial portion of the calcaneal **Su** facet.

I could not differentiate between the medial and lateral tibial facets, although these are relatively clearly delineated in *Oligokyphys* and mammals. The **UAJ** facets do not suggest significant load bearing and movement differences between the the tibia and fibula; the articular surfaces are nearly uniformly convex. Yet the astragalus and its contact with the navicular are among the most significant aspects of the foot. Articulation with the navicular is perceptibly sellar.

I have restudied the pedal remains of the tritylodontid *Oligokyphus* (Fig. 5.4) from the Liassic of England, previously described by Kühne (1956). There are significant differences in the astragalus and calcaneus from those of the Manda cynodont. These consist of the more highly differentiated, "mammal-like," body and head sections and a clearly demarcated **ATim** facet of the astragalus. The astragalar **CaA** articulation is unmistakably sellar, suggesting that considerable mobility could be achieved in the **LAJ** by the calcaneus, and therefore the whole foot distal to it. The astragalar **AN** articulation is slightly concave (? sellar), and the separation of the **UAJ** articulations with the tibia and fibula persists. While there is no clear demarcation of the lateral border of the **ATil** facet on the astragalus, it appears that the groove between the **AFi** facet and the more medial tibial articulation is so similar to *Ornithorhynchus* that it was almost certainly nonarticular. The **CaCu** facet is largely medially oriented, and, as in the Manda cynodont, the calcaneal **Su** facet is convex with only its distal portion concave. The arc of **UAJ** motion, as in the Tanzanian cynodont, is remarkably small, suggesting extensive **LAJ** adjustments.

One current consensus considers the Morganucodontidae as

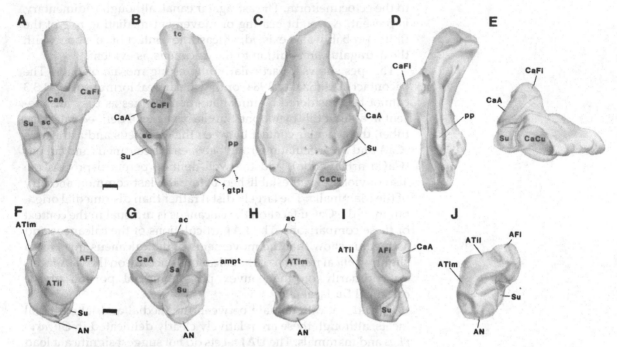

**Figure 5.4.** Oligokyphus, *based on BMNH R7503, 7504, 7500, and 7505, Early Jurassic, England. Left calcaneus (above) and astragalus (below). From left to right for the calcaneus: surface with facets of the* **LAJ** *and* **UAJ** *(A), dorsal (B), plantar (C), lateral (D), and distal (E) views. From left to right for the astragalus: surface with facets of* **UAJ** *(F), surface with facets of* **LAJ** *(G), medial (H), lateral (I), and distal (J) views. For abbreviations see Table 1.1. Scales = 1 mm.*

somewhat representative of the stem Mammalia in most of its known morphology. Its postcranial remains have been studied by Jenkins and Parrington (1976). The tarsal remains are particularly well represented (Figs. 5.5–5.6). The similarities to *Oligokyphus* of both the calcaneus and the two types of astragali are very great indeed. The convex portion of the calcaneal **Su** facet persists in being the more extensive part of the sustentaculum as compared to the distal concave portion. The **CaA** articulation is clearly sellar. On the larger astragalus, the astragalar canal is more dorsal than in *Oligokyphus*, and the astragalus is in contact with the cuboid, a condition which may or may not have been present in the tritylodontid. A neck and a head are well differentiated with a sellar (or at least concave) facet for navicular contact. As in the previously discussed groups, sustentacular support (i.e., a flangelike projection of the calcaneus under the astragalus) is minimal, probably less than seen in the Manda cynodont.

In his most recent accounts, Lewis (1983, 1989) employed monotreme foot morphology as a stage of evolution leading to the foot structure of the last common ancestor of living therians. This perspective was a direct consequece of the taxic outgroup proposition that monotremes are the living sister group of the therians, a prop-

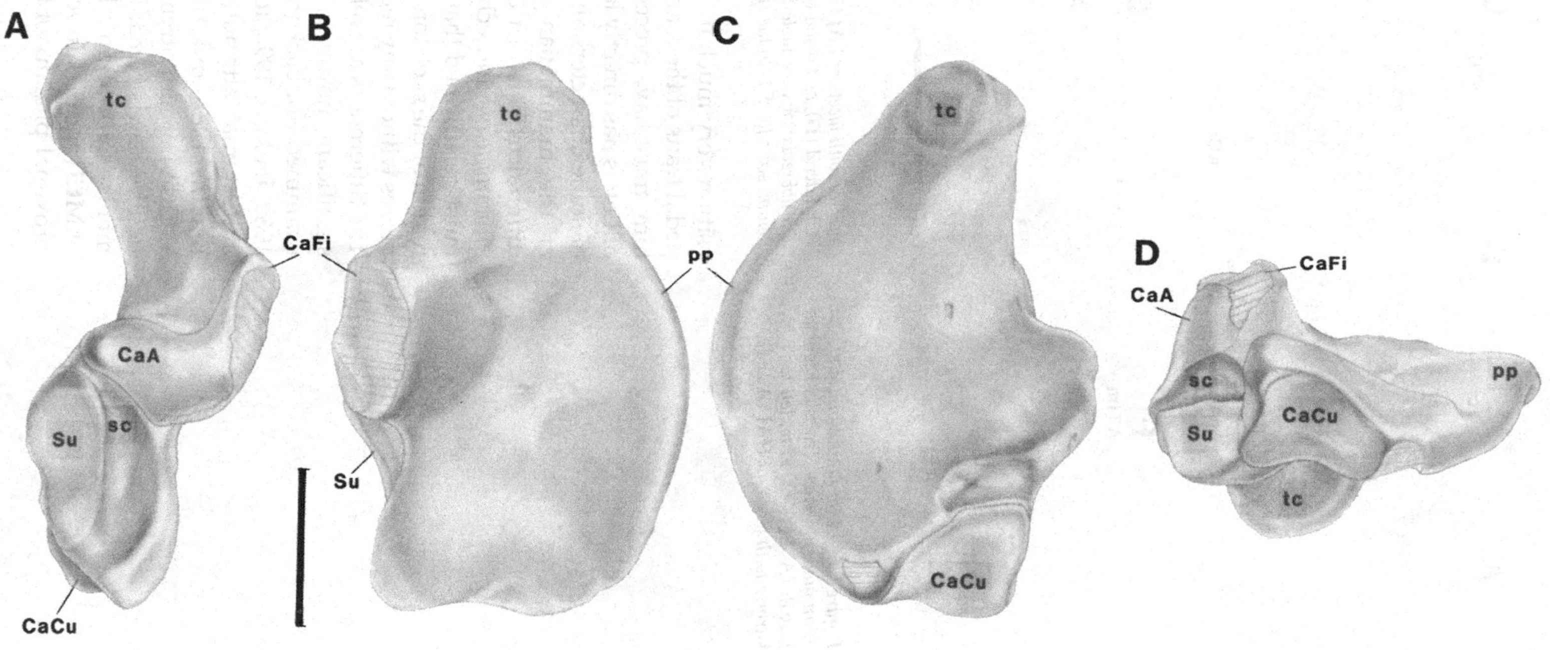

Figure 5.5. Morganucodontid, Pont Alun quarry, Wales, Triassic. Left calcaneus. From left to right: surface with facets of the **LAJ** and **UAJ**; (A), lateral (B), plantar (C), and distal (D) views. For abbreviations see Table 1.1. Scale = 1 mm.

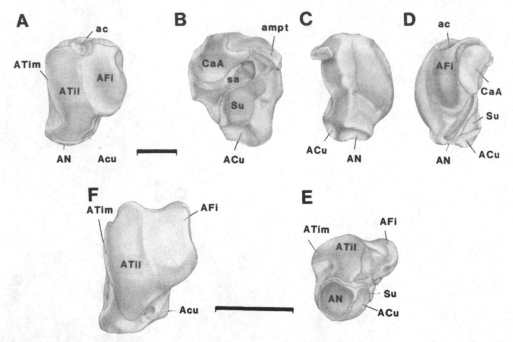

**Figure 5.6.** *Morganucodontid(s), Pont Alun quarry, Wales, Triassic. Right astragali (A–E, & F). From left to right: surface with facets of **UAJ** (A & F),* *surface with facets of **LAJ** (B), medial (C), lateral (D), and distal (E) views. For abbreviations see Table 1.1. Scales = 1 mm; scale for **F** is to the right.* *Size of the articular facets of this specimen (F) matches those of the calcaneus shown in Figure 5.5.*

osition very much in harmony with Hennig's deviation rule. The pedal traits of these animals, specifically of echidnas, according to him, must have preceded therian attributes in the foot. This view of Lewis has directly led to (1) a misidentification of the homology of some characters and (2) an unsupported view of the evolution of the mammalian pedal complex. There are some glaring differences between the foot structures and tarsal details of *Ornithorhynchus* and echidnas (see Figs. 5.2 & 5.3). Nevertheless, it must be affirmed that there are uniquely shared attributes in this region of the skeleton that unequivocally support their holophyletic ties without any recourse to the comparison of other characters. The differences as well as their shared unique similarities become significant, however, only when their ecological morphology is examined (for background summaries on monotremes see Grant, 1989, Jenkins, 1990; for phylogenetic assessment based on dental and cranial attributes see Kielan-Jaworowska, Crompton and Jenkins, 1987, and Archer, Murray, Hand, & Godthelp, 1993).

In living monotremes the calcaneal process previously identified by Lewis (1983, 1989) as the tuber is strongly bound by ligaments (**CaMt5l**) to a laterally and proximally attenuated process of **Mt 5** (Fig. 5.1). In echidnas this connection extends as well to the proximal phalanx of the fifth ray. In both groups of monotremes this area of the calcaneus is the peroneal process and not the tuber. The real tuber of the calcaneus is easy enough to identify in *Ornithorhynchus* both from the attachment of the tendon of the two headed gastrocnemius and from the position of the process on cleaned bones. In echidnas, however, an extreme transformation has taken place. The phyletic depression of the tuber flexad resulted in its becoming a tuberosity prominent on the edge of the extensad surface of the pes. In incomplete muscle preparations (e.g., Lewis, 1963, Fig. 1), it appears that the tendon attaches laterally to the most lateral bony projection, identified as the tuber by Lewis (1983, 1989), but in fact it loops to the flexad edge of the actual tuber, identified as the peroneal process by Lewis (1983, 1989). Even from Lewis's (1963) figure (republished in 1989) it is evident that the tendon does not extend to the laterally and distally visible peroneal process that is parallel with the fifth ray.

After having made his assessments of homologies on echidnas, Lewis has transferred the assumed topology to the platypus. The necessary explanation for the morphological change that I suggest (Szalay, 1993a) occurred in echidnas resides in the fact that the arc on the medial side of the astragalus, and the ability of the tibia to travel through it, is nearly 100 degrees. To implement such rotation on a transverse axis, the mechanical advantage of the UAJ is greatly increased by the flexad and distal migration of the insertion of the tendon of the gastrocnemius. This unique increase in the UAJ arc is probably related to the original digging or shoveling adaptation of the monotreme ancestry, carried to confusing extremes in echidna morphology.

Concomitant with the extreme flexion–extension ability of the foot in echidnas on a transverse axis is the additional supplemental motion that occurs along the virtually same transverse axis in the greatly stabilized LAJ (actually a sagittally oriented joint). The premium on mobility as well as stability in the LAJ is manifested in a highly conical, pivotlike, calcaneal CaA and in astragalar Su facets that move in the conarticular female facets. This arrangement provides additional flexion to the foot on approximately the same axis as the UAJ flexion–extension! A functional and adaptational perspective is clearly indispensable in testing this homology hypothesis.

There is an occasional "atavistic" remnant of the astragalar canal in the nonarticular groove between the ATil and AFi facets on some astragali of the platypus, but not in echidnas. As in therians the latter have closely abutted ATil and AFi facets but this condition is almost certainly a convergent attainment in the two major groups. Monotremes do not have a neck–head configuration on the astragalus, not even similar to *Oligokyphus* or morganucodontids. Furthermore, the ANJ of the platypus shows the unmistakable combination of a concave lateral channel on the astragalus and a convex medial component that may be considered a homologue of the "head." What is significant is that this joint is sellar, unlike the therian concave (navicular) and convex (astragalus) combination. Echidnas also retain a smaller concave astragalar component. The presence of prehallux and tibiale are well documented (Emery, 1901; see also Fig. 5.1). The outline of the bony plate (Os calcis; **Oc**) to which the horny perforated spur is attached, via a syndesmosis (Lewis, 1963) is apparent on the flexad view of the platypus's foot (Fig. 5.1).

It appears that Lewis (1983, 1989) misidentified some of these critical aspects of the tarsus in monotremes because he failed to take into account factors evident from ecological morphology. The platypus makes little or no propulsive use of its hind legs while swimming, and therefore most of its functionally significant pedal features may be related to its terrestrial, partly burrowing bioroles. It does not, however, depend for its livelihood and safety on digging and rapid "sinking" into the earth as echidnas probably do. The misidentification of the critical tarsal homologies of monotremes resulted in a general inability to differentiate among diagnostic monotreme apomorphy and the shared primitive synapsid, cynodont, and mammalian traits in the ancestor of the Monotremata, when compared to the unique attributes of either echidnas or the platypus.

The tentative articular relationship of the elements of the foot skeleton for the multituberculate genus *Eucosmodon* is shown in Figure 5.7. The mechanical shifts of the foot have been clearly described and illustrated by Krause and Jenkins (1983) and Jenkins and Krause (1983). Cretaceous material along with the evidence described by Krause and Jenkins (1983) testifies that the foot bones in this taxonomically diverse order were relatively very diagnostic (Figs. 5.8 to 5.11). I have not seen the morphological disparity displayed, for example, between the platypus and echidnas among multituberculates of vastly different sizes belonging to different suborders from different continents. I concentrate in this background account on the Cretaceous proximal tarsals and on *Eucosmodon* for the remainder of the foot (Figs. 5.9–5.11).

The **UAJ** articulation of the crus may support the inference of a calcaneofibular contact. As in multituberculate astragali I have seen, the medial articular condyle of the tibia is a surface largely in line (i.e., the tibial **ATim** facet; not on a recognizable medial malleolus), and does not form an angle with the **ATil** facet. Although in the Eocene *Eucosmodon* the two astragalar tibial facets are demarcated by a sharp diagonal crest, in the Cretaceous material shown here this crest is only a faint feature of the **UAJ** surface of the astragalus. Both this lack of medial buttressing and the sharp angle that the astragalar **AFi** facet forms with the *tibial articulation* are diagnostic of the multituberculate **UAJ**. In addition, the **AFi** is only contiguous with the **ATil** facet distally, suggesting that the wedge-shaped nonarticular area on the astragalus is the equiv-

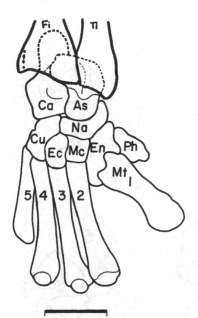
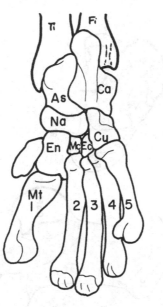
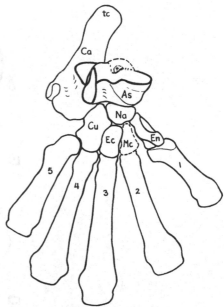

**Figure 5.7.** *Dorsal (left) and plantar (middle) views of the right distal crus, tarsus, and metatarsus in* Didelphis virginiana; *scale represents 1 cm. On the right dorsal view of right tarsus and metatarsus of the Eocene multituberculate* Eucosmodon. *Plantar process of calcaneus in the multi disarticulated from contact with* **Mt5** *(Kielan-Jaworowska and Gambaryan, personal communication). For abbreviations see Table 1.1.*

alent of the primitive synapsid separation of crural articulations. In my view there was probably calcaneofibular contact, as suggested by the faceting of the calcaneus and the configuration of the distal fibula, and by the documented presence of a sesamoid (in Krause & Jenkins, 1983; a lunula?) between it and the astragalus posteriorly.

The uniquely arched buttress around the astragalar canal of multituberculates eventually requires an explanation – it is highly diagnostic. The articulation of the astragalus with the navicular is equally diagnostic: it is an extreme modification of the primitive mammalian trait of a convex or sellar navicular **AN** facet. This **AN** facet is a curved and concave sulcus on the astragalus and a convex male equivalent on the navicular. This facet bears no resemblance to the well-established therian apomorphy, which is an astragalar head rather convex in all directions.

The relatively large peroneal process of the calcaneus of a Cretaceous sample shown in Figure 5.8 is much reduced compared either to Triassic–Jurassic mammals or to known early therians. [Kielan-Jaworowska and Gambaryan (personal communication after this book has gone to press) have strong evidence that in multituberculates the peroneal process articulated with **Mt5**, as in monotremes! This may well represent the primitive mammalian condition.] The size of the peroneal process in the Eocene

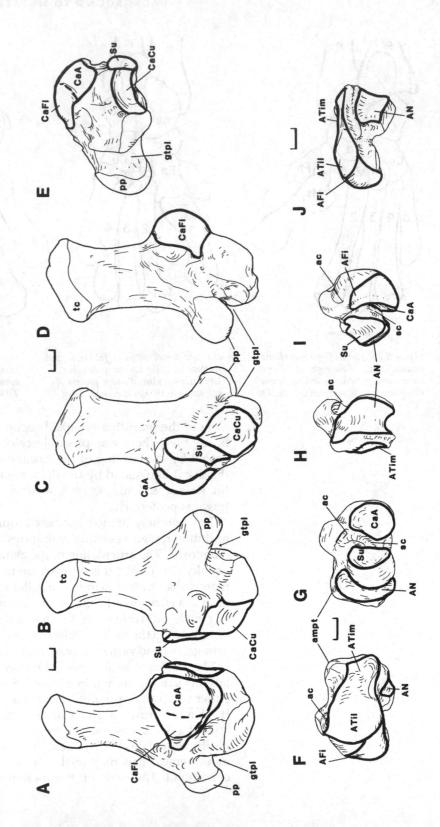

*Figure 5.8. Ptilodontoid, right calcaneus and left astragalus (reversed), TMP No. 8802030, Frenchman Formation, Gryde locality,, Alberta, Late Cretaceous. From left to right: dorsal, plantar, medial, lateral, and distal views. For abbreviations see Table 1.1. Scale = 1 mm.*

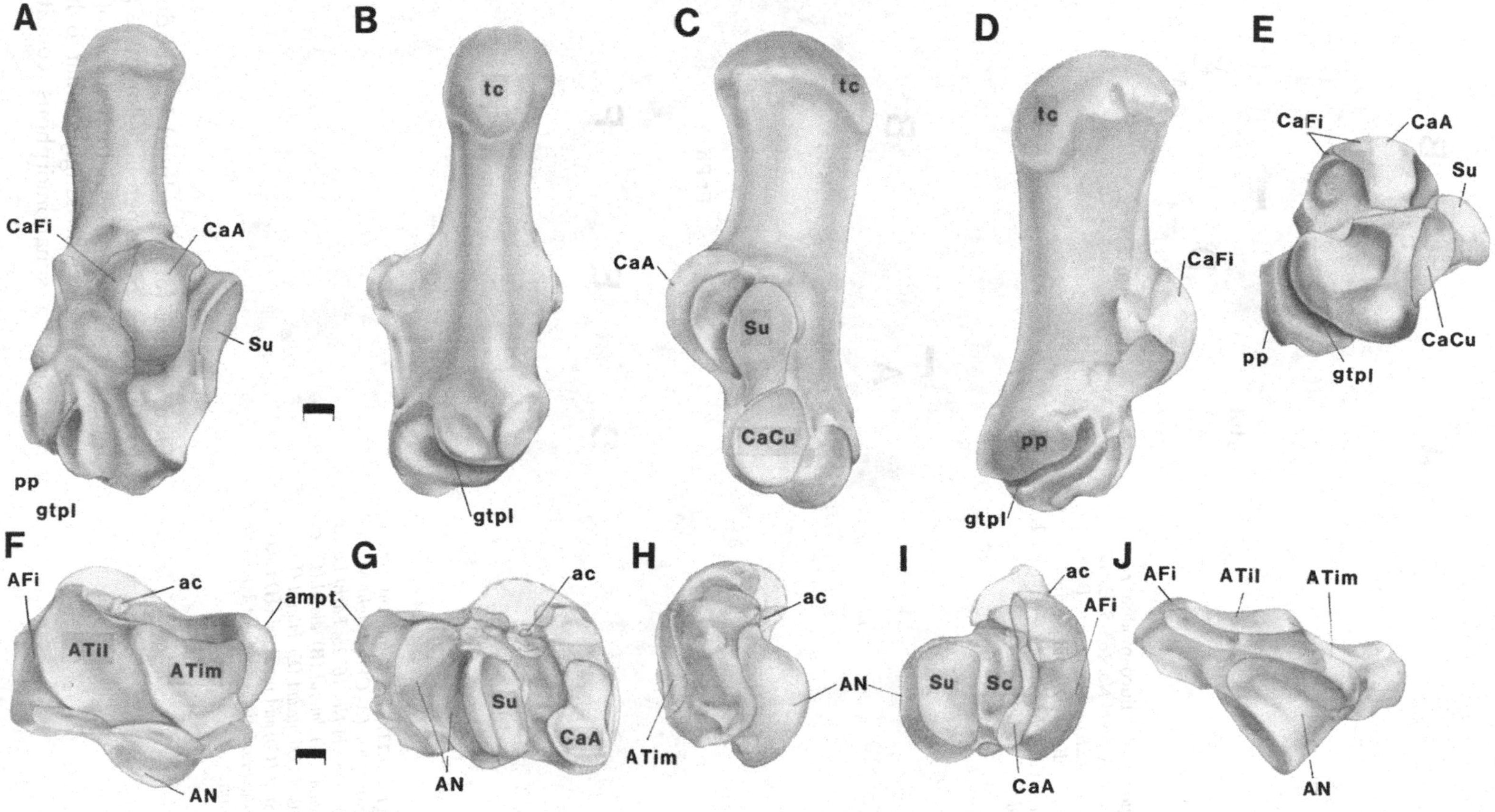

*Figure 5.9.* Eucosmodon, right calcaneus and astragalus, AMNH No. 16325, Early Eocene. From left to right as in Figure 5.8. For abbreviations see Table 1.1. Scales = 1 mm.

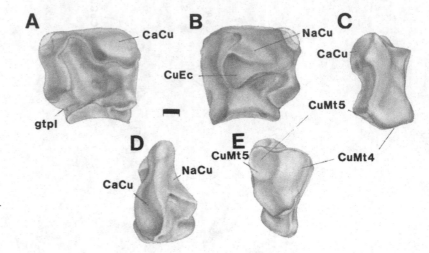

**Figure 5.10.** Eucosmodon, *right cuboid, AMNH No. 16325, Early Eocene. Lateral (A), medial (B), extensad (C), distal (D), and proximal (E) views. For abbreviations see Table 1.1. Scale = 1 mm.*

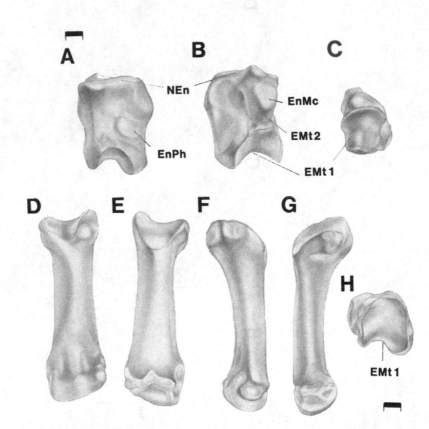

**Figure 5.11.** Eucosmodon, *right entocuneiform (A–C) and right Mt1 (D–H), AMNH No. 16325, Early Eocene. Medial (A), lateral (B), distal (C), extensad (D), flexad (E), lateral (F), medial (G), and proximal (H) views. For abbreviations see Table 1.1. Scale = 1 mm.*

*Eucosmodon* is further reduced, but the deep and confining groove for the tendon of the peroneus longus is preserved. The tuber is a long and slender proximal extension of the calcaneus in multituberculates and is strikingly similar to that of therians. This is not necessarily a synapomorphous condition exclusive to these

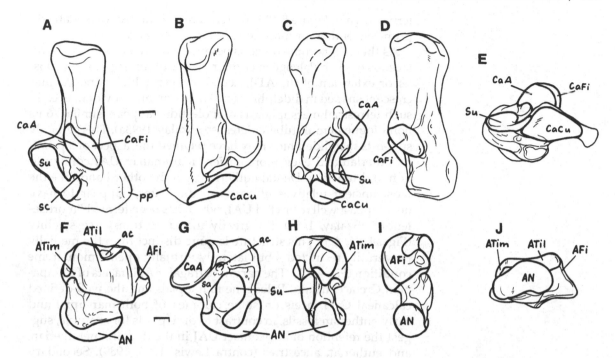

**Figure 5.12.** *While certainly not the tarsally most primitive, the arctocyonid* Protungulatum donnae *is a repre-sentative archaic eutherian; left calcaneus and astragalus, UMVP Nos. 1914 and 1915, Maastrichtian latest Cretaceous. From left to right as in Figure 5.8. For abbreviations see Table 1. Scales = 1 mm.*

groups, judging from the early appearance of the tuber in synapsids.

The sustentacular shelf so evident in tritylodontids, morganucodontids, and therians is noticeably absent from multituberculates, as it is from monotremes – but differently. These superficial similarities of multituberculates and monotremes are likely to be independently attained secondary conditions.

The cuboid and entocuneiform of *Eucosmodon* (Figs. 5.10 & 5.11) are different from the conditions known in therians or monotremes.

The therian common ancestor is difficult to reconstruct. The few known and available specimens from the Cretaceous that can be unequivocally allocated to marsupials (discussed in detail below in this chapter) and placentals add some additional insight and confirm expectations based on inductive and temporally based evaluation of the Paleogene evidence. The eutherian evidence reviewed by Szalay (1977b) is now dated, and an interim study (Szalay, 1984) has touched on the ancestral therian and marsupial traits of the crus and tarsus. The sundry complex topics that bear on the whole subject of eutherian pedal evolution and the origins of various ordinal modifications will be discussed elsewhere (in preparation). The example of a selected eutherian cruropedal pattern shown in Figure 5.12 is merely a representative of a relatively primitive eutherian conditions of this anatomical region.

As this will become obvious below, there are no described Cretaceous or even Paleogene crural remains of marsupials. The posterior extension of the ATil facet, however, which is helical and crescent shaped in didelphids (and this remnant is evident even in such terrestrial marsupials as caenolestids), appears similar to its homologue in multituberculates (see Szalay, 1993a). This certainly suggests that didelphids may have retained this aspect of a primitive therian and even more ancient mammalian UAJ configuration and function. Dental eutherians, on the other hand, like the most ancient relatives of various "ungulates," appear to have developed a well restricted UAJ, which has been termed "mortise-tenon" (Szalay, 1984), for greatly increased transverse stability (Figs. 4.4 & 4.5). This structure is quite distinct from the medially powerfully restricted – but laterally virtually open – monotreme condition (Fig. 4.5A). The narrow astragalar ATil facets of Campanian Cretaceous and Paleogene marsupials and the well-defined calcaneal CaFi facets on some calcanea of both marsupial and early eutherian fossils, in contrast to didelphids for example, suggest the retention of a tricontact UAJ in the therian, metatherian, and eutherian ancestries (contra Lewis, 1983, 1989). Secondary tricontact UAJs have evolved several times in both groups of therians (see Chapter 7 under Didelphidae, Peramelidae, etc.), but this fact does not axiomatically explain calcaneofibular articulations in the early fossil record of therians.

The question of astragalocuboid articulation is a problematic one. Transversely more compressed tarsi, as in "semi-cursorial" forms (in a broad sense to be applied to early, and probably terrestrial scampering eutherians; see in Chapter 7 under Didelphidae, and Szalay 1984), contact between the astragalus and cuboid is to be expected.

Contrary to statements in the general literature, the rounded head of the astragalus is a therian but not a mammalian character. As noted above, the neck is far less discernible in metatherians than in even the earliest eutherians. There is of course the danger of circularity in attempting to identify astragali as eutherian based using the criterion of a well-differentiated neck alone. Yet the shape and mechanics of the ANJ is a highly informative character for evaluating tarsal evolution.

While many intermediate nuances are entirely unknown, the condition of the last common ancestor of the therians can be reconstructed when we keep in mind the fossil record, the data from the living, and the ancestral constraints. Although the actual supporting role of the sustentaculum is already established, the astragalus is not completely superimposed, as this condition is probably attained independently in both eutherians and metatherians (see Szalay, 1993a). While the calcaneal sustentaculum was a supporting shelf proximally, its more distal extension was nearly as sharply angled with the **CaA** facet as the condition seen in multituberculates. The **LAJ** is in effect oblique, probably similar to the condition seen in Mesozoic and Paleogene marsupials.

The **UAJ** is a tricontact type but with crural components which articulate distally (**TFJ**). Unlike the condition in *Oligokyphus* and such mammals as morganucodontids, astragalar **ATi** and **AFi** facets are not separated in therians. Movement of the foot in the **UAJ** was relatively unconstrained. There is as yet no way to ascertain the presence or absence of a functionally significant meniscus in the **UAJ** in various taxa of therians. It may have been present and retained in marsupials, or perhaps it was acquired only by the first metatherians, ameridelphians, or even didelphids with their derived **UAJ** configuration. If it was present in the stem prior to the origins of Eutheria, then it was lost in the protoeutherians in the same way it was reduced in bandicoots and kangaroos.

It is important to note here that in a study unique among the comparative accounts of a mammalian postcranial system, Haines (1958) probed the question of arboreality in placental ancestry through the analysis of the forearm and hands, in particular, of therians and other vertebrates. Although not strictly related to the understanding of the hind foot, this far-reaching analysis has clear relevance to the mode of life of both ancient marsupials and placentals, and therefore it is of great interest for any future study dealing with early therians. Haines's analysis is one of the most complete "modules" to be tested and set in a phylogenetic framework. This latter is yet to be constructed for the phylogenetic analysis of carpal and relevant forelimb evidence for synapsids and mammals, and for the diversity within Mammalia. Haines's major conclusion, based on extensive comparisons of the contrahentes muscles in hands, was that his "convergent"-type hand, found primitively in placentals, was probably primitive in the Mammalia, whereas the "clasping" type displayed in didelphids was an advance built on the former condition. He clearly implied that full grasping arboreality seen in the generally regarded most primitive living marsupials, the opossums, was not present in the first therians or in the protoplacentals. The amazing and almost certainly derived saddle joint between the unciform and metacarpal 5 (**Mcp5**) in didelphids corroborates Haines's views on the evolution of the hand.

## Character transformations in the Mesozoic

The framework supplied by the Mesozoic pattern of tarsal character evolution is an important storehouse of information, a background against which are made the complex judgments required for explanations of specific polarity in various cruropedal groups, such as the metatherians. The potential number of characters that may be noted when describing the anatomy of organisms is great indeed. Yet from the accounts of the past descriptive, functional, and ecological morphology that have dealt with the comparative biology of the crus and the foot, a number of characters have emerged which make far more biological sense than those from less functionally and adaptively conceived comparative accounts. These characters form the bases of the various taxonomic properties.

### Calcaneal tuber

There is little doubt that the tuber is a synapsid–cynodont acquisition. One of its most extreme modifications, from a relatively short cynodont-like conformation, is seen in the monotremes, and its elongation (or rather its transformation into the posterior nonarticular half of the calcaneus) is documented for the therians and multituberculates. It may have been an independently lengthened feature of both multituberculates and therians from a common mammalian condition. It should be emphasized that the well defined tuber itself, unlike the highly modified condition in monotremes, is an ancient synapsid attribute. It may be viewed as the consequence of a drastic reduction of the peroneal shelf into a peroneal process which resulted in the slender shaft of the tuber (as for example the "elongate, square tuber calcis," the sole diagnostic character given by Rowe, 1988, for "Theriiformes"). It appears probable that this effect of peroneal shelf reduction has arisen independently several times. If this is the case, then this trait *alone* is a poor diagnostic feature indeed for a taxon. The shape of the cross section of the "shaft" and the end of the tuber are largely determined by loading not only by the tendon of the gastrocnemius but by the very important plantar flexors such as the flexor digitorum brevis (in therians). Such, or an equivalent, muscle is probably responsible for the great depth of the cross section of the "shaft" and the flexad extension of the calcaneal tuber (a plantar process on the tuber) in the multituberculates illustrated here.

### UAJ

The evidence appears unequivocal that a tricontact cruropedal joint (i.e., contact between fibula, astragalus, and calcaneus; Szalay, 1984; 1993a) is primitive for monotremes and all mammal groups that make their appearance in the Mesozoic. In fact articulation of the proximal tarsals of *Oligokyphus* and the Manda cynodont leaves no doubt that a large portion of the posterior calcaneal facet is a **CaFi** facet and not an extended crescent shaped **CaA** facet. In spite of Lewis' (1983, 1989) statements that therians are primitively bicontact in the **UAJ**, the secondary redevelopment of this condition from fully superimposed, bicontact ones (i.e., contact between fibula and astragalus only) in both metatherian and eutherian lineages is highly likely in light of the evidence (see Szalay, 1993a).

## LAJ

The early therapsid foundation of this joint, in which one cannot speak of any significant superposition, is most probably related to inversion and eversion accomplished by supination and pronation in the forefoot as a result of action in this joint. The **LAJ** does not seem to have been established to extend the foot on the lower leg. Detailed cineradiographic work on the feet of monotremes would be most welcome. The extreme sagittal orientation of the monotreme **LAJ**, rather than an ancient synapsid attribute, may be a consequence of the forces generated by the shortening, flexad, forward, and lateral migration of the tuber and the lateral twist of the peroneal process. It may have started from a condition seen in the morganucodontids. It is highly unlikely that it represents a pre-cynodont condition in which there was no "astragalar lean" on the calcaneus.

What has become evident (Szalay, 1993a) is that the evolutionary and directionally dynamic concept of "astragalar superposition" is a useful heuristic, the end of a particular transformation driven by mechanics, and that it has highly specific morphological and functional meaning when applied to a group of mammals. I do not believe that there is a phylogenetic constraint at the base of mammalian lineages that we can call "mammalian superposition." Although it undoubtedly began among early cynodonts, it was, I believe, achieved to different degrees in different lineages. In Paleogene fossil marsupials the presence of a relatively enormous astragalar medial plantar tuberosity (**ampt**), together with the equally robust, in-line navicular tuberosity just distal to the astraglar protuberance, in my view, constitutes evidence of substantial astragalar support resting on the substrate. These conditions, together with the helically oriented **Su** articulation, which is distally nearly sagittal, strongly support the idea of astragalar contact with the substrate. This, for example, could not be considered an equivalent degree of astragalar superposition compared to that which is seen in most eutherians. The foot structure of the earliest reproductively eutherian mammals, however, may have been identical to their reproductively marsupial sisters.

While the multituberculate astragalus laterally extends, almost horizontally, onto the calcaneus, the angle of sustentacular contact to the **CaA** articulation along its length is close to 90 degrees, and there is no noticeable shelf for the sustentaculum on the calcaneus. All this suggests that astragalar superposition in multituberculates was initiated by the lateral part of the astragalar facets. A condition like this may have been the source for the small sustentacular shelf discerned in the earliest known marsupials. In fact there is evidence from the archaic metatherians discussed below that the development of the shelf was initiated at the proximal (posterior) extremity of the sustentaculum. The distal segment is still nearly sagittal (at least in relation to the **CaA** facets) in orientation in these therians.

## *ACJ*

The significance of the morphological attributes of this joint is limited, I believe, without a full understanding of the ecomorphology of the animals being examined. If this articular contact is present, then it transmits loads of a relatively transversely constricted tarsus. Despite the assertions in much of the early literature concerned with the alternating versus parallel tarsal arrangements, presence or absence of this articulation among nonungulate therians appears to have more functional significance than immediate phylogenetic relevance. This condition fluctuates with the relationship of substrate usage and speed of travel, and could be different in close relatives that need a narrower scampering versus a broader climbing foot.

## *ANJ, and the astragalar neck and head*

The well-differentiated condition of therians and additional differences within the Theria, coupled with the distinctive multituberculate state, render this character very useful. The fossil evidence suggests that the sulcal (concave) astragalar "head" of cynodonts and Triassic mammals is primitive, and that this condition is further – and extremely – accentuated in multituberculates and retained in a modified form in monotremes. There is no doubt about a "neck," a narrowed distal portion of the astragalus, in tritylodontids and morganucodontids. The convex distal articulation of the astragalus, the "head," is a development in therians, and appears to be independent from the early appearence of the "neck." Multituberculates, without any "neck" or "head" on the astragalus, probably lost this premammalian trait as they expanded **ANJ** articulation transversely. The "neckless" appearance of the astragali of didelphids is a reflection of the large and medially distributed articular surface of this arboreal radiation. The powerfully grasping hallux generates great loads medially in this joint. Paleogene lineages of marsupials either retain, or more likely, develop independently (and secondarily) astragalar distal halves of relatively narrow proportions. Such a loading of the **ANJ** is probably indicative of nongrasping locomotion.

## *CCJ*

The nature of this articulation can be one of the most significant mirrors of pedal and tarsal movements in many groups of mammals. The movements and force transmission from calcaneus to cuboid tend to reflect flexion–extension, resistance to habitual abduction, and ability to rotate (pronate–supinate) the forefoot on this joint. While the stem mammalian and therian **CCJ** condition suggests forefoot rotational abilities, the limited data on multituberculates indicates great flexion–extension and abduction–adduction at the **CCJ** and **ANJ**, but shows a derived calcaneal restriction of forefoot rotation.

*Transverse tarsal arch*

Jenkins (1971) has discussed the presence of this character in the Manda cynodont. The wedging of the cuboid and the cuneiforms can be confirmed, and a similar condition is present in monotremes. It is almost certain that this predates the last common ancestor of morganucodontids and other mammals. It should not, therefore, be regarded as a mammalian feature.

*EMt1J*

The articular surfaces of this joint can be highly diagnostic, and their detailed transformation can be followed in metatherians and some grasping eutherians (Szalay & Dagosto, 1988). The diversity in morphology is clearly constrained and channeled by specific initial and boundary conditions in phylogeny, and therefore this is a potentially extremely useful character among some therians, particularly in the marsupials examined in the following chapters below. Little can be said of the diagnostic therian condition. While therian pedal graspers show distinct convergent solutions (see Szalay & Dagosto, 1988), the multituberculates exhibit a textbook example of a nearly unmodified sellar articulation (see Fig. 5.11). It does not resemble that of the most primitive living marsupial graspers, the didelphids. This potentially important area, therefore, offers very little evidence as yet for a particular transformation hypothesis for grasping related change of Mesozoic mammals. The ability to abduct–adduct slightly and rotate **Mt1** on the entocuneiform is not, by itself, adequate evidence for or against the arboreal or terrestrial nature of these animals.

**Mesozoic pedal evolution in mammals – An interim summary**

This highly preliminary synthesis serves only as background for understanding the therians. Bold hypotheses and the materials described by past studies of various cynodonts should be reexamined before a more synthetic story of mammalian pedal evolution can emerge from a series of well corroborated historical–narrative explanations. The recent brief review of Sues (1986) of early therapsid locomotion has not dealt specifically with the details of the tarsal evidence. As established by Jenkins (1971), the pedal attributes of the Manda cynodont are "mammalian" in a retrospective, albeit not formal taxonomic, sense. Not only because of their antiquity, but because of their structural and functional attributes, the morganucodontid tarsal material is in every way potentially antecedent to that of other mammals. The tritylodontid *Oligokyphus*, however, is extremely similar to the former, but not to specific therians, as has been stated in the literature. It follows that most pedal attributes, that have been considered mammalian, predate the traditional and reasonably well-justified taxonomic boundary between mammals and other synapsids.

Jenkins (1971, p. 197) is also convincing on the issue of the origins of *mammalian* plantigrady, particularly why "the pelycosaur calcaneum and astragalus probably did not have a plantar contact during the initial phase of propulsion, but remained off the ground throughout the entire stride. . . ." The early appearance of the calcaneal tuber is, therefore, unquestionably correlated with the beginnings of mobility in the **UAJ**. The subsequently advanced arguments, however, that the increasingly mobile **UAJ** could only be stabilized by a previous or synchronous evolution of plantigrady, rest tenuously on the lack of evidence for stabilizing malleoli in cynodont crural remains. Is it likely that a gradual shift from the distal ends of the proximal tarsus was followed by a plantigrade stance? The problem, corollary to the cynodont question, is that of ancestral therian plantigrady in the strict sense, meaning that the stride involves full contact with the skin under the calcaneal tuber. There are as yet no studies on living forms to probe and corroborate this likely primitive therian condition.

The pedal morphology of monotremes discussed above suggests derivation from a stage in which astragalar superposition was not achieved, the tibial and fibular facets of the **UAJ** were not contiguous, and the **ANJ** was sellar. I have demonstrated, however, an oblique sustentacular shelf that was present on the ancestral mammalian calcaneus. The medial buttressing of the **UAJ** by the tibial malleolus is probably a hypertrophied version of a well differentiated ancestral process. The level of this stage of evolution, or the level at which these may be considered shared attributes of monotremes with other mammalian groups, is that of the morganucodontids, or some postcranially unknown derivatives such as docodonts or pantotheres. Some of these attributes, however, are present in groups traditionally excluded from the Mammalia, such as *Oligokyphus*. From the retrospective of the Recent, and from extramammalian considerations, these traits are thus primitive, the retained traces of apomorphies of some lineage. The other attributes are the differences of monotremes from other pedally known mammals, and these traits probably represent independent evolution. None of these latter traits are shared with therians. The transformation of the tuber and the peroneal process, and the ligamentous attachment of the latter to the proximal **Mt5** (see the **CaMt5l** in both living genera of monotremes in Fig. 5.1) are specific to monotreme ancestry (but see notes under multituberculates above). Lewis (1983, p. 30–31) is thus wrong in his view that "the transition from a monotreme to a therian foot

architecture has, as its essence, the bending outward and back-
ward of the primitively downwardly projecting heel" and in his
assertation that "a consequence of this re-alignment is that the
massive flexor fibularis tendon now enters the sole undercutting
the distal talar facet." The tuber of monotremes, in order to increase the mechanical advantage for added flexion of the foot in
the UAJ, has migrated increasingly flexad. It reaches its extreme in
echidnas, in which it has come to be located distolaterally on the
calcaneus, while the peroneal process (Lewis's "tuber") is twisted
and laterally extended into alignment with Mt 5 to which it is
bound by the broad ligament I have described. The platypus is
clearly not as derived in its tuber and peroneal process morphology as are echidnas. In light of the monotreme calcaneal homologies Lewis (1983, p. 31) makes no sense in his statement that "Only
when this structural grade [i.e., the re-alignment he describes] is
reached is it truly accurate to speak of a sustentaculum tali and
sustentacular facet; attributions of these terms to the cynodont
condition is not, therefore, strictly appropriate." A flexor fibularis
tendon undoubtedly passed medial to the tuber since the first
synapsids developed this extension before the Manda cynodont.
Pedal attributes, then, do not cladistically link monotremes with
that area of any known group of mammals, but they do help focus
on an anagenetic stage from which the pedal characters of these
animals were evolved.

As might be expected, the multituberculates pose a considerable problem in interpretation. In spite of the seemingly near complete astragalar superposition (of the lateral part of the astragalus
only!) on the calcaneus, the combined pattern of LAJ, CCJ, and
ANJ articulations (but not the crural part of the UAJ, this chapter),
while structurally derivable from a morganucodontid articulation, are not really similar to the apomorphies attributable to the
first therian. In that regard the concept of "Theriiformes" is not
corroborated. The following independently derived differences
are particularly striking: large "headless" concave semicircular
articulation of the ANJ with its extensive sellar astragalar facet
(nearly the opposite of the therian condition) and its facet at
the large CCJ, where abduction–adduction and flexion–extension
of the foot was achieved. These are fundamental differences of
multituberculates from therians. While the long tuber is a shared
similarity with therians, its possible independent elongation from
a well established early mammalian tuber renders it at present
a poor character. In contrast, however, the UAJ offers some intriguing interpretations. The nearly horizontally oriented flat or
sulcal (concave) ATim facets in multituberculates are diagnostic of
this group – as is the peculiar astragalar canal. There is virtually
no arc in this UAJ, yet the configuration of the distal crus and the
motion possible on its surface, appear very similar to that of
didelphids and possibly of ancestral therians. The form–function

of the **UAJ** of multituberculates and therians, therefore, may well have originated from an exclusively shared ancestry, whatever the level of this ancestry. Perhaps it was the more primitive morganucodontid–like condition. Needless to say, additional material of crura of various Mesozoic groups may not corroborate this hypothesis.

# 6

# Mesozoic and Cenozoic: Fossil tarsals of ameridelphians unassociated with teeth

> *Are the Deltatheriidae marsupials? Their upper molar patterns, with broad stylar shelf and well developed stylocone, resemble some late Cretaceous Didelphidae more than Palaeoryctidae.*
> Butler and Kielan-Jaworowska (1973, p. 106)

> *The deltatheriid postcanine formula is logically irrelevant to refutation of the hypothesis of marsupial affinities for the Pediomyidae. . . . It is clear . . . that the reduction in the stylar [area] on upper molars of the Pediomyidae and 'dog-like' marsupials [i.e., borhyaenoids] was generated independently, from very different demands of natural selection in the two groups, and that the one pattern had nothing whatever to do with the other in either a functional or genealogical sense.*
> Fox (1979c, pp. 733–5)

## North America, Europe, Asia, and Africa

The convincing case made recently by Kielan-Jaworowska and Nessov (1990) for the metatherian affinities of the Deltatheroida of the Early to Late Cretaceous of Asia (and possibly also of North America) will be reviewed in Chapter 8 along with the less convincing suggestion of Kielan-Jaworowska (1992) for aegialodontian–deltatheroidan ancestor–descendant relationships. The postcranial anatomy of some deltatheroidan species is currently being described (Szalay & Nessov, in preparation), and a new group of Asian metatherians represented by the late Cretaceous *Asiatherium* (*nomen nudum*, Trofimov & Szalay, 1993) is discussed elsewhere (in preparation). A discussion of the relationships of the described taxa is presented in Chapter 8 under "Theria" and "Metatheria."

The earliest known marsupial remains from North America are three species of dentally *Alphadon*-like marsupials and a form similar to *Pariadens* from the Dakota Formation of the Cenomanian Late Cretaceous of Utah (Eaton, 1990; personal communication). Cifelli and Eaton (1987) and Cifelli (1990a–c) described additional marsupials from slightly younger deposits, the Kaiparowits Formation, and also from the Cenomanian of Utah (see also Eaton & Cifelli, 1988). Clemens (1966, 1977), Fox (1976, 1987) and others have discussed the distribution of North American metatherians. These relatives (for the most part undoubted) of the living marsupials were common in the faunas of the late Cretaceous of North America, but in the Cenomanian at least there are taxa that are better considered tribotherians (constituting the paraphyletic Tribotheria; see classification above in Chapter 2 and discussion in Chapter 8) without any clear cladistic ties with metatherians (see Cifelli, 1990b–e, 1992; Eaton, 1990). In the much younger Maastrichtian Late Cretaceous Lance and Hell Creek faunas, as many as thirteen species of metatherians – making up nearly 50 percent of the mammalian taxa at these collecting areas – might have been present, judging from the dental evidence (Clemens, 1966; Clemens & Lillegraven, 1986; Archibald, 1982, 1988; Archibald & Lofgen, 1990). While these species are known predominantly from parts of their dentition, from both the Cretaceous and Paleogene, some cranial material has been described and carefully analyzed (Archibald, 1979; Clemens, 1966; Wible, 1990). The dental remains along with limited cranial material are discussed by Archibald (1982), Cifelli and Eaton (1987) Clemens (1966, 1968a,b, 1977), Eaton and Cifelli (1988), Fox (1971, 1979a,b, 1981, 1983, 1987), Lillegraven (1969), Lillegraven and McKenna (1986), Reig et al. (1987), Rigby and Wolberg (1987), Sahni (1972), and most recently by Cifelli (1990a–c; 1993a). What is known of their available tarsal characteristics is discussed in this chapter.

To accommodate the generic diversity of these Cretaceous and Paleogene marsupials that are not holarctidelphians (discussed in Chapter 8 under the Archimetatheria), I include them in the families Pediomyidae (North America, Europe, Asia, Africa) and Stagodontidae (North America; the genera *Pariadens*, *Eodelphis*, *Boreodon*, and *Didelphodon*).

The Pediomyidae contains the Alphadontinae (with *Protalphadon*, *Alphadon*, *Turgidodon*, *Albertatherium*, *Ectocentrocristus*), the Glasbiinae (with *Glasbius*), the Pediomyinae (with *Iqualadelphis*, *Aquiladelphis*, *Pediomys*), and the Peradectinae. *Anchistodelphis*, as discussed by Cifelli (1990b) is problematical although I believe that it will prove to be a metatherian and will discuss it more fully later in Chapter 8. The genera listed above are North American and late Cretaceous, whereas the Peradectinae is composed primarily of Tertiary species from both sides of the North Atlantic; its origins, however, are probably from some *Alphadon*-like Cretaceous taxon.

*Peradectes* is allegedly present in South America (Crochet, 1980; Marshall & Muizon, 1992), in beds originally considered Cretaceous (but see Muizon, 1991; and Marshall & Muizon, 1992, for a reinterpretation and references therein). The Tertiary subfamily, however, is known from North America (Paleocene–Eocene–

early Oligocene *Peradectes*, early Eocene *Armintodelphys*, Eocene–Oligocene to early Miocene *Herpetotherium*, Early Oligocene *Alloeodectes*, Eocene–Oligocene to Miocene *Nanodelphys*, and ?Peradectinae: Early Eocene *Mimoperadectes*); Europe (Eocene *Peradectes*, Eocene–Oligocene *Peratherium*, and Eocene–Oligocene to Miocene *Amphiperatherium*; ?metatherian, Crochet, 1980); Africa (Tunisian early Eocene ?metatherian *Kasserinotherium*, Algerian early Eocene ?metatherian *Garatherium*, and Egyptian Oligocene *Quatranitherium*); and Asia (Early Oligocene Central Asian peradectine *Asiadidelphis* and the southeast Asian *Siamoperadectes*; see Gabunia, Shevyreva & Gabunia, 1990; Ducroq et al., 1992). The dental remains of the group are discussed in detail by Bown and Rose (1979), Crochet (1980, 1984–1986), Cifelli (1990a, b), Green and Martin (1976), Krishtalka and Stucky (1983a,b, 1984), Lillegraven (1976), L. S. Russell (1984), Setoguchi (1975, 1978), and Simons and Bown (1984), as well as in the additional references just cited. The studies of Crochet, Fox, and those of Krishtalka and Stucky stand out as the most comprehensive analyses of the dental morphology of the relatively abundant Tertiary metatherians.

While little can be said here of the dental attributes and adaptations of the sundry genera grouped under Pediomyidae, the glasbiines do represent animals that are more adapted for fruit, seed, and other plant feeding than any of the more omnivorous alphadontines, pediomyines, or peradectines. The stagodontids, however, are not only the physically largest of the known and undoubted metatherians of the American Cretaceous, with robust teeth, hypertrophid paraconids and metacristae and reduced metaconids, but, as I will demonstrate, their tarsal remains strongly suggest aquatic adaptations.

Didelphidae are first known on the North American continent with the appearance of *Didelphis* in the Plio–Pleistocene, and in South America in the Paleocene.

The known Mesozoic tarsal evidence (see Table 6.1) for the Ameridelphia (Figs. 6.1–6.8) all comes from North America. It is meager, but because of its antiquity and character diversity, it is especially precious and informative for metatherian phylogenetics. In spite of some of the remarkable polemics from some paleontologists (C. Patterson, 1981; Forey, 1982) about the irrelevance of fossils for "phylogeny reconstruction" (meaning clearly only the results of cladistic analysis), the fossil record is proving to be fundamental for phylogenetic analysis for any group that has a record (see a near concensus on this in the literature in Szalay, Novacek, & McKenna, 1993a, b). The descriptions that follow of the fossils that are not directly associated with dental data, and therefore are usually deprived of proper taxonomic allocation, are presented under the heading of the various geologic formations where they were found. The fossils are, of course, recognizably

Table 6.1. *Measurements of North American Cretaceous and Paleogene metatherian calcanea and astragali.*

| Formation | Museum no. | Calc. | ast. | LAJ / AAW | Length (mm) | Width (mm) |
|---|---|---|---|---|---|---|
| Oldman | UAVP 22793 | X | | (1.90) | . . . | 2.50 |
| | UAVP 8673 | | X | 1.90 | 3.00 | 2.40 |
| ═══════════════════════════════════════════════════ |
| | UAVP 22790 | X | | (3.00) | . . . | . . . |
| ═══════════════════════════════════════════════════ |
| | UAVP 22792 | X | | (5.50) | ?13 | . . . |
| | UAVP 22789 | X | | (5.60) | . . . | 8.70 |
| | UAVP 8674 | X | | (4.60) | . . . | . . . |
| | UAVP 22791 | | X | 5.50 | 8.50 | 8.50 |
| Frenchman | TMP 87.101.5 | X | | (5.10) | 10.60 | . . . |
| Lance | UCMP 111154 | X | | (4.40) | . . . | . . . |
| | UCMP 111156 | X | | (4.35) | 10.20 | . . . |
| | UCMP 111157 | X | | (4.00) | 8.60 | . . . |
| "Tiffany" | AMNH 29111 | X | | 1.50 | 3.60 | 2.10 |
| Wasatch, Four Mile (Sand Q.) | AMNH 89531 | | X | 4.00 | 6.20 | 4.90 |
| Wasatch, Bitter Creek Area | | | | | | |
| Loc. V-70214 | UCMP 111139 | X | | 1.80 | ?3.70 | 2.70 |
| | UCMP 113296 | X | | 1.70 | 3.40 | 2.35 |
| | UCMP 113304 | X | | 1.80 | . . . | . . . |
| Loc. V-71231 | UCMP 113307 | X | | 1.65 | . . . | 2.25 |
| Loc V-70246 | UCMP 113297 | X | | 1.85 | . . . | . . . |
| | UCMP 113303 | | X | . . . | 1.90 | . . . |
| | UCMP 113305 | | X | 1.40 | 1.95 | 1.80 |
| Green River, Loc. Powder Wash | AMNH 29124 | X | | 2.35 | . . . | . . . |
| Bridger | | | | | | |
| Loc. Sage Creek Main | AMNH 29137 | X | | 2.10 | 4.55 | . . . |
| | AMNH 29167 | X | | 2.55 | . . . | . . . |
| | AMNH 125774 | X | | 2.30 | . . . | . . . |
| | AMNH 29150 | | X | 2.10 | 4.45 | . . . |
| | AMNH 29148 | X | | 2.40 | 5.30 | . . . |
| | AMNH 29146 | | X | 2.00 | 3.35 | 2.70 |
| | AMNH 29149 | X | | 3.40 | . . . | . . . |
| | AMNH 125771 | | X | 2.75 | 4.40 | 3.30 |
| | AMNH 125770 | X | | 3.20 | 5.40 | 4.40 |
| | AMNH 29145 | | X | 2.95 | 4.40 | 3.55 |
| Loc. East Hill | AMNH 29141 | X | | 2.10 | . . . | . . . |
| | AMNH 29135 | | X | 2.25 | 4.80 | 3.15 |
| Loc. LSV | AMNH 125775 | X | | 2.35 | . . . | 3.20 |
| | AMNH 29110 | X | | 2.35 | 5.40 | 3.50 |
| | AMNH 125776 | X | | 3.00 | . . . | . . . |
| Loc. LTWL | AMNH 29134 | X | | 2.10 | 4.20 | . . . |
| | AMNH 125773 | X | | 2.75 | 4.50 | 3.70 |
| Tepee Trail | | | | | | |
| Loc. 5A | UCM 55904 | | X | 1.95 | . . . | . . . |
| | UCM 55905 | X | | 1.80 | 3.00 | 2.30 |
| ═══════════════════════════════════════════════════ |
| | UCM 55906 | X | | 2.90 | . . . | . . . |
| Loc. 20* | UCM 47200 | X | | 2.00 | . . . | . . . |
| | UCM 47201 | X | | 2.35 | . . . | . . . |
| | UCM 47202 | | X | 1.95 | 3.10 | 2.25 |

mammalian and therian, and as certainly metatherian in their allocation as the early marsupial dental remains can be. Because so much of the fossil mammalian taxonomic framework is based on dental evidence, I will attempt the tentative allocations of the tarsals to the known dental record from these beds.

### Oldman Formation, Alberta, Canada

There are three size groups (see Table 6.1) and two morphological categories within the group of seven tarsals, the five calcanea and two astragali. Fox (1987) recently reviewed all of the dental evidence originally described by him (Fox 1979a–c, 1981, 1987, 1988a,b; Fox & Naylor, 1986), an effort that allows some tentative correlation between the dental and tarsal data. There is no serious doubt that the larger specimens represent the stagodontid *Eodelphis*; the smaller specimens, however, may belong either to *"Alphadon"* or *"Pediomys"*.

### Pediomyidae (Figs. 6.1, 6.2 & 6.4)

UAVP Nos. 22793 and 8673, a calcaneus and an astragalus, respectively, have approximately matching widths of the articular areas in the **LAJ,** and therefore may represent samples from the same species. It must be kept in mind, however, that the measurement for UAVP No. 22793 is for the AAW dimension (see Table 6.1), or the combined "articular area width" which also includes the **CaFi** facet. Nevertheless these specimens could be conspecific.

CALCANEUS A.    The peroneal process is very large (Figs. 6.1 & 6.2), but considerably reduced compared to *Oligokyphus* or morganucodontids. The calcaneal **CaA** facet is nearly directly above (flexad) the **Su** facet, given the horizontal orientation of the calcaneus. This arrangement suggests a more medial and less superimposed position for the astragalus than that seen in any of the living didelphids. The **Sus** facet is pronounced and can be functionally correlated with the ribbonlike extension of the **Sus** on the astragalus AUVP No. 8673 (Fig. 6.4B,G). The **CaCu** facet is relatively very large, somewhat semicircular, transversely arcuate,

Notes to Table 6.1.

---

*Notes:* **LAJ** width is the combined width of the **CaA** and **Su** facets on either the calcaneus or astragalus. Articular area width (**AAW**) is the **LAJ** width and the **CaFi** facet (when present) on the calcanea; these measurements are shown in parentheses, and they indicate a tricontact **UAJ.** Width measurements of the calcanea themselves represent an oblique width between the most lateral point of the peroneal process and the most proximal and medial extent of the sustentaculum. Double broken lines represent possible separation of paleontological species samples in same formation. Museum abbreviations are as follows: AMNH, American Museum of Natural History; TMP, Tyrell Museum of Paleontology; UAVP, University of Alberta Vertebrate Paleontology; UCM, University of Colorado Museum; UCMP, University of California Museum of Paleontology.
*Three additional calcanea, UCM 55907–55909, are too damaged for measurement.

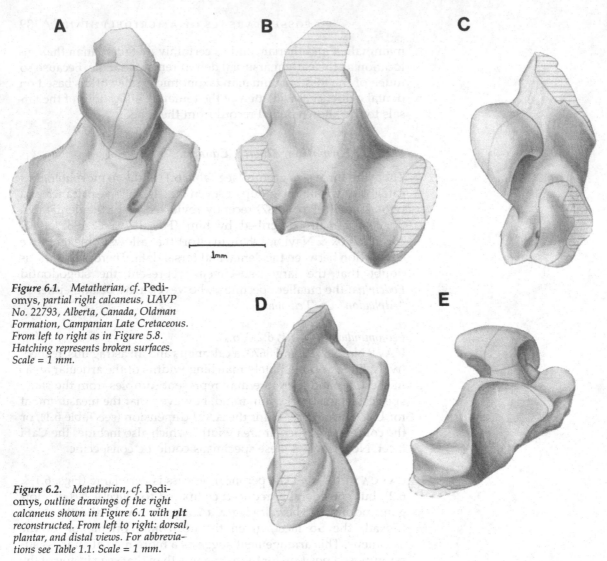

**Figure 6.1.** Metatherian, cf. Pediomys, partial right calcaneus, UAVP No. 22793, Alberta, Canada, Oldman Formation, Campanian Late Cretaceous. From left to right as in Figure 5.8. Hatching represents broken surfaces. Scale = 1 mm.

**Figure 6.2.** Metatherian, cf. Pediomys, outline drawings of the right calcaneus shown in Figure 6.1 with **plt** reconstructed. From left to right: dorsal, plantar, and distal views. For abbreviations see Table 1.1. Scale = 1 mm.

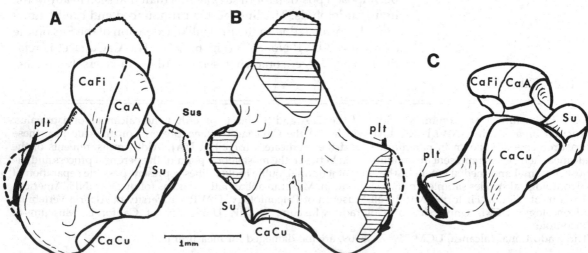

but without any sign of a **CaCup** facet, so completely diagnostic of all living Didelphidae irrespective of their specific adaptive modifications. In fact this facet is best compared to a segment of a cylinder, as it appears that cuboid–calcaneus rotation occurred through an axis that was dorsoplantar through the cuboid. This structure seems to suggest a greater ability of the forefoot to adduct and abduct (and pronate and supinate) than to flex on the calcaneus. The **CaFi** facet is very large; this structure is concordant with the small **AFi** facet of the astragalus AUVP No. 8673. This configuration unquestionably indicates, even without our actual knowledge of the crus in these forms, that the fibula was load-bearing directly through the calcaneus, an arrangement completely eliminated by the last common ancestor of living didelphids. The anterior tubercle (**at**) is all but absent and that distal area is slightly hollow. Although half of the tuber is missing, this part of the calcaneus was considerably deep.

CALCANEUS B.    Although AUVP No. 22790 is badly broken, it shows that this specimen was very similar to AUVP No. 22793 in the relationship of its articulations with the fibula and astragalus. The size of the **CaFi** is also similar. Both the peroneal process and the distal end bearing the **CaCu** facet are broken off, and the **Su** facet is damaged proximally. The relative width of the calcaneal sulcus may have been greater on this specimen than on the smaller calcaneus.

ASTRAGALUS.    AUVP No. 8673 (Figs. 6.3B, & 6.4A–H), the only astragalus known in this size range from the Oldman Formation, may be, as noted above, conspecific with the calcaneus AUVP No. 22793. The mediolaterally narrow **AFi** facet forms an angle of 100–110 degrees, relatively small compared to those of the Itoraí metatherians or the Didelphidae. The **ATil** facet is longer than wide with a medial distal extension onto the neck. The **ATim** facet extends to the medial and plantar (or volar or flexad) border of the astragalus. The astragalar sulcus is about as narrow as that on the calcaneus AUVP No. 22793, quite different from the stagodontid condition I will describe. The **ampt** occupies the proximal third of the plantar side of the bone, extending against the proximal border of the **CaA** facet. Signs of an astragalar canal are discernable on the distally facing surface of the astragalus. On a badly worn, but very similar astragalus from Bug Creek Anthills, USNM specimen (Fig. 6.4I–K), a small but unmistakable sign of an astragalar canal (**ac**) is visible on the proximal border of the plantar side. Eutherians appear to have a dorsal opening for the canal, whereas when the **ac** is present in metatherians, both of the openings appear on the distal or plantar side of the astraglus. The narrow **Su** facet is continuous with a facet which can be identified only as an **AN** facet, although the possibility exists that the **ACu**

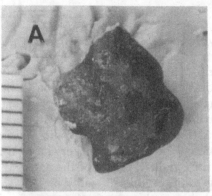
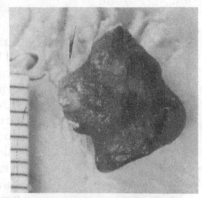
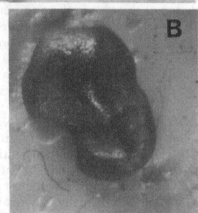
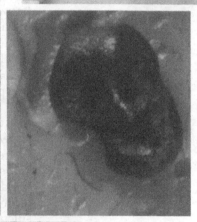
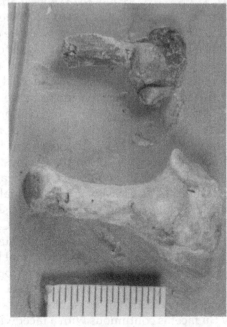
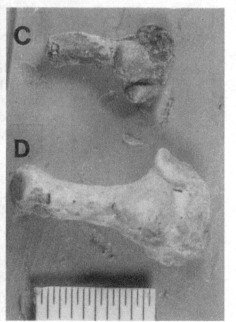

*Figure 6.3.* Stereophotos. **A.**
*Eodelphis sp., partial right calcaneus,
UAVP No. 8674, Alberta, Canada, Old-
man Formation, Campanian Late
Cretaceous.* **B.** *right astragalus, proba-
bly metatherian, UAVP No. 8673,
Alberta, Canada, Oldman Formation,
Campanian Late Cretaceous.* **C.**
*Didelphid (ICS 9), left calcaneus, DGM
No. 1.131-M, Itaboraí Fissures, Brazil,
Paleocene.* **D.** *Didelphid (ICS 23), right
calcaneus, DGM No. 1.180-M, Itaboraí
Fissures, Brazil, Paleocene. Subdivisions
on scales = 0.5 mm.*

facet was also present wedged between these facets. Slight round-
ing of the worn specimen in this area could easily disguise this
characteristic. The **AN** facet, deep in a flexad–extensad section,
appears to have facilitated extensive flexad and extensad motions.
To what degree this is indicative of substrate preference is difficult
to judge, but a comparison with the equivalent function of relevant living didelphids reveals important differences. The distal
outline of the **AN** facet does not show the curvature of this facet
characteristic of arboreal didelphids to the same extent. As in
highly terrestrial didelphids such as *Metachirus*, for example, the
curvature of the **AN** facet is oriented in an essentially flexad–
extensad direction in order to facilitate flexion of the forefoot on
the astragalus at the **TTJ**. The possibility exists that, in spite of all
the seemingly metatherian connections, these astragali may have
belonged to a eutherian.

### Stagodontidae (Figs. 6.3, 6.5, 6.6)

Three calcanea and one perfectly preserved astragalus from the
Oldman Formation, and one calcaneus from the Campanian of
Dinosaur Park give an excellent base for the morphology of these
two bones in this enigmatic group of relatively large Mesozoic
marsupials. The size association of these specimens with the dental dimensions from the same localities leaves little doubt that
these bones are indeed stagodontid: *Eodelphis* from the Oldman
Formation, and probably *Didelphodon* from the Dinosaur Park.

CALCANEUS. The three Oldman calcanea (see particularly
Figs. 6.5A–E & 6.6F–H) show the identical combination of features; these suggest a somewhat autapomorphic mix of characters
with what appears to be widely shared primitive mammalian and
apomorphic metatherian (or possibly therian) traits. As in the
pediomyids, the calcaneal **CaFi** facet is very prominent, correlated
with the small (but probably diagnostically modified) astragalar
**AFi** facet. This pattern is typically part of a tricontact **UAJ**. Although large, as in the ancestral therian, the peroneal process is at
the same time also highly diagnostic in that it terminates at the
level of the distal end of the **CaA** facet and has a prominent groove
for the peroneus longus tendon. Whether the **Su** facet, relatively
very small compared to pediomyids, is another diagnostic stagodontid attribute or a therian or pretherian retention is difficult to
determine, but I offer an interpretation under the discussion of the
stagodontid astragalus. The relatively very flexad position of the
**CaA** facet over the **Su** facet, however, appears to reflect an ancient
mammalian level of organization in which the astragalus is more
medial than extensad to the calcaneus. The conformation of the
calcaneal **CaCu** facet is so unique that I do not seriously doubt its
apomorphic nature, and that at the same time it is one of the best
indicators of some foot mechanics, at least based on this one ele-

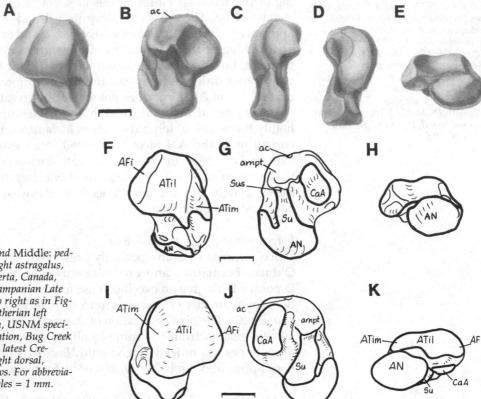

**Figure 6.4.** Above *and* Middle: *pediomyid metatherian, right astragalus, UAVP No. 8673, ALberta, Canada, Oldman Formation, Campanian Late Cretaceous; from left to right as in Figure 5.8.* Below: *?metatherian left astragalus, water worn, USNM specimen, Hell Creek Formation, Bug Creek locality, Maastrichtian latest Cretaceous; from left to right dorsal, plantar, and distal views. For abbreviations see Table 1.1. Scales = 1 mm.*

ment of the tarsus. The calcaneal **CaCu** facet is almost as close to an unmodified ovoid joint as are the diagnostic didelphid and some "edentate," archontan, or euprimate radiohumeral articulations. The slight lateral extension of the distal end of the calcaneus (as in pediomyids and didelphids) that does not have obvious articular contact suggests this condition to be derived in Metatheria, a diagnostic stagodontid apomorphy. The mechanics are reflected in the joint surface instead of the contiguous area, and therefore a direction is strongly indicated for the appropriate vertical comparison. A derivation from a primitive marsupial condition is likely; this shallow ball and socket joint, because of its circular conformation, is unlike other marsupial **CCJ** arrangements. In the latter the slight asymmetry of the articulation is almost invariably a clear indicator of a particular close-packed position. In the stagodontids only, the facilitation of mobility appears to be the overriding concern. This feature, together with the apparently closely bound (alternating) anterior tarsal region, may be the indicator of the rotational ability of the foot of an aquatic animal.

The well-preserved stagodontid calcaneus from Dinosaur Park (TMP No. 87.101.5; Fig. 6.6A–E) lacks only the peroneal process

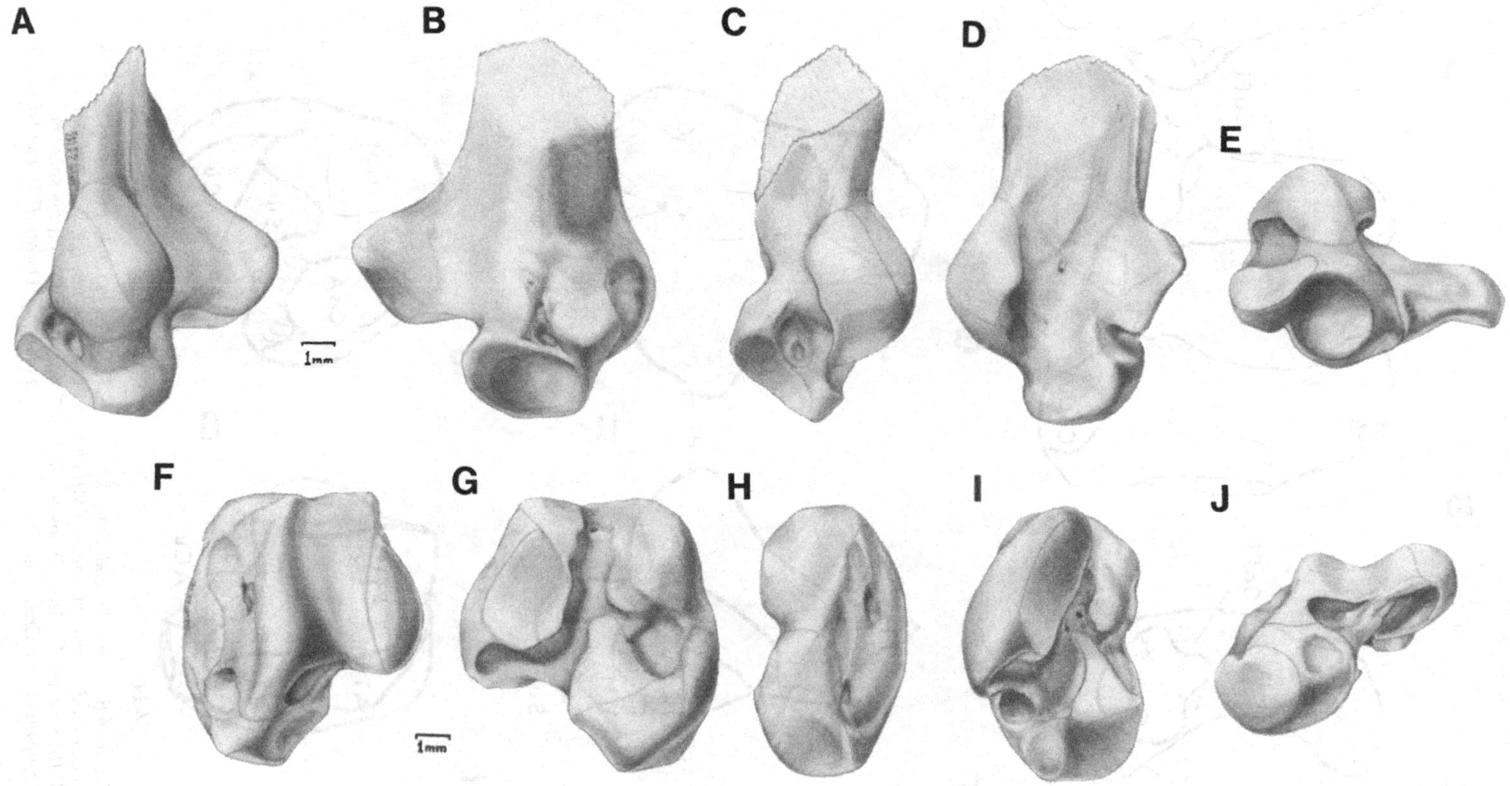

*Figure 6.5.* Eodelphis browni, partial left calcaneus (reconstructed from UAVP Nos. 22789 and 22792) and left astragalus (AUVP No. 22791), Alberta, Canada, Oldman Formation, Campanian Late Cretaceous. From left to right as in Figure 5.8. Scales = 1 mm.

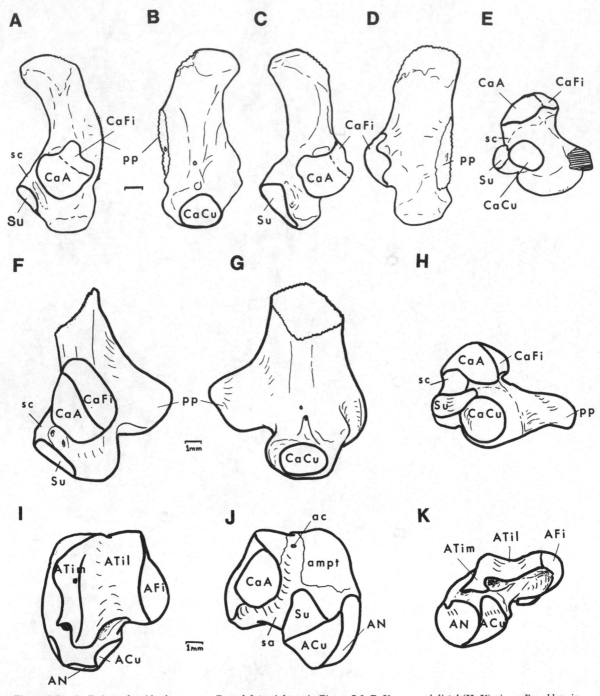

**Figure 6.6.** *A–E. Stagodontid, cf.* Didelphodus, *left calcaneus, TMP No. 871015, Steveville Area, Dinosaur Park, Campanian Late Cretaceous.*

*From left to right as in Figure 5.8. F–K. Outline drawings of calcaneus and astragalus of* Eodelphis *sp. shown in Figure 6.5. Dorsal (F, I), plantar (G, J),*

*and distal (H, K) views. For abbreviations see Table 1.1. Scales = 1 mm.*

**Figure 6.7.** *Metatherian, partial left calcaneus with peroneal process broken off, UCMP No. 111154; Wyoming, Lance Formation, Loc. V5620, Maastrichtian latest Cretaceous. A–C are labeled outline drawings of D–F. From left to right : dorsal, plantar, and distal views. For abbreviations see Table 1.1. Scales = 1 mm.*

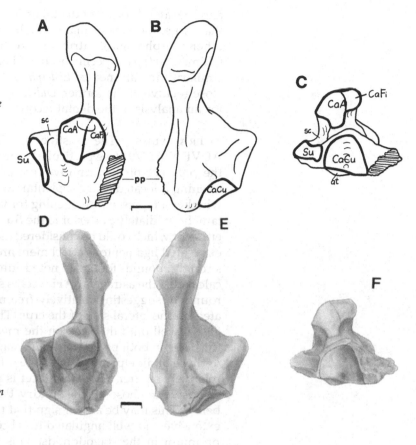

**Figure 6.8.** *Metatherian, left calcaneus with peroneal process broken off, Mason Pocket, Colorado, Tiffany beds of the Animas Formation, Tiffanian Late Paleocene; AMNH No. 29111. From left to right: dorsal (A), plantar (B), and distal (C) views. For abbreviations see Table 1.1. Scale = 1 mm.*

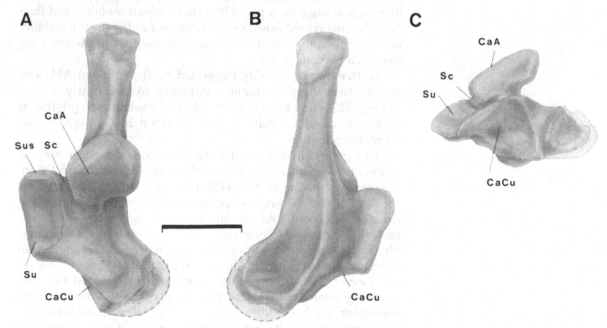

and the lateral corner of the tuber. The latter was not deep, unlike that of pediomyids or many didelphids, but rather rounded. All other morphological attributes are like those described for the Oldman *Eodelphis* specimens. It is likely that in spite of its extreme similarity to calcanea of *Eodelphis*, this specimen belonged to a close relative of the former, *Didelphodon* (see Fox & Naylor, 1986, for an analysis of the dental record).

ASTRAGALUS. The pristine specimen of a left astragalus, AUVP No. 22791 (see particularly Figs. 6.5F–J and 6.6I–K), shows the presence of the primitive mammalian astragalar canal. One opening is located near the plantar border of the distal surface, but whether it represents an opening for vessels is impossible to ascertain. Immediately posterior to the **Su** facet there are two foramina, either of which could be considered as the exit of the canal. Deeply excavated ligamentous attachment areas suggest an exceptionally strongly bound foot. As noted under the discussion of the calcaneus, the astragalar **AFi** facet is small but also diagnostically rounded, suggesting relatively free movement of the foot mediated on the lateral side of the crus. The **ATil** facet is greatly elongated, well onto the neck on the medial side. The elongation of this facet in both pediomyids and stagodontids stands in contrast to those of didelphids which appear to be (with some exceptions) relatively shorter. The **ATim** facet is relatively small, nowhere as extensive as those on the sundry Paleogene astragali discussed below. This may be a clear sign that the stabilizing function of an extensive and well-angulated tibial contact medially was not at a premium in the stagodontids. This functionally significant feature, again, suggests a pedal biorole in which mobility and flexibility were favored selectively, in contrast to the medial stability necessary in a terrestrial or arboreal animal that habitually traveled on a hard substrate.

The freedom of mobility suggested by the rounded **AFi** facet appears to be the reason for an astragalar **CaA** facet nearly as wide as long. This latter facet actually takes on a sellar configuration at its lateral edge, a condition just described in *Oligokyphus* and morganucodontids.

The very small sustentacular facet on both the calcaneus and astragalus is probably the result of the extensive **ACu** articulation. A large part of the area usually occupied by the sustentacular contact with the calcaneus appears to have been taken over by the contact with the cuboid. Stating this I am suggesting that the extensive **ACu** articulation of the stagodontids is a unique apomorphy of this group and not a retention from the metatherian ancestor. The **ACu** articulation of vombatiform diprotodontians, discussed below, is never this extensive. As suggested from the relatively small calcaneal **CaCu** facet, the ribbonlike medioplantar astragalar **AN** facet indicates a somewhat shared load resistance and transmission between the navicular and cuboid, suggesting that the direction of loading on the most mobile unit of the tarsus, the astragalus, was nearly equal distally both from the navicular and cuboid. The reduced **AN** facet and the large, wedge-shaped, and highly convex astragalar **ACu** facet, then, suggest extensive mobility between the cuboid and the astragalus. Perhaps this also reflects habitual use of the foot in water.

The astragalar sulcus clearly separates the **CaA** facet from the anteromedial plantar tubercle (**ampt**). This configuration may be also a diagnostic stagodontid feature (in combination with others), although what aspect of the mechanical alterations it mirrors is not clear. It is interesting to note that among the living Didelphidae it is *Chironectes* that also has an astragalar sulcus separating the **ampt** from the **CaA** facet. The biological (i.e., form–functional) significance of this convergence is entirely unclear to me.

### Lance Formation, Wyoming

The dental and cranial remains of the marsupials from these Maastrichtian Late Cretaceous beds have been chronicled in detail by Clemens (1966) and commented on by many in publications dealing with marsupial evolution (e.g., Lillegraven, 1969). The dental remains of *Alphadon, Pediomys, Glasbius,* and *Didelphodon* make up the taxonomic diversity among the most abundant elements, the teeth. The calcanea which I have studied (see Table 6.1) are not those of a stagodontid, and they are too big to belong to *Glasbius.* The closest size match between the length of either the first or second molars and the articular area width (the **CaFi** facet included) of the calcanea is in *Pediomys florencae.*

#### Pediomyidae (Fig. 6.7)
Like the pediomyid calcanea of the Oldman Formation, those from the Lance Formation have a large **CaFi** facet. So in the Oldman specimens, an angle between the horizontal mediolateral plane of the **CaCu** facet and the plane of the **LAJ** articulation is large, nearly 90 degrees, suggesting very slight asragalocalcaneal superposition. Although the peroneal process is broken off from all the specimens I have examined, it appears to have been relatively as large as the one on UAVP No. 22793 of the Oldman Formation. The relative dimensions, shape, and articular curvature of the **CaCu** facet are also similar to the Oldman specimen. What is quite remarkable is that the Lance specimens, so much larger in absolute size, are so similar in morphology and articular conformation to the Oldman pediomyids.

### Tiffany beds of the Animas Formation, Colorado
#### Peradectinae (Fig. 6.8)
A single, nearly perfectly preserved left calcaneus, AMNH No. 29111 (see Table 6.2) is the only North American Paleocene tarsal

element known to me, although others probably exist in collections. It is probably of *Peradectes elegans*. It clearly differs from the pediomyids and stagodontids described in that it lacks a **CaFi** facet. Like the pediomyids, it has a quasi-semicircular **CaCu** facet and a large peroneal process, as far as the latter can be judged from the area of breakage. The small angle between a mediolateral plane spanning the extreme widths of the **CaCu** facet and the articular surface of the **LAJ** facets is about 50 degrees, in marked contrast to the Oldman marsupial tarsals. There is no indication of a **CaCup** facet on this or on other calcanea in the North American Paleogene.

### Wasatch Formation, Colorado

*Metatheria, incertae sedis (Fig. 6.9A,B)*
A relatively large (see Table 6.1) single specimen, a left astragalus from the Four-Mile Sand Quarry, AMNH No. 89531, although heavily worn, is very distinct from other known North American marsupial astragali. The relatively long appearance of this astragalus is largely due to the sharply angled **ATim** facet on the medial side of the bone. The **AFi** facet is very narrow transversely, and the **ATil** facet extends far onto the neck of the bone. While the dorsal aspect is quite similar to the pattern most commonly encountered in Bridger and Tepee Trail peradectines, the plantar articulation as reflected by the **LAJ** facets is distinctive from both the archimetatherian peradectines, and the sudameridelphian itaboraiforms and sparassodontans. In the latter group the **Su** facet tapers into a **Sus** extension facet which continues into the cavity formed by the **ampt**, but in the Sand Quarry astragalus the laterally placed **Su** facet, without any sign of a **Sus** facet extension, is separated from the **CaA** facet by a very narrow astraglar sulcus. The orientation of the **AN** facet strongly suggests emphasis on dorsiflexion and extension. The possible significance of these observations is that they may point to a group of marsupials, perhaps dentally represented by *Mimoperadectes* (see Bown & Rose, 1979) which cannot be accommodated within the Peradectinae of the North American Eocene. This view must be tempered by the fact that it is supported by only a single and heavily worn astragalus.

### Wasatch Formation, Bitter Creek Area, Wyoming

*Peradectinae (Figs. 6.9C,D, 6.10, 6.21A)*
Under Dr. Donald E. Savage the field effort of the University of California, Berkeley, has produced from this area a modest collection of marsupial tarsals of what appears to be a single species based on size and certainly on morphological criteria (see Table 6.1). From the combination of characteristics there is little doubt

*Figure 6.9.* Stereophotos. *A, B.* Metatherian, worn left astragalus, AMNH No. 89531, Sand Quarry, Wasatch Formation, Four Mile fauna, Wasatchian Early Eocene. Dorsal (A) and plantar (B) views. *C.* Metatherian, left calcaneus, loc. V. 70214. *D.* Metatherian, left astragalus, UCMP No. 113305, loc. V. 70246. Both *C* and *D* are from the Wasatch Formation, Bitter Creek fauna, Wyoming, Wasatchian Early Eocene. Subdivisions on scales = 0.5 mm.

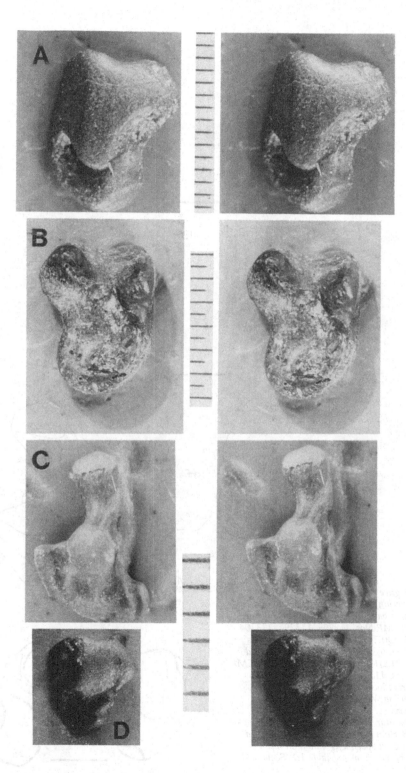

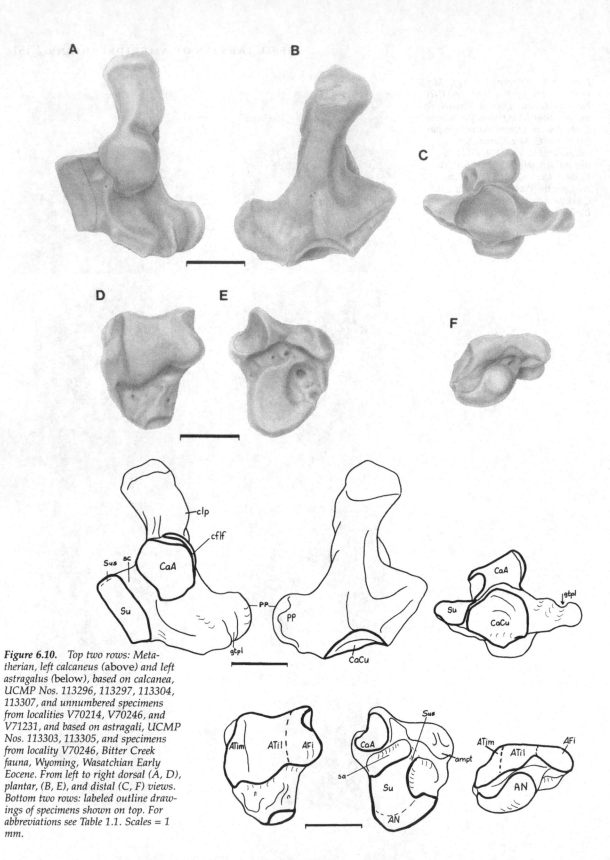

*Figure 6.10.* Top two rows: Meta-therian, left calcaneus (above) and left astragalus (below), based on calcanea, UCMP Nos. 113296, 113297, 113304, 113307, and unnumbered specimens from localities V70214, V70246, and V71231, and based on astragali, UCMP Nos. 113303, 113305, and specimens from locality V70246, Bitter Creek fauna, Wyoming, Wasatchian Early Eocene. From left to right dorsal (A, D), plantar, (B, E), and distal (C, F) views. Bottom two rows: labeled outline drawings of specimens shown on top. For abbreviations see Table 1.1. Scales = 1 mm.

that the most recent affinity of the animals was with the Mason Pocket and Bridger marsupials. This corroborates the detailed review by Krishtalka and Stucky (1983a, b) of the far more abundant dental evidence from the Bridger beds. There can be little doubt that the known Early Eocene peradectines were arboreally adapted. This is strongly supported by the skeleton of the Messel Eocene *Peradectes* from Germany (Storch, 1990) which has a long, muscular, and almost certainly prehensile tail.

CALCANEUS. The medially oriented tuber bears a well developed process, the **clp,** the significance of which is not clear. Nevertheless this feature, perhaps a guide to tendinous orientation on the lateral side of the back of the foot, is a rather ubiquitous trait in the Itaboraí assemblage also. As in other peradectines and didelphids, there appears to be no **CaFi** articulation. The narrow strip of articular contact lateral to the **CaA** facet is almost certainly the pressure facet of the calcaneofibular ligament (**cflf**) and not of the fibula. The peroneal process is large and offset laterally rather than plantarly. The **CaA** facet is mediolaterally and proximodistally extensive, and the calcaneal **Su** facet has the expected **Sus** extension facet, concordant with the conarticulating area of the astragalus. The **CaCu** facet permitted extensive rotation of the cuboid (and forefoot) on the calcaneus. The quasi-semicircular morphology with a slight central depression of this facet stands in telling contrast to the more cylindrical orientation of this facet in the small pediomyid described above from the Oldman Formation. The superposition angle, as measured by the angle enclosed by the plane of the **LAJ** articulation and the horizontal plane between the extremes of the **CaCu** facet, is approximately 45–50 degrees.

As far as we can tell from the cuboid facet, the considerable ability of peradectines to rotate the forefoot on an axis that traversed the calcaneus in distoextensad and proximoflexad direction, resulting in pronation and supination, is similar to the pattern shown by didelphids and some carnivorans, primitive archontans, and early primates. Flexad a slightly retroflexed articular rim suggests a stabilized articulation with the cuboid in the close-packed position. This area of the **CaCu** facet cannot be homologized with those of the living didelphids (and of the two categories of didelphid calcanea of Itaboraí; see ensuing discussion). Didelphids share important morphological and functional similarities in this area that are not encountered in these peradectines.

ASTRAGALUS. The **UAJ** articular area is smoothly continuous, suggesting great mobility in the cruropedal joint but relatively poor mediolateral stability. This in turn strongly implies an arboreal substrate preference, terrestrial scamperers might display a greater **UAJ** stability in their morphology, as do caenolestids for example. Specifically, the **ATim** facet is not sharply angled with the **ATil** facet, as it is, for example, in the later Eocene peradectines or in many of the Sudameridelphia. The **AFi** facet, horizontally continuous with the tibial articulation, is not as wide as one usually encounters in didelphids, but it is not as transversely narrow as in itaboraiforms, caenolestids, borhyaenids, or late Eocene peradectines.

In all of the North American Paleogene astragali known to me, the persistently pediomyine-like sustentacular extension, the **Sus** facet, and its intertaxonal variation in Itaboraí taxa, as well as its derived absence in didelphids, all suggest the phyletic distinction of peradectines known by tarsal evidence from the origin of the Sudameridelphia. The **Su** facet is broadly continuous with the **AN** facet, and if there was any cuboid contact, I find it impossible to discern. This broad distal **LAJ** continuity with the **ANJ** is characteristic of pediomyines, peradectines, and some itaboraiforms and borhyaenids, but not of didelphids, in which this continuity of facets, if retained, is relatively narrow. The **ampt** in the Bitter Creek astragali is shallow, as in didelphids; it does not curl under (i.e., flexad) the **Sus** facet. The **AN** facet suggests a relatively transverse orientation and habitual pronation and supination of the forefoot rather than the increasing flexion–extension orientation of the geologically younger peradectines, caenolestids, and borhyaenids, and the more ancient itaboraiforms.

### Bridger Formation, Wyoming

*Peradectinae (Figs. 6.11A–D, 6.12, & 6.14)*
Of all the Bridger samples I have examined, in collections made by Dr. Robert M. West, there is a recurrent pattern of a smaller and a larger tarsal phenon, *Bridger Metatherian Group I and II* (see Fig. 6.11). To what formally named taxa these tarsals actually belong is at present not possible to assess. Yet it appears certain to me that these phena belong to a very closely related species group. If intermediate size parameters existed, I would hesitate to assign these specimens to different species. As it is, they may still represent larger- and smaller-sized clusters of sympatric species. I find no meaningful differences worth describing to distinguish between the two samples, so they are compared as essentially one phenon for the calcaneus and for the astragalus. The splitting into two families of the dental morphology of these species represented by tarsals based on minor dental differences alone is entirely unacceptable to me, if the zoological concept of a family in therian mammals is to connote either a genuinely tested phyletic distinction or an adaptively significant morphological distance.

CALCANEUS. Like the Bitter Creek calcanea, the Bridger specimens lack a **CaFi** facet but have a well defined **cflf**. The **CaCu**

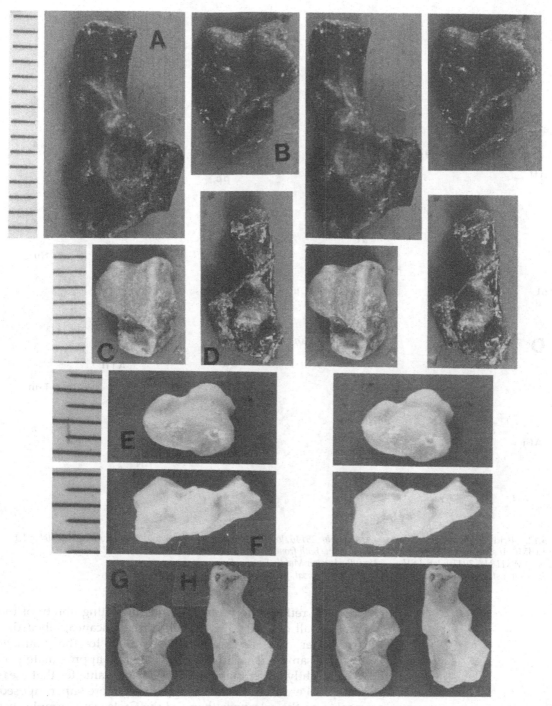

**Figure 6.11.** *Stereophotos. **A-B** and **C-D** represent two species of metatherians, a smaller, Bridger Metatherian Group I (BMG I), and a larger, Bridger Metatherian Group II (BMG II), from the Sage Creek Main quarry of the Bridger Formation, Wyoming, Bridgerian Middle Eocene. **A.** Right calcaneus of BMG II, AMNH No. 29149. **B.** Left astragalus of BMG II, AMNH No. 29145. **C.** Right astragalus of BMG I, AMNH No. 29146. **D.** Left calcaneus of BMG I, AMNH No. 29148. **E–H** represent one species from the Tepee Trail location 5a, Tepee Trail Formation, Wyoming, Uintan Late Eocene. **E, G.** UCM No. 55904, right astragalus, dorsal (E) and plantar (G) views. **F, H.** UCM No. 55905, broken left calcaneus, dorsal (F) and plantar (H) views. Subdivisons on scales = 0.5 mm.*

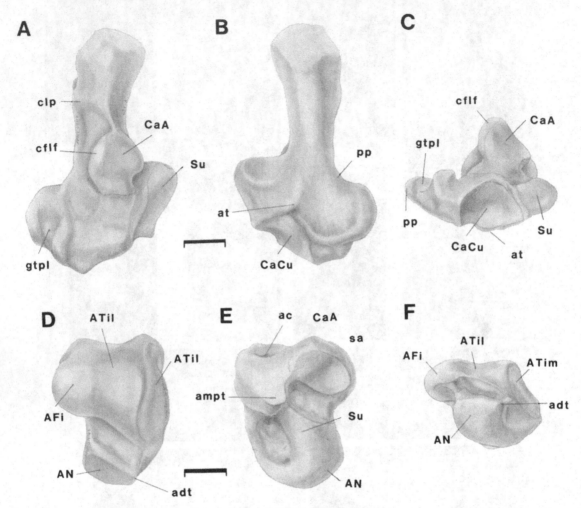

**Figure 6.12.** Bridger Metatherian Group I (BMG I). Above: *right calcaneus, AMNH No. 29110, loc. LSV-A.* Below: *right astragalus (AMNH No. 29146, locality is Sage Creek Main. Both from Bridger Formation, Bridgerian Middle Eocene. From left to right: dorsal, plantar, and distal views. For abbreviations see Table 1.1. Scales = 1 mm.*

facet clearly reflects the pronating and supinating ability of the forefoot. In all of their other details, these calcanea, albeit distinctly larger, are also essentially similar to the younger Wasatchian sample. The superposition angle is approximately 55 degrees, slightly larger than the Early Eocene value for that meager sample. Peradectines had an apparently more superimposed astragalus on the calcaneus than did the Cretaceous samples yet described.

ASTRAGALUS. The differences of this sample from the Bitter Creek sample are greater than those differences from the calcaneal samples from the same localities. This variation is largely due to

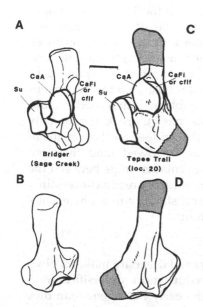

*Figure 6.13. Comparison of calcanea in dorsal (above) and plantar (below) views from the Bridger Middle Eocene (A, B) and the Tepee Trail Late Eocene (C, D). Shaded area is reconstructed. For abbreviations see Table 1.1. Scale = 1 mm.*

changes in the **UAJ** and also in the **ANJ**. Slight consequences of all these changes were reflected in the construction of the **LAJ** as well.

The modifications of the **UAJ**, even without the critical evidence from the crus, appear to be reflected in the greater angulation of the **ATim** facet with its lateral tibial articulation and slightly furrowed conformation. In addition, the medial extension of the **ATil** facet onto the neck may mirror an increased range of translational cruropedal movements.

The distinctly more extensad–flexad motion in the Bridger **ANJ** than in the Bitter Creek sample is indicated by the **AN** facet, which is transversely extensive ("dorsally"). The more plantar components of this facet are essentially tucked flexad (under) the astragalus. This pattern, a ubiquitous selective reponse to flexion–extension at the **TTJ** in marsupials, is probably closely tied to similar axial mechanics at the **UAJ**. This pattern occurred entirely independently in both sparassodontans and in some primarily terrestrial didephids (see below). The incidental but related phenomenon of astragalar distal tuber (**adt**) is a reflection of width increase of the astragalar **AN** facet in the close-packed position when the navicular is most flexed on the astragalus. This position most likely occurred at the point of toeing off during locomotion.

The **LAJ**, while very similar to the older peradectine pattern, is different in one significant way. The **ampt** is hypertrophied, and it curls under the **Sus** extension. This configuration had the effect of locking the **LAJ** more securely against the loads related to substrate contact either during toeing off or landing. It may also reflect loading of the astragalus from the substrate itself, hence the hypertrophied tuberosity.

### *Tepee Trail Formation, Wyoming*

*Peradectinae (Figs. 6.13 & 6.14)*
Although the tarsal sample known to me of the Late Eocene North American marsupials is meager (see Table 6.1), it is of particular importance. It attests to a change in the closely clustered stock of the Peradectinae, without any suggestions from the entire span of the Eocene in North America of the presence of didelphids. The Tepee Trail tarsals are slightly more modified than the Bridger sample – but in exactly the same manner as the latter differ from the Bitter Creek sample. Small as they may have been, these marsupials appear to have improved their scampering–cursorial abilities in a manner exactly parallel to the living caenolestids.

CALCANEUS. Given the meager sample, the differences between the Tepee Trail and the Bridger calcanea are minor. The **CaCu** facet may be slightly more narrowed mediolaterally, and therefore higher flexad–extensad.

ASTRAGALUS. The differences between the Tepee Trail sample and the Bridger one are simply extensions of the changes perceived between the latter and the older Bitter Creek tarsals: these changes represent a temporal and morphological trend. The flex-ad **ampt** is as fully developed as this character is expressed in itaboraiforms and borhyaenids. It is a robust stop for the sustentacular wing of the calcaneus and possibly also an important substrate contact point along with the navicular. A functionally correlated minor modification from the Bridger level of organization consists of the somewhat more furrowed (and therefore pulleylike) **ATil** facet and the slightly smaller angle between the **ATim** and the former facets. Both alterations are almost certainly the result of selection for mediolateral stability in a phyletically increasingly scampering–cursorial foot.

## South America

The early dental, cranial, and postcranial diversity makes it obvious that the metatherians have become more diversified morphologically in South America by the early Paleogene than they were in North America in the latest Cretaceous and Paleogene. Their absence from the late Cretaceous Los Alamitos fauna of Patagonia (Bonaparte, 1987) is of critical importance (see Chapter 9). Although the long known riches of marsupial species from the early Paleogene Itaboraí fissures hinted at discoveries to come in earlier periods, it was Sigé's (1972a, b) reports on the mammals of Laguna Umayo from the the Peruvian Andes that gave the first substantial glimmer of the possible diversity of marsupials in South America, perhaps as early as the latest Cretaceous. The subsequent discoveries of Muizon, Marshall & Sigé (1984), Marshall, Hoffstetter & Pascual (1983), and Marshall and Muizon (1988) showed a bursting and substantial marsupial radiation from Tiupampa in south central Bolivia. This rich locality, with at least ten species of marsupials and seven (or eight ) species of placentals, was originally suggested by the authors to be of late Cretaceous age. The evidence for chronology rested primarily on charophytes (algae). It appears to be early or medial Paleocene in age, slightly older than the fauna from Itaboraí (see also Van Valen, 1988a, b; Muizon, 1991, Pascual & Ortiz Jaureguizar, 1990).

### Itaboraí Formation, São José de Itaboraí, Brazil

The long-known limestone fissures at Itaboraí, now under water, which yielded the largest variety of early South American marsupials are of medial Paleocene Itaboraian (Francisco & Souza Cunha, 1978; Pascual & Ortiz Jaureguizar, 1990) or Riochican (Late Paleocene) age (Marshall et al., 1983; Cifelli, 1983). Coupled with the dental diversity, the relatively abundant foot bones described here give us an unparalleled picture of some aspects of marsupial

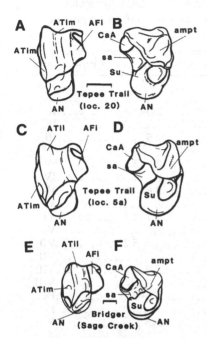

*Figure 6.14. Comparison of astragali in dorsal view (left) and plantar view (right) from the Bridger Middle Eocene (E, F) and the Tepee Trail Late Eocene (A–D). For abbreviations see Table 1.1. Scales = 1 mm.*

postcranial diversity at that time (see Tables 6.2 & 6.3). These also allow direct comparisons of northern and southern groups which have been contentiously compared on dental grounds. They also permit ecological comparisons as these aspects of the taxa relate to locomotor adaptations (e.g., Figs. 6.15 & 6.16). A relatively minor concern is that, in spite of the many first-rate efforts that have gone into the (species level) microtaxonomy of the dental evidence, the actual diversity is still somewhat disputed. I, for example, do not recognize "*Robertbutleria*" recently erected by Marshall (1987), but consider it a junior synonym of *Protodidelphis*, and find the comparison and published distinctions between the genera *Guggenheimia* and the more recently described *Procaroloameghinia* (Marshall, 1982b) either inadequately discussed or unconvincing. Some uncertainty, therefore, as to the exact number of species present at Itaboraí still exists. Marshall (1987) omitted from his Itaboraí summary *Patene colhuapiensis* (Simpson, 1935a), presumably because this taxon is clearly a borhyaenid.

The list of the dental taxa is presented in Table 6.3, arranged in an order of the increasing length of either lower or upper first or second molars, or both when known. This list is the result of my examination of most of the relevant dental specimens, many of them epoxy casts, and I have also taken into account the extensive published literature on the Itaboraí marsupials. This microtaxonomy was described, significantly emended and discussed, and illustrated by Paula Couto (1952a–c, 1959, 1961, 1962, 1970), Simpson (1947), Marshall (1982b), and Marshall and Muizon (1984). The "opossum-like" forms were recently summarized by Marshall (1987). The marsupials described were exclusively jaw and dental elements, with the exception of a skull of *Epidolops* (Paula Couto, 1952b; 1961).

I accommodate the Itaboraí marsupials, as seen in the classification of Metatheria in Chapter 2, in the groups Itaboraiformes, Polydolopimorphia, Sparassodonta, Paucituberculata, and Didelphimorphia, all of these being South American endemics with the exceptions of the dispersal of the didelphids into North America in the Plio-Pleistocene, and the occurrence of the Eocene Antarctic polydolopids. I see no clear evidence for the presence of either pediomyines, peradectines, or microbiotheriids in the tarsal or dental evidence, although microbiotherians may be represented dentally. The dental traits cited by Reig et al. (1987) and Marshall (1987) for various taxa are not convincing synapomorphies shared between Itaboraí forms and the Holarctic groups. I provide here a detailed account of the tarsal evidence, treated under various *Itaboraí Metatherian Groups* (*IMGs*), which are expediently size- and trait-determined tarsal phena. While the tarsals represent sudameridelphidans and some probable glirimetatherians, only two of the species-level tarsal phena may be considered didelphimorphian didelphids in the strict sense.

Table 6.2. *Mediolateral measurements of* **LAJ** *or* **AAW** *(articular area width which includes the calcaneal* **CaFi** *facet as well, when present) for the Itaboraí phena of marsupial tarsals.*

| ICS/IAS No. | N | Astragalus (mm) | Calcaneus (mm) | IMG |
|---|---|---|---|---|
| 1 | (1) | | 1.75 | I |
| 2 | (2) | 1.3–1.5 | | I |
| 3 | (5) | | 2.0–2.3 | II |
| 4 | (3) | 1.9–2.3 | | II |
| 5 | (6) | | 2.5 | III |
| 6—(IAS 8, below, may be part of IAS 6)—III | | | | |
| 7 | (2) | | 3.1–3.5 | IV |
| 8 | (1) | 2.0 | | IV |
| 9 | (2) | | 3.6–4.0 | V |
| 10 | | | | V |
| 11 | (3) | | 4.1–4.5 | VI |
| 12 | | | | VI |
| 13 | (14) | | 4.0–5.2 | VII |
| 14 | (2) | 4.1–4.5 | | VII |
| 15 | (8) | | 5.5–6.0 | VIII |
| 16 | (4) | 5.2–5.6 | | VIII |
| 17 | (7) | | 6.5–7.2 | IX |
| 18 | (1) | 5.6 | | IX |
| 19 | (1) | | 6.5 | X |
| 20 | | | | X |
| 21 | (4) | | 7.2–7.7 | XI |
| 22 | (1) | 6.2 | | XI |
| 23 | (1) | | 8.4 | XII |
| 24 | | | | XII |
| 25 | (3) | | 8.8–9.9 | XIII |
| 26 | (2) | 7.5–7.6 | | XIII |

*Note:* Because of the inclusion of the width of the **CaFi** facets in the measurements of calcanea, **LAJ** widths for the latter are likely to be greater than those of the conspecific astragali.

Ideally, consecutive odd- and even-numbered samples (of calcanea and astragali, respectively) combined, should represent at least one species. These numbered samples, designated as *Itaboraí Calcaneus Sample* and *Itaboraí Astragalus Sample,* are abbreviated as *ICS* and *IAS,* followed by the sample number. They are first and foremost size divisions, but qualitative considerations help define the boundaries. For this reason there may be more species phena represented by the two kinds of tarsals than half of the total sample number.

The inferred morphospecies are designated as *Itaboraí Metatherian Group,* abbreviated as *IMG* and designated by Roman numerals. The assignments of these sample and group numbers, and the suggested association of the taxonomic names attached in the text to dental samples and groups, are clearly no more than moderately supported hypotheses.

The samples, and therefore the groups also, are numbered from the smallest (1 or I) to the largest (26 or XIII).

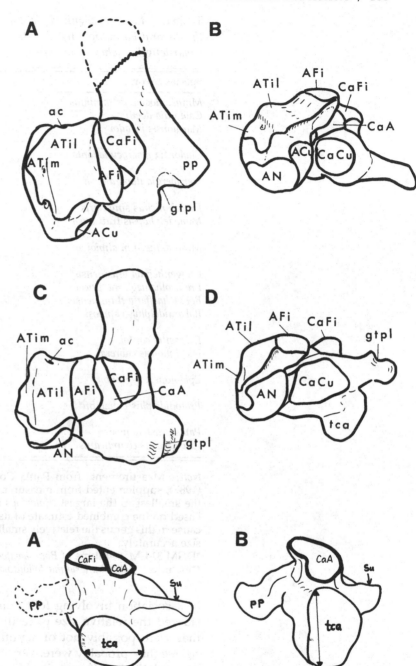

**Figure 6.15.** *Comparison of articulated calcaneus and astragalus of the Campanian Cretaceous Oldman* Eodelphis, *a stagodontid, (above) and the Itaboraian Paleocene IMG XIII (a probable borhyaenid) (below). There are no special similarities between these two groups. Dorsal views to the left and distal views to the right. For abbreviations see Table 1.1.*

**Figure 6.16.** *Distal view of two Itaboraí calcaena to show the relationship of tuber to inferred function and substrate preference.* **A.** *Represents ICS 7 and shows a relatively transversely wide* **LAJ** *(sustentacular facet is not visible), very large* **CaFi** *facet, and a tuber which is transversely wider than taller.* **B.** *Represents ICS 5 and shows a relatively narrow* **LAJ**; *it lacks a* **CaFi** *facet, and its calcaneal tuber is taller than wide. These two complexes probably reflect adaptive optima for primarily terrestrial (?) vs. small branch-grasping arboreal locomotion in Itaboraí marsupials. See text for details. For abbreviations see Table 1.1.*

My probable and often questionable tentative allocations of tarsal remains to described dental taxa from Itaboraí are based primarily on assumed size associations. I have struggled for a long time to find the most convincing correlates between the available tooth parameters and some measures of the tarsals at hand. Unquestionably, some size related factors probably affect any system

Table 6.3. *List of recognized, dentally based and named, species from the Itaboraí Formation, listed in the order of their mean first or second molar lengths (when a sample of more than one is known).*

| Species taxon | Tooth size (mm) |
|---|---|
| *Minusculodelphis minimus* | 0.7 (M/1) |
| *Gaylordia doelli* | 1.2 (M/1) |
| *Marmopsis juradoi* | 1.5 (M1/) |
| | 1.3 (M/1) |
| *Gaylordia macrocynodonta* | 1.8 (M1/) |
| | 1.7 (M/1) |
| *Sternbergia itaboraiensis* | 1.8 (M1/) |
| | 1.7 (M/1) |
| *Derorhynchus singularis* | 1.8 (M2/) |
| *Monodelphopsis travassosi* | 1.9 (M2/) |
| | 1.9 (M/2) |
| *Mirandatherium alipioi* | 2.1 (M1/) |
| | 2.2 (M/1) |
| *Guggenheimia brasiliensis** | 2.4 (M1/) |
| *Procaroloameghinia pricei* | 2.5 (M/1) |
| *Bobbschaefferia fluminensis* | 3.0 (M/1) |
| *Itaboraidelphys camposi* | 3.9 (M2/)) |
| | 3.5 (M/1) |
| *Eobrasilia coutoi* | 4.0 (estimated for M1) |
| *Didelphopsis cabrerai* | 4.4 (M1/) |
| | 4.0 (M/1) |
| *Epidolops ameghinoi* | 3.5 (M1/1; estimated) |
| | 7.2 (P3/3; estimated) |
| *Protodidelphis vanzolini*** | 4.6 (M1/) |
| | 4.7 (M/1) |
| *Patene castellanosi* | 5.0 (M/1) |
| *Zeusdelphys complicatus* | 8.3 (M3/) |

*Notes:* Measurements from Paula Couto (1952a–c, 1962) and Marshall (1987), supplemented from measurements taken from epoxy casts, from the smallest to the largest. *Epidolops* is placed in its sequential position based on the combined estimate of its molar and premolar dentition because in this genus the relatively small molars are unlikely to reflect body size accurately.
*DGM 314-M, paratype of *Bobbshaefferia*, is *Guggenheimia*.
**includes *"Robertbutleria mastodontoidea"*

of association involving foot bones and teeth, even if one considered the relative size of teeth as feeding related adaptations that were possibly not of significant consequence, although of course they probably were. Very small mammals almost certainly have relatively different sized foot bones than their larger cousins. However, short of a major biometric study employing scaling and ecological measures coupled with shape analysis, this very real biological problem cannot be solved. Such a study would not only serve mammalian paleontologists but it would undoubtedly yield insights into biological similarities or differences between the two major living groups of therians.

Although the rough correlation I use here is not as precise as one would wish, some of the tarsal groups may be associated with dental measurements with some degree of certainty. More importantly, the morphological pattern of the tarsals can be analyzed as an independent character complex or taxonomic variable. In some fossils, and of course in the living animals, important connections can be made between teeth and tarsals, and this line of evidence has some consequences on the interpretation of these phena. In living didelphids, both large and small bodied taxa, the width of the **LAJ** (including both the **Su** and **CaA** facets) falls within approximately the same range as the observed range for the lengths of **M/2** and **M2/**. Given the nature of somewhat greater ontogenetic range of "post-juvenile" (i.e., nearly fully ossified) bone size variation (based on tooth eruption criteria) in marsupials than there is in eutherians, this imprecise, albeit handy, correlation will have to suffice until more careful studies using large ontogenetic samples are made. For a priori biological reasons (until proven otherwise), I consider those measurements of footbones most significant (i.e., potentially diagnostic on the species level) that have the closest articular associations with surrounding bones. These areas are the joint surfaces (as defined earlier), and therefore in my metric characterization of the available tarsals I have used **LAJ** widths of both the calcaneus (including, however, when present, the **CaFi** facet which is part of the **UAJ**) and astragalus. The calcaneal **CaFi** measurements are included for sheer expediency; this facet, when present, is difficult to delineate exactly from the **CaA** facet. So for the sake of easier repeatability of measurements on new specimens that will be undoubtedly described, this minor difference from the astragalar counterpart parameter is used. As expected for facets which articulate, the calcaneal and astragalar **LAJ** measurements should give the best chance for associating these two bones of the same species. This is closely equivalent, and replete with the same, but surmountable, difficulties as associating upper and lower teeth based on occlusal assessment.

In the next section I describe (or designate) and compare with other homologous elements 26 size samples of calcanea (*Itaboraí Calcaneus Sample; ICS*) and astragali (*Itaboraí Astragalus Sample; IAS*). The calcaneal and astragalar samples are combined, and judged to be conspecific (or at least practically and meaningfully further indivisible at the present) into groups, each group being a species "taxon" (unnamed) based on a *tarsal phenon*. This process yields, as a minimum, thirteen species (*Itaboraí Metatherian Group #I-XIII*; abbreviated as *IMG I, II*, etc.). As the dental record indicates, more species are known from Itaboraí (18 groups; see Table 6.3). In a few instances particular samples (i.e., either an *ICS* or *IAS* cluster) have no specimens allocated to them; they do not appear to be present in the known collections but are presumed to exist based either on astragalar or calcaneal samples that may be associated with them. At best, these phena correspond to the size categories of **M1s** or **M2s** of the described dental taxa from Itaboraí and therefore allow some level of attempted comparison, or at least food for taxonomic thought and further opportunity for testing.

*Itaboraí Metatherian Group I (Fig. 6.17)*

This is the smallest of the tarsal samples, and based on the available measurements it could represent either *Minusculodelphys*, *Gaylordia doelli*, or *Marmopsis*. As it should become obvious in the account below, this animal did not have the diagnostic combination present in the didelphid common ancestor.

ITABORAÍ CALCANEUS SAMPLE 1. The calcaneus is very similar to those described for the peradectines, with the notable exception of the **CaA** facet. In peradectines this facet is distally elongated, ribbonlike and somewhat helical in appearance, but the peradectine facet is quite distinct from the already helical **CaA** facets encountered in various eutherians. In *ICS 1* this facet is cylindrical, almost hingelike in appearance. There is no articulation for the fibula. The **clp** is closer to the **CaA** facet, in effect more distal, than in peradectines. This lateral process may be related to the ligamentous binding of the fibula to the calcaneus. As in the peradectines, the superposition angle is approximately 45 degrees. The peroneal process is moderate in size; it essentially retains the support for the **gtpl**. The **CaCu** facet is quite circular and dished out. There is no indication of a **CaCup** facet seen in the didelphids.

ITABORAÍ ASTRAGALUS SAMPLE 2. Two perfectly preserved specimens indicate an astragalar form–function very similar to that of *Late* (!) Eocene North American peradectines and the extant caenolestids. This close similarity to the American fossils is almost certainly convergence or parallelism; this critical tarsal closely reflects locomotion in small, probably rapidly scampering, "cursorial" animals. Did it derive from an ancestry (structural or actual) similar to the astragali from Bitter Creek of Wyoming, as the younger peradectines probably did? This is quite likely. The primitively small **AFi** facet persists and the strongly flared **ATim** facet gives secure support for tibial articulation. What is interesting both for *IMG I* and for the North American peradectines is that in spite of an obvious lack of articulation of the calcaneus with the fibula, the contact of the latter with the astragalus remained relatively small.

The **LAJ** articulations are very similar to the peradectine and pediomyine patterns. The **ampt** is exceptionally robust, with the apparent correlation of securing articulation with the calcaneal sustentaculum. (In didelphids this is largely accomplished by the

*Figure 6.17.* *Top two rows:* Itaboraí Metatherian Group I (IMG I). *Above: left calcaneus (ICS 1; based on DGM No. 1.106-M) Below: left astragalus (IAS 2; based on DGM No. 1.105-M),* Itaboraí, Brazil, Paleocene. *In the text and legends to follow the consequetive odd and even numbers of ICS and IAS designations refer to the calcaneal and astragalar samples, respectively, of the same hypothetical species that are designated as Itaboraí Metatherian Group plus a Roman numeral (see also Table 6.3). From left to right: dorsal (A, D), plantar (B, E), and distal (C, F) views. Bottom two rows: labeled outline drawings of specimens shown in top two rows. For abbreviations see Table 1.1. Scales = 1 mm.*

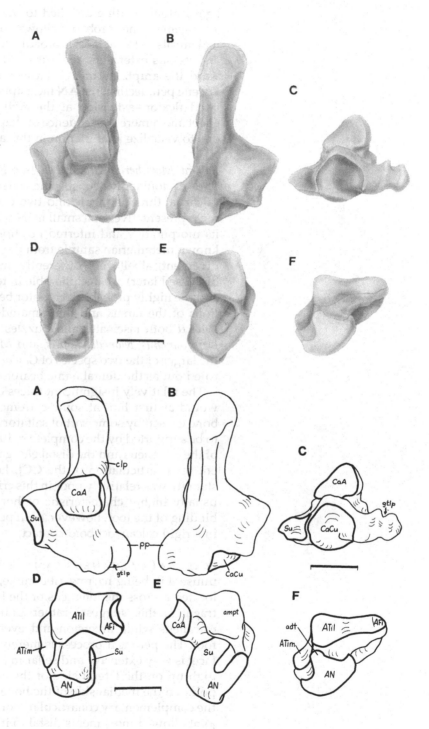

ligamentous binding attached to the relatively smaller **ampt,** an arrangement that probably allows greater flexibility of the proximal tarsals.) The **Su** facet is broadly continuous with the **AN** facet, and its **Sus** extension facet runs well extensad and proximal toward the **ampt.** As in living caenolestids and middle and late Eocene peradectine, the **AN** facet appears primarily to have facilitated flexion–extension at the **ANJ.** The characteristic **adt,** an effect and a mere consequence of this hingelike articulation, helps one to visualize joint motion in this area.

*Itaboraí Metatherian Group II (Figs. 6.18, 6.19, & 8.10)*
The undoubtedly conspecific tarsal elements (twenty four calcanea, three astragali, and two cuboids) are without question the representatives of a small marsupial more peculiar in terms of its morphology and inferred positional behavior than any other known metatherian sample from the Americas. While the equally divergent, albeit very differently, modified argyrolagids (to be discussed later) are decipherable in terms of the likely mechanics and the highly probable locomotor behavior reflected by the functions of the tarsus and the remainder of the postcranial bones, *IMG II* both fascinates and puzzles. It falls in the size range of *Derorhynchus, Monodelphopsis,* and *Mirandatherium.* Nevertheless, the larger of the two species of *Gaylordia* and *Sternbergia* cannot be ruled out as the dental name bearers of this animal.

The relatively insignificant traces of a fibula in **UAJ** mechanics would at first hint at some extreme reduction or fusion of this bone caused by some sort of saltatory behavior. This would seem to be supported by the complete reduction of the peroneal process of the calcaneus and the distal elongation of this bone. The apparent sellar articulation of the **CCJ,** however, would suggest that stability was relatively poor in this critical area, a stability which is usually highly characteristic of hopping animals. Ligamentous binding of the foot, however, as in peramelids, may have resulted in a rigid calcaneocuboid contact.

ITABORAÍ CALCANEUS SAMPLE 3. The **CaA** facet is highly unusual in being narrow, ribbonlike, and somewhat transversely located accross the long axis of the bone. As noted under the astragalus, this suggests rather unusual rocking ability of the calcaneus while inversion and eversion was performed by the foot. The peroneal process is merely rudimentary, and the **CaCu** facet is very extensive and sellar in configuration. It extends from high up on the lateral side of the calcaneus to the nearly equal point on the medial side of the bone. The isolated cuboids reflect the complementary conarticular morphology of the **CCJ.** This foot joint allowed movements distal to it that permitted extensive abduction and adduction of the forefoot and considerable flexion and extension, as well as great latitude of motion in between these

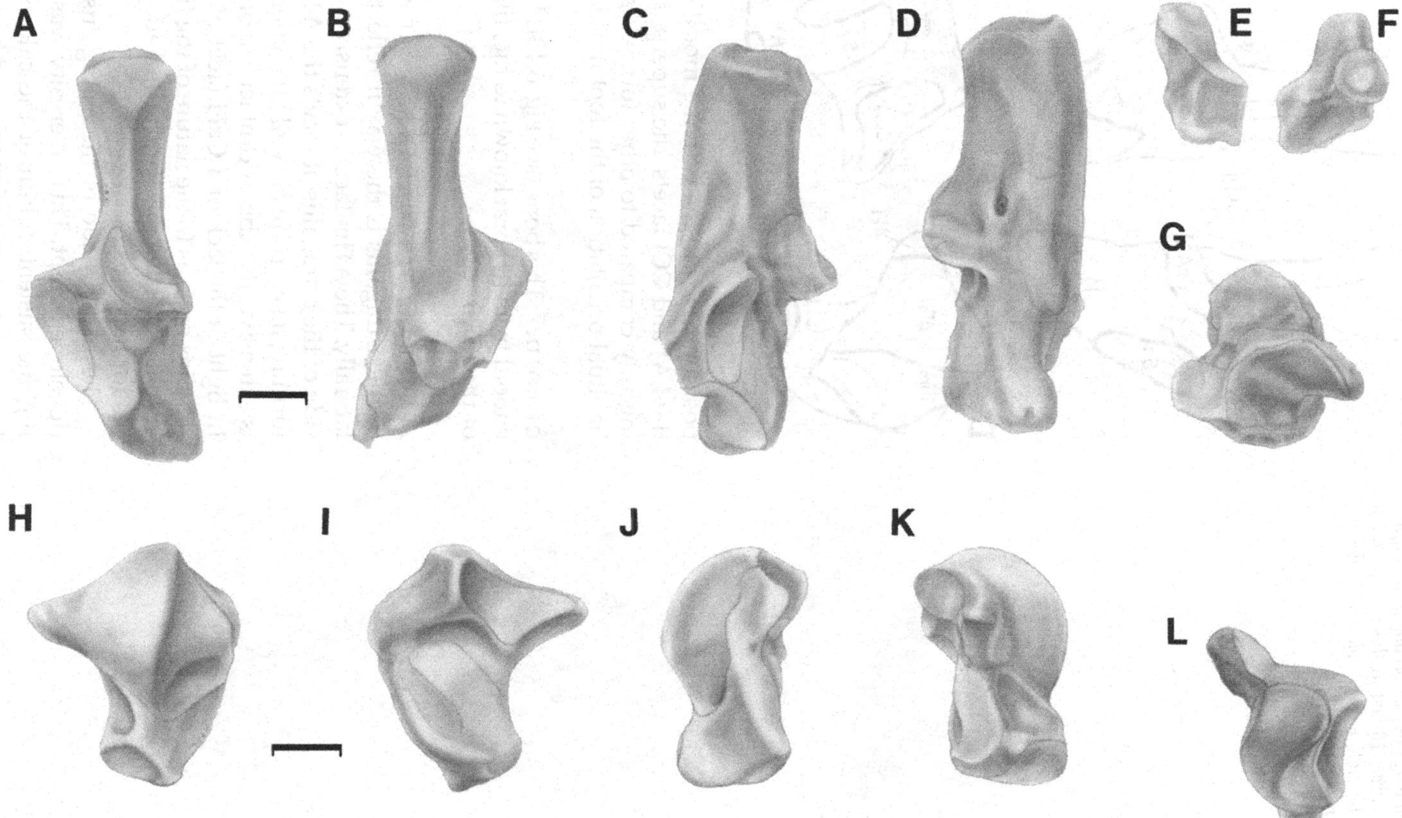

*Figure 6.18.* IMG II. Left calcaneus above (ICS 3; based on DGM Nos. 1.107-M and 1.104-M), right astragalus below (IAS 4; based on DGM No. 1.112-M), and left cuboid to the extreme right on top (based on DGM Nos. 1.118-M and 1.119-M). From left to right, depictions for the calcaneus and astragalus are given as in Figure 5.8. From left to right the cuboid is shown in dorsal (E) and plantar (G) views. Scales = 1 mm.

**Figure 6.19.** IMG II. *Labelled outline drawings of the specimens shown in Figure 6.18 (astragalus reversed). For abbreviations see Table 1.1. Scales = 1 mm.*

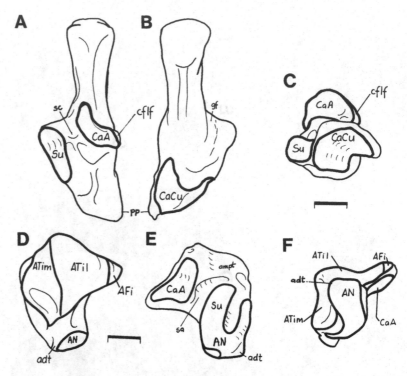

positions. Due to the extreme modifications of the orientation of the **LAJ** and **CCJ** facets, the superposition angle cannot be meaningfully compared to other forms. I am quite uncertain as to the habitual orientation of the foot in this animal.

CUBOID. This bone (see Fig. 6.18E,F), the only type of marsupial cuboid from Itaboraí known to me, fits perfectly with the calcanea of the *ICS 3.*

ITABORAÍ ASTRAGALAR SAMPLE 4. The most baffling aspect of the astragalus is the way it reflects the narrowing of the **UAJ** laterally. The **ATim** facet is extensive and spread out on the medial side of the astragalus; it meets the **ATil** facet at a 90 degree angle, forming a sharp crest. While the lateral articulation with the tibia is extensive, the fibular contact can only be described as marginal. In light of the lack of a **CaFi** facet on the calcaneus, the question remains open as to the nature of the fibula: was it very reduced or absent, and what mechanical and adaptive reasons were responsible for its state?

The plantar side of the astragalus shows a relatively weakly developed **ampt.** This tuberosity was not instrumental in restricting the sustentaculum of the calcaneus, although undoubtedly ligamentous connections existed between it and the lateral side of the plantar surface as well as the calcaneus. The astragalar **CaA** facet, as its calcaneal counterpart, reflects the ability of the calcaneus to rotate under the astragalus on an axis that traversed the calcaneus approximately 35 degrees from the tuber and exited mediodistally on the plantar side. This allowed considerable rotation of the calcaneus under the astragalus. The fact that the astragalar **CaA** facet tapers further nearly to a point laterally suggests that very little load was placed on the fibular portion of the crus.

In telling contrast to *IMG I*, living caenolestids, or peradectines, the **AN** facet in this animal is very narrow distally, suggesting more movement facilitation than compressive load bearing by the forefoot. This inference is reinforced by the observations above on the **CCJ.** Neither of the two moieties of the transverse tarsal joint suggest adaptations for either well stabilized load distribution from the toes usually seen in small or large cursors, metatherians or eutherians. Nevertheless, judged from the **CCJ** of peramelids, a narrow foot tightly bound by ligaments for hopping cannot be ruled out. These joints of the **TTJ** show no similarities to the various arboreally mediated mechanics of the living didelphids or to the fossil groups that may well have been capable arboreal climbers.

*Itaboraí Metatherian Group III (Figs. 6.20 & 6.21B,D)*
Only calcanea are known for this group. The possibility cannot be dismissed that the astragalus assigned to *IAS 8* may be conspecific with this species (i.e., it is *IAS 6*).

ITABORAÍ CALCANEUS SAMPLE 5.    Six excellently preserved calcanea show a morph virtually identical to *ICS 1*, the smallest of the calcaneal samples from Itaboraí. The pattern is also very similar to that shown in the Eocene peradectines of North America. The shaft of the tuber calcanei is medially deflected like the pattern in peradectines, but unlike the straighter pattern seen in didelphids or in *ICS 9* of *IMG V*. The superposition angle is approximately 45 degrees.

*Itaboraí Metatherian Group IV (Fig. 6.22)*
Although the size difference of this group from *IMG III* may not appear significant, the morphology is. In fact this species in many ways retains what may be the largest number of primitive attributes in the tarsus of the ancestors of known ameridelphian marsupials.

ITABORAÍ CALCANEUS SAMPLE 7.    A decidedly diagnostic difference is the presence of a well developed **CaFi** facet in both the adult (Fig. 6.22A–C) and what is judged to be a juvenile specimen (Fig. 6.22D–F). The **LAJ** appears to be relatively wide transversely, as it is in the specimen of the astragalus assigned to the

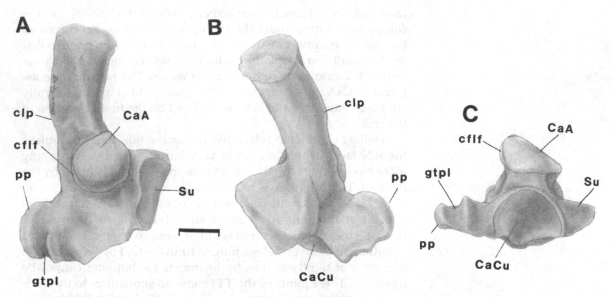

**A** clp CaA cflf pp Su gtpl

**B** clp pp CaCu

**C** cflf CaA gtpl Su pp CaCu

***Figure 6.20.*** IMG III. *Right calcaneus (ICS 5; based on DGM No.* 1.123-M). *From left to right dorsal, plantar, and distal views. For abbrevia-* tions *see Table 1.1. Scale = 1 mm.*

same species (see the following discussion of *IAS 8*). The lack of a calcaneal **Sus** facet should be mirrored in the reduction or absence of any astragalar **Sus** facet. In fact this feature is lacking in *IAS 8*. The peroneal process is moderate in size. The **CaCu** facet is semicircular, without any indication of a **CaCup** extension facet.

ITABORAÍ ASTRAGALUS SAMPLE 8. The only astragalus (Fig. 6.22G–I) assigned to this sample (or possibly a representative of either *IAS 6* or *IAS 10*) is an interesting mixture of attributes, some of which can be seen in the earliest Eocene peradectines from Bitter Creek. There are some additional features that may be either more primitive or more advanced than are the shared similarities with peradectines. The **UAJ** is characterized by a relatively small **AFi** facet, suggesting the presence of **CaFi** articulation. The angle of the prominent **ATim** facet with the relatively narrow **ATil** facet is about 125 degrees, relatively very obtuse compared to 105 degrees in the Bitter Creek sample and 95 degrees in the *IAS 2*. This suggests a strategy aimed at relative transverse stability by a widening of the **UAJ** and **LAJ** articulations. The **LAJ** articulations are relatively far apart, and the **Su** lacks a **Sus** extension facet. For this reason the large **ampt** appears to be a retained primitive, but clearly functionally important attribute. The **Su** facet is not as broadly continuous with the **AN** articulation as that found in the earliest Eocene peradectines. The **AN** articulation appears to reflect moderate emphasis on flexion and extension but with ample ability retained for supination necessary in extreme inversion.

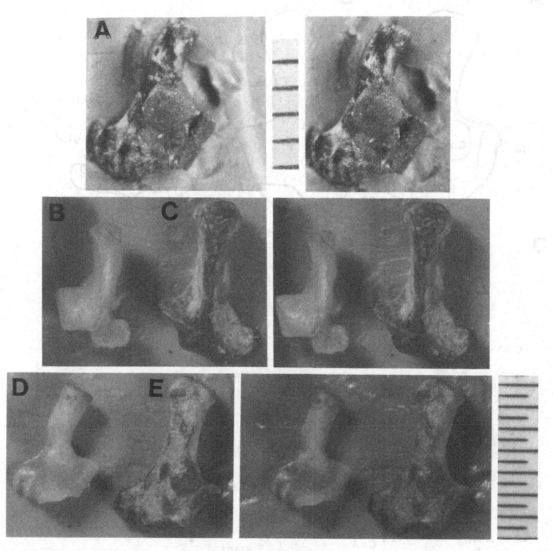

**Figure 6.21.** *Stereophotos.* **A.** *Metatherian, right calcaneus, unnumbered UCMP specimen, Bitter Creek fauna, Wyoming, Wasatchian Early Eocene.* **B,**
**D.** IMG III (ICS 5; *right calcaneus, DGM No. 1.123-M).* **C, E.** IMG V (ICS 9; *right calcaneus, DGM No. 1.131-M). A, D, and E are dorsal, and*
*B and C are plantar views. Subdivisions on scales = 0.5 mm.*

*Itaboraí Metatherian Group V (Figs. 6.21C–E & 6.23D–F).*
Two calcanea that attest to the existence of this group, along with a single calcaneus making up *IMG XII* (see below), represent the only unquestionable tarsal remains that share the synapomorphous calcaneal attributes of the living family Didelphidae. It is unfortunate that this group allocation to the very clearly diagnosable living family cannot be better corroborated by the appropriate features seen in astragali. There are no astragali that may be assigned to either one of these two Itaboraí

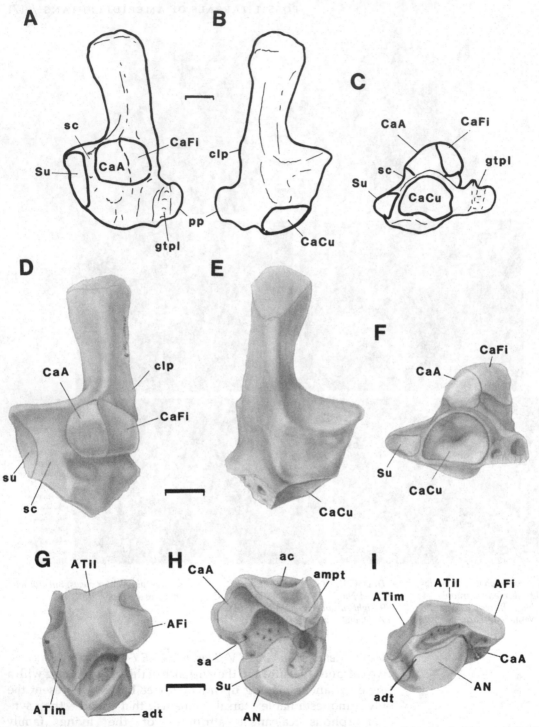

**Figure 6.22.** IMG IV. Above: *adult left calcaneus, (ICS 7 based on DGM No. 1.128-M). Middle: juvenile left calcaneus, (ICS 7, based on DGM No. 1.127-M, with the peroneal process broken off). Below: left astragalus (IAS 8, but may be IAS 6 or IAS 10; based on DGM No. 1.129-M). From left to right: dorsal, plantar, and distal views. For abbreviations see Table 1.1. Scales = 1 mm.*

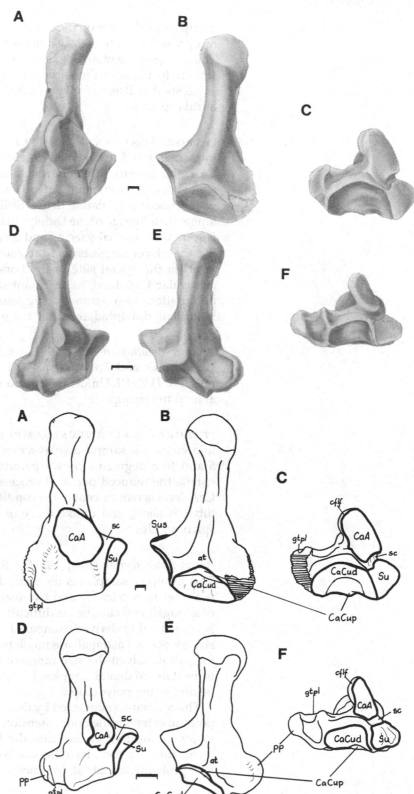

*Figure* 6.23. *Didelphidae. IMG V, below, and IMG XII, above in two top rows; right calcaneus with peroneal process broken off (ICS 9; based on DGM No. 1.131-M), and right calcaneus (ICS 23; based on DGM No. 1.180-M). From left to right dorsal, plantar, and distal views. Bottom two rows: Labeled outline drawings of specimens shown in top two rows. For abbreviations see Table 1.1. Scales = 1 mm.*

groups, based either on the appropriate size range or on the well-recognized morphological criteria of living didelphid astragali. The astragalus tentatively assigned to *IAS 8* may be structurally similar to those of the earliest didelphids. The size correlation suggests that *Itaboraidelphys camposi* may represent the dentition of this species.

ITABORAÍ CALCANEUS SAMPLE 9.   The barely noticeable **clp,** the straight and not medially recurved tuber in the proximity of **CaA,** the absence of **CaFi** articulation, all taken together with a well-defined, prominent **CaCup** facet plantar to a transversely wide cuboid articulation, leave little doubt that this sample belonged to a lineage of the Didelphidae. That the calcaneal **Su** facet is somewhat medially facing and is separated by a narrow **sc** from the **CaA** facet suggests that the astragalus may have had its **Su** facet on the lateral side of the bone and relatively close to the astragalar **CaA** facet, as in extant didelphids. The lack of **CaFi** articulation also strongly suggests another advanced ameridelphian, didelphid, feature – the widened **AFi** articulation.

*Itaboraí Metatherian Group VI (Fig. 6.24).*
*Didelphopsis* and *Protodidelphis* fall approximately into the size range of *IMG VI*. Unfortunately no sample of astragali is assignable to this group.

ITABORAÍ CALCANEUS SAMPLE 11.   The three calcanea known for this sample display a morphology rather similar to *ICS 5* and to a degree to known peradectines. The major difference entails the reduced peroneal process. The distal extension of the **CaA** facet is reminiscent of the condition seen in peradectines. The tuber is short, and unlike those of peradectines, it lacks a conspicuous **clp.**

*Itaboraí Metatherian Group VII (Fig. 6.25)*
This group of samples is the second best represented at Itaboraí with fourteen calcanea and two astragali. As always, association of astragali and calcanea is difficult. In this instance the calcaneus is considerably derived compared to the other samples, but the size associated astragali are much more conservative in their conformation. Given the size range of these bones and information from Itaboraí dental samples, it appears to me that these remains belong to the polydolopid *Epidolops*.

The mechanics suggested by the calcaneal morphology strongly point to extensive flexion–extension with relatively limited ability to invert the foot or supinate the forefoot. It appears that this form–function complex reflects a bounding–cursorial, and therefore probably terrestrial, locomotor habit.

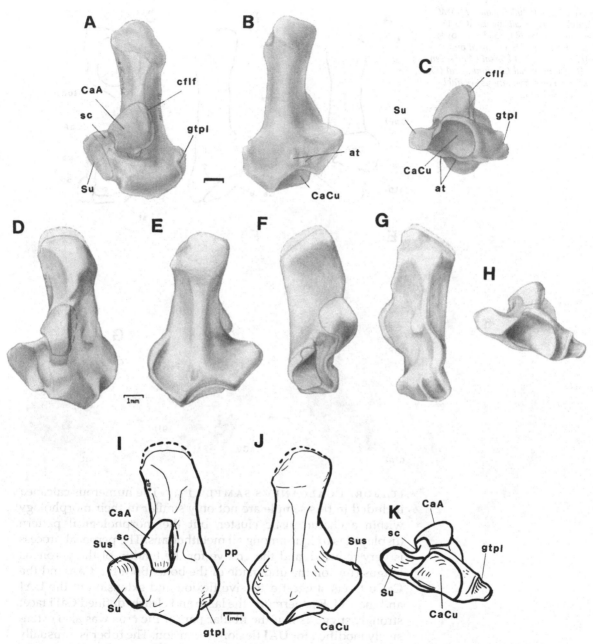

**Figure 6.24.** IMG VI. **A–C.** Left calcaneus (ICS 11, based on AMNH No. 55371). **D–K.** Left calcaneus (ICS 11; based on DGM No. 1.132-M). I-K: Labelled outline of left calcaneus shown in D–H. A–C and I–K dorsal, plantar, and distal views. D–H, from left to right as in Figure 5.8 For abbreviations see Table 1.1. Scales = 1 mm.

*Figure 6.25.* *Polydolopidae (?), IMG VII.* Below: *right calcaneus (ICS 13; based on AMNH No. 55370.* Above: *right calcaneus (ICS 13; based on AMNH No. 53372) Dorsal (A, E), ventral (B, F), proximal (C), and distal (D, G) views. For abbreviations see Table 1.1. Scale = 1 mm.*

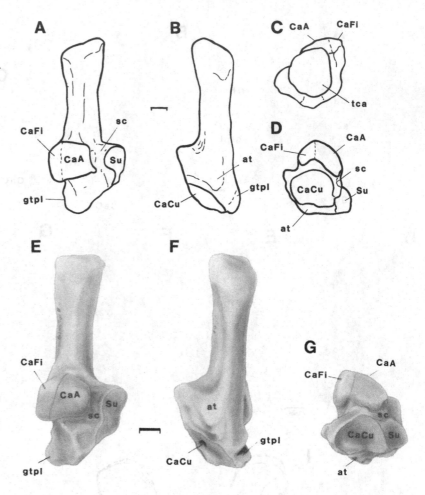

ITABORAÍ CALCANEUS SAMPLE 13. The numerous calcanea included in this sample are not only similar in their morphology within a characteristic cluster, but the morphological pattern displayed is unique among all metatherians. The peroneal process is very reduced, and the groove for the tendon of the peroneus longus runs on the underside of the bone. Both the **CaA** and the **CaCu** facets suggest extensive flexion and extension in the **UAJ** and the **TTJ**. Furthermore, the large and clearly defined **CaFi** facet strongly suggests that the fibular part of the crus was also extensively modified for **UAJ** flexion–extension. The tuber is unusually long and its cross section is not deep, in sharp contrast, for example, to the deep tuber of *ICS 11.*

ITABORAÍ ASTRAGALAR SAMPLE 14. The morphology of the two astragali (DGM Nos. 1.148-N and 1.151-M) making up this putative sample is extremely close to that of all the other Itaboraí samples, except of course that of *IAS 4* of *IMG II,* the most modi-

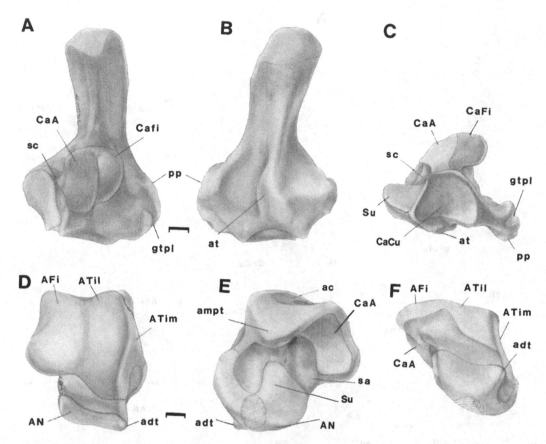

**Figure 6.26.** IMG VIII. Above: *left calcaneus (ICS 15, based on DGM No. 1.160-M). Below: right astragalus* (IAS 16; based on DGM No. 1.161-M). *From left to right: dorsal, ventral, and distal views. For abbreviations see Table 1.1. Scales = 1 mm.*

fied metatherian tarsals of the fauna. In fact this "Itaboraí" morphology of the astragali is also similar to that of the peradectines. It does not have any of the diagnostic features of didelphid astragali described below.

*Itaboraí Metatherian Group VIII and IX (Figs. 6.26, 6.27, & 6.29B,C)* This large sample falls within the range expected for *Patene*, in the upper size range of the dental taxa described from Itaboraí (see Table 6.3). It is relatively very well represented by eight calcanea in *ICS 15*, seven calcanea in *ICS 17*, and four astragali in *IAS 16*, and one astragalus in *IAS 18*. The number of species represented by the tarsals, however, becomes increasingly difficult to ascertain in this group. Although I discuss these under one heading, as the subdivision of this sample suggests, based on the persistent similarities of the characters encountered within these samples (namely those of the astragali), I believe that there may be two species represented.

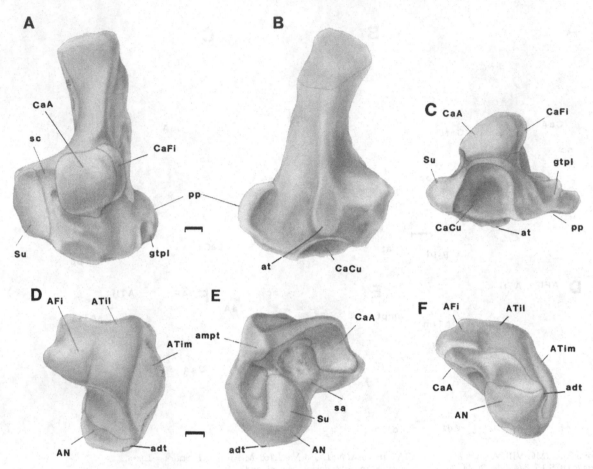

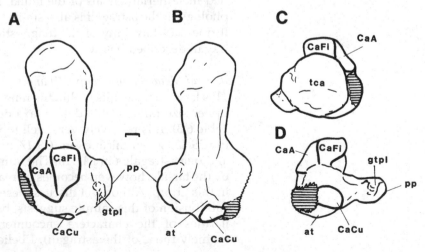

*Figure 6.27.* IMG VIII *and* IX. Below: *based on size criteria applied to the available samples, right astragalus (IAS 16, based on DGM No. 1–164-M). Above: left calcaneus (ICS 17, DGM No. 1.171-M. From left to right: dorsal, plantar, and distal views. For abbreviations see Table 1.1. Scales = 1 mm.*

*Figure 6.28.* IMG X. *Left calcaneus, (ICS 19, based on DGM No. 1.173-M). Dorsal (A), plantar (B), proximal (C), and distal (D) views. For abbreviations see Table 1.1. Scale = 1 mm.*

ITABORAÍ CALCANEUS SAMPLE 15.   In remarkable contrast to the specimens of *ICS 11*, the tuber calcanei of this larger sample are rounded and relatively thick mediolaterally compared to the deep tuber of the *ICS 11*. This shape reflects a great decrease in the load placed on the tuber by the tendon of the flexor of the toes, the flexor digitorum brevis. Fibular contact with the calcaneus is reflected by the long **CaFi** facet. The anterior tubercle is well developed in this sample, and the **CaCu** facet shows no signs of a proximal extension plantarly (**CaCup** facet). The **CCJ**, like those of the larger Itaboraí calcanea, shows a very stereotypic ball and socket configuration.

ITABORAÍ ASTRAGALUS SAMPLE 16.   The four astragali generally display the by-now familiar (see above) **UAJ** and **LAJ** articular patterns of putative basal sudameridelphian and undoubted borhyaenids. The **ANJ** is characteristically indicative of primarily flexion and extension. The slight medial excursion of the **AN** facet suggests a derived increase in mediolateral stability in the **LAJ** in this lineage.

ITABORAÍ CALCANEUS SAMPLE 17.   This calcaneus is quite similar to *ICS 15*. The size is larger but this may be an artifact of sorting. There appears to be a relatively smaller **CaFi** facet, but the significance of this size is impossible to establish. The validity of this sample is in some way dependent on the distinction of the astragalus (DGM 1.165-M) as the basis for *IAS 18*.

ITABORAÍ ASTRAGALUS SAMPLE 18.   While the astragalus assigned to this sample is similar to those comprising *IAS 16*, it is mediolaterally more spread out. The angle formed between the **ATim** and **ATil** facets is about 10 to 15 degrees greater – yet this feature may not be significant; without a larger sample this structure cannot be made into an adaquate diagnostic feature. The **Su** facet is characteristically ribbonlike, relatively narrower than those on the astragali in *IAS 16*. As a result of the greater mediolateral spread of this bone, the **ANJ** appears to have been aligned more mediolaterally. The ability of the **ANJ** to allow greater pronation and supination of the forefoot along with the increased mediolateral mobility of the **UAJ** may reflect functionally significant differences between *IMG 15* and *IMG 16* – if these groups are real!

*Itaboraí Metatherian Group X (Fig. 6.28)*

ITABORAÍ CALCANEUS SAMPLE 19.   The two calcanea representing this group are diagnostically distinct from the other samples. The calcaneus has a robust, rounded, and bulbous end on its tuber and a very extensive **CaFi** facet. Although the area is broken, it appears that the partly preserved **CaCu** articulation was of a ball-and-socket (slightly modified ovoid) variety. Judged from the tuber, this animal was primarily terrestrial, but not particularly fast on the ground.

### Itaboraí Metatherian Group XI

ITABORAÍ CALCANEUS SAMPLE 21. This sample (DGM Nos. 1.175-M, 1.176-M, 1.178-M, and 1.179-M) is quite similar to *ICS 15*, except significantly larger. The **CaFi** facet is discernable along with a well developed **clp**. The **CaCu** facet is deep and of a ball-and-socket type in its form–functional configuration. There is no indication of the extension facet seen in didelphids.

ITABORAÍ ASTRAGALUS SAMPLE 22. This sample is represented by a single well-preserved specimen (DGM No. 1.177-M) that is similar to the astragali of *IAS 18* and *IAS 26*, and to the advanced late Eocene peradectines in North America. Yet it is very likely that the northern character combination was acquired entirely independently of the complex seen at Itaboraí. The astragalar canal persists, as in the other Itaboraí taxa, and the **LAJ** pattern is a familiar, primitive one, similar to the others already described.

### Itaboraí Metatherian Group XII (Fig. 6.29A–C)
In addition to *IMG V*, this group includes the other undoubted didelphid at Itaboraí. As I discussed at length in Chapter 3 and 8, I consider the identification of the Didelphidae on molar grounds alone to be unreliable as yet.

ITABORAÍ CALCANEUS SAMPLE 23. Like *ICS 9*, the single calcaneus representing this group has the diagnostic **CaCup** facet shown in all the living didelphids, in addition to the apparent absence of the **CaFi** facet, a condition so prevalent among the Itaboraí marsupials.

### Itaboraí Metatherian Group XIII (Figs. 6.29A, D & 6.30, 6.31)
This is the largest of all the marsupials known at Itaboraí, yet in morphological attributes it parallels the majority of the samples just described.

ITABORAÍ CALCANEUS SAMPLE 25. Like most of the previously described Itaboraí samples, the **CaFi** facet is prevalent, the peroneal process persists, and the **CaCu** facet is relatively deep and facilitates the rotation of the forefoot.

ITABORAÍ ASTRAGALUS SAMPLE 26. As in the other samples, the emphasis on the astragalar facets of the **UAJ** falls on the

**Figure 6.29.** *Stereophotos. IMG XIII. (A, D) and IMG IX (B, C). **B, C.** Right calcaneus (ICS 17, DGM No. 1.170-M). **A, D.** Left calcaneus (ICS 25; AMNH No. 55375). A and B, dorsal views; C and D, plantar views. Subdivisions on scales = 0.5 mm.*

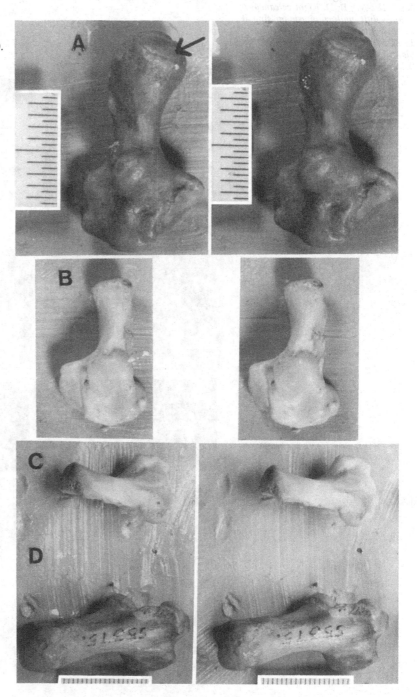

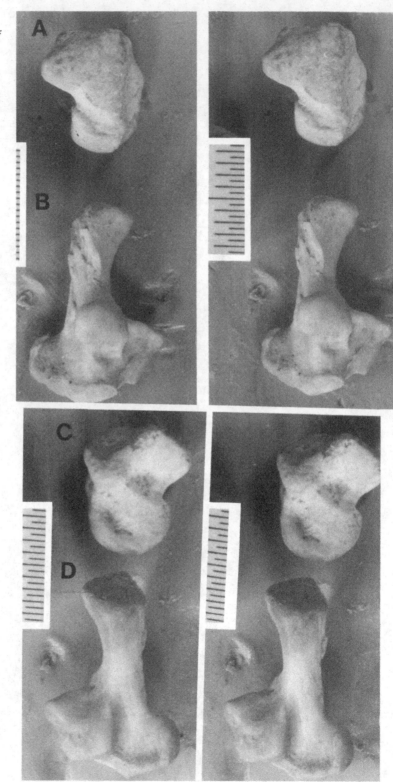

**Figure 6.30.** Stereophotos. IMG XIII. **A, C.** Right astragalus (IAS 26; DGM No. 1.182-M). **B, D.** Right calcaneus of young adult, with epiphysis missing at the end of tuber (ICS 25; DGM No. 1.181-M). A, B: dorsal views; C, D: plantar views. Subdivisions on scales = 0.5 mm.

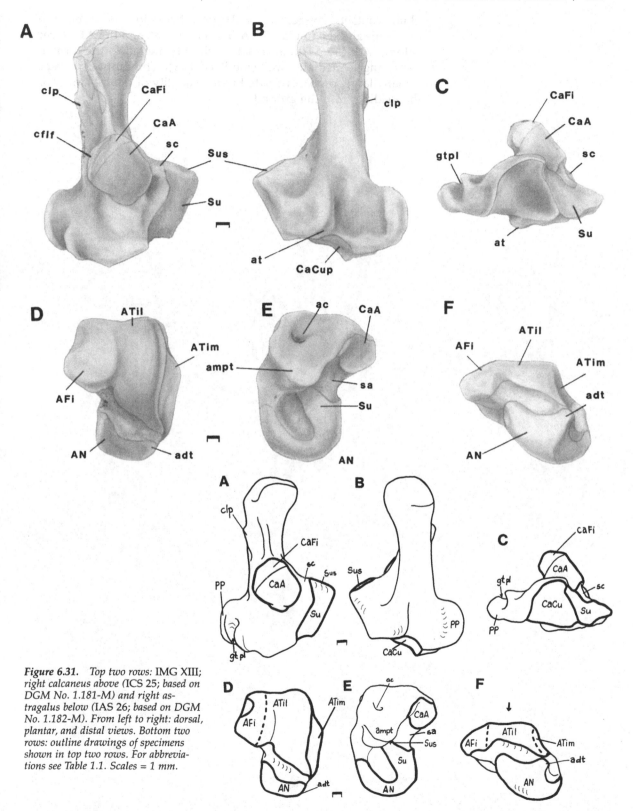

*Figure 6.31.* Top two rows: IMG XIII; right calcaneus above (ICS 25; based on DGM No. 1.181-M) and right astragalus below (IAS 26; based on DGM No. 1.182-M). From left to right: dorsal, plantar, and distal views. Bottom two rows: outline drawings of specimens shown in top two rows. For abbreviations see Table 1.1. Scales = 1 mm.

tibial contact, suggesting a large articulation between the calcaneus and the fibula. The **AFi** facet is small, and the **ATil** facet is long with the **ATim** articulation sharply angled to the former, indicating a medially well-stabilized **UAJ.** The **ampt** is exceptionally large, probably related to the stability of the **LAJ** and / or the loading from the ground.

# 7

# Cruropedal attributes of living and fossil families of metatherians

> It is that osteology must after all constitute the core of the true theory of mammalian history. It is only by means of the skeleton that we are able to correlate the knowledge of living with that of fossil mammals and thus to synthesize the results of palaeontology, systematic mammalogy and comparative anatomy.
>
> Gregory (1910, p. 112)

> It will, perhaps, be helpful for an understanding of marsupial conditions if it be explained that Huxley, Dollo, Bensley, and other authorities on this group have been convinced that the Metatheria were derived from an ancestor sufficiently specialized for an arboreal existence for the latter adaptation to have left a lasting impression upon the foot structure. This is in contrast to the protoplacental ancestor, which, although presumably to some extent arboreal in habit, was hardly modified in this direction to a very definite degree.
>
> A. B. Howell (1944, p. 28)

**Didelphidae (Figures 3.1, 3.2, 4.1–4.5, 6.23, 7.1–7.12, & 8.11–8.18)**

## Ecomorphology

Bock's (1991) and Goldschmid and Kotrschal's (1989) recent reviews (see additional references therein) explore and discuss in detail the role of functional and ecological morphology in systematic biology. Because the fossil record of the early differentiation of marsupial postcranial morphology primarily takes the form of proximal and distal ends of long bones, and calcanea and astragali, in addition to occasional skulls and the relatively abundant teeth, it would be ideal to emphasize and document here the consistency of the phylogenetic constraints and habitus-related variation exhibited by members of this extant family in the entire postcranium. Such lofty aims, however, are seriously constrained by the relative lack of understanding of the field biology of these animals. Within such limits I will attempt to point out some functional modifications in the tarsus which are probably adaptive. While a number of predictions could be made from the morphological evidence alone, this is not a satisfactory way to achieve the understanding of living animals – or their phylogeny. Detailed behavioral and ecological studies of these marsupials are urgently needed.

A recent volume by Lee and Cockburn (1985) concisely summarizes some of the information available on the behavioral and evolutionary ecology of didelphids and other living marsupials. The reviews by Collins (1973), Hunsaker (1977a, b), and Streilein (1982) that summarize the scattered information on the ecology and assorted habits of New World marsupials are important sources for what was known up to that time about the natural history of these animals. But in order to interpret the ecology and behavior of fossil forms known from their scantily preserved hard parts, one needs highly focused studies on aspects of the behavioral and ecological morphology of the living species. Such studies dealing with the evolutionary morphology of living marsupials are very few and are sometimes based on inappropriate assumptions. The knowledge needed about living species concerns their feeding preferences, methods, and regimes, as well as exact locomotor and postural behaviors, in addition to habitual substrate and habitat preferences. At best, most of this information is still anecdotal.

A notable exception, an examplary team study in mammalogy, is the recent analysis of Charles-Dominique et al. (1981; see also Charles-Dominique, 1983) of a nocturnal community of nine frugivorous mammals in French Guiana, five of which were didelphids: *Caluromys philander, Marmosa cinerea, Marmosa murina, Philander opossum,* and *Didelphis marsupialis.* The species were found to be less and less arboreal in the order listed, with *Philander* and *Didelphis* spending the most time on the ground. No other study on South American didelphids has been as detailed as the latter (see also Atramentowicz, 1982), but it is generally well established that *Metachirus nudicaudatus* is a capable terrestrial runner (probably largely scansorial), and that *Lutreolina crassicaudata,* in spite of its name lutrine opossum, is a clumsy and reluctant swimmer, and it is mainly terrestrial. *Didelphis, Philander,* and *Monodelphis* are terrestrial, yet not cursorial to the degree that *Metachirus* is known to be. *Lestodelphys,* a pampa inhabitant, is probably more terrestrial than arboreal. All of these terrestrial forms are also capable climbers, but there are virtually no data on speed, technique, or leaping behavior on any of these animals, information that is critical in the analysis of the comparative morphological evidence. The "megagenus" *Marmosa* (morphologically not clearly diagnosed and contains several genera such as the relatively large *Micoures*) is taxonomically diverse but very poorly known, in spite of numerous attempts at taxonomic revision (see Reig et al., 1987). The genera *Caluromysiops* (see Izor & Pine, 1987) and *Glironia,* like *Caluromys,* are probably arboreal. The species of *Caluromys,* highly arboreal, are reported to be particularly agile and fast in the trees (Nowak & Paradiso, 1983), in contrast to such well observed animals as the North American *Didelphis. Chiro-nectes* is semiaquatic, highly modified for this existence, yet even this opossum does climb, if unwillingly. For taxonomic summaries of species of didelphids see Gardner (1973, 1982), Marshall (1977b, 1978a, d, e) and McManus (1974), and references therein.

At this point a note should be added concerning the terminology used in the literature and in this study. Grand (1983), who has recently discussed locomotor-related terminology in small mammals, did not differentiate between commonly used terms such as scansorial and scampering. Böker's (1927) concept of "Klammerklettern" (which I designate as **graspclimbing**), in addition to the concepts (defined below) such as **scansorial** and **scampering,** represent, I believe, at least *three distinctive types of locomotor adaptations* in small mammals. The dictionary definitions of the adjectives "scansorial" and "scampering" should not be taken as the full meanings of these behavioral categories because these definitions are not precise and overlap. I therefore cannot follow Grand (1983) who suggests a similar descriptive meaning for the three locomotor behaviors.

Graspclimbing describes some arboreal activities of didelphid, microbiotheriid, and phalangeriform marsupials, as well as cimbing in most euprimates but not in other eutherians.

Scansorial is often a synonym for climber, but I will here restrict the use of climbing with the aid of claws to the term **clawclimbing,** behavior that is seen essentially in squirrels, and many other rodents, and in the dasyurid climbers, which do not make grasping use of the hallux. Didelphids will also clawclimb on larger trunks when they ascend or descend. Scansorial is also used to describe **arboreal scampering** (specified with an adjective; **arboscansorial** would be a precise designation), descriptive of a locomotor regimen which excludes grasping as the primary means of climbing. Jenkins and McClearn (1984), for example, have given a detailed functional–adaptive account of the hindfoot reversal practiced by both many marsupials and eutherians when hanging or descending trees.

The term scampering (to run, or to go hurriedly or quickly) should ideally be restricted to the terrestrial locomotion of small mammals through leaf litter, undergrowth or open space. I will use the term **terrestrial scampering** (**terrascansorial** to distinguish it from **arboscansorial**), which is in fact a near synonym of **semicursorial** or **subcursorial** without the past implications of the latter to the seemingly ungainly larger "subungulates." The terms "semicursorial" or "subcursorial" are more or less applicable to terrestrial fast-moving small mammals that do not have the kind of advanced mechanical adaptations found in the protoperissodactyls, protoartiodactyls, lagomorphs, many groups of South American native ungulates, or hyraxes. The various levels of morphological advances that can accompany this behavior are probably impossible to distinguish unless one resorts to morphological–mechanical criteria. Such designations, however, defeat the purpose of behavioral descriptors. Clearly, many open country cursors, however, have returned to scamper in the undergrowth, and there are no form–function barriers of a qualitative nature to prevent a scamperer (e.g., a eutherian viverrid or a metatherian caenolestid) to be a more open-country cursor. A case in point is *Metachirus*.

In an important recent study of body weight distribution of several didelphids (the same ones that were studied by Charles-Dominique et al., 1981) bearing on the evolutionary analysis of the skeleton, Grand (1983) instructively singled out *Metachirus*. He suggested (p. 309) that *M. nudicaudatus* "has undergone a complete transformation through the high proportion of muscle; the concentration of muscle in the lumbar region, the thighs, and upper arms; and the elongation of the distal limb segments." He further cited other sources for the terrestrial running ability of the animal and the lack of tail prehensility in the genus compared to most other didelphids. Grand (1983) also suggested, based on his body weight distribution studies, that the fully terrestrial *Monodelphis* is not a fully active, "high-speed" animal. According to Grand (1983), whose assessment of *Metachirus* is undoubtedly correct in this regard, the body proportions of this genus are in fact similar to such osteologically well recognizable eutherian cursors as *Dolichotis* and *Lepus*. This adaptation manifests itself in a higher percentage of musculature in the body rather than in the extremities. Compared to the arboreal didelphids with heavily muscled arms and thighs, *Metachirus* is relatively less muscled in the pes and manus. Yet, interestingly, when one assesses the osteological attributes of the brown "four-eyed" opossum, it becomes relatively difficult, but feasible, to identify features that suggest the extremes of subcursoriality documented by body proportions or observed locomotor behavior.

A question can only be posed but not answered here about *Metachirus* in particular and the nature of the problem in general: Is a lack of pronounced osteologically convincing evidence of the departure of this genus from the ancestral didelphid pattern (which is probably that of a graspclimber and arboscansorial form) due to the complete, perhaps less appreciated, locomotor pattern of this animal, or did the constraints of didelphid morphogenesis not yet permit some patterns to be available for selection? This and similar questions pose a problem that looms large when dealing with the interpretations of fossil remains. Another, similar example not pursued in any detail in this book concerns the phalangeroid *Gymnobelideus*, the astonishing jumping abilities of whom I have observed (undoubtedly influential in the development of gliding in its sister taxon *Petaurus*). Its jumping ability is not reflected in the pedal hard anatomy to a degree comparable to the bones and joints of rodent or primate arboreal leapers.

Given the fact that all Didelphidae, with the notable modification of *Chironectes*, share a hindfoot constructed around the antecedent constraints of a powerful grasping mechanism, I must assume that the most constraining heritage, the diagnostic attribute, of this taxon is related to graspclimbing arboreal habits. This assessment has consequences for the perspective employed on the pedal data (see Fig. 7.1 for representative didelphid osseous skeletons).

### Caluromyinae (Figs. 7.2–5, 7.8, 7.9, & 8.11–8.13)

A possible difference between members of this group and most other didelphids, one that may reflect not a habitus adjustment of the individual species but rather a shared phylogenetic constraint unique either to the whole family or to this subfamily, is the relative dorsoplantar flattening of the **CaCud** facet in these taxa. The facet, although higher laterally, is quite narrowed medially. Again, in the few species of *Marmosa* sampled a similar pattern seems to emerge.

### Didelphinae (Figs. 7.1, 7.6, 7.7, 7.10–7.12, 7.15C, & 8.11–8.13)

*Metachirus* (Figs. 8.11–8.13) should be singled out here because it is the most committed terrestrialist in this primarily arboreal radiation. As expected, the form–function of the facets reflecting the movements of the **TTJ** mirror this. There is also a secondarily acquired **CaFi** facet. What is surprising is that the other, thought to be critical, areas of the tarsus do not show equally clear modifications that should be due to the selective forces generated by this subcursor. The **AN** facet, however, is oriented much more vertically rather than transversely, and this is accompanied by an increase in depth of the **CaCud** facet as well. Both of these form changes reflect the increased functional importance of flexion–extension rather than the substrate-generated supination–pronation of the forefoot. Such characters by themselves make unreliable features to serve as bases of groupings. Because they often change independently from other areas of the tarsus, their reliability as meaningful tests of a phylogenetic hypothesis is very low indeed.

*Monodelphis*, which is probably generally terrestrial (Anderson, 1982), has the most reduced peroneal process among the didelphids. Equally significant is the fact that the angle made by the astragalar **ATim** facet with the **ATil** facet is fairly sharp, not unlike the one we see in most of the fossil American groups and the Caenolestidae. *Monodelphis* also displays the relatively vertical

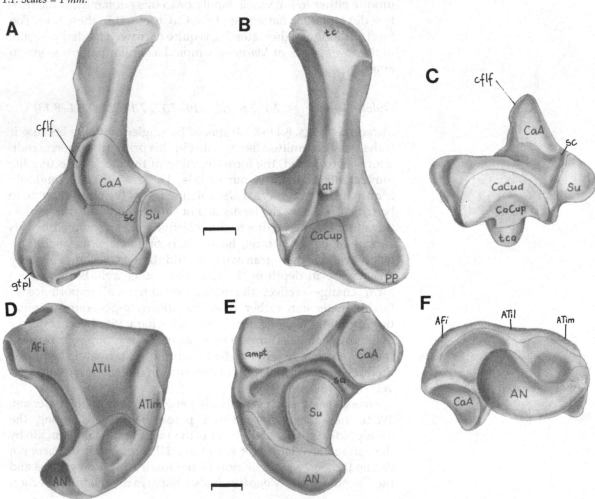

**Figure 7.1.** *Osseous feet of three didelphids: Philander opossum, Chironectes minimus, and Didelphis virginiana (phalanges are not shown in the latter). For abbreviations see Table 1.1. Scales = 1 cm.*

Ph

Ph

Ph

flattened distal phalanx

**Philander o.**    **Chironectes**    **Didelphis v.**

**Figure 7.2.** Caluromys derbianus, *AMNH No. 75512. Right calcaneus (above) and right astragalus (below). From left to right: dorsal, plantar, and distal views. For abbreviations see Table 1.1. Scales = 1 mm.*

A

cflf
CaA
sc  Su
gt pl

B

tc
at
CaCup
PP

C

cflf
CaA
sc
CaCud
Su
CaCup
t ca

D

AFi
ATil
ATim
AN

E

ampt
CaA
sa
Su
AN

F

AFi
ATil
ATim
CaA
AN

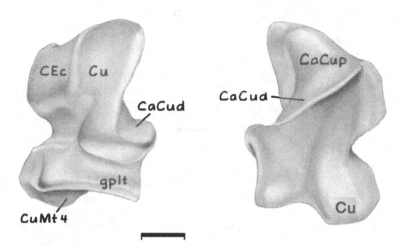

*Figure 7.3.* Caluromys derbianus, AMNH No. 75512. Right cuboid. Plantar (left) and dorsal (right) views. For abbreviations see Table 1.1. Scale = 1 mm.

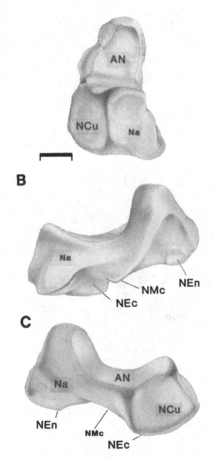

*Figure 7.4.* Caluromys derbianus, AMNH No. 75512. Right navicular. Dorsolateral (above), dorsal (middle), and plantar (below) views. For abbreviations see Table 1.1. Scale = 1 mm.

orientation of the **AN** facet noted under the earlier discussion of *Metachirus*. These features may be related to the need to brace the ankle medially in order to resist the relatively greater loads generated by fast galloping. We know virtually nothing of the locomotion-related behavioral ecology of the eleven recognized species of this genus.

The only semiaquatically well-adapted living marsupial, *Chironectes*, is unique indeed. From the more focused perspective of phylogenetics, *Chironectes* is additionally significant in that it shares those features, unequivocally and in spite of its unique adaptations, that unite the Didelphidae. Webbed hind feet and reduction of hallucial grasping notwithstanding, the attributes of the **EMt1J** (Fig. 7.7) and the **CCJ**, as well as the **LAJ**, remain fully didelphid. The modification of the **UAJ** are clearly made on the diagnostic didelphid base. It is reasonable to assume that a terrestrial lifestyle probably preceded the aquatic one. Its **UAJ** (Fig. 7.12) in uniquely modified among the didelphids in that the angle of the attachment of **AFi** facet to the **ATil** facet is sharp. The **ATim** facet is sharply offset medially – it is "stepped" off the **ATil** one – yet the surface of the former is parallel to the latter in a manner we see in *Metachirus* and *Philander*. The tightly bracing fibula on the lateral side also contacts, as in *Didelphis* and *Metachirus*, the surface lateral to the calcaneal **CaA** facet through the meniscus (Fig. 7.12). This is obviously a secondary development of indirect fibular contact, but it may be related to lateral bracing of the **UAJ** and not to load bearing originating from ground contact, as in many terrestrial forms. It is not unlikely, however, that this structure is simply a heritage feature derived from an ancestor that was a species of either *Didelphis* or *Metachirus*, taxa that have this secondary calcaneofibular contact.

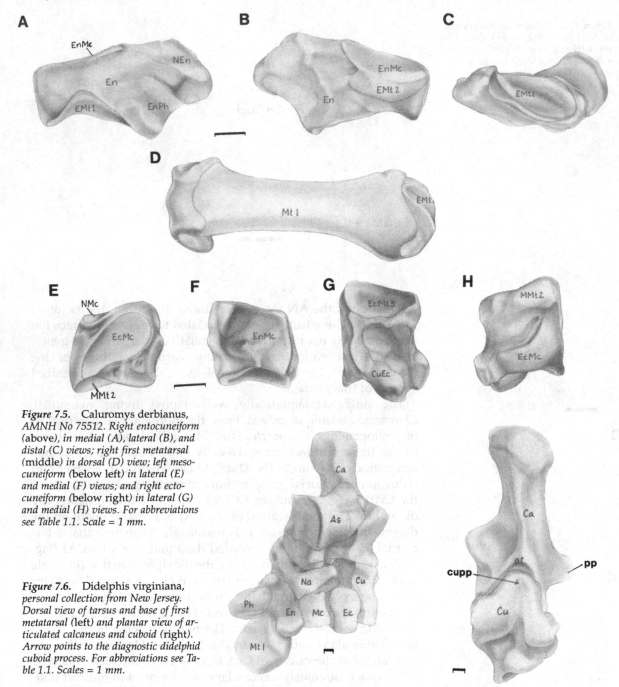

**Figure 7.5.** Caluromys derbianus, AMNH No 75512. *Right entocuneiform (above), in medial (A), lateral (B), and distal (C) views; right first metatarsal (middle) in dorsal (D) view; left meso-cuneiform (below left) in lateral (E) and medial (F) views; and right ecto-cuneiform (below right) in lateral (G) and medial (H) views. For abbreviations see Table 1.1. Scale = 1 mm.*

**Figure 7.6.** Didelphis virginiana, *personal collection from New Jersey. Dorsal view of tarsus and base of first metatarsal (left) and plantar view of articulated calcaneus and cuboid (right). Arrow points to the diagnostic didelphid cuboid process. For abbreviations see Table 1.1. Scales = 1 mm.*

*Some general features of the didelphid tarsus*
The opossums are the most abundant of the three living groups of American marsupials. They have served in zoology and paleontology as a living stand in, the best available appropriate model

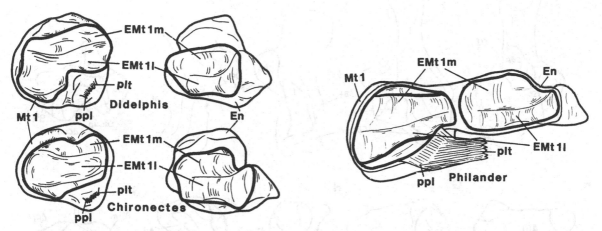

**Figure 7.7.** *Conarticular surfaces of the left EMt1J in three largely terrestrial genera of didelphids* (Chironectes *is also aquatic). The form–function of this joint is stereotyped and diagnostic of the family. This* joint is only known in the borhyaenids among fossil ameridelphians (see Fig. 7.23). For abbreviations see Table 1.1.

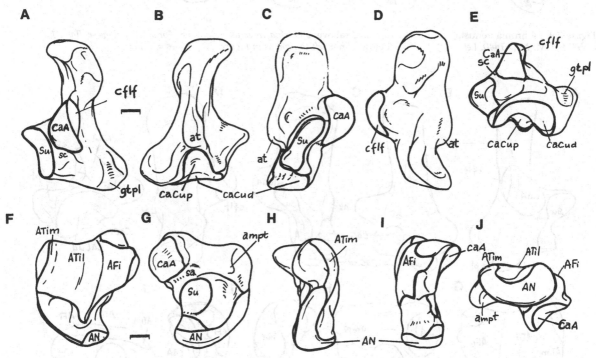

**Figure 7.8.** Caluromys philander, AMNH No 234989. Left calcaneus (above) and left astragalus (below). From left to right as in Figure 5.8. For abbreviations see Table 1.1. Scales = 1 mm.

for the quintessentially primitive marsupials, and their didactic use and fascination for students of mammalian evolution justifiably continues. But in a phylogenetic analysis aimed at understanding evolutionary history, preoccupation with what often amounts to systematic "non-statements" about which animals are "primitive" ("living fossils") and which are "advanced" can often

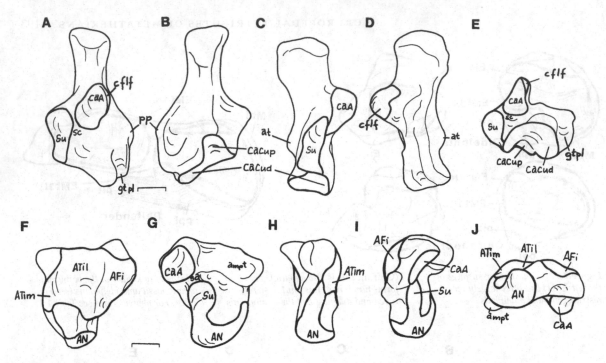

**Figure 7.9.** Glironia venusta, CNHM (DM) No. 41440. Left calcaneus (above) and left astragalus (below). From left to right as in Figure 5.8. For abbreviations see Table 1.1. Scales = 1 mm.

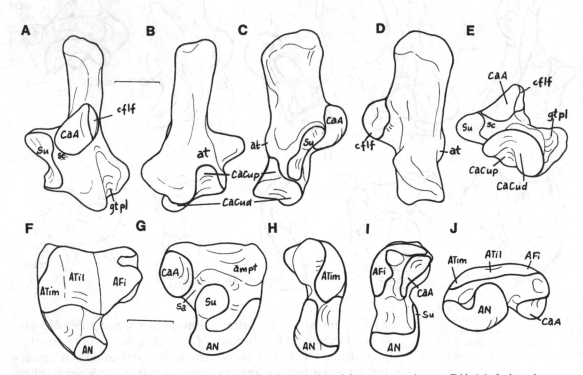

**Figure 7.10.** Marmosa sp., AMNH No. 99983. Left calcaneus (above) and left astragalus (below). From left to right as in Figure 5.8. For abbreviations see Table 1.1. Scales = 1 mm.

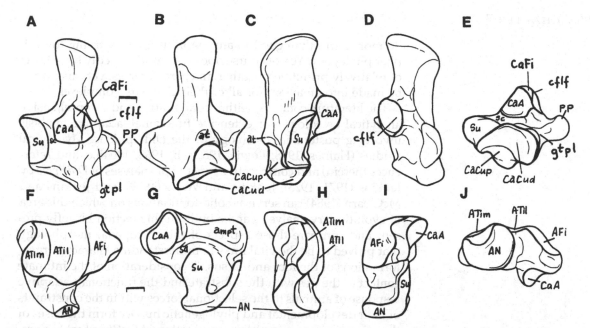

*Figure 7.11.* Didelphis patagonicus,
AMNH No. 132905. Left calcaneus
(above) *and left astragalus (below).*
*From left to right as in Figure 5.8. For*
*abbreviations see Table 1.1. Scales = 1*
*mm.*

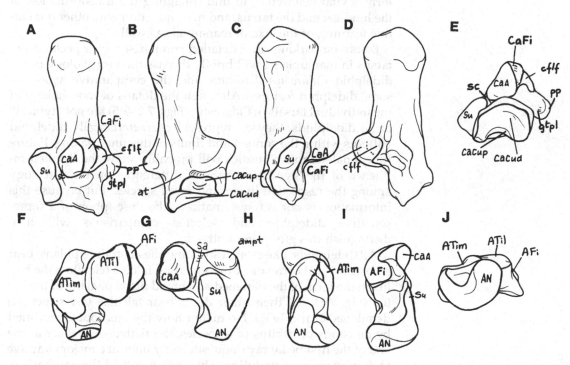

*Figure 7.12.* Chironectes minimus,
AMNH No. 148720. Left calcaneus
(above) *and left astragalus (below).*
*From left to right as in Figure 5.8. For*
*abbreviations see Table 1.1.*

do more harm than good to any rigorous attempt trying to decipher phylogeny. Yes, of course opossums undoubtedly have hosts of relatively primitive metatherian characteristics, yet this cannot be made into an axiom for all of their character complexes.

The literature contains neither comparative and comprehensive analytical osteology nor extensive functional–adaptive analysis involving postcranial osteology of the Didelphidae. Myological studies (Haines, 1958; Lewis, 1962a, b, 1963, 1964b,c; and references therein) and comparative functional analyses of *Didelphis* by Jenkins (1971, 1973), Jenkins and Weiss (1979), and Jenkins and McClearn (1984) can serve as solid foundations on which this area of evolutionary analysis can be built. In this section I briefly present and compare those aspects of the cruropedal morphology that proved to be the vital links in understanding character transformations of the foot and tarsus. A consideration of the intimate contact of the pes with the substrate and the functional–adaptive response of animals to the selectional forces within the constraints of their developmental and phylogenetic history form the bases of the recognition of a complex combination of features that are present in all living Didelphidae, and uniquely so. These features form a vital connection in understanding the transformation of the hindfeet and the tarsus, and in conjunction with other qualitative features, of the taxa of marsupials as well.

Before embarking on a detailed comparison of aspects of the tarsus in marsupials I will briefly discuss the morphology in the didelphid *Caluromys derbianus* and the comparative aspect of some didelphid features. Although the details of morphology of the individual tarsals of *Caluromys* (Figs. 7.2–7.5) are not "typical" of all didelphids, they do represent characteristically didelphid patterns within the context and limits of this undertaking. Future functional–adaptive studies will probably show the detailed influence of the significant habitus differences on the osteology among the respective living and fossil species. But because this information is not yet systematized, the description of a representative didelphid and selected comparisons will help distinguish this group from others.

All didelphids, like caenolestids and the australidelphians, bear an osteological correlate of their missing falcula (claw) on the hallux in the form of the flattened and broad distal phalanx of the big toe (Fig. 7.1). All their other digits bear falculae (for structural details see Clark, 1936). No matter how the hindfoot is modified by demands of habitus of a species, the flattened phalanx at the end of the first pedal ray unquestionably hints at a major adaptive shift in marsupial evolution. This shift involved the emphasis of hallucial grasping to the point where the falcula of the big toe became reduced and eventually disappeared. Such is a tantalizing hint from the living species, a phyletic marker with its far reaching phylogenetic significance, without any clue as yet from the described fossil record.

The prehallux (see Fig 7.1) is present in didelphids, a derivative of the early segmentation of the cartilaginous blastema of the tibiale in all amniotes that have it. It does not form a synovial contact like those found between the other tarsals. Its presence in any mammal probably does not signify the kind of articular and force-transmitting relationships that exist between other tarsals or carpals.

The second ray of the foot is always slightly more proximally articulated with the tarsus than the first and third rays, specifically with the mesocuneiform. This configuration may well be a primitive therian or metatherian pattern preserved in didelphids. The fourth and fifth rays articulate with the cuboid, slightly more proximally than the EcMt3 contact.

DISTAL CRUS (SEE *METACHIRUS* IN FIG. 4.4). The presence of a meniscus in the form of a semicircular and ligamentous ring between the fibula and the proximal tarsus is unique to marsupials, as far as this is known (but see earlier discussion of the Multituberculata). Judged by its great similarity to the relatively unmodified feet of most australidelphians, the didelphid condition is probably primitive, considering what we know of the meniscus in the living metatherians. As discussed in greater detail above, one or two sesamoids, the lunulae of the meniscus, brace the UAJ posteriorly and provide for smooth articular translation between the conarticular surfaces. The origins of these UAJ mechanics may well be related to selection for UAJ mobility in an arboreal pretherian, therian or metatherian lineage that had a tricontact UAJ (Szalay, 1984). Because of its close connection with fibulo-astragalar function, the meniscus is greatly diminished or lost in the stem lineage of such terrestrial groups as the macropodids or bandicoots (see details below). The articulation between the distal tibia and fibula is interestingly different from eutherians. The conjunctly rotated crural elements accommodate each other in a quasi tongue-in groove, curved articulation. The elongated concave tibial TiFi facet guides and slightly constrains the elongated and convex fibular TiFi facet.

The distal articular surface of the tibia shows the characteristic posterior elongation of the ATim facet, labeled as the ATip facet on the various figures. Judged from the slightly forward extending astragalar ATim facet in pediomyids, this elongated tibial ATim facet is probably primitive in the Metatheria. If the didelphid condition of a long posterior extension of the medial articulation of the tibial ATip facet (see Fig. 4.4) is a primitive metatherian condition, then this has significant bearing on the mechanics and consequent direction of phyletic change in marsupials in that area. The arcuate path of the tibial **ATip** facet strongly suggests that the conjunct medial rotation of the tibia during dorsiflexion and its reverse during plantarflexion were the major accommodators of the foot during inversion and eversion (Jenkins & McClearn, 1984; Szalay, 1984). During dorsiflexion the opossum foot is abducted and slightly everted, and during plantarflexion the reverse, adduction and inversion, take place. While supination and pronation of the distal tarsals and the rays of the foot are significant, the **UAJ** is the major area of accommodation of these animals (and possibly in the metatherian ancestry) for substrate inclination and irregularities. The extended tibial **ATip** facet may represent the apomorphous adaptation of some pretherians, first therians, or protometatherians, for transferring loads when climbing large vertical trunks. The **UAJ** adaptations, however, may not be closely linked with the grasping modification of the didelphids.

The slightly stepped separation between the tibial **ATil** and **ATim** surfaces is distinct, unlike the more derived condition in the putative ancestral australidelphian, a point discussed below. The didelphid pattern, regardless of the habitus of a particular species, still mirrors the condition of the ancestral tibial medial malleolus, which is relatively sharply angled against the astragalus. On the tibia itself the medial malleolus is not very prominent in arborealists when compared with the conformations of either the borhyaenids, bandicoots, or eutherians. The articular surface of the tibial **ATim** facet is nearly parallel to the **ATil** facet.

ASTRAGALUS (FIGS. 7.2 & 7.8).    The derived metatherian aspect of the didelphid **UAJ**, which is decidedly not diagnostic for the Metatheria, is well reflected on this bone. Although the **AFi** facet is broad, indicating the considerable superposition of the astragalus on the calcaneus and the exclusive weight bearing contact with the fibula, the **ATim** facet is relatively small. The latter forms a very obtuse angle with the **ATil** facet, and in general mirrors the relatively less restricted medial crural contact in the ancestral didelphid compared either to the caenolestids or to any of the fossil metatherian groups described from the Americas. This combination of **UAJ** features stand in sharp contrast to known stagodontids and pediomyids. Nevertheless, didelphids retain the distinctive ridge formed by the **ATim** and **ATil** facets, a clear heritage of the common ameridelphian metatherian ancestry.

The contact with the calcaneus in the **LAJ** is made through the separate synapsid **CaA** and **Su** facets, the latter distinct from the **AN** facet. All the known Cretaceous and Paleogene astragali possess confluent **Su** and **AN** facets, as I have demonstrated. Not only is the didelphid **Su** facet separated from the **AN** one, but on the lateral side of the plantar surface it also approaches the **CaA** facet.

The astragalar medial plantar tuberosity (**ampt**) is located well medial and only slightly plantar to the **Su** facet, and is bound by a thick band of ligaments to the posterior edge of calcaneal sustentaculum. This configuration is in diagnostic contrast to the the usually relatively much larger, more robust, and deeper **ampt** of the more ancient non-didelphid taxa in which usually (but see some of the exceptions above) the **ampt** hangs over the calcaneal sustentaculum posteriorly, presumably in addition binding it with the homologous ligament. This changed condition represents a decisive shift in didelphids towards a more mobile **LAJ**.

The **ANJ** is aligned somewhat transversely, primarily facilitating supination and pronation as forced by the substrate. In some terrestrial genera discussed below this function is reoriented into additional flexion and extension of the forefoot for the biological role of accommodating to conditions of relatively rapid scampering on a terrestrial substrate.

As a rule, didelphids do not retain an astragalar canal, although it is regularly present (or its trace is clear) in the Mesozoic and Paleogene marsupial material described. This absence is particularly significant in light of the fact that flexion–extension of the **UAJ** of didelphids is far more limited than in the other ameridelphians described.

CALCANEUS (FIGS. 7.2 & 7.8). The peroneal process is exceptionally large in this genus, comparable in its relative dimensions to many Paleogene taxa, with the groove for the tendon of the peroneus longus well offset laterally. The anterior tubercle is well developed and proximally retracted because of the spread of the diagnostic didelphid **CaCup** facet. The calcaneus is excluded from the **UAJ** – there is no **CaFi** facet. The narrow shiny band next to the **CaA** facet is for the calcaneofibular ligament (**cflf**). The semiconical **CaA** facet is transversely wide yet short, and it appears to permit relatively limited range of additional inversion in the already habitually inverted foot, as the calcaneus rotated forward and medially, and back and laterally. The **Su** facet, particularly its distal half, faces medially. The **CCJ** articulation is highly diagnostic in didelphids, and this undoubtedly derived ameridelphian metatherian pattern is reflected by the subdivision of the **CaCu** facet into the **CaCud** facet, which is crescent shaped, and the **CaCup** facet, which is proximally projecting, conical, and tongue shaped. As noted below, the broad plantar **CaCup** facet fully accommodates the projecting conical facet of the cuboid.

CUBOID (FIG. 7.3). The dominating feature of the didelphid cuboid, so clearly reflected on the calcaneus, is its projecting and conial articulation with the latter (see also Fig. 7.6B). This facet is easily, and for phylogenetic purposes usefully, divisible into the proximal and distal components (**CaCup** and **CaCud** facets). While this pivot facilitates the rotation of the forefoot on an axis nearly parallel to the foot and going through both the calcaneus and the cuboid, the structure probably also resists forces generated by the powerful grasp. From the anterior tubercle on the plantar side of the calcaneus to the entire plantar surface of the cuboid projection there is a massive bundle of ligaments that resists the bending moments that would tend to open up the **CCJ** plantarly. Rapid mediolateral forefoot accommodation and powerful grasping, together with possibly some leaping generated landing loads, may have been the functional requirements for the bioroles that were originally selected for and determined the ancestral didelphid solution. The groove for the tendon of the peroneus longus is prominent; it deeply notches the bone medially. Distal to this groove is the **CuEc** facet. The articulation with **Mt4** and **Mt5** is stepped.

NAVICULAR (FIG. 7.4). This bone is cradle shaped. The fact that its **AN** facet wraps extensively around the large head of the didelphid astragalus may be a testimony to the relative importance, both in terms of mobility–stability and also loading, of the medial part of the foot – probably tied intimately to the forces generated by the powerful didelphid grasp.

ENTOCUNEIFORM (FIGS. 7.5 & 7.7). On its slightly medial and distal end this bone fully reflects the diagnostic articular relationship with **Mt1**. The **EMt1J**, joint to be discussed fully later, is sellar but with a distinct ridge within it. The ridge is the entocuneiform **EMt1p** facet which articulates with the troughlike **Mt1** conarticular facet (see Fig. 7.5C, D). This pattern appears universally in didelphids, irrespective of the habitus (Fig. 7.7). The transformation of this complexly articulating bone with its surrounding units is one of the many unmistakable changes which occurred from a didelphid structural ancestry in the evolution of the protogondwanadelphian.

The lateral side shows the contiguous **EnMc** and **EMt2** facets. The **NEn** facet faces medially, proximal to the prominent trough, but not a real facet for the **EnPh** contact. The prehallux does not form a synovial articulation with the entocuneiform, in spite of its close association with it. A small strip of the **EnMc** facet is also visible on the medial side of the bone.

MESOCUNEIFORM (FIG. 7.5E, F). As expected in a bone sandwiched in the middle of the foot, the respective sides of the rectangular bone are each dominated by facets for the **Mt2** (**MMt2** facet), navicular (**NMc** facet), entocuneiform (**EnMc** facet), and ectocuneiform (**EcMc** facet).

ECTOCUNEIFORM (FIG. 7.5G, H). The lateral side is in contact with the cuboid through an elongated ribbonlike **CuEc** facet, whereas the medial side contacts the mesocuneiform via a triangular **EcMc** facet on the proximodorsal surface. The **MMt2** facet is distal on the medial side.

**Caenolestidae (Figs. 7.13, 7.14, 7.15B, & 7.23A, B, C)**

This small family, a "diprotodont" group inasmuch as they possess enlarged and procumbent central incisors, is a relic of a once flourishing radiation (known by at least up to ten recognized fossil genera; see especially Marshall, 1980a). Yet the living taxa – the terrestrial *Caenolestes*, *Orolestes*, and *Rhyncholestes* – display such an intriguing combination of osteological (Osgood, 1921; Gregory, 1922), ecological, and chromosomal (Hayman et al., 1971) features that they may eventually help our understanding of some of the adaptive attributes of many South American fossil marsupials.

Based on the available literature, Marshall (1980a, p. 21) has summarized the caenolestid habitat preferences as follows:

> Living caenolestines prefer densely vegetated, cold, and wet forest habitats, ranging from sea level to elevations exceeding 14,000 ft. They prefer moist, often moss-covered slopes and ledges, well protected from cold winds, mist, and rain. In the high cold paramos of Colombia, Ecuador, and Peru they are typically found in the scrub adjacent to meadows.

In spite of their obligate terrestrial, semicursorial, and galloping locomotor style (yet they can climb well), the hallux of these animals occasionally bears only a tiny vestige, a mere flake, of a "nail." This certainly suggests that the ancestry of these terrestrialists was a marsupial from a group that eliminated the claw on the hallux and perhaps made hallucial adaptations similar to the Didelphidae. Their tail is nonprehensile in contrast to the common didelphid pattern.

The caenolestid astragalus has a narrow **AFi** facet, unlike the derived broad didelphid condition. Similarly, the angle between the **ATim** facet and the medially long **ATil** facet is sharp, as in many of the Itaboraí metatherians but not in didelphids. The proximally ribbonlike and distally broad **Su** facet is confluent with the **AN** facet. The latter displays the characteristic orientation described repeatedly before, convergently similar to *Metachirus*, a superior-plantar curvature that primarily facilitates flexion and extension of the distal tarsus and the foot. As expected in such a primitive metatherian astragalus, the tuberosity for the posterior astragalocalcaneal ligament (**ampt**) is very large. In spite of its elongation distal to the **CaA** facet, the caenolestid calcaneus has a large peroneal process, a well-defined **CaFi** facet, and the primitively elongated metatherian sustentacular facet. There is not the slightest suggestion that the **CCJ** was ever similar to that of the diagnostic didelphid condition.

**Figure 7.13.** *Comparison of left foot skeleton of extant caenolestid* Lestoros inca *and partial left foot skeleton of the Santacrucian Miocene borhyaenid* Sipalocyon gracilis. *Scales = 1 cm.*

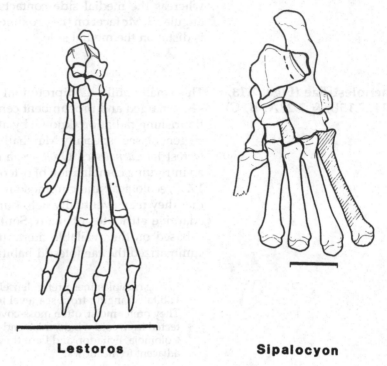

**Lestoros**                    **Sipalocyon**

**Figure 7.14.** Caenolestes *sp., AMNH No. 62915. Left calcaneus (above) and left astragalus (below). From left to right as in Figure 5.8. For abbreviations see Table 1.1. Scales = 1 mm.*

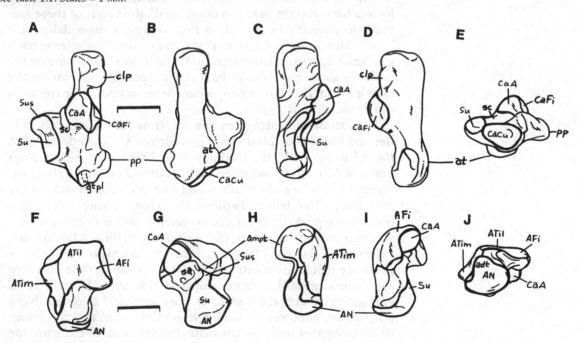

**Figure 7.15.** *Distal views of the right crus in* Trichosurus *(A), personal collection from Victoria, Australia;* Caenolestes *(B), AMNH No. 47178; and* Metachirus *(C), MNRJ No. 4000. Tibia is to the right; fibula is to the left. Anterior edge is at top of figures. For abbreviations see Table 1.1. Scales = 1 mm.*

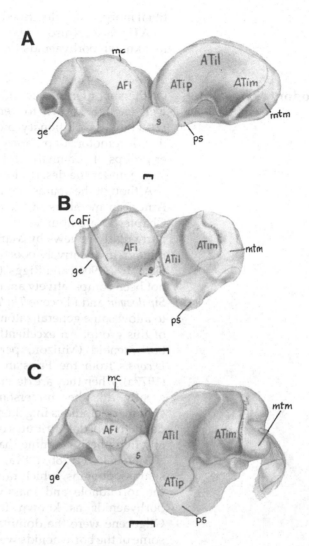

The distal halves of the caenolestid tibia and fibula (Fig. 7.15B) are closely bound together, yet the articular surfaces of the **UAJ** are distinct and separate. The presence of a posterior shelf (**ps**) *without* a significant tibial posterior extension facet (**ATil** and **ATip** facets; see Fig. 7.15C for the more primitive didelphid condition) is a potentially important fact about the evolution of these marsupials. As I have demonstrated, this facet mirrors the extreme of conjunct rotation of the tibia during foot inversion and dorsiflexion at the **UAJ**. Although without this posterior extension of the facet, the cursorial caenolestids appear to have, significantly, a bony area which may be a remnant of this eliminated articulation. The absence of a falcula from the hallux of caenolestids, as I have shown, also implies a grasping ancestry and therefore highly arboreal adaption. The primitive metatherian nature of the distal tibial morphology described for the didelphids, the significance of an **ATip** facet, is also supported by the condition of this area in the few known borhyaenids (Figs. 7.16 & 7.17).

**Sparassodonta (Figs. 7.16–7.25)**

Of the suborder Sudameridelphia only the sparassodontans are well enough known to venture a preliminary outline of their postcranial evolutionary morphology as it may have related to their locomotor adaptations. Whatever brief comments on the other groups of sudameridelphians were possible, have been made above under the description of the Itaboraí tarsals.

Although the sparassodontans (the name restricted here to the American members of the original concept; see classification in Chapter 2) are abundantly represented in the fossil record (see particularly reviews by Marshall, 1978b; 1981a), their postcranial remains are relatively poorly known; beyond Dollo's (1900, 1906), Sinclair's (1906), and Riggs' (1934) accounts, these specimens have not been comparatively analyzed. Only the Santacrucian Miocene *Sipalocyon* and Pliocene *Thylacosmilus* are sufficiently well studied to allow some generalizatons to be made about the tarsal features of this group. An excellently preserved skeleton of a Tiupampa borhyaenoid (Muizon, personal communication) and that of *Lycopsis* from the Friasian late Miocene reported by Marshall (1977a), when they are described and compared in detail, will add greatly to the understanding of the Borhyaenidae and sudameridelphians in general. As discussed in Chapter 8, neither dental nor pedal attributes of known borhyaenids lend support to any hypothesis claiming that they are derived from Didelphidae.

Marshall's (1978b, 1981a) analyses made it clear that among the sparassodontans, which range from small weasel-sized forms to the formidable and massive saber-toothed thylacosmilids, the borhyaenids as known from the Paleocene until the early Oligocene were the dominant carnivores. It is not unlikely that some of the borhyaenids were omnivorous. At that time, the large lineages at least, began their decline probably due to the success of the predatory ground birds, the "phororhacoids." This decline persisted into the Pliocene when the last sparassodontans disappeared.

Marshall (1978b, 1981a; see also Villarroel & Marshall, 1983) suggests that the small- to medium-sized Hathlyacyninae, which ranged from the Paleocene until the end of the Pliocene, have had a history of competitive interaction with carnivorously adapted didelphids. This well-argued hypothesis will undoubtedly receive rigorous tests both from the taxonomic refinements determining which groups are or are not didelphids, and from future functional–adaptive studies. Although much of the cranial and dental material for all sparassodontans has been carefully

*Figure 7.16.* Sipalocyon gracilis, YPM No. PU15702, Santacrucian Miocene. Distal (above), anterior (middle), and posterior (below) views of distal right crus. For abbreviations see Table 1.1. Scale = 1 mm.

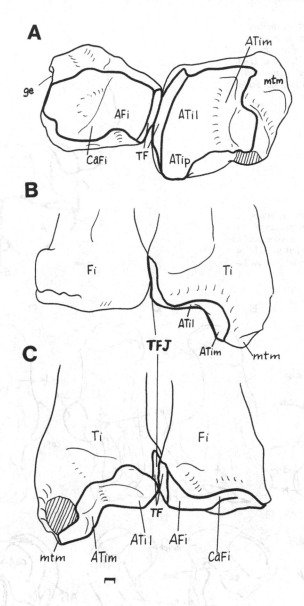

described, there are no comparably detailed documentations and functional–adaptive assessments of the postcranial evidence. In addition to the potential value of such studies for establishing the history of the group, the analysis of the postcranial remains is necessary to test any hypothesis relating to substrate preferences. For animals in the smaller size ranges, which was arboreal, semi-arboreal, or terrestrial cannot be fully ascertained from the usual dental and cranial evidence. As yet much postcranial material has not been analyzed in order to answer such important questions

*Figure 7.17.* Sipalocyon gracilis, YPM No. PU15702, Santacrucian Miocene. Anterior view of loosely articulated (in order to show **LAJ** and **UAJ** facets) left distal crus, calcaneus, and astragalus. For abbreviations see Table 1.1. Scale = 1 mm.

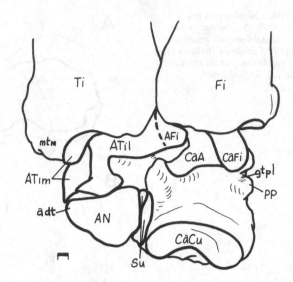

*Figure 7.18.* Sipalocyon gracilis, YPM No. PU15154, Santacrucian Miocene. Left calcaneus (above) and astragalus (below). From left to right as in Figure 5.8. For abbreviations see Table 1.1. Scales = 1 mm.

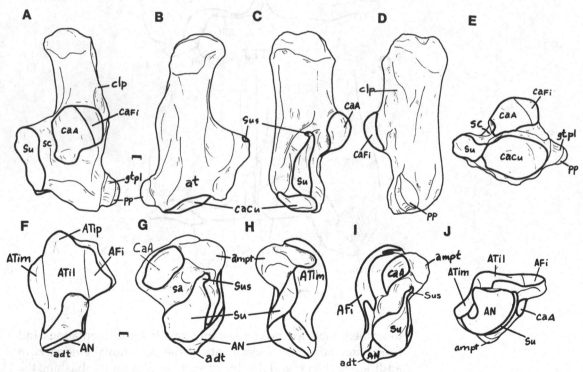

about these animals. As far we know, none appears to have developed the long-legged cursorial specialization of canids, a habitus approached by the dayuromorphian thylacinids. Consequently, the evaluation of the complex web of interactions of South American metatherian and eutherian carnivores from vari-

*Figure 7.19.* Sipalocyon gracilis, YPM No. PU15154, Santacrucian Miocene. Dorsal view of articulated left calcaneus, astragalus, navicular, and cuboid (A), and plantar view of left calcaneus and cuboid (B). For abbreviations see Table 1.1. Scale = 1 mm.

*Figure 7.20.* Sipalocyon gracilis, YPM No. PU15154, Santacrucian Miocene. Dorsal (A), lateral (B), and medial (C) views of left navicular. For abbreviations see Table 1.1. Scale = 1 mm.

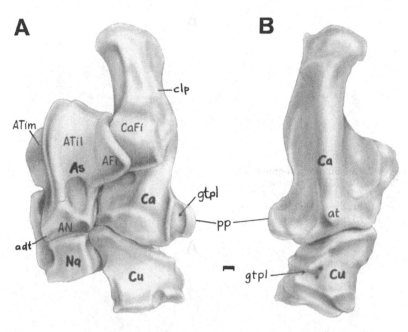

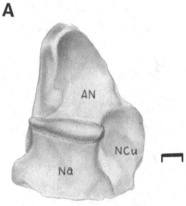

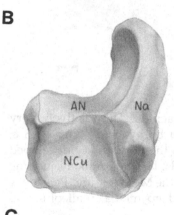

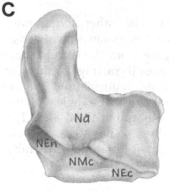

ous lineages remains difficult until this work is accomplished. The pedal remains of the Hathlyacyninae described here do not appear to support the hypothesis that these forms were partly arboreal, although undoubtedly some of the dental taxa described were probably capable climbers, as are some of the members of the Dasyuridae. Marshall believes that the large Prothylacyninae were terrestrial and omnivorous. The Proborhyaeninae, which became extinct in the early Oligocene, had reached skull sizes of up to 2 feet, and therefore it is safe to assume that they were terrestrial. After the disappearance of these latter giants, the Borhyaeninae also became very large. The saber-toothed sparassodontans – the thylacosmilids – are the last of the group to disappear, perhaps as a direct result of the appearance of the felid saber-tooths.

Unlike the caenolestid **TFJ** discussed above, the one known for *Sipalocyon* is somewhat less derived (Fig. 7.16). It preserves a considerably mobile articulation between the two bones (see also the condition described for didelphids). The **ATip** facet, a posterior continuation of the **ATil** facet, is strongly suggestive that extremes of dorsiflexion required or retained close tibioastragalar conarticular contact dorsally as well as medially. The distinct possibility exists that the posteriorly attenuated tibial **ATip** facet, although primitive for the known ameridelphian metatherians, is related to peak loading during a dorsiflexed and abducted position of the foot in borhyaenids. This *may* well represent the primitive climbing adaptation in ameridelphian metatherians, still possibly pre-

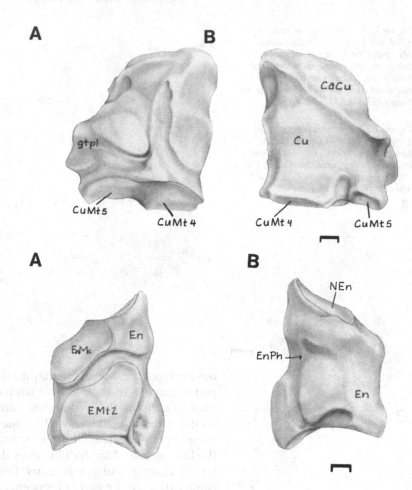

**Figure 7.21.** Sipalocyon gracilis, YPM No. PU15154, Santacrucian Miocene. Plantar (A) and dorsal (B) views of left cuboid. For abbreviations see Table 1.1. Scale = 1 mm.

**Figure 7.22.** Sipalocyon gracilis, YPM No. PU15154, Santacrucian Miocene. Lateral (A) and medial (B) views of left entocuneiform. For abbreviations see Table 1.1. Scale = 1 mm.

served in many of the small borhyaenids. The relatively very prominent medial tibial malleolus of borhyaenids (Figs. 7.16 & 7.17) stands in interesting contrast to the very poorly developed one of the largely terrestrial dasyurids. It is not unlikely, judging by the astragalar conarticular facet of the Mesozoic and Paleogene fossil marsupials, that a large and deep medial malleolus was primitive in the Metatheria.

Unlike the astragalus of didelphids or other living non-caenolestid marsupials, the astragalus of *Sipalocyon* is nearly adjacent to the calcaneus (Fig. 7.17). Judging from the earliest marsupials, therefore, the lateral orientation of the astragalar **Su** facet (Fig. 7.18) is a retention of the primitive conformation. The astragalar **ATim** facet is extensive and makes a sharp angle with the **ATil** facet, as expected from the corresponding tibial area. The **ATil** facet, extremely elongated medially, also has a posterior extension that can be appropriately called, at least in borhyaenids,

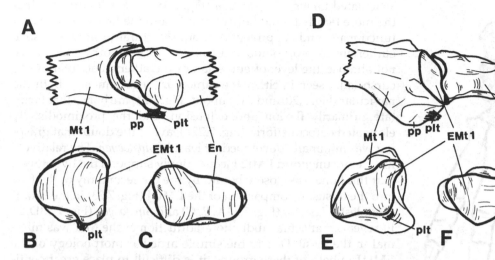

*Figure 7.23. Comparison of the left EMt1J of Lestoros inca, AMNH No. 42685 (A, B, C) and Sipalocyon gracilis, YPM No. PU15154, Santacrucian Miocene (D, E, F). Dorsal view of articulated entocuneiforms and proximal end of first metatarsals (A, D), proximal view of first metatarsals (B, E), and distal view of entocuneiforms (C, F). For abbreviations see Table 1.1.*

the astragalar **ATip** facet. The **AFi** facet is small as found in the more ancient marsupials. The **Su** facet has the ribbonlike extension leading above the **ampt,** a condition repeatedly noted for the Paleogene forms. This facet is helical and, given the orientation of the **CaA** facets, the **LAJ** probably facilitated parasagittal movements between the two proximal tarsals. The astragalar **AN** facet possesses the previously described extreme development at its medial and plantar extremity, the **adt,** which occurs in a number of taxa. The net effect of these articular contacts was primarily a flexion–extension in the **UAJ** and **TTJ.** The **adt,** therefore, reflects the fact that astragalonavicular contact was the most extensive during dorsiflexion of the distal part of the foot and distal tarsals on the proximal tarsals. The **adt** is the effect of the repeatedly attained dorsoplantar orientation of the curvature of the **AN** facet, which, as discussed above, is related to primarily flexion and extension, and minimally to pronation and supination. Its usefulness, in spite of its visual prominence, is virtually nil for phylogenetics.

The calcaneus has only a small peroneal process. The fibular contact is extensive, and this feature, together with what I judge to be the nearly vertical orientation of the sustentacular facets of the proximal tarsals, strongly suggests that the tricontact nature of the **UAJ** is the ancestral therian condition preserved in borhyaenids. The mediolaterally elongated **CaCu** facet shows no sign of derivation from the didelphid condition. In addition to the mechanics of the **UAJ** and **ANJ,** the amorphously rounded end of the tuber strongly suggests, as previously discussed, that the known borhyaenids were primarily terrestrial scamperers and bounders.

A comparison of a didelphid tarsus (Fig. 7.6) with the partially

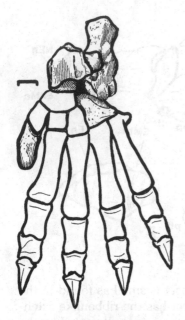

*Figure 7.24.* Thylacosmilus atrox, *FMNH (DVP) P14344, Pliocene of Argentina. Reconstructed skeleton of left pes (slightly modified after Riggs, 1934) in dorsal view, showing extent to which foot skeleton is known (note hallux reduced to Mt1). Scale = 1 cm.*

## Argyrolagidae (Figs. 7.26–7.30, & 8.10)

articulated tarsus of *Sipalocyon* (Fig. 7.19) makes it apparent that the more tightly bound borhyaenid foot in the few known taxa is functionally and adaptively (but not structurally) similar to such eutherians as procyonids and some viverrids. Borhyaenids have not attained the level of eutherian or metatherian adaptations to cursoriality seen in either the canids or the peramelids. Both the navicular (Fig. 7.20) and the cuboid (Fig. 7.21) are bones that facilitate primarily flexion and extension, and the proximodistally elongated entocuneiform (Fig. 7.22) leaves little doubt that grasping was minimally developed, at least in *Sipalocyon*. The relatively huge entocuneiform **EMt2** facet strongly supports the hypothesis that this bone was closely bound by the the second ray of the foot into the tarsus. A comparison of the **EMt1J** (Fig. 7.23) in the extant caenolestid *Lestoros* (Fig. 7.23A, B, C) and *Sipalocyon* (Fig. 7.23D, E, F) reveals that actual abduction–adduction of the **Mt1** was minimal in the fossil. Due to the simple articular morphology of the **EMt1J** in both of these groups, it is difficult to place great confidence in the possibility that the similarity of this complex is homologously derived. In either case, in addition to the articular similarity, the elongation of the entocuneiform in the two families is quite similar.

Although the much larger and younger Pliocene *Thylacosmilus* is less well known (Figs. 7.24 & 7.25), the proximal tarsal morphology of the large marsupial saber-tooth is not very different from that of its more ancient borhyaenid relative. What may have been a minimal (and therefore nearly impossible to detect) contact between the astragalus and the cuboid in earlier borhyaenids is a fully developed and well-defined **ACu** facet in *Thylacosmilus*.

The earliest evidence for the existence of this now-extinct family of marsupials comes from the Colhuehuepian Miocene of Argentina, but the material on which most of our knowledge is based comes from the Montehermosan Pliocene of the same region (Simpson, 1970a, b; see also Villaeroel & Marshall, 1988).

In the hindlimb-dominated saltatorial argyrolagids, the distal half of the crus is completely fused, and the resulting articulation of the **UAJ** mirrors the diagnostic attributes of the proximal tarsal conditions. Contact of the crus and its load-bearing role in relation to the tarsals are nearly equally divided between the fibula and tibia. In addition to the enclosure of the astragalus in a mortise to which both elements of the crus contribute, the fibula has an extraordinarily broadly curved band of contact with the calcaneus. The tibial **ATil** and **Atim** facets are separated from one another and the former only preserves a hint of a posterior extension.

Unfortunately, the specimen of a primitive Colhuehuepian argyrolagid (courtesy of R. Pascual) is only known postcranially from a calcaneus (Fig. 7.28A–C), so the comparisons that I make

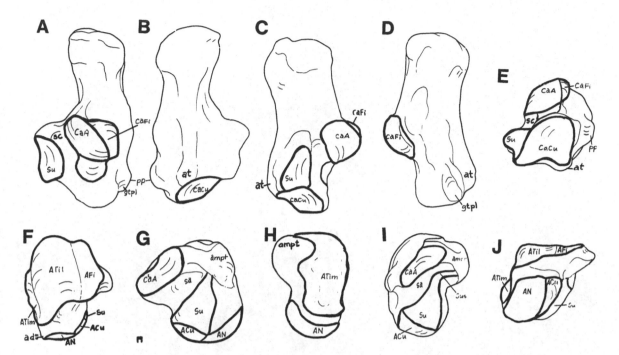

**Figure 7.25.** Thylacosmilus atrox, FMNH (DVP) P14344, Pliocene of Ar- gentina. Left calcaneus (above) and as- tragalus (below). From left to right as in Figure 5.8. For abbreviations see Ta- ble 1.1. Scales = 1 mm.

below are limited. The astragalus of *Argyrolagus* is dominated by the mechanics of the large and slightly trochleated **ATil** facet and the transversely cylindrical **AN** facet. The exclusively flexion–extension movements limited to the **UAJ** and the **TTJ** are severely constrained mediolaterally, and are facilitated dorsoplantarly by the **ATim** and **AFi** facets. These are obvious from the illustrations. Yet in spite of the dominant influence of mechanics related to a saltatorial biorole, the small **AFi** facet is distinctly unlike that of any didelphid. One of the apomorphic attributes of the pro-todidelphid, as noted, was a widening of the **AFi** and the probably concurrent elimination of the calcaneofibular contact. The **Su** facet of the argyrolagid has the ribbonlike extension of the ambigu-ously primitive (and also caenolestid and borhyaenid-like) condi-tion which runs above the **ampt**. This facet also possesses a transverse extension proximally, very similar to facets in some macropodid taxa (see below).

Although restricted only to two taxa, the calcaneal record is extraordinarily interesting. The more ancient Miocene form (Fig. 7.28A–C) is exactly intermediate in morphology between the Pliocene *Argyrolagus* and between the pattern predominant in the Itaboraí radiation of marsupials, with the exception of derived Itaboraí didelphids, polydolopids, and *Itaboraí Metatherian Group*

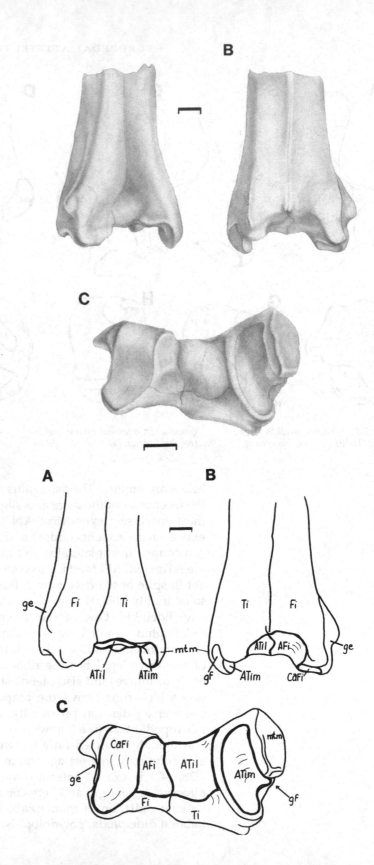

**Figure 7.26.** Top two rows: Argyrolagus scagliai, MMMP No. 785-S, Montehermosan Pliocene of Argentina. Distal end of right crus. Anterior (A), posterior (B), and distal (C) views. Bottom two rows: labeled outline drawings of specimens shown in top two rows. For abbreviations see Table 1.1. Scales = 1 mm.

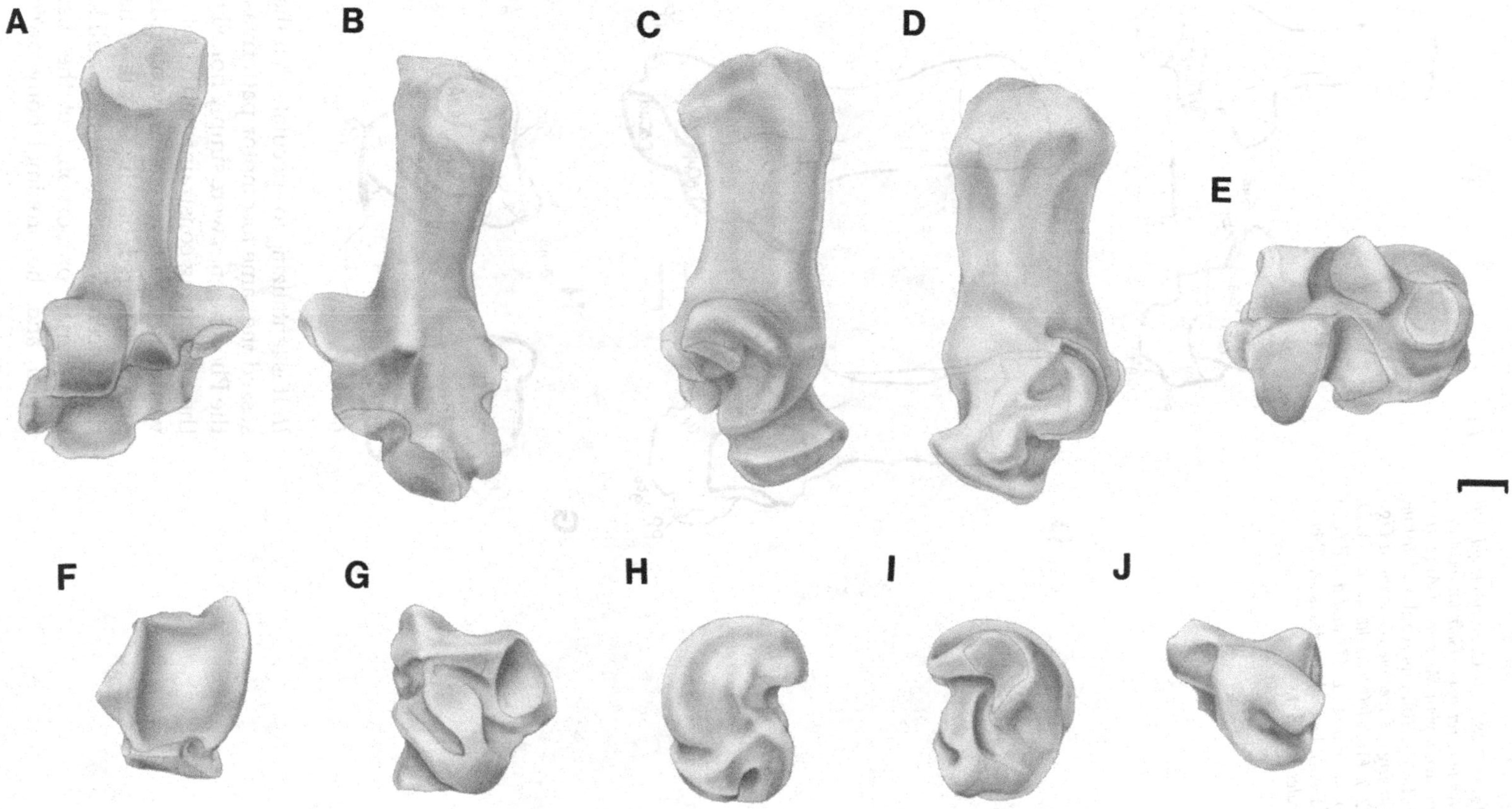

**Figure 7.27.** *Argyrolagus scagliai, MMMP No. 785-S. Montehermosan Pliocene of Argentina. Right calcaneus (above) and astragalus (below). From left to right as in Figure 5.8. Scale = 1 mm.*

*Figure 7.28. A–C. Argyrolagid, left
calcaneus reversed, Gaiman locality,
Colhuehuapian Miocene of Argentina.
D–I. Argyrolagus scagliai. (outline
drawings of specimens shown in Fig.
7.27). From left to right dorsal (A, D,
G), plantar (B, E, H), and distal (C, F,
I) views. For abbreviations see Table 1.1.
Scales = 1 mm.*

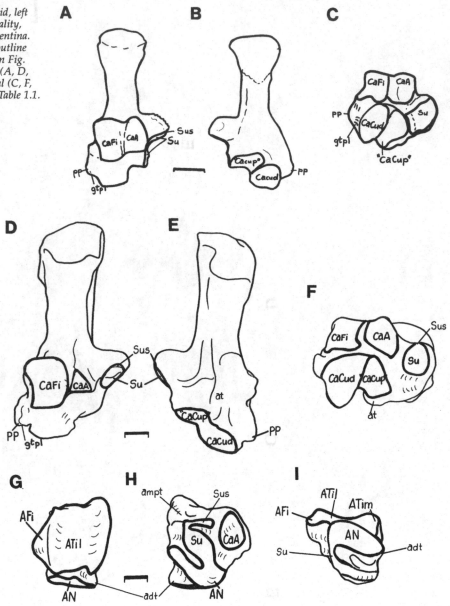

*II.* It is tantalizing to speculate that the Miocene argyrolagids possessed the same locomotor pattern as the Pliocene ones, but that
the Pliocene ones were simply honed to the extremes permitted by
the underlying constraints of the total available structural framework. In the Oligocene argyrolagids the **CaFi** facet is already
greatly enlarged, but the **CaA** facet has not become as small as in
the Pliocene form. The **Su** facet still keeps its elongated primitive
metatherian proportion, yet the distal portion is already atrophied, and the proximal bony projection indicates a mechani-

*Figure 7.29.* Argyrolagus scagliai, *MMMP No. 785-S, Montehermosan Pliocene of Argentina. Right navicular, in dorsal (A), lateral (B), and medial (C) views. For abbreviations see Table 1.1. Scale = 1 mm.*

*Figure 7.30.* Argyrolagus scagliai, *MMMP No. 785-S, Montehermosan Pliocene of Argentina. Left cuboid. Medial (A), plantar (B), dorsal (C), and proximal (D) views. For abbreviations see Table 1.1. Scale = 1 mm.*

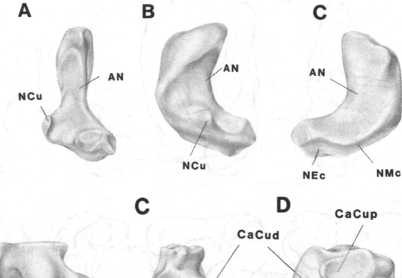

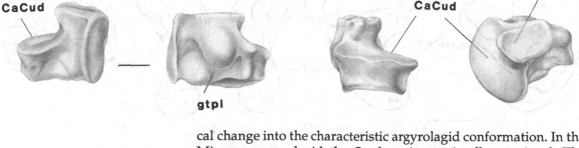

cal change into the characteristic argyrolagid conformation. In the Miocene argyrolagid the **Su** facet is proximally restricted. The **"CaCup"** (in quotes to distinguish it from the nonhomologous proximal development characterizing the Didelphidae) is already proximally retreated from the more distal facet, resulting in a "stepped," highly stable **CCJ** arrangement, closely paralleling the structure found in macropodids. In the Pliocene form these two facets, while no less "stepped," are clearly continuous.

The navicular of *Argyrolagus* (Fig. 7.29) is a crescent-shaped bone which supports the astragalus distally and posteriorly (plantar side). It is relatively far more slender and proximodistally less deep than its homologue in either known borhyaenids or caenolestids, or in the far less closely related dasyurid hopper *Antechinomys*, for that matter. The cuboid (Fig. 7.30), as expected, mirrors the diagnostic separation of the **"CaCup"** and **CaCud** facets. The larger cuboidal **CaCud** facet compared to its articulating equivalent on the calcaneus strongly indicates that these animals were capable of considerable distal foot adjustment – more specifically, rotation on a long axis of the foot and crus at the **TTJ**.

## Microbiotheriidae (Figs. 7.31, 7.50, 8.17 & 8.18)

The Valdivian temperate rain forests of south–central Chile and areas of adjacent Argentina are the home of the only living microbiotherian, *Dromiciops australis*. This easily climbing small marsupial has a prehensile tail, like didelphids, in addition to its

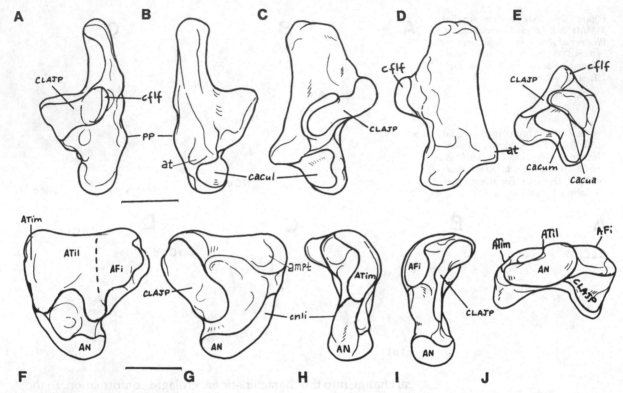

***Figure 7.31.*** Dromiciops australis, AMNH No. 97746. *Left calcaneus* (above) *and astragalus* (below). *From left to right as in Figure 5.8. For abbre-* *viations see Table 1.1. Scales = 1 mm.*

superbly adapted grasping foot. Kelt and Martinez (1989) report that when disturbed these marsupials show the same open-mouthed, undoubtedly archaic, threat display familiar in opossums. Jimenez and Rageot (1979) and Mann (1978) note that *Dromiciops* constructs spherical nests 1–2 meters above ground from plant materials that resemble those of birds. These same authors report litter sizes of three to four young.

Mann's (1955) studies of *Dromiciops* are the most comprehensive of the pioneering studies done on the morphology and physiology of this genus. I will merely list a few of the characters noted by him: (1) dense hair cover on both the dorsal and ventral surfaces of the tail; (2) small hairy ear lobes; (3) nipples never exceeding four in number, in a small pouch; and (4) third median vagina in a small proportion of multiparous females. Mann made the interesting suggestion that the enlarged bulla of the animal may be a compensation for the reduction of its ear lobes. The latter condition is undoubtedly the result of the cold and humid climate of southern Chillean forests. Mann commented that the animal is specialized for climbing large trunks, a behavior reflected in the large pads on its relatively large hands and feet and in its short and broad bones. According to Mann, the diet of *Dromiciops* consists of the larvae and imagos of great variety of arthropods, mostly insects. Grant and Temple-Smith (1987) added new observations to the growing wealth of information on the biology of *Dromiciops*. Marshall (1978c) has summarized much taxonomic information on the genus.

Similarly to most other American marsupials, this important little animal is all but unstudied from the perspective of behavioral and ecological morphology, except for the observations of Mann (1955) and my initial efforts concerning the hand and foot. Postcranial remains of this family are not known in the fossil record (I am currently studying the whole skeleton of *Dromiciops*), and the described cranial and dental remains of *Microbiotherium* itself are only slightly different from the living *D. australis*. Marshall (1981b) has reviewed the fossil record of the family, and Segall (1969) and others have studied the basicranium. *Microbiotherium*, the only other valid genus recognized in the family, is known by seven species samples (Marshall, 1981b), going back to the Colhuehuepian late Oligocene and last appearing in the Santacrucian Early Miocene.

The details of the distal articular surfaces of the crus are indistinguishable from the petauroids and phalangeroids, particularly from those of the small burramyines and petaurines. The sharp distinction between the tibial **ATim** and **ATil** surfaces in didelphids is smoothly and seamlessly continuous in *Dromiciops*, as in the large or small petauroids or the inferred protophalangeriform.

The astragalar articular surfaces also reflect the smooth continuity of all the **UAJ** facets. They also possess relatively very small **ATim** and very large **AFi** facets, a derived ameridelphian pattern seen in didelphids. This didelphid taxonomic property clearly forms the foundation of the minor but mechanically significant **UAJ** modifications in the protogondwanadelphian: the further reduction of the **ATim** facet and the relatively smooth articular contact between the crus and the entire dorsal surface of the astragalus. The distal extension of the **ATil** facet laterally, both in *Dromiciops* and some burramyids, is reminiscent of the condition encountered in some didelphids. The head of the astragalus is similar in shape and orientation to that of arboreal didelphids. The **AN** facet is transversely oriented. The plantar side of the relatively flat astragalus has a small **ampt,** restricted to the posteromedial corner of the bone, a condition again similar to the didelphid one, the latter derived in the Ameridelphia. The **LAJ** articular relationships, however, are distinct, shared with all australidelphians regardless of the differences in the size and habitus of that ecologically and behaviorally varied radiation. The **CLAJP,** consisting of the confluent **CaA** and **Su** facets, is a taxonomic property of the Gondwanadelphia, the stem of the Australidelphia, in addition to the other characters of the astragalus, calcaneus, cuboid, and **EMt1J**, to mention only tarsal attributes here.

The calcaneus of *Dromiciops* is also representative of the pro-togondwanadelphian and protosyndactylan structures, and the protoaustralidelphian as well. In addition to the **CLAJP**, the cuboid contact has been diagnostically modified. The proximal, crescent-shaped, distally facing **CaCum** facet, is bordered by the continuous but sharply angled **CaCul** facet which faces medially. The **CaCua** facet is best explained as a new addition. The similarity of the cuboid (Fig. 8.18) to that of phalangeriforms is a shared pattern, derivable from a didelphid one. The medial wall of the large **cump** that portrudes out is both part of the lateral tunnel for the enormous flexor complex going to the rays of the foot, and also part of the lateral tunnel for the peroneus longus. This feature is almost certainly related to what I assume to be the hypertrophied pedal grasping musculature of the first aus-tralidelphian. This myological assumption, however, needs careful comparative investigation. The critical transformation hypothesis of the **CCJ** is discussed below in Chapter 8.

The **EMt1J** conformation (Fig. 8.17) and both the entocuneiform and **Mt1** morphology, like the morphology and function of the other tarsals, are unquestionably most similar to other primitive australidelphians. While the articular surfaces are sellar and can be divided into an **EMt1d** and **EMt1p** facets, these structures are much more similar to those of phalengeriforms than to those of didelphids.

## Dasyuromorphia (Figs. 7.32–7.40, 8.16, & 8.18–8.21)

In their exemplar book on the evolutionary ecology of marsupials, Lee and Cockburn (1985) have selected *Antechinus* since it is one of the best studied of dasyurids and in many ways a representative of the primitive life history strategies of that family. The dasyurids, although they cover a size range from some of the smallest (4 g) to middle sized (7–10 kg) forms, are crepuscular and nocturnal insectivorous–carnivorous forms. They have hairy and nonprehensile tails, even in the most arboreal species. Not all of them have well-developed pouches; in some it only appears as a rim during the breeding season or it is even absent. These carnivores are all but unstudied postcranially from the perspective of evolutionary morphology. In fact, I should repeat, the adaptive significance of many features is poorly understood because of a lack of unified evolutionary approach to historical–narrative explanations. For that reason alone undoubted holophyletic clades like the Dasyuromorphia are ideal for such future studies.

Archer (1984b, p. 636) notes that "Many dasyurid groups show independent evolution of arid adapted features such as evacuated palates, long tails, fat tails, granulated soles on the feet . . . ." The exact distribution of striations versus granulation on the pedal

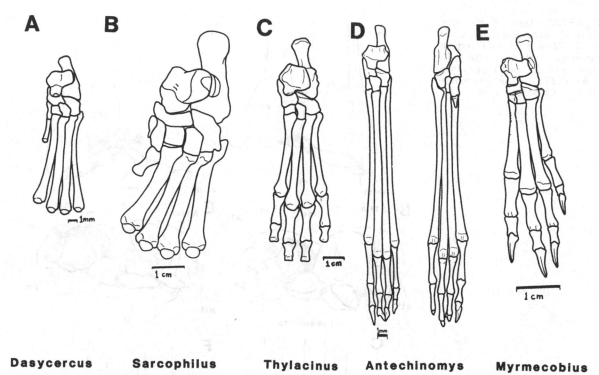

**Dasycercus**   **Sarcophilus**   **Thylacinus**   **Antechinomys**   **Myrmecobius**

*Figure 7.32.* *Osseous left foot showing tarsus and metatarsus in dorsal view in five dasyuromorphians:* Dasycercus, Sarcophilus, Thylacinus, Antechinomys, *and* Myrmecobius. *Phalanges are shown in* *the four figures to the right, and plantar side of foot is also shown in* Antechinomys.

pads and their correlations with habitat and other factors, and the exact morphology of the phalanges and falculae (claws) are some of the several topics of study that would yield potentially significant information for understanding dasyurid evolution. Tate (1947, p. 108) wrote that

The normal, moderately broad dasyurid foot has striated pads. The narrowly elongated foot often has granular pads. The former may indicate scansorial, the latter terrestrial, habits. The pads of the two highly specialized families, the Myrmecobiidae and the Notoryctidae, both relatives of the Dasyuridae, are without striae. On the other hand, more than twice as many of the genera of true dasyurids have striae as lack them. Progressive diminution of the striated condition in dasyurid genera is shown (table 2) in contrast to the high degree of uniformity present in most other marsupial families.

Tate (1947, pp. 108–109) has made an important contribution to the beginnings of the analysis of marsupial external foot morphology. Specific details and the significance of this type of information are as yet inadaquately explored. Dental characters of the dasyurids have been assessed in considerable detail by Archer

**Figure 7.33.** *Distal left crus of the dasyurids* Sarcophilus harrisi, *AMNH No. 65672 (A, B, C), and* Dasyurus viverrinus, *SAM M2086 (D, E, F). Anterior (A, D), distal (B, E), and posterior (C, F) views. For abbreviations see Table 1.1. Scales = 1 mm.*

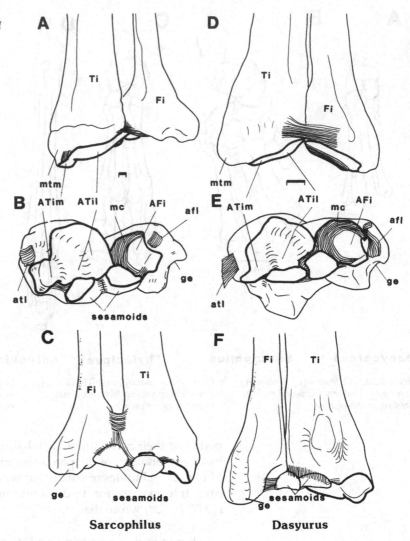

Sarcophilus          Dasyurus

(1976b, 1977b, 1978, 1982b, c), but phylogenies of taxa based on tooth analysis and molecular studies alone simply cannot adaquately test and therefore reflect the history of lineages without supplementary information concerning other character complexes. These non-dental traits must be evaluated on their own merits.

The only known living representative of the Myrmecobiidae (see Calaby, 1960), the numbat is rapidly disappearing, yet its morphology has not been adaquately studied from either behavioral or ecological perspectives. It is a largely crepuscular, nocturnal, and solitary termite- and ant-eating animal. There is a prominent chest gland on this animal, and it lacks a pouch. Its tail is nonprehensile; it is terrestrial, but can climb very well.

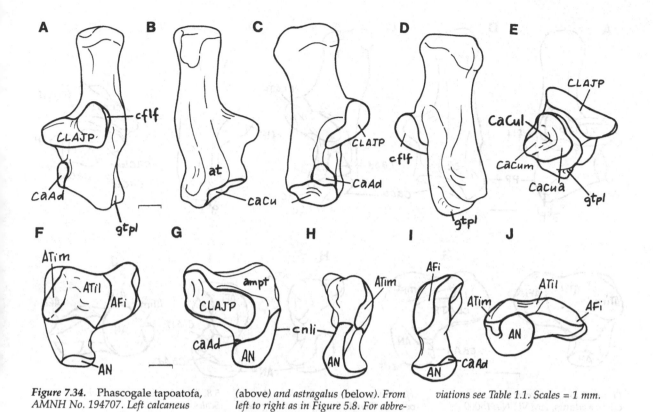

**Figure 7.34.** Phascogale tapoatofa, AMNH No. 194707. Left calcaneus (above) and astragalus (below). From left to right as in Figure 5.8. For abbreviations see Table 1.1. Scales = 1 mm.

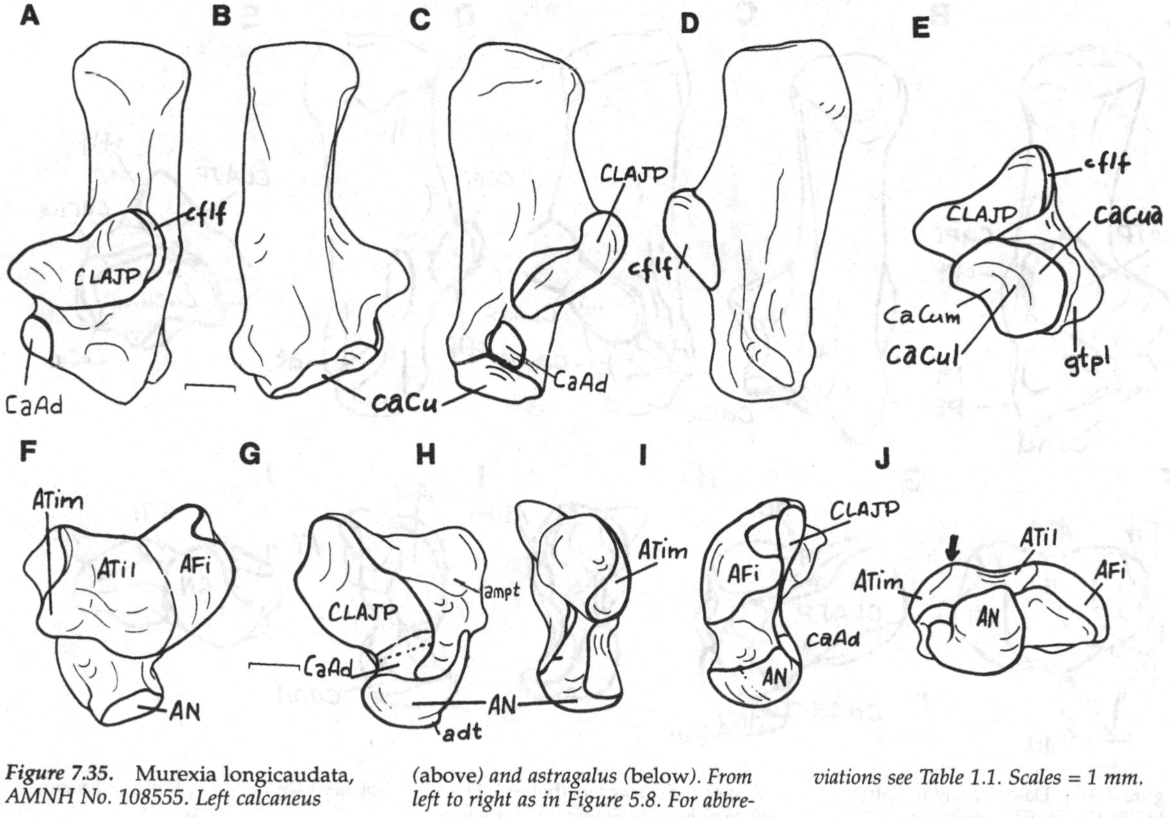

**Figure 7.35.** Murexia longicaudata, AMNH No. 108555. Left calcaneus (above) and astragalus (below). From left to right as in Figure 5.8. For abbreviations see Table 1.1. Scales = 1 mm.

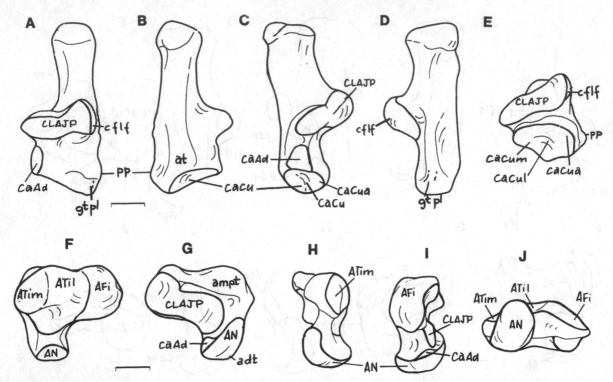

**Figure 7.36.** Dasycercus cristicaudatus, AMNH No. 15009.

Left calcaneus (above) and astragalus (below). From left to right as in Figure 5.8. For abbreviations see Table 1.1. Scales = 1 mm.

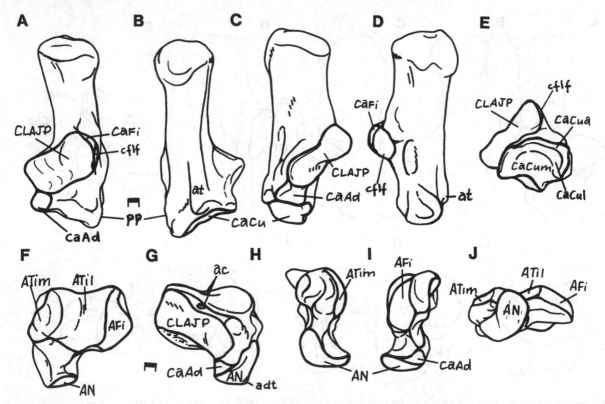

**Figure 7.37.** Dasyurus maculatus, AMNH No. 66192. Left calcaneus (above) and astragalus (below). From left to right as in Figure 5.8. For abbreviations see Table 1.1. Scales = 1 mm.

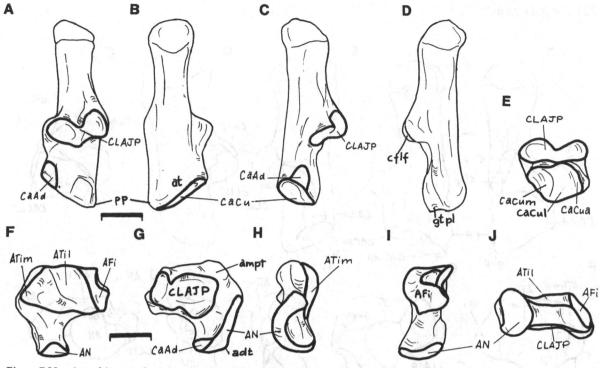

**Figure 7.38.** Antechinomys laniger, SAM No. M3908. Left calcaneus (above) and astragalus (below). From left to right as in Figure 5.8. For abbre- viations see Table 1.1. Scales = 1 mm.

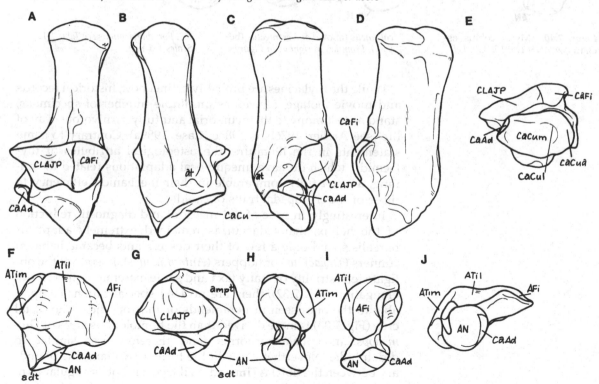

**Figure 7.39.** Thylacinus cynocephalus, AMNH No. 35504. Left calcaneus (above) and astragalus (below). From left to right as in Figure 5.8. For abbreviations see Table 1.1. Scale = 1 cm.

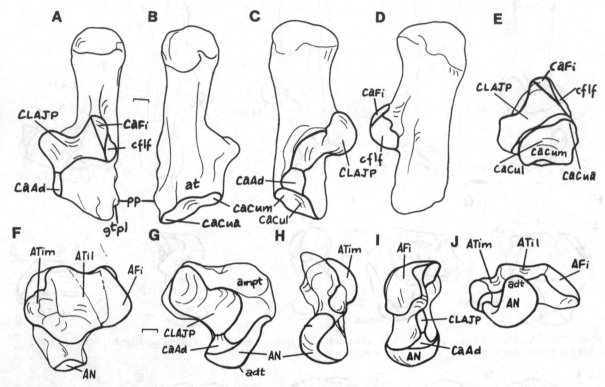

**Figure 7.40.** Myrmecobius fasciatus, AMNH No. 155330. Left calcaneus (above) and astragalus (below). From left to right as in Figure 5.8. For abbreviations see Table 1.1. Scales = 1 mm.

While the thylacines are probably extinct now, historical records and movie footage, as well as an ample number of specimens, stand as testimony to their cursorial and fully carnivorous way of life (see Archer, 1976b, c, 1982c; Case, 1989a). Contrary to some statements in the literature, the osteological attributes of thylacinids, which do have unequivocal adaptations related to running, are not fully comparable in their mechanical efficiency to those of the cruropedal traits of canids.

Interestingly, in spite of the ancestral and diagnostic reduction of the hallux, many dasyurids remained extremely adapt arborealists, and only a few of their descendants became habitual runners (*Thylacinus*) or hoppers (*Antechinomys*). In spite of the obligate terrestriality of many taxa and some extremes of metatarsal elongation (Fig. 7.32), there are no **UAJ** specializations comparable to the ones seen in peramelids or macropodids. The distal crus (Fig. 7.33) is more derived than the homologous areas of *Dromiciops* and phalangeriforms (and therefore less similar to didelphids). Nevertheless, as in the former two taxa, the boundary between the tibial **ATim** and **ATil** facets is not as angular as in didelphids. In dasyurids the tibial **ATil** facet no longer retains the posterior extension (the **ATip** facet) characteristic of the phalangeriforms. The double sesamoids posterior to the **UAJ** are also typically australidelphian. The dasyurid condition of the **UAJ**, like the other joints of the foot, is derivable from the structure seen in *Dromiciops* and the phalangeriforms, rather than a didelphid, as this patristic relationship will be explored in detail next and in Chapter 8.

The proximal tarsals illustrated give a representative sample of dasyurid astragali and calcanea. The **UAJ** morphology, like that of didelphids and *Dromiciops* as well as phalangeriforms, consistently displays the relatively large **AFi** facet. In some taxa such as *Antechinus, Antechinomys,* and *Thylacinus,* the **ATim** facet forms a more acute angle with the **ATil** facet than in others. This consequence of a well-developed downwardly projecting tibial medial malleolus might be expected to occur in the stem of this primarily terrestrial radiation, yet it is not the rule. As expected, however, the orientation of the **AN** facet is almost invariably in a dorsoplantar (extensad–flexad) arc, primarily facilitating flexion and extension at the **ANJ.**

In addition to the joined facets of the **CLAJP,** dasyurids display a secondary contact between the astragalus and calcaneus on the lateral side of the plantar surface of the neck. This **CaAd** facet is contiguous with the **AN** facet; this contact may vary in its proportion but it is invariably present. Explaining this contact is an important thread in the analysis of the transformational history of this group. In a postulated arboreal radiation that acquired the **CLAJP,** no buttressing was observed between the astragalus and calcaneus distally. In animals which possessed an extremely abducted hallux, the mobility of the two bones distally was the critical factor in the adjustment of the grasping foot. Loads that would have required osseous contact were not generated at the distal ends of these two bones. However, in the dasyurids (primarily scamperers on the ground or large branches) the loads are generated in a more tightly bound and nongrasping foot and require the additional support provided by the **CaAd** contact. This explanation is corroborated by the occurrence of an entirely independently acquired facet, in a nearly identical position, in the terrestrial and hopping Macropodoidae.

As in all other australidelphians, the calcanea possess only a rudimentary peroneal process. **CaFi** articulation is present only in the large *Thylacinus* and the numbat. The articulation of the calcaneus with the cuboid and the morphology of the latter is critical in resolving the transformation sequence of crurotarsal attributes in the australidelphians and will be discussed in detail below. The morphology of the cuboid in dasyurids shares important characteristics with other australidelphians that are not shared with didelphids. The process on the cuboid (**cump**), described and illustrated also for *Dromiciops, Burramys*, and didelphids (Fig. 8.18), is prominent, unlike the condition in didelphids. Given the relationship of this feature to the hypertrophied pedal flexors in forming a lateral wall of the tunnel for its passage, and given the reduced hallux in dasyurids, it is highly probable that **cump** expression reflects the ancestral hypertrophied australidelphian condition. This hypertrophy was probably initially related to the lateral rotation of the calcaneus in the stem australidelphian (the first gondwanadelphian), together with the cuboid, and this in turn resulted in the drastic reorientation of the **CCJ**. I believe that the evidence compellingly suggests that this ancestral australidelphian pattern is the base or initial condition from which the dasyurid **CCJ** was modified. The faint subdivision of the dasyuromorph **CCJ** into three identifiable surfaces homologous to the ones prominent on the microbiotheriid (and phalangeriform) pattern (Fig. 8.18) is almost certainly a derived gondwanadelphian (and australidelphian) condition compared to the latter. No matter how poorly delineated they may appear, the articular facets follow the pattern of three tiers displayed in the **CCJ** by the microbiotheriids and phalangeriforms. Although dasyurids possess only a suggestion of the new and apomorphous expression of the **CaCum** facet of the protoaustralidelphian, the total tarsal pattern, taken with the secondary **CaAd** facet already discussed, is derivable from the structurally microbiotheriid condition and not from a didelphid condition.

*Myrmecobius fasciatus*, like dasyurids, has no functional hallux. An interesting aspect of the myrmecophagous numbats is that their small cheek teeth are not only simple but that they also vary both in number as well as in morphology. The selective agents responsible for occlusion and mastication have been all but eliminated, and the previously shearing-related bioroles of the teeth appear to have become nearly incidental to survival. Compared to the dasyurid astragali, that of *Myrmecobius* displays an **ATim** facet that is distinctly trochleated with a longitudinal groove. As in all dasyurids, the **CaAd** facet is present. The calcaneus, with the minor difference of a **CaFi** contact, has essentially a dasyurid configuration. Judging by the obligate terrestriality of these animals, the calcaneofibular contact, as in *Thylacinus*, is a secondarily derived metatherian condition, not the primitive one in the Australidelphia. The articulation of the calcaneus with the cuboid shows the three barely distinct facets (see especially Fig. 7.40B, C). This feature, in addition to the overall similarity of the tarsus to dayurids, strongly corroborates the phyletic proximity of *Myrmecobius* to dasyurids. It also gives indirect support to the hypothesis that the primitive gondwanadelphian and australidelphian

CCJ is represented not by the dasyurid, but by the microbiotheriid condition.

## Notoryctidae (Figs. 7.41 & 7.42)

Almost all of the peculiarities in notoryctids may be related to their burrowing way of life (see especially Winge 1923, 1941; Stirling, 1891a, b, c, 1894; Sweet, 1904, 1906, 1907). The functional and adaptive aspects of ecological morphology of digging in quadrupeds are treated by Hildebrand (1985), and any evolutionary hypotheses about *Notoryctes* should be examined from such a perspective. These astonishing animals sink out of sight rapidly and "swim" through the soil without leaving a burrow. This unique marsupial, more than any other, is in need of a comprehensive functional–adaptive analysis of its anatomy, particularly its musculoskeletal morphology.

In the marsupial mole there are no manifestations of the external ears (pinnae), and the opening to the middle ear is deep within the fur; the nose is covered with thickened skin, while the nostrils are only slits; the tail is short and stubby, with a knob at the end. The atrophied and vestigial eye and lost pinna of the external ear, the cresting of the cranium for nuchal muscle insertion, and the enlarged and fused spinal processes of the second to sixth cervical vertebrae are some of the numerous fossorial specializations. These animals possess strong zygomatic arches and a highly inflated bulla. As expected, the optic foramen is lost.

The cervical vertebrae behind the atlas appear to be fused, and the coalesced neural arch and spinous process portions of these vertebrae prevent effective dorsiflexion of the atlas and the skull. This adaptation appears to be effective for the resistance of forces during digging when the head may be snapped upwards if the nose is used in some of the tunneling activities. Another adaptation of the axial skeleton is the buttressing of the sixth cervical vertebra with the first rib, the latter effectively serving as a strut between the sixth cervical, first thoracic, and the manubrium. This manubrium is very large with a highly characteristic ventral keel. The slender clavicle is well buttressed against the scapula. The highly diagnostic scapula, with its second scapular spine on the axillary border and a tunnel for the infraspinatus formed by the two scapular spines, like the rest of the skeleton, deserves a careful and detailed analytical description within the framework of an ecomorphological study. The remaining portion of the pectoral girdle is also unique among diggers. In spite of its unique peculiarities, the manus deserves special consideration. The carpus (Fig. 3.2) should be compared with other Australasian forms beyond what is documented here. Its ecological morphology undoubtedly holds the key for its appropriate comparison with other australidelphians.

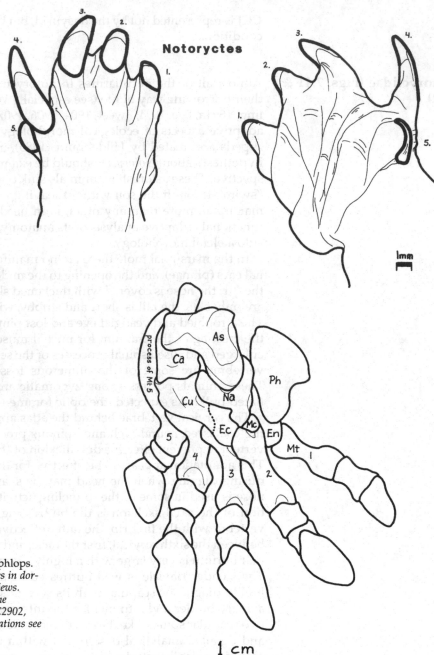

**Notoryctes**

*Figure 7.41.* Notoryctes typhlops. Above: MUZ No. 193, left pes in dorsal (left) and plantar (right) views. Numbers next to toes depict the cheiridia. Below: NMV No. C2902, right osseous foot. For abbreviations see Table 1.1.

The sacrum is massive, being composed of six vertebrae. The strong reduction of the epipubic bones is not adequately explained by fossorial activity. Winge (1941, p. 76) notes that

the hip bone and its surroundings at least are modified so that they resemble a shovel-like implement; the spinal processes and the

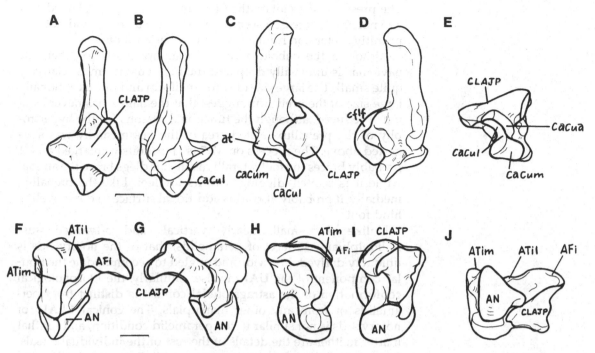

**Figure 7.42.** Notoryctes typhlops, NMV No. 200241. Left calcaneus (above) *and astragalus (below). From left to right as in Figure 5.8. For abbre-* viations see Table 1.1. Scales = 1 mm.

*Prc. mammillares* of the sacral vertebrae and of the anteriormost caudal vertebrae, as well as all salient crests of *Ilium* and *Ischium* are flattened, their upper face expanded, and they have partly fused together.

Although the manus is greatly modified, the third and fourth claws are enlarged and elongated for *digging*, not unlike those of the eutherian *Myospalax* (zokor) and Chrysochloridae (golden moles). It is the peculiar conformation of the external morphology, as well as osteological features of the *shoveling* pes (Figs. 7.41 & 7.42), that is the key to understanding the evolutionary morphology of notoryctids, and therefore to establishing their possible phylogenetic ties. Curiously, the enlarged fourth digit is set off from the first three, which are bound together by skin down to the base of the flattened digging falculae (perhaps more properly called nails, although histologically this has not been investigated). The fourth ray, therefore, is not bound to the others except at its base. Although the manner of use of the foot may help to elucidate these attributes, the conformation of the fifth ray adds to the peculiarity of the pes. This toe is plantar and is not excessively reduced – both conditions are entirely unlike those found in other "moles." The enormously enlarged prehallux forms the base of an additional lobule of the shoveling foot, and the first, second, and third rays are bound together to form the central scoop of the pes.

The presence of a nail on the first ray can only be explained as a neomorph, or at least a secondary adult occurrence, and not as a primitive retention from ancient metatherian roots.

Although the cuboid and navicular are fused together, the navicular is unusually deep and the mesocuneiform is relatively quite small. The large size of entocuneiform and the exceptionally large size of the prehallux suggest that the hallux has never been much reduced, although the functional reasons for the hypertrophy of the prehallux-related area of the pes medially was suggested above. The process on **Mt5** is enormously enlarged and it probably braces the foot laterally, as the hindlegs brace the animal while it is scratch–digging with its forefeet. Like the prehallux medially, it probably also adds additional surface to the shoveling hind foot.

Unlike the small, nearly vertical, and advanced australidelphian **AFi** facet of peramelids, that of the notoryctids is uniquely derived. It is wide and angled to accommodate the fibular component of the **UAJ** mortise. Similarly, the morphological conformation of the astragalotibial contact is distinct from peramelids or from any other marsupials. The conical **CLAJP** of notoryctids is not similar to the peramelid condition, and for that matter neither are the details of the rest of the individual tarsals. The tuber of the calcaneus is exceptionally long and slender, and the **CaCu** articular area is not excessively modified from a primitive syndactylan condition. All three of the angled subfacets of this contact are distinct; the **CaCum** is proximal, the **CaCul** is nearly aligned with the long axis of the bone, and the **CaCua** is distal and narrow. The **CLAJP,** not centrally pinched as in peramelids and macropodids, dominates the dorsal surface of the calcaneus. Unlike the condition in peramelids, there is no fibular contact with the calcaneus.

**Peramelidae (Figs. 7.43–7.49, & 8.23–8.25)**

This family, including *Macrotis* as well which was recently classified by several workers into its own family, the Thylacomyidae, is completely terrestrial, hindlimb dominated, largely nocturnal, and omnivorous. In spite of their retention of the near-ancestral, probably metatherian and ameridelphian, incisor numbers ("polyprotodonts") and relatively unmodified cheek teeth in *Peroryctes*, bandicoots represent one of the more highly modified holophyletic clades of australidelphian marsupials, both osteologically and reproductively. Their snouts are elongated and pointed to varying degrees, and this configuration is well reflected on the skull. Their tail is not prehensile and they have a well developed pouch which opens posteriorly or posteroventrally (in *Macrotis*). Peramelid life history strategies have been reviewed by Lee and Cockburn (1985; see also Gordon, 1974).

The bandicoots are not hoppers in the sense that kangaroos are

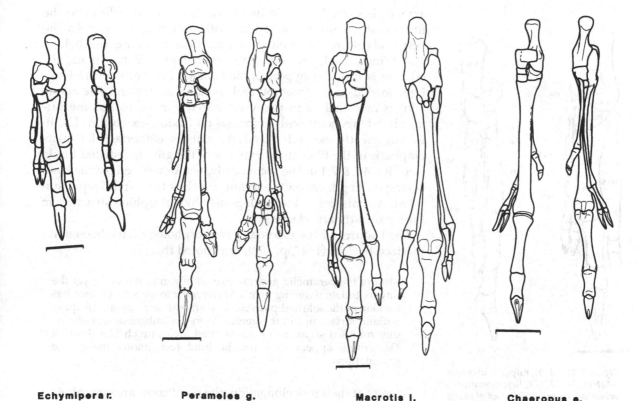

**Echymipera r.**  **Perameles g.**  **Macrotis l.**  **Chaeropus e.**

*Figure 7.43. Comparison of representative left foot skeletons of the Peramelidae. From left to right first ray is increasingly reduced and finally lost,* and fourth ray is increasingly elongated. From left to right: Echymipera rufescens (dorsal and medial views), Perameles gunnii (dorsal and plantar views), Macrotis lagotis (dorsal and plantar views), and Chaeropus ecaudatus (dorsal and plantar views). Scales = 1 cm.

(see below). For a more complete understanding of their musculoskeletal adaptations, more in-depth comparative studies are needed (e.g., see Filan, 1990). Their loss of clavicle has been probably mistakenly attributed to digging. If anything, a clavicle helps to buttress the pectoral girdle while digging. This apomorphic trait of bandicoots, I believe, can be associated with what may be called a cursorial–bounding type of locomotion. Pridmore's (1992) recent cinematographic study of the didelphid *Monodelphis domestica* strongly suggests that, after the ability to trot, the next stage of locomotor evolution may have involved the adoption of half-bounding or bounding locomotion, and all the necessary modifications that such a shift entailed. So in spite of the obvious digging propensities of bandicoots, the loss of the clavicle is probably the result of modifications related to speed of travel on the ground. These quasi-cursorial adaptations are most pronounced in *Chaeropus*.

As Winge (1941, p. 81) described it, the fossorial hand of per-

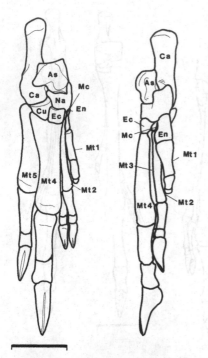

**Figure 7.44.** Echymipera rufescens, AMNH No. 104572. Representative primitive peramelid right foot of living taxa in dorsal (A) and medial (B) views. For abbreviations see Table 1.1. Scale = 1 mm.

amelids is strongly clawed on rays 2–4, while the pollex and the fifth toe are reduced. Converging with bounding lagomorphs, the vertebral column has been greatly modified (see diagnosis below; also Winge, 1941). As noted in other mammalian cursors, the clavicle is lost, the hypertrophied radius has become the long-bone loaded in the forefoot, and the ulna is reduced. The elbow joint is modified as in many advanced cursors to become primarily a hinge joint with emphasis on flexion–extension. The ilium is greatly expanded and the crus is eutherian-like in its emphasis of the tibia at the expense of the greatly reduced fibula (see Figs. 4.5 & 7.45). The secondarily well-developed patella has a corresponding facet on the femur which is long and deeply sulcated, reminiscent of the analogously hypertrophied structure in the small early artiodactyls.

The features of the external foot morphology have been summarized by Tate (1947, p. 108), who noted that

> Most of the peramelid genera have wholly granular pads; yet the tiny mountain dwelling genus *Microperoryctes* of New Guinea has the skin of the sole and pads soft and smooth and apparently quite without striae; and in the genus *Peroryctes*, subgenus *Ornoryctes*, tiny rounded striated or semi-striated areas, much like those of *Dasyurinus*, appear only on the hind feet among the coarse granulations.

In spite of their foot elongation, the bandicoots are nevertheless plantigrade – the entire sole of their pes touches the ground resting in a stance. This plantigrade position is changed to a ditigrade one (as in cats or dogs) when locomoting. While facultative unguligrady during running is likely, the extent to which an unguligrade toe-off is significant in bandicoot running has not been studied, and therefore functional inferences about these marsupials are limited. What is obvious is that the fourth ray of the foot is the weight bearing one (Winge, 1941), and that the toe-off at the end of this digit greatly increases the arc through which the animal is accelerated while the foot is still on the ground. The momentum of the body steadily increases as the load of the animal at the ankle joint moves across an arc of motion, from the point touched on the ground by (presumably) the terminal phalanx of the fourth ray. These animals display the complex mechanics from plantigrady, digitigrady, and to probably unguligrady during their positional repertoires. They deserve careful functional study.

Wood-Jones (1923–1925, p. 150) described an aspect of locomotor behavior of *Perameles myosura* as follows: "It is an animal of astonishing activity, its powers of jumping being all the more remarkable from its habit of rising vertically into the air. When alarmed on its evening excursions it will pause, and then in an instant, spring into the air and vanish in the most remarkable

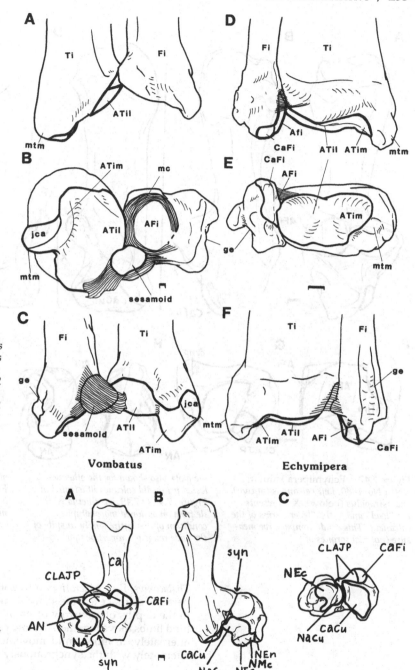

**Figure 7.45.** Comparison of distal crus in two independently terrestrial lineages of syndactylans. Peramelid Echymipera rufescens, AMNH No. 104572 (D, E, F, right leg; right), and vombatid Vombatus ursinus, AMNH No. 65622 (A, B, C, left leg; left). Anterior (A, D), distal (B, E), and distal (C, F) views. For abbreviations see Table 1.1. Scales = 1 mm.

**Figure 7.46.** Echymipera kalubu, ANM No. 1880. Left calcaneus and navicular in articulation. From left to right dorsal (A), plantar (B), and distal (C) views. Syn = syndesmosis. For other abbreviations see Table 1.1. Scale = 1 mm.

manner." The proportions of the tarsus, metatarsus, and phalanges of this species are approximately in between those of *Echimipera r.* and *Perameles g.* as illustrated on Fig. 7.43. The genus *Macrotis* is the most fossorial of the peramelids. Wood-Jones (1923–1925, pp. 165–166) notes that

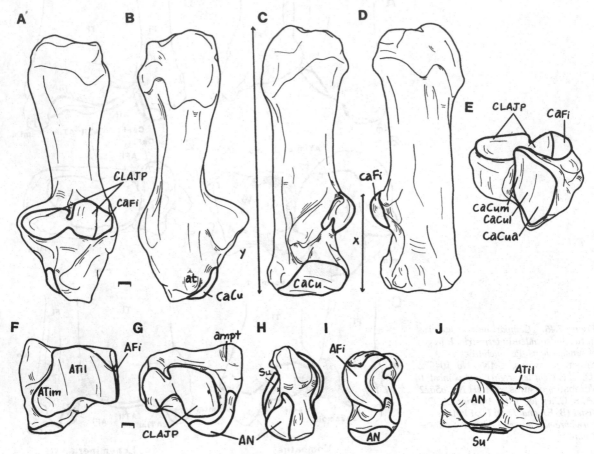

**Figure 7.47.** Echymipera kalubu, ANM No. 1880. Left calcaneus (above) and astragalus (below). X represents the "load" and Y the "lever" arms of the calcaneus. This crude comparative measure (not real torques of the foot), also shown for the other selected peramelid calcanea illustrated in Figures 7.48 & 7.49, has fluctuated little from an assumed morphotypic condition of this entire family in spite of the wide range of proportions for other pedal components in peramelids. From left to right as in Figure 5.8. For abbreviations see Table 1.1. Scales = 1 mm.

*Thalacomys*, [i.e., *Macrotis*] on the other hand, [compared to other peramelids that excavate shallow runways] passes most of its time in the depths of a burrow of its own making. . . . In any gait the hind limbs move together. In slow progression the fore limbs move alternately; in more rapid movements they move in unison, but alternately with the synchronously acting hind limbs.

*Chaeropus*, although probably extinct, was undoubtedly the most cursorial of the bandicoots. It has extremely elongated and slender limbs, a two-toed condition of the manus similar to artiodactyls, and a functionally one-toed condition of the hind limb elongated far beyond that seen in the other genera. Wood-Jones's (1923–1925, p. 170) account of Krefft's description of the gait of *Choeropus*

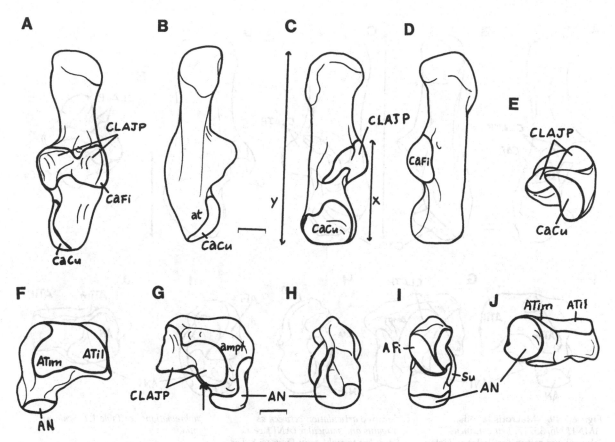

*Figure 7.48.* Microperoryctes murina, AMNH No. 221709. Left calcaneus (above) and astragalus (below). Arrow points to *Su* facet, indicating the similarity of this proximal extension to that of *Su* facet in Macrotis. In the latter the facet is extended and makes contact with the navicular (*ANl* facet, see Fig. 7.49). From left to right as in Figure 5.8. For abbreviations see Table 1.1. Scales = 1 mm.

is worth quoting here, as it colorfully describes the movement of this animal "like that of a 'broken down hack in a canter, apparently dragging the hind quarters after it.' I know of no more precise account of sequence of movement of the limb. . . ."

The extent of the relative elongation of the fourth metatarsal and the reduction of the hallux in the living bandicoots (Fig. 7.43), compared to the length of the tarsus itself, is a rough measure of the relative adaptation for cursoriality among the Peramelidae. In spite of the propensity of these animals for burrowing, and the genuine commitment of *Macrotis* to a fossorial life, the hind legs appear to be honed for speed and not for digging. The long nailed manus, on the other hand, is more in line with the mechanical requirements for digging. For these animals, the form–function solutions for cursorial bioroles must also incorporate their fossorial attributes as well. This intrafamilial comparison suggests that the inferred protoperamelid was in fact probably highly

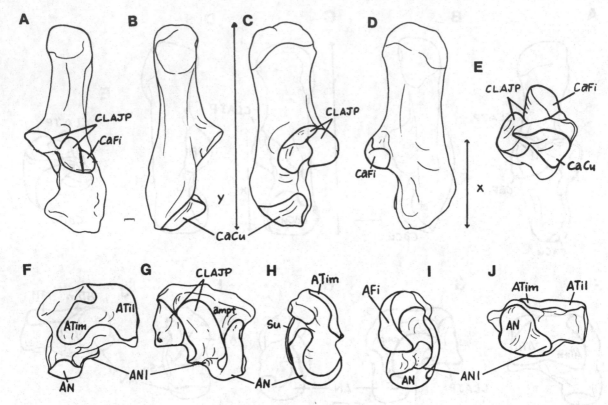

**Figure 7.49.** Macrotis lagotis, AMNH No. 35717. Left calcaneus (above) and astragalus (below). Note de novo *articulation between astragalus and navicular (ANl facet) From left to right as in Figure 5.8. For abbreviations see Table 1.1. Scale = 1 mm.

adapted to a running–bounding terrestrial locomotion. The differences among known peramelid tarsal features are relatively small, and the extreme crural and tarsal adaptations of the family are rather stereotyped, with the noted additional minor modification in *Macrotis*.

The polytypic genus *Echymipera*, in spite of its minor specialization of its loss of **I5/** and advances in cheek tooth morphology over those of *Peroryctes*, for example, is of critical significance in the study of peramelid carpal and tarsal morphology. I will therefore discuss it as essentially the closest known representative of the protoperamelid condition in the morphology of the hind foot, and in the details of the tarsals and metatarsals. This syndactyl foot has robust fourth and fifth rays, so that **Mt4** and **Mt5** represent the dominant mass of the foot. The diagnostic peramelid double contact of the **Mt4** with the ectocuneiform in addition to the cuboid is fully achieved (Fig. 7.44). Although the hallucial ray does not participate in any obvious bioroles of which I am aware, it is not greatly reduced; in fact, it is less reduced than that of any known dasyuromorphian. The contact between the **Mt4** and the cuboid and ectocuneiform is not parallel to the ground but rather angled, in direct contrast to the condition in *Chaeropus* in which a more transversely horizontal articulation between these bones is achieved (Fig. 7.43). While the reduced hallux (beyond the condition seen in *Echymipera*) persists in *Perameles*, it is merely vestigial or eliminated in forms like *Macrotis* or *Chaeropus*. The latter genus also achieves the drastic reduction in size of the fifth ray of the foot. In the other peramelids this ray maintains the relative proportions of a putative ancestor, displayed structurally by *Echymipera*. How the fifth toe functions in the locomotion of peramelids is yet to be elucidated. Judging by the distal articular morphology of the cuboid, however, it would appear that that the **Mt5** supplies a critical lateral buttress to the cuboid in the virtually immobile, tightly articulating tarsal complex. The distal surfaces of the navicular, cuboid, and ectocuneiform have exceptionally deep projections that lock these bones tightly not only with one another, but the latter two with the proximal surfaces of **Mt4** and **Mt5**.

In addition to the specializations of the bony dimensions, the **UAJ** of peramelids (Fig. 7.45D,E,F) is one of the four known mortise–tenon configurations in Metatheria. Like the **UAJ** of the last common ancestor of the Eutheria, and obviously independently of argyrolagids and macropodoids, this construction achieves the mechanical advantage of mediolateral stability by the additional constraint supplied from the fibula. As in many living eutherians but not most marsupials (except macropodoids), the fibula is a relatively slender bone but it serves as a nearly vertical lateral buttress for the astragalus, and it rests on the calcaneus as well. The peramelid crus has the derived syndesmosis between the tibia and fibula in place of a synovial **TFJ**. This condition is relatively rare in mammals, and it is even more unusual among marsupials. The contrast between the cruropedal morphology of the vombatoid *Vombatus* and *Echymipera* on Figure 7.45, both exclusively terrestrial syndactylans, is unquestionably largely related to the adaptive differences in relative loads generated between a relatively slow moving animal versus the stem peramelid, which was cursorial.

As expected in a semicursorial animal even in a relatively pedally primitive peramelid such as *Echymipera*, the tarsals are exceptionally well locked together. The calcaneus is tightly bound by a syndesmosis to the navicular, a rare condition in mammals (Fig. 7.46). This area, because it is nonarticular, would be very difficult to detect in a fossil calcaneus if the navicular was not known. Although this is the only intertarsal syndesmosis, the remaining joints between the seven tarsals are articulated in such a restrictive way that significant readjustments of the foot are not possible between the cuboid and navicular, or between the latter and any of the cuneiforms. The mesocuneiform is exceptionally reduced, possibly because of lack of any significant loads on the grooming claws. Movements between the calcaneus and cuboid are very limited, and this range of motion would be difficult, if not impossible, to deduce from the isolated bones alone.

The calcaneus of *Echymipera* and other peramelids has a secondary constriction between the **Su** and **CaA** facets of the **CLAJP**. The **CaA** components of both calcaneus and astragalus are exceptionally curved, the arc covering nearly 180 degrees, whereas the medial **Su** facets are less curved. This combination provides excellent lateral stability while the foot rotates, with the astragalus being held relatively stationary in the crus during dorsi- and plantarflexion. Movement of the foot at the **LAJ**, like that of the **UAJ**, may involve rotation on a nearly completely transverse axis with very little wobble. This critical issue of peramelid foot mechanics, or that of any other marsupial for that matter, has never been investigated. Nevertheless taken together, the consistently larger astragalar **Su** facet in contrast to its conarticular mate on the calcaneus and the rigid interlocking of the remaining tarsals strongly suggest that most of the mobility in the joints of these cursorial animals, in addition to that in the **ANJ**, is in the **LAJ**, in a manner partly similar to the condition in artiodactyls.

The narrowing of the **CLAJP** in the **LAJ** greatly facilitates the stability of this highly mobile joint. Yet the **LAJ** *immobility* in macropodoids, to be described later is another different solution which also manifests itself in a "pinched" **CLAJP**. Both conditions are unquestionably derived from the **CLAJP** displayed by gondwanadelphians, nonmacropodoid phalangeriforms, and the vombatiforms. The advanced condition in kangaroos results in transversely offset facets which completely prohibit **LAJ** motion in all axes. It is nevertheless clear, given the unbroken connection of these facets, their exact transverse alignment on the calcaneus, the conformation of the **CCJ** and the **UAJ**, and syndactyly, that this condition of narrowing in the middle of the **LAJ** is derived independently and for different functional (mechanical) ends in both the peramelids and macropodoids.

Unlike in the pedally more advanced peramelids, the **CaFi** facet of *Echymipera* is not as vertical as in the rest of the family (compare Fig. 7.47 with the other peramelid conditions). As in all other australidelphians, the **CaCu** facet faces for the most part distomedially, with the characteristic medial orientation of not only the **CaCul**, but also the **CaCum** and **CaCua** facets. Although these facets cannot be as clearly delineated as those in microbiotheriids or in other syndactylans, they are nevertheless discernible. In order to account for such facet orientations in the **CCJ** of the obligate terrestrial peramelids, it makes sense to postulate stem australidelphian origins rather than evolution directly from didelphids. In the stem australidelphian condition, the habitual extremes of lateral rotation of the calcaneus and cuboid resulted in the remodeling of this joint described in the discussion of the Microbiotheriidae.

The proportions of the calcaneus including and distal to the **LAJ** articulation as compared to the entire length of the calcaneus are among the most conservative aspects of the tarsus and metatarsus in the family. This feature, as in *Echymipera* for example, reflects a relatively powerful role for the triceps surae, given the relatively short tarsus and forefoot anterior to the cruropedal articulation. While the calcaneus remains similarly built, this proportion of the calcaneus compared to the entire length of the foot is considerably altered, however, in forms with an elongated foot (see Fig. 7.43). In the species with long feet a relatively short calcaneal tuber (compared to overall foot length) generates enough force to keep the relatively enormous torque developed by the arc of the foot distal to the **LAJ** as a result of the animal's momentum.

The body of the astragalus of *Echymipera*, as that of all peramelids, is characteristically quadrate in appearance with large **ATim** and **ATil** facets. The relative size of these facets is indicative of the importance of the tibial side of the **UAJ** mortise. As in many eutherians, the **AFi** facet in all peramelids, forms an approximately 90-degree angle with the **ATil** facet. The "pinched" **CLAJP**, as noted, is not only secondary, but the **Su** facet component is invariably longer proximodistally than the **CaA** portion, suggesting the derived nature of the extensive **LAJ** motion. Although the navicular accommodates this motion of the astragalus, the movement at the tarso-metatarsal articulations, judging from the nature of articulations of the tarsals with the metatarsals, is probably minimal. Detailed functional–adaptive studies are needed to understand these relationships more fully.

The only feature that significantly differentiates the tarsus of *Macrotis* from the other peramelids sampled is the distal extension of the **Su** facet into a laterally extended astragalonavicular (**ANl**) facet, creating a de novo contact between the astragalus and navicular (Fig. 7.49). As noted, the **LAJ** and **ANJ** in the peramelid foot are the areas where, in addition to the transfer of great loads, significant movements take place. This lateral facet extension makes a secondary connection of the two joints that enhances the mobility of these critical areas. The cylindrical surfaces of the articulating bones in the modified **ANJ** further facilitate both the flexion–extension of the foot on the astragalus and also stabilize motion in this area mediolaterally.

**Petauridae, Tarsipedidae, and Phalangeridae (Figs. 7.50–7.61 8.26–8.31)**

The living petaurids, the petaurines (Petaurini, Dactylopsilini, and Pseudocheirini) and burramyines (Burramyini and Acrobatini), represent a varied arboreal radiation of nocturnal insectivores–gumivores, nectarivores, graminivores, and foli-

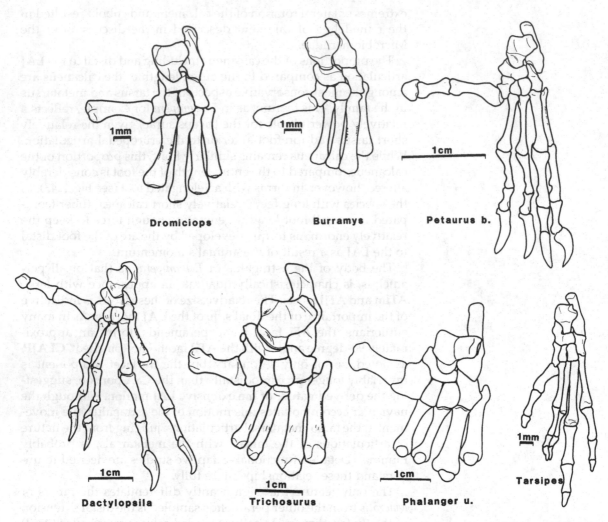

**Figure 7.50.** *Representative left foot skeletons of* Dromiciops *and phalangeriforms. Only tarsus and metatarsus are shown for* Dromiciops, Burramys, Trichosurus vulpecula, *and* Phalanger ursinus. *Scales apply to feet under which they are positioned.*

vores (see Smith & Lee, 1984). The arboreal glider *Petaurus* has a patagium which stretches from the hand (along the fifth ray) to the ankle on the foot, while its apparent sister taxon *Gymnobelideus* lacks any sign of this gliding adaptation. The genera *Dactylonax* and *Dactylopsila*, with enlarged fourth hand rays and hypertrophied upper incisors, are similar to the petaurins, although they are still poorly known ecomorphologically. The pseudocheirin radiation is a folivorous one, and although *Pseudocheirus* is a climber, the greater glider *Schoinobates* (=*Petauroides*) has a well-developed patagium running from its elbow to the tarsal region. The genus *Hemibelideus* appears to be in the incipient

**Figure 7.51.** *Distal left crus and astragalus of* Phalanger maculatus, *QMJ No. 9946, showing interrelationship of components of* **UAJ.** *Anterior view of distal crus and astragalus (A), dorsal view of astragalus schematically showing the joint capsule and the meniscus (B), and distal view of crus (C). For abbreviations see Table 1.1. Scales = 1 mm.*

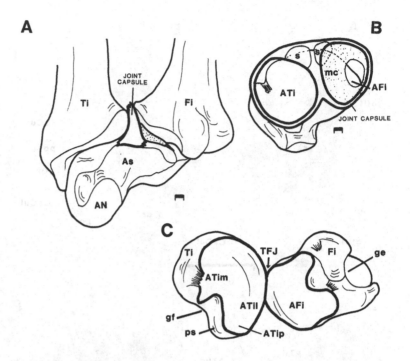

stage of gliding adaptation (see Johnson-Murray, 1987), a condition from which *Schoinobates* has evolved. The ecomorphology of the skeleton of these forms is all but unstudied. In general the appendicular skeleton of *Schoinobates* compared to that of its close nongliding relatives *Pseudocheirus* and *Hemibelideus* is strikingly elongated, as is often the case in gliders (see especially the placental colugos; but not in tiny *Acrobates*). The greater glider also has hypertrophied cheiridia and claws compared to the ringtail possums. The comparative myology of all marsupial gliders has been studied by Johnson-Murray (1987; see references therein), in addition to those of *Hemibelideus* and *Pseudocheirus. Hemibelideus*, while not a glider, is a leaper, and it apparently has small skin flaps close to the body wall in the axillary and inguinal areas. Very little is known of the functional bony correlates of such transformations. All these animals provide ideal models to study the evolutionary transformations involved in gliding. The differences and similarities between the musculoskeletal strategies of placental and marsupial gliders, once carefully delineated, could potentially reveal much information on one of the most interesting adaptive transformations in mammalian evolution: the change from quadrupedalism to flying.

The Burramyinae are mouse-sized forms which are mostly insectivorous and nectar feeding. They may be scansorial or fully arboreal forms, agile leapers or gliders, as in the case of *Acrobates*, with prehensile tails. *Burramys* with its plagiaulacoid premolars

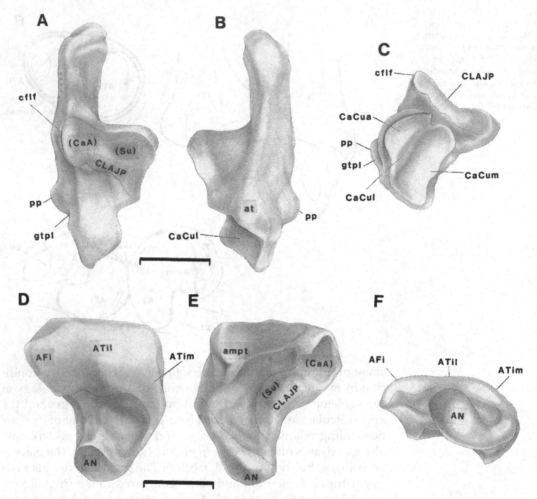

**Figure 7.52.** Burramys parvus, MUZ No. 4446. Right calcaneus (above) and astragalus (below). From left to right: dorsal, plantar, and distal views. For abbreviations see Table 1.1. Scales = 1 mm.

appears to be specialized for seed feeding, and the remaining forms have mixed diets which can include invertebrates and fruits as well as pollen and nectar (Smith, 1980; Smith and Hume, 1984). *Acrobates* appears to have evolved gliding independently from the other petaurid gliders; in addition to its narrow plagiopatagium between the limbs (from the elbow to approximately below the knee), it has a laterally projecting stiff fringe of hair on its tail which appears to aid gliding. The latter trait is shared with the nongliding *Distoechurus* of New Guinea. Given the constructional differences of the gliding membranes (Johnson-Murray, 1987), gliding has evolved probably independently from that seen in *Petaurus* and *Schoinobates*.

The flower- and stalk-climbing, mouse-sized, and pollen- and

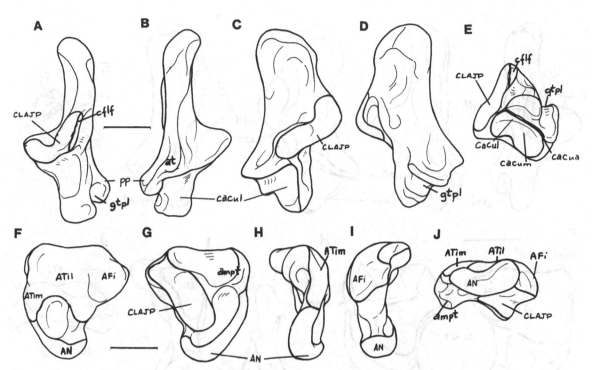

**Figure 7.53.** Distoechurus pennatus, *AMNH No. 105938. Left calcaneus (above) and astragalus (below). Arrow pointing to flexad surface of astragalus in the various*

*Figures 7.54–7.84, of diprotodontians indicates approximate boundary between the ATim and ATil facets; this surface is usually rounded in arboreal forms but becomes increasingly angular*

*and stabilized in more terrestrial groups. From left to right as in Figure 5.8. For abbreviations see Table 1.1. Scales = 1 mm.*

nectar-feeding *Tarsipes* (see Russell & Renfree, 1989) has a highly modified cranial form–function commensurate with its feeding specializations (Archer, 1984b, and references therein). This little animal has an extremely elongated skull, with a beaklike rostrum resulting in a palate greatly narrowed anteriorly. *Tarsipes* has a protrusible and greatly elongated tongue with hypertrophied "hair-like" papillae on its tip, and only a slightly modified digestive system for pollen digestion (Richardson, Wooller, & Collins, 1986). As a consequence of its pollen-adapted diet its dentition is vestigial, and the areas associated with chewing are atrophied. Its mandible is reduced in the extreme with just traces of a coronoid process and an uninflected angle. Like that of most of the other phalangeriforms, the postcranial morphology of this genus has not been studied. Its foot (Figs. 7.50 & 7.61) is elongated and slender – highly derived within the Phalangeriformes.

The living phalangerids are relatively larger than most of the petauroids discussed in this book. They are strongly arboreal with a prehensile, glabrous tail. *Trichosurus*, which overran New Zealand after its introduction in the last century, is also the most suc-

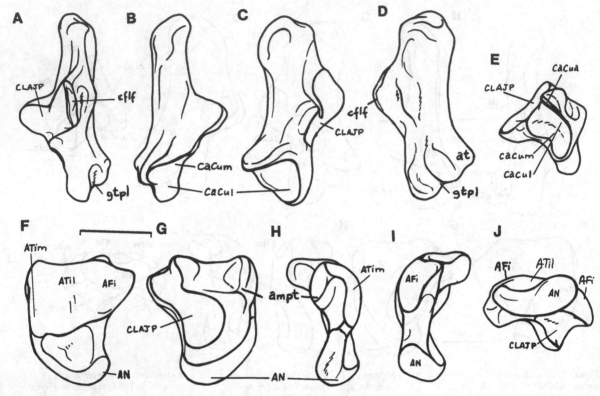

**Figure 7.54.** Acrobates pygmaeus, AMNH No. 858. Left calcaneus (above) *and astragalus (below). From left to right as in Figure 5.8. For abbre-* *viations see Table 1. Scale = 1 mm.*

cessful Australian possum to coexist with human settlements. In general the phalangerids can be characterized as relatively slow and cautious climbers, making full use of their prehensile tails. *Trichosurus* (personal observations), and probably other phalangers as well, use a fully reversed (hyperinverted) foot position when descending a tree trunk. When on the ground *T. caninus* is capable of considerable speed using a half bounding type of progression. There are no functional–adaptive studies regarding the postcranial osteology of the phalangers. The osteological patterns displayed by the foot of phalangeroids are built on the exceptionally mobile and widely grasping hallux of the australidelphian ancestor. This has been described and emphasized above in the section entitled "Microbiotheriidae." The **UAJ** patterns are essentially similar in phalangeroids, although future surveys will undoubtedly discover additional important taxon-specific functional–adaptive differences. While the proportions of the tibial **ATi** facets are similar to that of ameridelphians, they are smooth and uninterrupted, as in the living microbiotherians.

While some species- and genus-specific differences were appar-

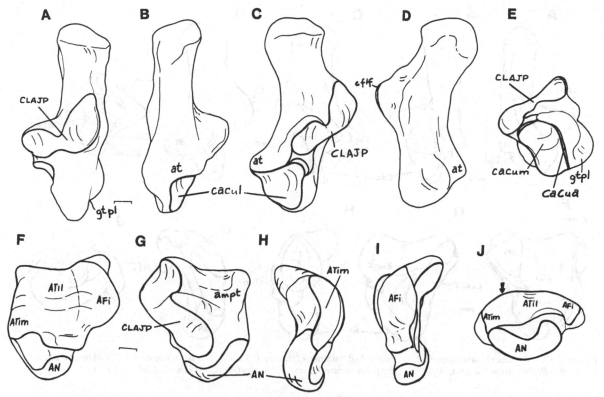

*Figure 7.55.* Petaurus breviceps, AMNH No. 42993 and MUZ No.

4787. Left calcaneus (above) and astraglaus (below). From left to right as

in Figure 5.8. For abbreviations see Table 1.1. Scales = 1 mm.

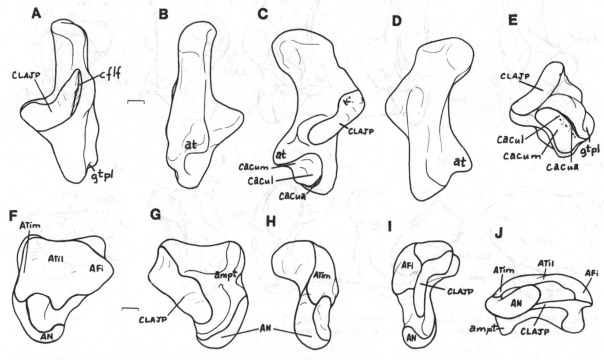

*Figure 7.56.* Dactylonax palpator, AMNH No. 79778. Left calcaneus

(above) and astragalus (below). From left to right as in Figure 5.8. For abbre-

viations see Table 1.1. Scales = 1 mm.

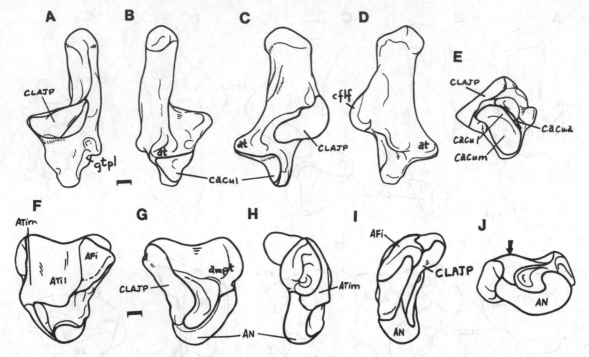

**Figure 7.57.** Pseudocheirus archeri, personal collection from Victoria, Australia. Left calcaneus (above) and astragalus (below). From left to right as in Figure 5.8. For abbreviations see Table 1. Scales = 1 mm.

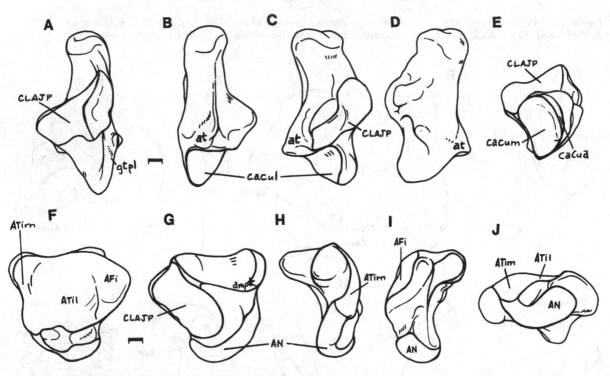

**Figure 7.58.** Schoinobates volans, AMNH No. 65364. Left calcaneus (above) and astragalus (below). From left to right as in Figure 5.8. For abbreviations see Table 1.1. Scales = 1 mm.

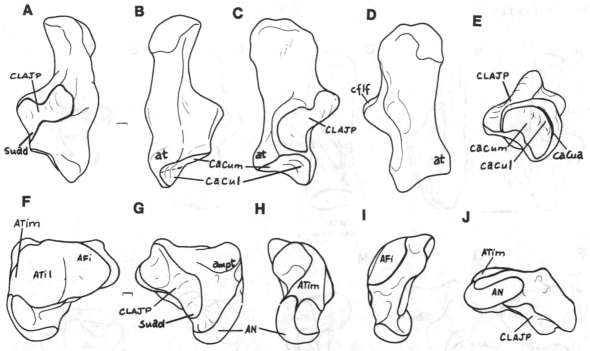

**Figure 7.59.** Trichosurus vulpecula, AMNH No. 160348. Left calcaneus (above) *and astragalus (below). From left to right as in Figure 5.8. For abbre-* viations see Table 1.1. Scales = 1 mm.

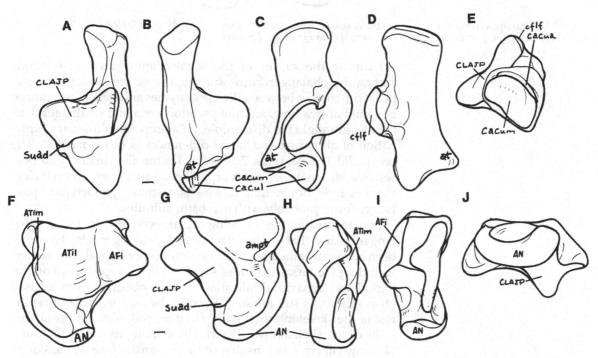

**Figure 7.60.** Phalanger gymnotis, AMNH No. 104089. Left calcanieus (above) and astragalus (below). Note in particular the relatively acute angu- *lar relationship between ATim and ATil facets. This is similar to the conditions seen in macropodids and vombato- morphians. From left to right as in* Figure 5.8. For abbreviations see Table 1.1. Scales = 1 mm.

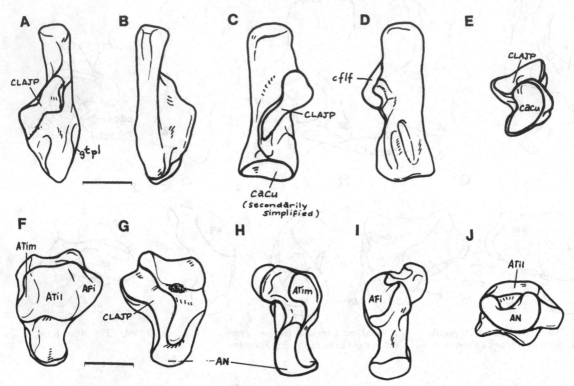

**Figure 7.61.** Tarsipes spencerae, AMNH No. 119714. Left calcaneus (above) and astragalus (below). From left to right as in Figure 5.8. For abbreviations see Table 1.1. Scales = 1 mm.

ent during the survey of the tarsals, family level distinctions among the phalangeriforms are much more tenuous. Diagnostic pedal differences between the burramyines and petaurines cannot be found at this time, and the pseudocheirin and phalangerid attributes are slight but discernible, yet a proper evolutionary explanation of all carpal and tarsal differences is much needed. The tarsipedid *Tarsipes* (Figs. 7.50 & 7.61) is tarsally distinct and possesses additional aberrant morphological attributes. Nevertheless there is little doubt that *Tarsipes* is a small syndactylous phalangeriform, probably with acrobatin affinities.

In conclusion, the osteological aspects of the living phalangeriforms, that is, petauroids and phalangeroids, have not been adequately analyzed from an ecomorphological perspective, including the tarsal attributes surveyed here. A great deal of ecological and behavioral information, to be obtained from focused observations, is still missing. Because the carpus, tarsus, and the rest of the skeleton are so poorly understood, some of the shared traits of the skeleton reveal virtually nothing as yet of the intraphalangeriform relationships of these animals (see discussion in Chapter 8).

**Macropodoidea (Figs. 7.62–7.73, & 8.36–8.41)**

Although *Hypsiprymnodon* is a living genus, in most (but certainly not all) of its morphological attributes and probably in much of its habitat preference and behavior, it appears to represent a somewhat frozen model in time from the beginnings of the macropodoid radiation. This quadrupedal bounder has nearly equally long front and hidlimbs and a simple stomach. The diverse Potoroinae, on the other hand, are hoppers like the macropodines and also have the sacculated stomach used in pregastric digestion. The macropodines, dendrolagines, and the extinct sthenurines represent the most advanced locomotor diversification within this large superfamily. In their feeding ecology the kangaroos are the marsupial equivalents of ungulates in Australasia (Janis, 1990; Sanson, 1980).

The capsule summary of Flannery (1984) that gives a flavor of this group cannot be easily improved on. He states (p. 817) that

> few people realize that there are approximately 60 living species of kangaroos and an equivalent number of extinct ones, with new living and extinct species being discovered every year. While all kangaroos follow the same basic body plan, possessing two powerful hindlimbs, weaker forelimbs and a strong tail, they are otherwise an amazingly diverse lot, ranging in size from . . . around 1 kg, to . . . as much as 90 kg. Kangaroos are adapted to an astounding variety of lifestyles.

The latter is a reference to both the return to an arboreal habitat by dendrolagins and to the burrowing of bettongs, in addition to the occupation of diverse terrestrial habitats.

The locomotor adaptation of the group is unquestionably the key to its origins. Postcranial traits, therefore, particularly those of the hindlimb, may be far more critical in the eventual recognition of the earliest kangaroos in the fossil record than other remains. The foundations of the "big footed" (macropod) adaptations have been well understood for centuries, even though some of the fine points concerning the energy-storing ability of the tendon of the triceps surae are only a recent addition to our knowledge (Dawson, 1977). Balanced evolutionary accounts, particularly ecomorphological ones, however, are lacking in the literature. The preexisting syndactyly of the phalangeroid-like ancestry of kangaroos has certainly channeled change in a mechanically necessary direction. The quadrupedal bound, as Flannery (1982, 1984, 1987, 1989) referred to it, was probably antecedent to an increasingly hind-limb-dominated hop. Yet in all known macropodoids, the hind-limb dominance as a locomotor engine compared to the modifications of the forelegs is obvious, in spite of the fact that *Hypsiprymnodon* and *Potorous* do make extensive use of a qaudrupedal locomotion. In evaluating the change to a rapid terrestrial locomotor mode, one cannot help but be influenced by the

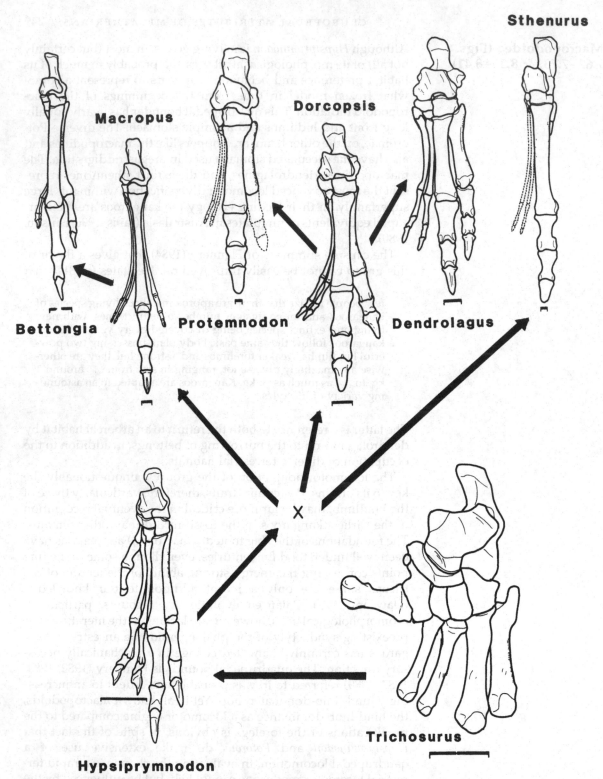

**Sthenurus**

**Dorcopsis**

**Macropus**

**Dendrolagus**

**Bettongia**

**Protemnodon**

**Hypsiprymnodon**

**Trichosurus**

*Figure 7.62.* *Representative left foot skeletons of a phalangerid and various* *macropodoids, showing path of possible structural transformation of the pes. All* *views are dorsal. This is not a phylogeny of the taxa shown. Scales = 1 cm.*

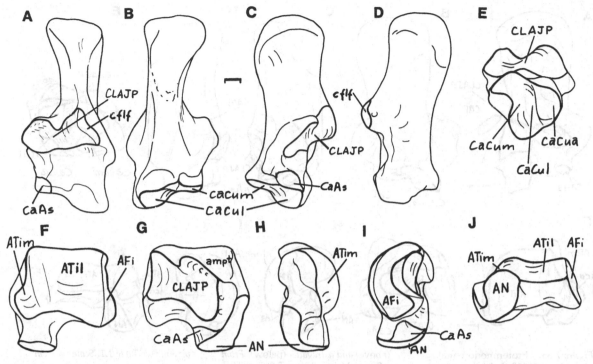

**Figure 7.63.** Hypsiprymnodon moschatus, *AMNH No. 184580. Left* calcaneus (above) and astragalus (below). From left to right as in Figure 5.8. For abbreviations see Table 1.1. Scale = 1 mm.

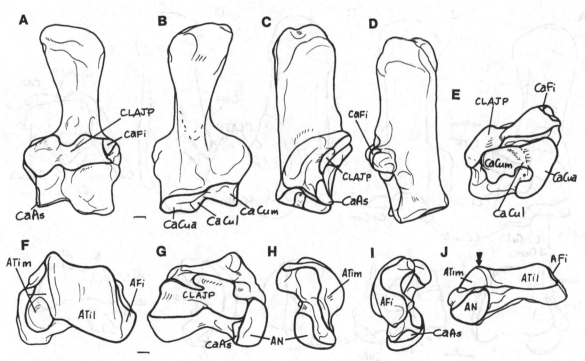

**Figure 7.64.** Potorous tridactylus, *AMNH No. 65295. Left calcaneus* (above) and astragalus (below). From left to right as in Figure 5.8. For abbreviations see Table 1.1. Scales = 1 mm.

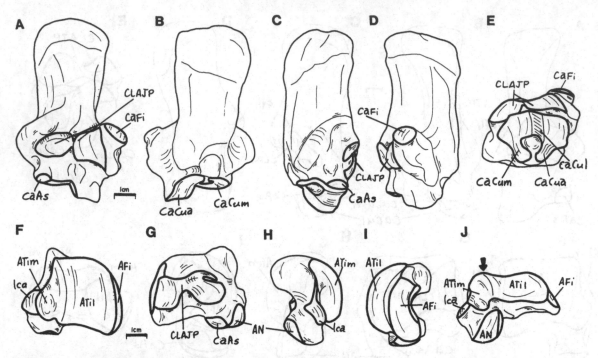

**Figure 7.65.** Protemnodon anak, NMV No. P39118. Left calcaneus (above) and astragalus (below). From left to right as in Figure 5.8. For abbre- viations see Table 1.1. Scales = 1 cm.

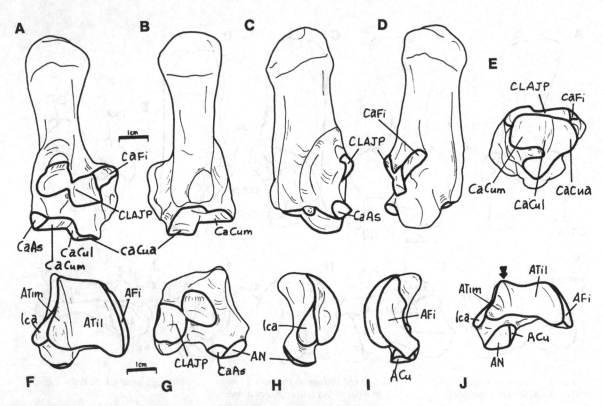

**Figure 7.66.** Prionotemnus rufus, personal collection from Victoria, Aus- tralia. Left calcaneus (above) and as- tragalus (below). From left to right as in Figure 5.8. For abbreviations see Table 1.1. Scales = 1 cm.

*Figure 7.67.* Dendrolagus lumholtzi, AMNH No. 35731. *Left distal crus in anterior (A), posterior (B), and distal (C) views. For abbreviations see Table 1.1. Scale = 1 mm.*

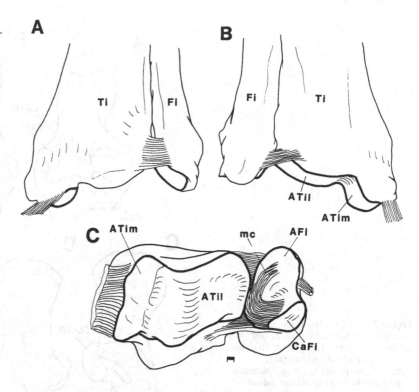

perception that the adaptive potential of the forelimbs of marsupials is restricted by the biorole of neonatal crawling to the teat.

No adequate account can possibly be given in this book of the intricate evolutionary morphology of the foot within this, the probably most successful diprotodontian radiation. The interplay of habitat and the adaptive modification of morphology and its relevance to kangaroo evolution need to be explored in much greater detail than is possible at present (Szalay & Flannery, in preparation). Nevertheless the major outlines of this origin and radiation (Fig. 7.62) are some of the most instructive available for the understanding of character evolution. Kangaroo pedal diversity supplies us with incomparably excellent, repeated examples of the interplay of habitat, behavior, and morphological solutions constrained by a given heritage.

As the protomacropodoid bounder committed itself to force transmission through the fourth ray of the foot, the direction of the transformation followed a predictable pattern. The speed and distance which kangaroos are capable to attain and cover are difficult to imagine without an extremely and transversely stable **UAJ.** This joint is entirely the same type as that of eutherians (see *Hypsiprymnodon* in Fig. 4.4). The surprising lack of mobility of the **LAJ,** however, is entirely unlike that of peramelids or artiodactyls, but it closely parallels the perissodactyl condition. Still unex-

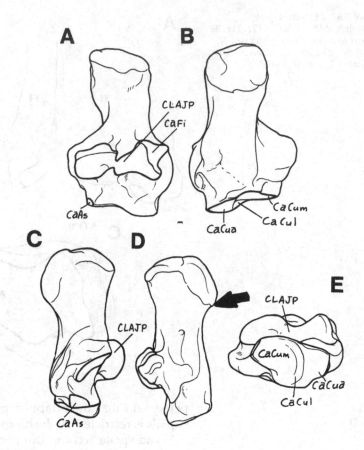

**Figure 7.68.** Dendrolagus lumholtzi, AMNH No. 65258. Left calcaneus in dorsal (A), plantar (B), medial (C), lateral (D), and distal (E) views. Arrow points to the exceptionally strong development of plantar side of tuber for flexors of rays in this tree kangaroo. For abbreviations see Table 1.1. Scale = 1 mm.

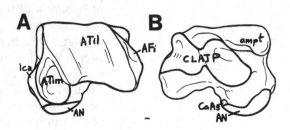

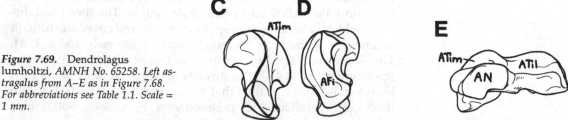

**Figure 7.69.** Dendrolagus lumholtzi, AMNH No. 65258. Left astragalus from A–E as in Figure 7.68. For abbreviations see Table 1.1. Scale = 1 mm.

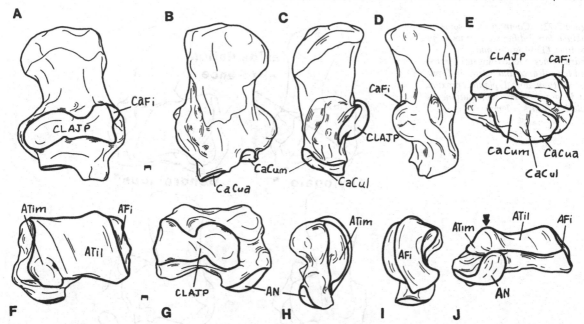

*Figure 7.70.* Dendrolagus matchei, NMV No. C5582. Left calcaneus (above) and astragalus (below). From left to right as in Figure 5.8. For abbreviations see Table 1.1. Scales = 1 mm.

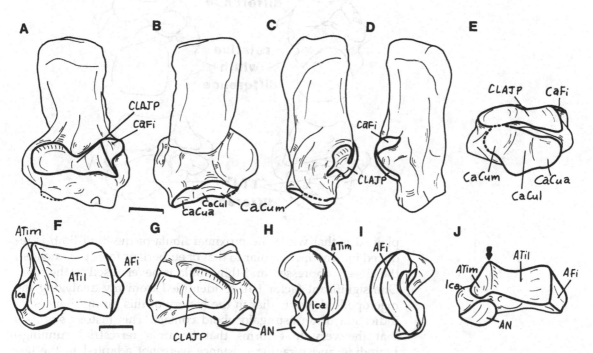

*Figure 7.71.* Bohra paulae, AM Nos. F62099 and F62100, Pleistocene, Wellington Caves. Left calcaneus (above) and astragalus (below). From left to right as in Figure 5.8. For abbreviations see Table 1.1. Scales = 1 cm.

*Figure 7.72.* *Comparison of some proximal tarsal differences between macropodines (Thylogale) and dendrolagines (Dendrolagus). Articulated left calcanea and astragali. From top to bottom: dorsal, plantar, proximal, and distal views. For abbreviations see Table 1.1. Scales = 1 mm.*

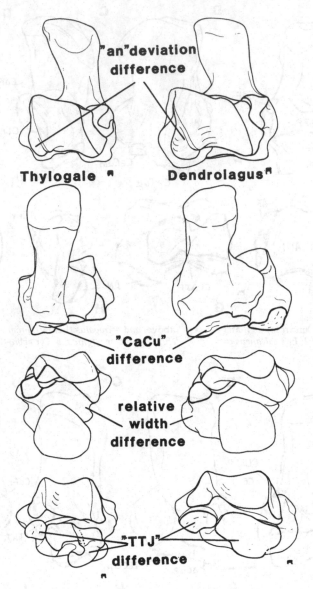

plained is that while the proximal fibula of the hind limb is reduced in a manner similar to that of eutherians (and peramelids), the closely appressed and thin fibula is never fused to the tibia! One significant factor I have identified from my analyzing the macropodoid feet is that in the history of this group there is no indication of an astragalocuboid contact. This history suggests that the very first forms that began a terrestrial running–bounding and foraging existence were not adapted to the terrestrial existence seen in a number of phalangeriform-derived lineages. In the latter groups which are habitually plantigrade with-

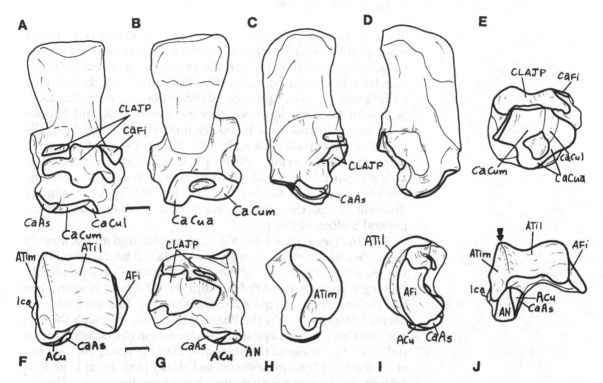

**Figure 7.73.** Sthenurus occidentalis, SAM No. P20811, Pleistocene of South Australia. Left calcaneus (above) and astragalus (below). From left to right as in Figure 5.8. For abbreviations see Table 1.1. Scales = 1 cm.

out digitigrade abilities to run, the cuboid tightly articulates with the astragalus. Like the peramelids, which also used the fourth toe as the sole digitigrade contact when running, macropodoids lack the astragalo-cuboid articulation. But unlike the bandicoots, kangaroos all but eliminate all movement in the **LAJ**, an extreme adaptation in this nonmobile area of contact between the calcaneus and astragalus to spread the great loads transmitted.

Dismissing the validity of the transformation series obtained through functional–adaptive analysis can lead to fundamental misreading of critical information. Specifically, a rigorous cladistic perspective on character analysis (e.g., as in Reig et al., 1987) prevented the acceptance of the *secondary* nature of the separation of the articular facets of the ancestral australidelphian **CLAJP**, particularly as seen in occasional specimens of some macropodoids and peramelids. The secondary waisting of the transversely aligned **CLAJP** is related to independently attained critical **LAJ** stability in open country rapid locomotion, related to different tarsal mechanics.

### Hypsiprymnodontidae

The only living representative of this pedally primitive and early macropodoid radiation, *Hypsiprymnodon*, displays all the diagnostic pedal and tarsal features of the protomacropodoid, including its relatively unreduced hallux (Fig. 7.62). In spite of the retention of the latter and the acquisition of the critical macropodoid adaptations in the foot, however, this macropod runs and bounds quadrupedally and leaps bipedally instead of leaping bipedally only. It retains a prehensile scaly tail which it also uses for counterbalancing and carrying. All the phylogenetically and mechanically diagnostic macropodoid attributes of the tarsus are present, however (see also Flannery & Archer, 1987a, for a slightly different perspective; see also Johnson & Strahan, 1982, on the general biology of this genus).

The **UAJ** (see Figs. 4.4 & 7.63) is, as emphasized above, already of the "eutherian type"; namely the fibula is a lateral splint bracing the astragalus as part of the crural mortise. Correspondingly, the angle between the **AFi** facet and the **ATil** facet is sharp, just like the angle between the **ATim** facet and the **ATil** facet. Both are derived features within the Phalangeriformes, although they do not invariably covary, as the earlier discussion of sparassodontans indicates. But unlike in the Macropodidae, however, the **CaFi** facet is absent in *Hypsiprymnodon*, and thus in this regard the **UAJ** retains the more primitive phalangerid conformation. The astragalar **ATil** facet, however, is gently sulcated for enhanced flexion–extension and transverse stability, in contrast to the more primitive phalangeriform condition.

The **LAJ** has developed the characteristic distal **CaAs** facet seen in all macropodoids. This is not a sustentacular homologue, as the interpretation of tarsal morphology by Reig et al. (1987), or the general misunderstanding of marsupial **LAJ** evolution by Hershkovitz (1992) mistakenly imply. This phyletically new facet simply stabilizes the astragalus on the calcaneus and does not allow any motion. The **CLAJP** of all macropodoids is ridged to prevent **LAJ** movement, and this strategy often results in further fragmentation of that joint in the more derived macropodines and sthenurines.

The **CCJ**, while it clearly reflects the ancestral diprotodontian pattern, already expanded the small strip of the **CaCua** facet into an additional platform distal to the parallel one of the **CaCum** facet. Thus the well-stabilized, step-locked **CCJ** of macropodoids is already fully formed in the living hypsiprymnodontid. The diagnostically enlarged **CaCua**, however, is not extended plantarly and medially as in the Macropodidae. The neck of the astragalus is brought distal to and under the body, and the **AN** articulation is of the flexion–extension variety so invariably present in any living marsupial (and most eutherians) that are rapid terrestrial locomotors.

## Macropodidae

The slow locomotion of kangaroos, described as pentapedal in reference to the use of the tail as a fifth prop, allows the synchronous moving forward of the hypertrophied hind limb of these animals. As the animal increases speed, the characteristic bipedal hop comes to predominate, during which the tail plays an important balancing role. As expected in animals with a hind-limb dominated locomotor system, there has been no reduction of the clavicles or any noticeable modification of the shoulder girdle.

In spite of the locomotor differences of the two families, there is no objective evidence, as far as I am aware, that would support an independent loss of the hallux and an independent acquisition of the **CaFi** facets in the non-hypsiprymnodontid kangaroos, that is, the potoroines and macropodines. In all species of the macropodids, without exception as far as I know, the astragalus is relatively more transverse and its body less deep than in *Hypsiprymnodon*, in addition to having the derived, well-developed concave mortarlike **ATim** facet for the distal articular surface of the medial malleolus. For the obvious phylogenetic and subsequently adaptive causes I have outlined earlier, the ancestral **CLA-JP** is both extended transversely and proximally, and is also pinched; very rarely is it fragmented along the border of the ancestral **LAJ** facets. To varying degrees the selection for **LAJ** stability has produced an extreme syndesmotic connection between an attenuated astragalar medial plantar tubercle (**ampt**) and the calcaneus, resulting in an almost ossified locking contact of the **LAJ** (see also Flannery & Archer, 1987b).

While I do not contest the phyletic distinctness of the Potoroinae (e.g., Case, 1984), there are no significant attributes of the foot (proximal morphology of **Mt5** notwithstanding) that would support such a monophyly independent of a macropodid ancestry. In fact potoroines and most macropodines are remarkably similar to one another. Nevertheless the genus specific features of the tarsus and the rest of the foot of these sundry individual taxa cannot be denied even after a brief comparison of not only the representative figures but also of the specimens.

*Prionotemnus rufus*, the red or plains kangaroo, has evolved a distinctive tarsal morphology in contrast to the selected macropodid genera depicted in the other figures. The sustentacular support of the **CLAJP** is moved laterally and, in contrast to the previously discussed taxa, it lies on the long axis of the tuber of the calcaneus. This narrowing of the calcaneus results in the derived loss of the medial sustentacular shelf. This change is also reflected

in the astragalus with its equally derived and functionally correlated elongated **UAJ** articular areas, and in the relatively transversely narrowed astragalar body.

It is in *Protemnodon* (Fig. 7.65), *Prionotemnus* (Fig. 7.66), and the
Sthenurinae (Fig. 7.73) that the exceptionally developed syndesmotic binding of the astragalus and the calcaneus is strikingly
manifested in a hypertrophied **ampt** and the corresponding area
of contact on the calcaneus. These areas come close to being ossified, and thus effectively prohibit any **LAJ** movements.
Sthenurine kangaroos, while pedally distinctive, also have the
sustentacular portion of the **CLAJP** located more laterally, as in
*Prionotemnus,* compared to other macropodoids. The mechanical
significance of this modification is clear: It loads the foot over the
the most robust area of the calcaneus.

## Dendrolagini

The modification of the tree kangaroos is a classic example of
successful application of ecological morphology to evolutionary
explanation, a vivid illustration of the importance of tested patristic ties with enormous empirical content, a historical–narrative
explanation that successfully accounts for the direction of transformation of these animals from their terrestrial relatives. The
details of the foot morphology of this transformation are of exceptional interest for understanding the practice of functional–
adaptive analysis for phylogenetic (evolutionary) reconstruction.
What makes the assessment of the direction of evolutionary
change (the testing of transformation series) so immediately obvious is an easy understanding, even by students who are not specialists in morphology, that tree kangaroos exhibit the ancestral
specializations of terrestrial hoppers modified into the needs of
arborealists (see also Ganslosser, 1977, 1980; Groves, 1982).

The **UAJ** of tree kangaroos (Fig. 7.67) reflects the relatively
greater transverse spread of this articular area compared to terrestrial kangaroos. The most characteristic attributes of the tarsus,
however, are reflected in the form-function of the transverse tarsal
joint. Both the **ANJ** and **CCJ** have undergone transformation in
conjunction with the movements and the stability requirements of
the narrow and uneven substrate of the arboreal environment in
which tree kangaroos live. Especially instructive is a comparison
of this substrate to the ancestral terrestrial substrate on which the
adaptations for high-speed hopping occurred. The stepped **CCJ** of
the macropodoid ancestor is de-emphasized (Figs. 7.68–7.72), although the ancestral macropodoid taxonomic property is well reflected. The morphology of the head of the astragalus and the
navicular mirrors the stability requirements and movements in
this joint. Rather than the flexion–extension of terrestrial kangaroos, emphasis has shifted to mediolateral mobility in the **TTJ**.

Concomitant with these demands are the relatively more deviated astragalar head and the greater width of the calcaneus in tree kangaroos. These differences are summarized in Fig. 7.72.

Only in one tree kangaroo, *Dendrolagus lumholtzi* (Fig. 7.68), is there a clear modification of the calcaneal tuber. As in slow climbers and hangers, such as some anthropoids and tree sloths (Sarmiento, 1983) for example, the plantar heel process is well developed in this form. This condition suggests an adaptive modification of the pedal flexor originating on the tuber, and probably of the plantar aponeurosis originating on the plantar process of the tuber for flexing the digits. This configuration is an advantage for a quasi-grasping stance of an arboreal kangaroo.

### Sthenurinae

The Plio-Pleistocene fossil kangaroos that have been allocated to this subfamily are characterized by a functional pedal monodactyly, since the **Mt5** is all but atrophied (Tedford, 1966, 1967; Flannery, 1983b). In these forms the astragalar trochlea is laterally rounded and the separation of the distal and proximal components of the **CLAJP** is well established (Fig. 7.73). This configuration is occasionally found in some specimens of *Onychogalea* and probably also in other kangaroos as well. It may be a consequence of extreme **LAJ** rigidity or nonmovements – lack of selection for any movement or selection for ligamentous synostosis to enhance stability. In these forms there is ossification of the ligaments binding the astragalus distally to the calcaneus. Unlike the configuration found in the macropodines, the **CaCua** and **CaCum** facets are plantarly continuous.

**Vombatiformes (Figs. 7.45, 7.74–7.84, & 8.42–8.46)**

The two living families, the Vombatidae and Phascolarctidae, represent only a small part of the once flourishing radiation of these middle- to large-sized marsupials.

In spite of the fact that the koala is arboreal, its arboreal locomotion stands in striking contrast to the large phalangers. The reduction of the tail in the ancestry of koalas has necessitated, so to speak, the hypertrophy of the hands and feet, and the development of curved claws associated with these. When koalas descend tree trunks, they do not proceed head first, and therefore the reversal of hindfeet does not come into play. Lee and Carrick (1989, p. 742) give the following succinct account of the positional behavior of koalas:

> *Phascolarctos* is often viewed as a clumsy and sluggish creature, which clings to branches by its modified feet. . . . This view is conditioned by the most common experience of the animal in captivity or occasional glimpses in the wild, which are mostly individuals resting high up in the trees during daylight hours. While

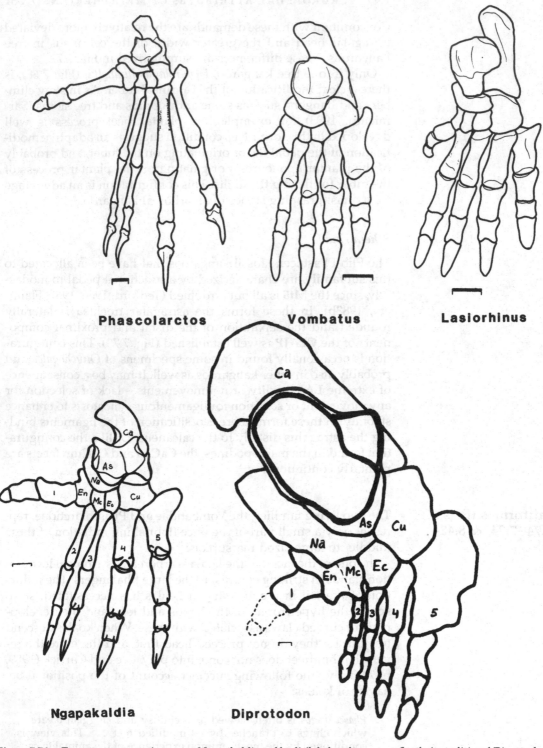

**Phascolarctos**

**Vombatus**

**Lasiorhinus**

**Ngapakaldia**

**Diprotodon**

*Figure 7.74.* Top row: representative left foot skeletons of three extant vombatiforms. Bottom row: reconstructed left feet of the diprotodontiforms

Ngapakaldia tedfordi (left; based on cast reconstruction by Carol Munson, courtesy of M. Woodburne; specimens from UCMP Tirari Desert localities,

South Australia) and Diprotodon optatum (right; from sundry specimens of the Pleistocene Lake Callabonna sediments, South Australia). Scales = 1 cm.

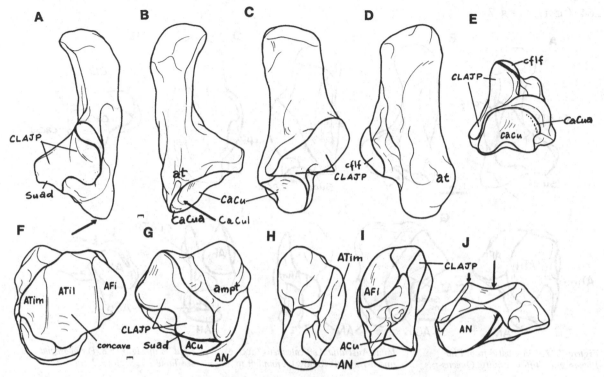

*Figure 7.75.* Vombatus ursinus, AMNH No. 65622. *Left calcaneus (above) and astragalus (below). From left to right as in Figure 5.8. Second arrow* to right points to flexad surface of astragalus, showing extent of trochleation of **ATil** facet in this form and other vombatomorphians below. Arrow to the left, as noted in Figure 7.53 for diprotodontians, shows angulation between **ATim** and **ATil** facets. For abbreviations see Table 1.1. Scales = 1 mm.

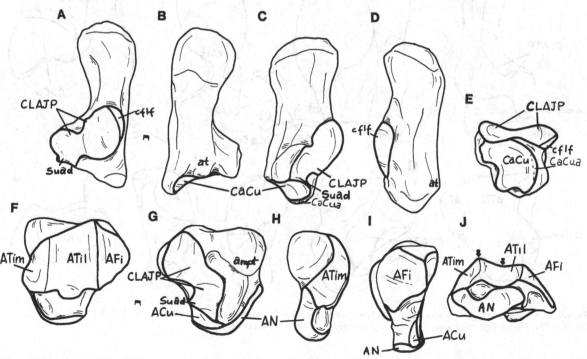

*Figure 7.76.* Lasiorhinus latifrons, AMNH No. 6708. *Left calcaneus (above) and astragalus (below). From left to right as in Figure 5.8. For abbre-* viations see Table 1.1. Scales = 1 mm.

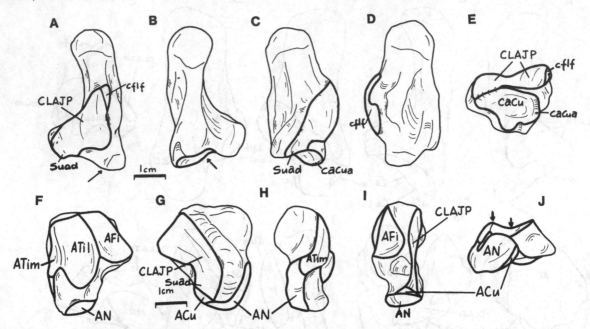

**Figure 7.77.** Vombatid from Plio-Pleistocene Clifton locality, Queens-land, Australia. Left calcaneus (above) and astragalus (below). From left to right as in Figure 5.8. For abbreviations see Table 1.1. Scales = 1 cm.

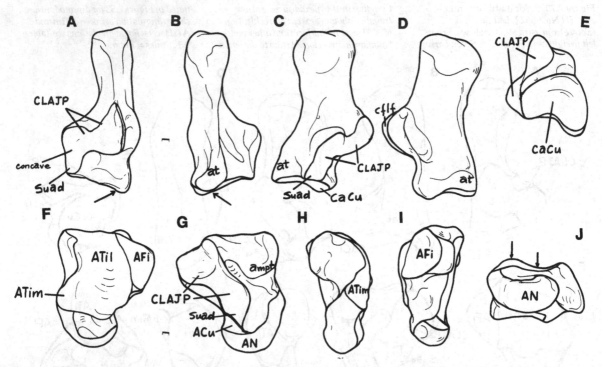

**Figure 7.78.** Phascolarctos cinereus, AMNH No. 65608. Left calcaneus (above) and astragalus (below). From left to right as in Figure 5.8. For abbre-viations see Table 1.1. Scales = 1 mm.

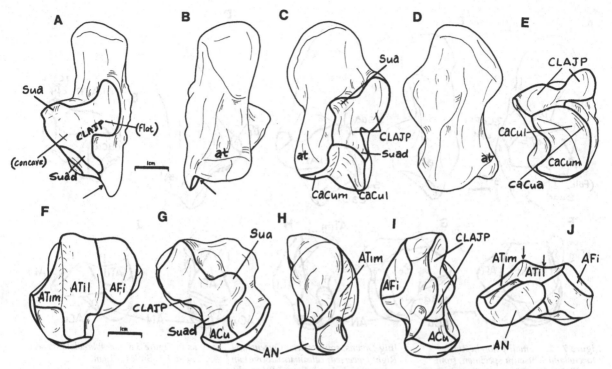

**Figure 7.79.** Thylacoleo carnifex, SAM No. P20806–7, from Pleistocene *Naracorte Cave deposits. Left calcaneus (above) and astraglaus (below). From* *left to right as in Figure 5.8. For abbreviations see Table 1.1. Scales = 1 cm.*

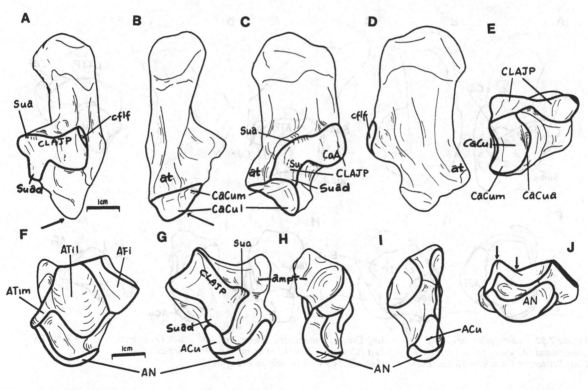

**Figure 7.80.** Ngapakaldia tedfordi, UCMP No. 69813, from middle Miocene UCMP locality V6213, South *Australia. Left calcaneus (above) and astragalus (below). From left to right* *as in Figure 5.8. For abbreviations see Table 1.1. Scales = 1 cm.*

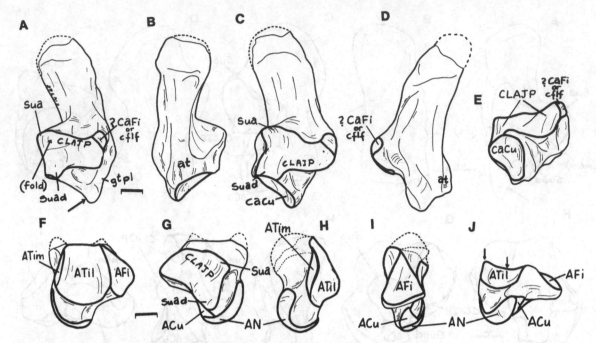

**Figure 7.81.** *Small palorchestid, Queensland Museum specimen, from Plio-Pleistocene Chinchilla Fauna, Darling Downs, southeastern Queensland. Right (reversed) calcaneus (above) and astragalus (below). From left to right as in Figure 5.8. For abbreviations see Table 1.1. Scales = 1 cm.*

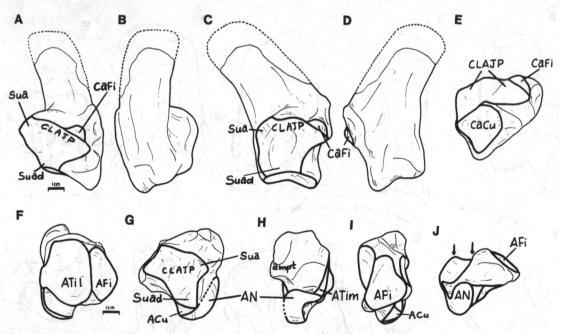

**Figure 7.82.** *Large palorchestid, Queensland Museum specimen, from Plio-Pleistocene Chinchilla Fauna, Darling Downs, southeastern Queensland. Left calcaneus (above) and astragalus (below). From left to right as in Figure 5.8. For abbreviations see Table 1.1. Scales = 1 cm.*

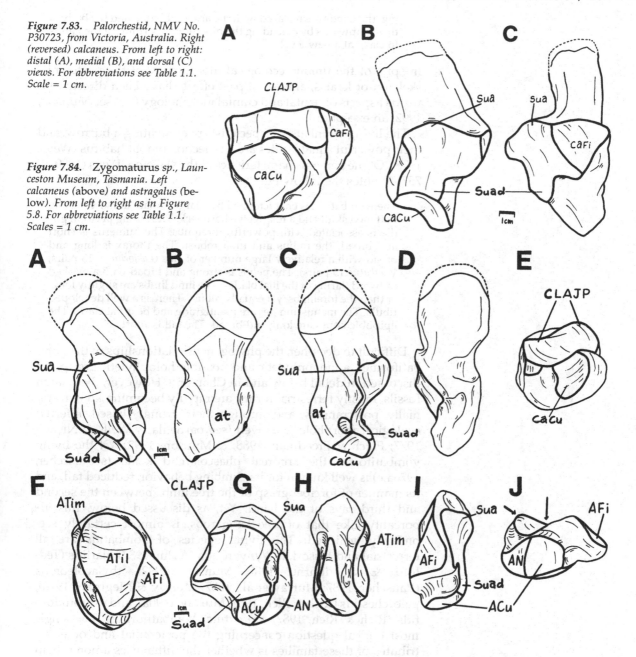

**Figure 7.83.** *Palorchestid, NMV No. P30723, from Victoria, Australia. Right (reversed) calcaneus. From left to right: distal (A), medial (B), and dorsal (C) views. For abbreviations see Table 1.1. Scale = 1 cm.*

**Figure 7.84.** *Zygomaturus sp., Launceston Museum, Tasmania. Left calcaneus (above) and astragalus (below). From left to right as in Figure 5.8. For abbreviations see Table 1.1. Scales = 1 cm.*

*Phascolarctos* usually moves at a sedate pace, the proportionately very long limbs can propel the animal rapidly over the ground or up the trunks of small and very large trees alike. *Phascolarctos* walks by moving the diagonally opposite limbs alternately and runs by moving the forelimb and then the hind limbs in unison. The extended digits of the palm adjacent to the digits touch the ground as does the entire hind foot, with the toes extended and the hallux held at right angles to the axis of the foot. . . . When climbing, the hands are released with the arms extended and the body is thrust upwards by extending the hind limbs, permitting the hands to clasp at a new level.

In spite of the unique ecological morphology of the postcranial skeleton of koalas, almost all past efforts have been directed towards aspects of dental and cranial morphology (but see Munson, 1992, an exception).

The living wombats are specialists in excavating a burrow, and their powerful squat bodies are a reflection of that habitus (Wells, 1978). Of the general characteristics of the skeleton, Wells (1989, p. 757–8) notes the following:

> The wombat . . . is characterized by a large, broad, dorso-ventrally flattened skull and a relatively short neck. The strong pectoral girdle is associated with powerful forelimbs. The humerus is short and broad, the radius and ulna robust. The thorax is long and broad with a relatively large number of ribs (*Lasiorhinus* 13 pairs, *Vombatus* 15 pairs). The pelvis is strong and broad with a marked outward flaring of the ilial blades. The hind limbs are slightly longer than the forelimbs, yet equally robust. There is a well developed fibula. The manus and pes are plantigrade and bear flat claws. The epipubic bones are long and broad. The tail is short.

Difficult to decipher, the phylogenetic relationships of the Vombatiformes, a monophyletic and possibly holophyletic group, are discussed in detail below and in Chapter 8. However, these large fossils, mostly terrestrial forms, are rapidly becoming known cranially, postcranially, and through their strongly based, inferred ecological morphology as well (see particularly Wells & Nichol, 1977; Finch & Freedman, 1988; & Munson, 1992). Of the living vombatiforms the arboreal phascolarctid koala (see Archer, 1976a,e) is well known for its climbing behavior, reduced tail, and its manner of forcipet grasp of the tree limbs between the second and third rays of the hand. Yet, as discussed below, this arboreality, like that of tree kangaroos, is almost certainly secondarily achieved. The living species of vombatids are all terrestrial and fossorial. Wynyardiids (Aplin, 1987), ilariids (Tedford & Woodburne, 1987; Munson, 1992), thylacoleonids (Rauscher, 1987; Murray et al., 1987; Murray & Megirian, 1990), palorchestids (Rich & Rich, 1987; Murray, 1990a), and diprotodontids (Rich & Rich, 1987) are extinct Vombatiformes. The single most critical question concerning the postcranial and other attributes of these families is whether the differences among them merely reflect divergence from a single terrestrial ancestor or are indications of independently attained terrestrial adaptations. For some characteristics, as for some of the taxa, these queries can be answered with greater confidence than for others. The question is particularly difficult to answer about the vombatomorphians (*sensu lato*, including phascolarctids, vombatids, and ilariids) on the one hand, and the diprotodontiforms (including the wynyardioids, diprotodontoids, and thylacoleonoids) on the other. These two groups, I believe, are monophyletic clusters within a possibly natural Vombatiformes. The questions remain whether or not it is likely that some of the vombatiform or even vombatomorphian or diprotodontiform traits are homoplasies, independently derived from an arboreal ancestry that could not be included in the suborder because of its cryptic phalangeriform conformation. Answers to such questions will likely come from the analysis of new fossils and from new functional–adaptive efforts like those of Munson (1992).

One of the most obvious shared attributes of a former, and commonly used, concept of the Vombatiformes is the robusticity of the fifth metatarsal for making the foot more suitable for a terrestrial existence. At present, this relatively simple trait is perhaps better viewed as adaptive, and therefore, because of its relative structural and functional simplicity, a homoplastic similarity rather than a phylogenetic one. The slight emphasis on the robusticity of Mt5 in arboreal forms such as phalangeroids might have laid the foundation for the great hypertrophy of this bone in the groups of vombatiforms. Whether this has occurred once or twice, once in a vombatomorphian and once in a diprotodontiform ancestor, is one of the numerous related phylogenetic questions.

In spite of the grasping foot of *Phascolarctos* (Fig. 7.74) enough details are known of its surface biology and osteology to corroborate the hypothesis that these relatively slow and clumsy arboreal animals are the descendants of a primarily terrestrial, vombatiform, or more specifically vombatomorphian, ancestry. The craniodental and postcranial remains of the vombatoid ilariids, with relatively primitive diprotodontian dentitions (Tedford & Woodburne, 1987), coupled with their wombat-like postcranial adaptations (Munson, 1992), strongly support the special relationship of ilariids and vombatids. According to Munson (1992) the postcranial attributes of the Middle Miocene *Ilaria*, particularly the manus and the pes, are "nearly identical" to wombats. Ilariids retain the phalanger-like pollex and the primitive syndactylous (slender) appearance of the second and third metatarsals, and share a posteriorly *Thylacoleo*-like vertebral column that Munson convincingly attributes to similar mechanical functions serving different biological roles. Given a more primitive vombatiform lower back construction, incipient adaptations to digging may have either contained the adaptations necessary to anchor the hind quarters while dealing with a struggling prey, or else the similarities are convergent.

Although according to Munson, *Ilaria* had a unique magnum and scapula, these elements taken together with the ulnar remains indicate an animal with robust forelimbs. She believes that these traits do suggest some degree of fossorial behavior, although less developed than in wombats. Because the pes is somewhat less modified than that of wombats, Munson questions its effectiveness as a digging or shoveling tool, and further suggests that ascribing to *Ilaria* the fossorial behavior of its fossorial living relatives that dig deep burrows is not supported by the habitus-reflecting vertebral column. The sacrum of the fossil genus, unlike that of wombats, was deep and transversely narrow, and, taken with the relatively stocky and deep centra of lumbars, this evidence suggests much less flexibility than characteristics of living fossorial wombats. The body size of the known species of *Ilaria*, which appears to have been calf-sized or larger, makes a fully fossorial habitus less than likely (Munson, 1992).

As expected in secondarily terrestrial mammals, the stem vombatiform changed its **UAJ**, as seen in wombats. Interestingly, some of these apparently homologous modifications are even recognizable in the arboreal koalas! A mortise–tenon **UAJ** form is approached in vombatiforms in that the **AFi** facet is considerably more angled to the tibial articulation (Fig. 7.45) than in phalangerids. The tibial restraint of the **UAJ**, the internal surface of the medial malleolus, is in fact relatively quite large and restricting in vombatids (but not as much in *Phascolarctos*), almost as fully as in many eutherians. Also, interestingly, the proximal tibio-fibular joint **(TFpJ)** in wombats slightly approaches the eutherian condition in that the fibula articulates more distally (rather than laterally) with the tibia compared to non-peramelid and non-macropodoid marsupials. The **UAJ** in wombats, however, unlike this joint in terrestrial marsupials capable of cursorial speed, is not as much altered laterally as it is medially from a putative phalangeriform condition. In species of *Phalanger*, although not in *Trichosurus*, the **ATil** facet is concave, forming a sulcus for stability. This same condition is somewhat exacerbated in vombatids, but not any more than in *Ngapakaldia* or *Thylacoleo*, and certainly not to the degree seen in the large palorchestids. Partly because of its common recurrence in unrelated groups but primarily because it also occurs in *Phalanger*, this **UAJ** sulcal configuration is not a convincing vombatiform synapomorphy, a point of disagreement between this study and that of Munson (1992).

The fibula continues to be an important element, either being directly loaded as a weight-bearer or functioning as a bracing column, and both the meniscus and the mobile **TFJ** are retained. While terrestrial locomotion clearly demanded different roles from the foot in different lineages, the mechanics of digging were contained within the capabilities of the mobile phalangeroid-like **UAJ**. The more unidirectional and stability-dependent needs of terrestrial cursors and jumpers, such as the bandicoots and kangaroos, required much more limiting (i.e., restraining) form–function solutions than those seen in the slow moving vombatoids. The wombats, in addition to greatly reducing, but not losing, the grasping hallux, show a significantly altered **CCJ**. What was previously a stout bracing wedge of bone, the calcaneal **CaCul** facet, is remodeled to a less restricting articulation between the two bones. In addition, the articulation is transversely widened for stability while allowing new mobilities. Significantly, this secondarily simplified condition, is also present in koalas namely the reduction of the inverted L-shaped **CaCu** facet, in spite of the original arboreal role of this character related to grasping and inversion in the protoaustralidelphian. The terrestrial locomotion of wombats is likely to be responsible for a far wider **LAJ**, a restructured **UAJ**, and a more tightly articulating tarsus than the one in koalas; all of these properties are probably due to the advantages of improved terrestrial load distribution.

Hallucial deviation, so necessary for grasping, is generally correlated with an astragalar head that is medially removed from the proximity of the cuboid both for widening the pedal palmar surface as well as for receiving the loads from the first ray. A sizable astragalocuboid contact, a derived syndactylan condition, however, is present in all known animals discussed under this heading. This astragalocuboid contact is unique in koalas among arboreal syndactylans, and it is very probably retained from a vombatimorph ancestry, in spite of the obligate arboreality of *Phascolarctos*. This vombatiform contact between astragalus and cuboid reaches the extreme in the large *Diprotodon* (Fig. 7.74) where the navicular, the calcaneus, and the cuboid form a large ovoid mortar for the entire rounded head of the astragalus. *Phascolarctos* possesses only a small **ACu** articulation, probably secondarily reduced from a larger one. Vombatomorphians, like diprotodontiforms, share with some species of phalangers a distal extension of the sustentacular facets, the **Sud** facet.

The size and morphology of the entocuneiform of *Lasiorhinus* is more transformed from a phalangeroid-like, primitive syndactylan condition than the entocuneiform of *Vombatus* or the very primitive (phalangeroid-like) entocuneiform of *Ngapakaldia*.

An important point must be made here concerning the loading of the rays of the hand and foot, particularly the distal ends of the metapodials, as these are often interpreted in connection with habitus evaluation. The relative length of these rays is highly adaptive, and tends to reflect either arboreal or terrestrial habitus. The arboreal forms clearly tend to have relatively longer rays and metapodials, and the plantigrade terrestrial forms have shorter ones, for obvious reasons. It is a common mistake, however, to assume that most arboreal forms have less distally expanded metapodials than terrestrial forms – a difficult point to assess visually since the terrestrial forms, such as wombats, have considerably shorter bones. The expansion is significant in terrestrial forms on the distal articular condyles of the metapodials. While arboreal graspers will tend to have more spherical distal metapodial articulations, those of the shorter rayed terrestrial forms are more transversely expanded. In fact, while in the latter, the articular condyle may span the entire distal width of the metapodial, in arboreal phalangers, which also possess greatly expanded distal metapodials, the articular area itself will only span a little more than half of the distal width of the bone.

In addition to providing a detailed description of the skeletal remains of *Ilaria*, Munson (1992) has also described and extensively compared the excellent postcranial fossil remains of the Middle Miocene palorchestid *Ngapakaldia*. In an unpublished master's thesis, Waters (1967) has assessed the osteology of the Pleistocene *Diprotodon* and also described aspects of the postcranials of this genus. In many ways the Miocene palorchestid is one of the most important animals the study of which will expand our perspective and add to our understanding of the evolution of diprotodontiforms and vombatiforms. Munson made extensive comparisons of non-vombatiforms with *Trichosurus* among the phalangerids. Given the large, sheep-sized dimensions of *N. tedfordi*, the best preserved and smallest of the known species, it would be additionally valuable to continue making comparisons with the largest of the living phalangers. The generally primitive vombatiform nature of the skeleton of *Ngapakaldia*, very phalangerid-like in many respects, is well conveyed by Munson's account. From the casts of a composite manus and pes generously supplied by her I could also verify her comparisons independently. Our minor differences of opinion relate only to the significance, either phylogenetic or functional, of some of the objectively presented information in her paper. I am in agreement with Munson (1992) that while *Ngapakaldia* probably was an able climber, it spent most of its time on the ground. Occasional, or may be systematic, exploitation of the leaf resources of the canopy of forests, however, was well within the ability of the species of this genus. Given the size of these animals, and their probable habitual terrestrial locomotion, the deep central sulcus on the astragalus for the tibia (**ATil** facet) is a simple functionally and adaptively significant trait. Because of the simplicity of a groove on the **Atil** facet, its widespread expression in other groups, however, is likely to be convergent. This condition is also well developed in kangaroos, virtually all eutherian cursors, and even occasionally in the very large living cuscuses. Munson draws attention to the fact that while *Ngapakaldia* had strong and robust arms, it was almost certainly not adapted to a fossorial existence. She concurs with others (Flannery, 1983a) who suggested that the larger, some cattle-sized, palorchestids used their arms and hands to bend forage down from trees and bushes, and states that such bioroles may well account for the pectoral limb morphology of *Ngapakaldia* as well. I believe that these assesments probably represent the most strongly corroborated hypothesis for palorchestid hand and arm use for activities other than locomotion. The assessment of the manner of quadrupedal locomotion for the diprotodontiforms has been advanced by Winge (1941), and it also clearly applies to *Ngapakaldia*, but in proportion, however, to the lesser degree of morphological modifications of the latter.

There is nothing in the manus or pes of *Ngapakaldia* and other palorchestids and diprotodontids that might suggest fossorial behavior. The claws, judging from terminal phalanges that are very deep and therefore relatively short, did not resemble those of wombats – in fact they are almost at the opposite end of terminal phalanx construction. Nevertheless the cheiridia might have been too short for such large animals to be effective at habitual climbing.

A plantar and medially twisted tuberosity at the bottom of the medially curved calcaneal tuber appears to raise a functional problem. In some primates, edentates, and other marsupials such a process signifies the hypertrophy of the area of attachment of the digital flexors. In diprotodontoids as in hominids, however, this tuberosity is almost certainly a point of physical support for the plantigrade foot. An accentuated plantar process of the navicular immediately posterior to the entocuneiform, together with medial and posterior extension, may represent additional points of support for a plantigrade terrestrial pes. The latter features on the navicular and entocuneiform, however, may be primitive, or at least found in some phalangerids where it can be tied to "large branch plantigrady," and therefore their significance for habitus postdiction in this case may be doubtful. *Ngapakaldia* shares this character complex with *Thylacoleo*, along with another complex discussed later, except for the entocuneiform which is not known in the latter. While *Ngapakaldia* may or may not have had **CaFi** articulation, the other tarsally known palorchestids possessed an increasingly larger **CaFi** facet – and so did *Zygomaturus*. Unlike these animals, however, the very large and pedally extremely modified *Diprotodon* completely lacks calcaneal contact with the fibula.

Particular note should be taken of the degree to which the **UAJ** articulation was developed for graviportality in the large tapir-like palorchestids and *Zygomaturus* on the one hand, and *Diprotodon* on the other (Figs. 7.82–7.84). The extreme (probably primitive) concavity of the sulcus on the astragali of palorchestids and *Zygomaturus* and the absence of this in *Diprotodon* (Fig. 7.74) in which a nearly flat **ATil** facet can be seen, is a telling testimony to two separate adaptive solutions. These differences, however, are seen in animals which may have been donkey- to large tapir-sized (such as the larger palorchestids), and rhinoceros-sized (like *Diprotodon*). The **UAJ** morphology seen in the latter giant, therefore, may be a reflection of the extreme slowness of these very large animals. They may not have had their limbs fully positioned under the body to relieve the greatly increased loads. In an animal of the size of *Diprotodon* the sharp stabilizing keel between the **ATil** and **ATim** facets seen in the smaller palorchestids, *Zygomaturus*, thylacoleonids, and vombatoids might not have accommodated the significantly greater loads.

A number of genera of other diprotodontoids, were, of course, already well known postcranially before *Ngapakaldia*. Owen's profusely illustrated studies of diprotodontids (*Diprotodon*), the review of the family by Stirton et al. (1967), and popular accounts of the South Australian Lake Callabonna diprotodontids (Tedford, 1973), as well as Waters's (1967) study have increased our understanding of the general ecology of these extinct forms. Given the phylogenetic constraint of syndactyly and the functional constraint of large body size, the diprotodontids have evolved one of the most bizarre of mammalian feet (Fig. 7.74). Not only is the tuber of the calcaneus enormously expanded for weight support, particularly the medial extension discussed above for *Ngapakaldia*, but the **Mt5** is hypertrophied beyond that seen in any other mammal. These graviportal animals had only insignificantly small toes and walked in a pigeon-toed manner. Winge's (1941, pp. 86–87) account of the extremities of *Diprotodon* succinctly sums up their adaptive significance.

> The hand and foot are peculiarly modified as a result of the manner in which they have been used: not for climbing but for walking on the ground. They have developed certain characters remarkably similar to those of megatheriids, which likewise were primarily climbers that have become adapted for ground-living. They trod more upon the wrist and ankle than upon the fingers and toes, holding the hand and foot edgewise in such a manner that the weight of the body fell mostly upon the outer margins. The bones of the wrist and ankle, as well as the fifth metacarpal and metatarsal, became rounded, whereas the fingers and toes diminished in size; least reduced are the fifth finger and toe which take on a small part in carrying the weight.

Comparison of the feet of *Ngapakaldia* and *Diprotodon* (Fig. 7.74) shows that the extreme specializations so well described by Winge are not yet developed in the older genus (see also Archer, 1977c). Nevertheless, as Munson (1992) notes, *Ngapakaldia* probably favored the outer edge of its feet for weight support.

The thylacoleonids are becoming increasingly better known in the fossil record and have been expertly studied (Wells & Nichol, 1977; Wells, Horton, & Rogers, 1982; Murray et al., 1987). The partial foot, along with the hand, of *Thylacoleo* was described and functionally analyzed by Wells and Nichol (1977). Wells and Finch (1982) and Finch and Freedman (1988) have added detailed understanding to the natural history of one of the most interesting extinct Australasian marsupial groups in their study of postcranial morphology. Early in her studies of the genus, specifically *Thylacoleo carnifex*, Finch (1982) has suggested that the nearly equal lengths of the relatively long limbs indicate cursorial abilities in the large mammal. In her later work with Freedman, many of the original points were greatly augmented (Finch & Freedman, 1988, p. 251):

> In the forelimb the radius was clearly longer than the humerus (115%), and [in] the hindlimb the tibia was considerably shorter than the femur (82%). Among the marsupials, the main *Thylacoleo*

indices were more similar to those of *Sarcophilus*, but with some significant differences, notably in propodial/epipodial length ratios. Compared to *Panthera leo* there were many marked similarities. Morphologically, the *Thylacoleo* scapula conforms to that found in walking, trotting, rather than climbing viverrids; the pelvis similarly agrees with that of ambulators and cursors.

Finch and Freedman conclude (1988, pp. 270–71) that:

*Thylacoleo* appears to have been a slow to medium cursor on the evidence of its limb indices and its scapular and pelvic morphology. The animal was probably capable of leaping, as evidenced by the long femur and the strengthening of the dorsal and anterior walls of the acetabulum. The potential for abduction of the limbs, especially of the forelimb, might suggest scansorial ability. However, the animal was large and heavy and it did not have the well-developed subscapularis minor muscle needed to resist the anterior pull on the scapula during climbing. Also, the small claws on digits II-V (Wells & Nichol, 1977) do not appear sufficiently powerful to support a weighty animal when ascending a tree. . . . Thylacoleonid features clearly related to the carnivorous feeding habit are: long neck, a lengthened radius, the facility for axial rotation of the forelimb (possibly also a grooming adaptation), the huge terminal phalanx on digit I of the manus and retention of the abductive capacity of the forelimbs. The stout olecrenon process and the unusually expanded fifth metacarpal suggest a powerful strike, a necessary accompaniment to efficient use of the claws in catching prey.

A few tarsal attributes discussed below appear to add weight to the hypothesis of terrestrial existence in the large thylacoleonids. Given the relatively small and not particularly robust phalanges of *Thylacoleo* compared to its metapodials, however, I do not believe that the family was habitually digitigrade. While these predators could obviously run, their structure did not remotely approach the locomotor functional complex seen in the skeleton of cats, particularly in their tarsal and metapodial mechanics. Given what I can assess, these animals were probably incapable at maintaining running speeds anywhere near those attained and sustained in kangaroos. Diprotodontoids and vombatoids, however, were a widely available prey for the dentally and cranially (and undoubtedly behaviorally) supercarnivorous thylacoleonids.

Although quite modified from a phalangerid-like pattern, the tarsals of *Thylacoleo* (Figs. 7.79, 8.45, & 8.46), like the palorchestid *Ngapakaldia*, retain, the australidelphian calcaneal and cuboid **CaCul** facets, but have considerably modified **UAJ** and **LAJ** articulations. It cannot be objectively stated that these animals are less modified than vombatimorphians – they are merely differently transformed. The distal extension of the astragalar **ATil** facet onto the neck of the bone in *Thylacoleo* is reminiscent of the squatting facet one finds in the sparassodontans and in living mammals such as some primates that habitually keep the foot flexed onto the crus. It is not implied that such mechanics are arboreally linked. In fact this condition is primarily related to a crouching position. The channel under the sustentacular portion of the calcaneus that shunts the **fft** and **fdbt** is relatively narrower in *Thylacoleo* and *Ngapakaldia* than in phalangerids, phascolarctids, or vombatids. The most significant difference of the **LAJ** of known thylacoleonids from those of phalangeroids and other australidelphians (except the large diprotodontids) that I notice is the degree to which the critical articular areas lost their concavity–convexity. The arcs of these **CaA** and **Su** facets are far smaller than the corresponding values in the more concave and covex facets of other diprotodontians. Drastic reduction of the arcs of the **LAJ** suggests a considerable shift toward translation in that joint from the antecedent transverse axial rotation of the astragalus and calcaneus on one another and of the astragalus on the crus. This **LAJ** attribute is probably related to the troughlike channeling of the **ATil** facet and the extension of this facet distally on the astragalus. These attributes may have facilitated greater transverse stability in the foot while allowing for anteroposterior adjustments.

Thylacoleonids and diprotodontoids (diprotodontids and palorchestids) appear to share some characters to the exclusion of other known syndactylans. One such attribute of the tarsus is the conformation of the **LAJ**. The **Sua** facet, a unique extension of the sustentacular articulation among syndactylans, is present in both of the *differently* adapted diprotodontoids and thylacoleonids, suggesting that this feature is a shared apomorphy of the last common ancestor of the families, rather than a convergence. In addition to a well-developed **ACu** contact, the narrowed channel for the pedal flexors and the medial extension of the sustentaculum, that is, the **Sua** facet, may link thylacoleonids (Fig. 7.79) with the best known of the primitive palorchestids, *Ngapakaldia* (Figs. 7.74 & 7.80), with all other tarsally known palorchestids (Figs. 7.81–7.83), and with the diprotodontids (Fig. 7.84). *Thylacoleo* shares with both vombatids and palorchestids the plantar projection of the navicular. For the reasons offered during discussion of the palorchestid evidence, this attribute suggests a plantigrade stance. The foot of *Ngapakaldia*, then, cannot be considered as to generally represent a more primitive condition than that of vombatoids – it displays critical apomorphies some of which it shares with thylacoleonids. Although certainly not fossorial, these shared diprotodontiform attributes are differently derived than their homologues in the vombatoids.

# 8

# *Taxa and phylogeny of Metatheria*

> *it is necessary to emphasize the distinction between definition and the evidence that the definition is met. . . . Another way to put the matter is to say that categories are defined in phylogenetic terms but that taxa are defined by somatic relationships that result from phylogeny and are evidence that the categorical definition is met.*
>
> Simpson (1961a, p. 69)

> *That is the basis of the challenge for the incorrigible phylogenetic detective: most puzzle pieces gone and the ones that remain [are] lying their heads off (may be) about their original position in the phylogenetic tapestry. With handicaps of that sort the best we can hope to achieve is an hypothesis of relationships involving the fewest* improbable *assumptions about the missing parts of the puzzle.*
>
> Archer (1984b, p. 760)

The known groups of metatherians as I recognize them (and they most likely were not the only marsupial mammals) are classified at the beginning of the book in Chapter 2, and these taxa are diagnosed here. In the diagnoses I cite and discuss taxonomic properties, those well tested ones in which I have confidence, inasmuch as they are the derived features probably present in the last common ancestor of each group (see Rowe, 1987; Szalay et al., 1987). These are the hypothesized (and corroborated to varying degrees) apomorphies of the postulated ancestors in contrast to their respective antecedents. The scientific validity of groups so diagnosed is dependent on all the objective information against which such concepts can be tested. I should add here that providing definitions, where all or most members of a group possess a common trait, is not always the hallmark of a genuinely phylogenetic enterprise. Note that the quote from Simpson, above, refers to a concept of definition that is identical to the concept that I shall continue to refer to as diagnosis. Many homologous, *nonidentical,* traits of members of a group are often considered "synapomorphies." Yet, as I have developed this argument, it is the various explanations that support and allow assumptions that such homologies were indeed derived and subsequently modified from the designated ancestral apomorphy of a group. These latter taxonomic properties are used to diagnose a monophyletic group at its root (see Fig. 2.2).

Given the practical necessities of gaining access to the testing of H-N Es against well-corroborated taxonomic properties, the groups recognized here cannot be isomorphic equivalents of the most corroborated cladograms. Enormous gaps remain in the factual knowledge of Cretaceous therian diversity, hampering its evolutionary understanding. Whenever probabilities of cladistic relationships are not sufficiently convincing, morphological divergence is not well established, or the evolution of one well-diagnosed group from another equally well delineated one is highly probable, paraphyletic (horizontal, stage, but certainly not "grade") groupings are a necessity. This approach is not merely pragmatic; its sound theoretical foundations are based on evolutionary realities. Therefore such an approach is preferred here. Autapomorphies (or so designated apomorphies) of groups are probably often the antecedent conditions for the apomorphies of other groups. The Tribotheria, which includes all tribosphenic forms that are not likely to be cladistically either Metatheria or Eutheria, is such a monophyletic taxon. This group maybe viewed as the formal, higher level "stem group" of the other Theria.

In addition to the diagnoses, I also attempt to present below the supportive and corroborative historical–narrative explanations of taxonomic properties and the taxa. Some of these were detailed in Chapters 6 and 7. Some, but not many, of the traits in the diagnoses are soft anatomical features. Given my area of emphasis, and because many fossil groups either represent or at least hold the promise of an eventual uncovering of critical missing information, I concentrate primarily on those cranioskeletal traits that have been studied in both the living and fossil samples. Clearly I do not imply that soft anatomical and molecular evidence is not important for discovering and testing relationships, and therefore grouping living organisms. Yet it is a fact that a large part of the mammalian record, in this case the marsupial one, is known only as fossils. It is for the reason of potential temporal connectedness, without any desire to imply a "hierarchic" perspective for different kinds of evidence, that I, like others before me, firmly believe that evolutionary morphology of hard tissues will remain the central focus of all those phylogenetic efforts that strive for a *complete* account of evolutionary history. Unlike the more controversial debates about reproduction, metabolic rates, and encephalization, it is the patently obvious simple fact that hard anatomy has been the connecting link between the extant and extinct, and therefore has the potential for linking the greatest span of the available evidence. I also hold, I have already emphasized, that we have not even begun the causal (i.e., mechanical, ecological, and behavioral) linkage of soft anatomical evidence with hard morphology, and therefore we have not yet begun to realize fully the potential of all the available osteological evidence for taxonomy.

Evolutionary morphology of the mammals after the late nineteenth century and early-to-mid-twentieth century blossoming has not soared, as one might have hoped, after the solid outlines of the Evolutionary Synthesis were fitted together. In general, paleomammalogists, some mammalogists, and functional anatomists carried on the tradition of evolutionary analysis of form, but Davis's (1964) exemplar monograph showed its great strengths (as well as its avoidable pitfalls) and the enormous potential of evolutionary morphology.

Still waiting to be widely appreciated, however, is Bock's (1989) recent call (see also Bock, 1990, 1991) for functional–adaptive research on organisms which can connect structural and functional attributes and lead to new understanding of the taxonomist's most important data – the characters. Quantitatively reassuring fixes using algorithms are held out as the promise for purely cladistic (classificatory) attempts at resolution of the phylogenetics of mammals and other groups. Variously manipulated cladograms are often claimed to "win" when they have the greatest number of putatively synapomorphous characters at nodes, and thus various problems are said to be either resolved or not. Strangely, a pessimistic attitude about the potential of morphology as the central column of data for deciphering evolutionary history is most manifest among a number of paleontologists who seem to have turned with increasing enthusiasm toward not only algorithms but the promise of molecular phylogenetics (e.g., McKenna, 1987, and references therein). In fact, in Gould's (1985, 1986) widely read statements (one of which was astonishing in its extreme espousal of the idea of a pure genealogy to be lifted from the molecular evidence, the "perfect homology"; and as such akin to the braggadocio of a few students of molecular genealogy), he has claimed that, morphological evidence is an "inherently flawed measure of genealogical relationships" (Gould, 1986, p. 68). Yet this view astutely captures the mood and the implicit beliefs of a number of leading taxonomists and paleontologists whose respect for technique has sometimes overshadowed the potential of the morphological evidence. Given either their technique-cum-methodological, stratigraphic, or taxic perspectives or some other agendas, these workers have come to view the enterprise of evolutionary morphology as a practice which should *follow* the "establishment of genealogies" (i.e., cladistic statements) in order to add "scenario" to the "genealogy." The functional morphologist Lauder (1990) is also a recent proselytizer of the perspective which would give a taxic cladistic analysis primacy in determining the most probable transformation series of characters. Apparently, the nineteenth-century practice of enumerating traits from skeletal remains in order to sort them into phenons and link them using various techniques, as practiced by most taxonomists, has probably reached its zenith. It may well be that an implicit pessimism concerning the multifaceted research required for evolutionary morphology is the reason for the panacea-like appeal of taxic cladistic techniques, even though they are often based on false evolutionary assumptions and on highly questionable taxonomic properties.

Of course the practice of taxonomy means different methods and subsequently different techniques and procedures to different students, although, presumably, all aim for the best understood phylogeny and its most consistent expression in a usable classification for nonspecialists (see Szalay & Bock, 1991). Yet paradoxically, the overwhelming taxic focus tends to rob taxonomic practice of its firmest foundations: the evolutionary morphology of those characters that are complex enough to allow for the ruling out of convergences and even parallelisms. It is morphological evidence, *interpreted* from the fully integrated perspective of evolutionary causality, evolutionary mechanisms, developmental, and functional–adaptive processes, that holds the promise for the understanding, and therefore testable interpretation, of the single greatest continuous line of evidence – the anatomical record as reflected in the hard morphology.

## Subclass Theria Parker and Haswell, 1897

### *"Diagnosis," and additional cynodont and mammalian attributes of the last common ancestor of therians*

The last common ancestor of *living* therians probably had: a didelphid-like pattern of reproduction (oviparity for some tribotherians cannot be ruled out), with a bifid reproductive tract, and male with the scrotum anterior to a bifid penis; neonates which after birth became attached to fully formed nipples and performed "pumpsucking," driven by the tongue and the floor of the mouth; and probably well-developed facial vibrissae. The first therians (in the sense employed here, the first mammals with tribosphenic molars) had: five upper and four lower incisors and at least seven postcanine tooth families; a tribosphenic molar morphology with relatively small, but clearly established protocone, a well defined talonid basin (not only a high buccal crest) with a relatively large hypoconid, and the presence of a smaller hypoconulid and probably an entoconid, a fossa distal to the trigonid into which this protocone occluded, and the primitively smaller (not subequal) metacone than paracone retained from the nontherian ancestry, without a reduction in the size of the paracone as seen in some archaic metatherians; a lower jaw which opened by digastric and hyoid complex of muscles; septomaxilla retained; broad nasal-lacrimal contact; the floor of the tympanic cavity formed by fibrous membrane, possibly with ventral fenestra in embryo in contact with the mandibular angle, probably somewhat inflected; ectotympanic was an incomplete narrow bony C-shaped curved plate, possibly causally related to fibrous attachment of angle; internal carotid artery, entering through a posterior carotid foramen, split into stapedial and promontory branches, the latter entering the cranial cavity through the anterior carotid foramen; superior sagittal sinus drained through sigmoid sinus through the foramen magnum and through the prootic sinus which runs into the lateral head vein; the lateral head vein which drained through the postglenoid foramen; epitympanic recess either retained or independently evolved; bicrurate, stirruplike stapes that accommodated stapedial artery; palate possibly fenestrated; pattern of cynodont cranial articular area in which the jugal anteriorly borders the dentary-squamosal articulation retained; epipubic ("marsupial") bone of cynodont ancestry retained; possible presence of a small patellar sesamoid ossification; scapula, with well developed supraspinous fossa, developed from the lateral part of the coracoid-scapular anlage retained from ancestry; coracoid process fused to the scapula, developed from the anlage of the metacoracoid; a procoracoid embryonic anlage probably expressed in the adult as a "praeclavium"; dermal clavicle, and a manubrium of the sternum that is partly an ossification rostrally of the anlage of the interclavicle, and also partly formed from small portions on either side of the coracoid anlage (whereas the rest of the manubrium and the sternum developed from the paired embryonic sternal bands); metacarpals, metatarsals, and phalanges with two sets of ossification (on **Mcp2–5** and **Mt2–5** distal epiphyseal ossification only and proximal one for body; on **Mcp1** and **Mt1** with proximal epiphysis only; all phalanges with proximal epiphyses only); prehallux and prepollex retained; carpus with independent (unfused) centrale and probably unreduced lunate; large fibula in close synovial contact distally with tibia, and proximally with both the tibia and femur (the so called fabella or parafibula associated with the head of the fibula); possible retention of sacral ribs in adults (present in opossum embryos); foot which was probably semi-plantigrade; **UAJ** of the tricontact, primitive mammalian type, but with the three articular contacts (**CaFi, AFi,** and **ATi** facets) contiguous with one another; **AFi** facet transversely narrow (the presence of a meniscus, with lunula, between astragalar and fibular **AFi** and **CaFi** facets uncertain, but not unlikely); cuboid distal rather than distomedial to the calcaneus; probably no astragalocuboid contact; head, if not the neck, of astragalus well differentiated from body; articulation with navicular through a convex astragalar head; all digits falcula (claw) bearing.

### Distribution

Valanginian (Early Cretaceous; 141–135 MYBP?) to Recent.

### Included taxa

Infraclasses Tribotheria, Metatheria, and Eutheria.

## Discussion

Gauthier, Kluge, and Rowe (1988), in their discussion of the importance of fossils in amniote phylogeny (see also Donoghue et al., 1989), provide a large list of soft anatomical and osteological features putatively present in the protomammal of their definition. Employing features of the soft anatomy, but also osteology, the exact delineation of the character states for a particular taxon are replete with difficulties. Although the literature on the mammalian background of the therians (all tribosphenic mammals) is enormous and cannot be covered here (see Kemp, 1982; Kermack & Kermack, 1984; Szalay et al., 1993a, b; and references therein), all these contributions are relevant to the understanding of the therian mammals.

There is no clear agreement in the literature on what might constitute a widely accepted concept of the Theria. In an important recent paper, for example, Kielan-Jaworowska (1992) included the Symmetrodonta, Eupantotheria, and Monotremata in the concept of Theria, in addition to the Peramura, the latter as a stem group whence both the Metatheria and Eutheria originated independently. Her phylogenetic hypothesis deserves careful scrutiny; however, I can only deal here with the tribosphenic animals with relevance to the concepts of Theria and Metatheria that are espoused here.

In spite of the profuse taxonomic literature on the history of concepts surrounding Metatheria and Eutheria, as well as the various hypothesized and named nodes below the concept of Theria, the latter, as used here, is a widely accepted holophyletic group, even with the inclusion of the stem taxon Tribotheria. The infraclass Tribotheria, which by necessity remains paraphyletic, is a useful classificatory entity. It is at the very root of the subclass Theria for those taxa that we cannot allocate with confidence to either Metatheria or Eutheria. Tribotherians may have been reproductively and physiologically marsupial, even "looking like" a didelphid (see also Archer, 1984a, pp. 593–5, for essentially identical views on this), although the causally linked osteological correlates of ovoviviparous versus oviparous reproduction have not been unequivocally recognized in mammals (see Chapter 3 on epipubic bone bioroles). The Aegialodontia is the earliest known and most primitive stage of what may be considered a (nearly) fully developed tribosphenic molar bauplan. In spite of the suggestion of the wear facets on the lower teeth, whether or not it had a protocone, is not known in the published literature. I only tentatively include it in the Tribotheria. The Pappotherida, on the other hand, represents an assemblage of fully tribosphenic mammals of as yet unspecifiable ties to either the metatherians or eutherians.

It is certain (as indicated where I believe it to be corroborated) that in characterizing the first therians in the "diagnosis" I have included characters that were not apomorphies of this stem, but rather plesiomorphies retained from a remote cynodont, mammalian, and atribosphenic background. Nevertheless, because my account focuses on the Metatheria, I consider it important to present as fully as I can the structural and particularly the osteological, background from which the Metatheria might have evolved. This approach I hope, gives a fuller perspective on the various hypotheses of evolution detailed in the diagnoses of taxa below. Gregory (1951, pp. 366–7), for example, has compiled a long and valuable list of attributes characterizing what he considered metatherian, and therefore possibly therian, modifications to the cranium seen in an ancestral form. He used the didelphid condition as a representative of the cranial morphology of this ancestry. The recent functional–adaptive review of therian cranial development by Maier (1993) is also of particular significance for future assessments of cranial characters. Maier pursued the thesis that the development of skull of the altricial pumpsucking marsupial neonates reflects an ancient therian neonate adaptation, still exhibited in marsupials. He considered middle-ear development to be primitive therian in didelphid marsupials, and derived therian in eutherians, due to the prolonged intrauterine development of the latter (see Chapter 3 under the discussions of developmental and reproductive biology).

Based on careful analytical examination of developmental evidence, the contributions of Presley (1979; 1993) have simplified and sharpened our understanding of basicranial arterial circulation. There appears little doubt that he is correct in noting that the mammalian pattern, retained in monotremes and in the first marsupials and protoplacentals, consisted of a single internal (medial) carotid splitting into the stapedial and the medial branch, which I will call here the promontory artery. There is no reason to assume that the insectivoran or other eutherian promontory arteries are any other than the distal medial segments of the original medial branch of the internal carotid.

The problem of reconstructing the cranial blood vessels in mammal-like reptiles, early mammals, and various Cretaceous atribosphenics and therians has been pursued with some vigor, as these patterns are potentially important taxonomic properties. As stated by Gow (1991, p. 142),

> "Early cynodonts have separate prootic and opisthotic bones. The position of the suture between these bones is fundamental to understanding the prootic canal because it is in this suture that the canal is formed. In tritylodontids the prootic and opisthotic bones fuse to form the periotic bone, and in mammals the compound bone develops a new structure, the promontorium, and is then called the petrosal bone.

Although there is relatively little disagreement on the homologies of the bony elements of the base of the skull, the terminology applied to the cranial arteries and veins through the years has varied greatly and obviously complicated communication and consensus concerning not only a reconstruction of morphological patterns but also the establishment of these in extant forms.

The extensive studies of Wible (1984, 1990, 1991) and Rougier et al. (1992) on the ontogeny of the mammalian circulatory patterns resulted in much valuable information and analysis, particularly in clarifying difficult issues of development in the circulatory system in monotremes and therians. The ontogeny and inferred phylogeny of development of the chondrocranium and the fate of the sagittal and lateral head veins in monotremes, marsupials, and eutherians are depicted in Figure 8.1 (courtesy of J. R. Wible). Wible (1989, p. 44a) reports the following relevant comparisons and conclusions concerning a dichotomy in circulation inferred from the petrosal bone topography in mammals:

> In the monotreme *Ornithorhynchus* and presumably in morganucodontids, triconodontids, and multituberculates, two well developed vessels, the stapedial artery and lateral head vein, run through the middle ear on the surface of the petrosal. Whereas the stapedial artery is lacking in marsupials, the lateral head vein disappears in eutherians. . . . the major vein draining the cranial cavity, the postglenoid vein, is not homologous in marsupials and eutherians . . . [and] sulci for the inferior petrosal sinus, stapedial artery, and facial nerve are not unique to marsupials and eutherians among mammals (see particularly Figs. 8.1–8.4).

In their analysis of the basicranium, particularly the petrosal, of the atribosphenic mammal *Vincelestes*, a eupantothere, Rougier et al. (1992, p. 213), consider the following combination of craniovascular features to be part of the protomammalian (and also nonmammalian cynodont) pattern:

> Included are well-developed posttemporal, ascending, and orbitotemporal channels whithin which we reconstruct the arteria diploetica magna, ramus superior, and ramus supraorbitalis, respectively. . . . This pattern has been modified to varying degrees in all Recent mammals. Though monotremes retain what we believe is the primitive venous pattern, both the platypus and echidna lose the connection between the arteria diploetica magna and the ramus superior. Both the arterial and venous systems are drastically altered in marsupials, with the reduction of the stapedial system, arteria diploetica magna, and lateral head vein. The venous pattern is further altered in placentals, with the total loss of the lateral head vein. Though the primitive arterial pattern is modified within the placental orders, all of the major components of that pattern in different proportions are retained in some lipotyphlans (Wible, 1984, 1987).

The difficulty, recognized early by van Gelderen (1924), is the lack of clear homology between the veins exiting through the postglenoid foramen in the two extant therian groups. Wible

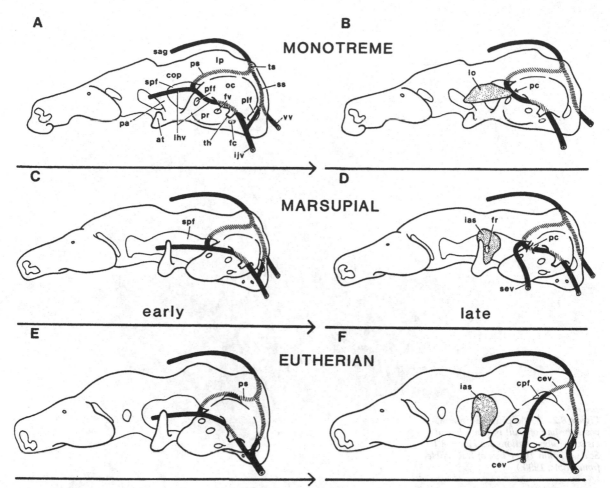

A

**MONOTREME** B

C

**MARSUPIAL** D

early

E

**EUTHERIAN** F

late

*Figure 8.1.* Schematized chondrocranial developmental patterns depicting selected venous drainage in early (A, C, E) and late (B, D, E) developmental stages in the three major groups of living mammals (courtesy of J. R. Wible, from Wible, 1990). See text for discussion. Abbreviations: **ac,** aqueductus cochleae; **acf,** anterior carotid foramen; **as,** alisphenoid; **asc,** gyrus of anterior semicircular canal; **astp,** alisphenoid tympanic process; **at,** ala temporalis; **av,** aqueductus vestibuli; **bo,** basioccipital; **bs,** basisphenoid; **cc,** crus commune; **cev,** capsuloparietal emissary vein; **cf,** condyloid foramen; **cop,** commissura orbitoparietalis; **cp,** crista parotica; **cpf,** capsuloparietal foramen; **cr,** crista petrosa; **ctpp,** caudal tympanic process of petrosal; **e,** ectotympanic; **eo,** exoccipital; **er,** epitympanic recess; **et,** groove for eustachian tube; **fc,** fenestra cochlea; **fi,** fossa incudis; **fm,** foramen magnum; **fn,** facial nerve; **fnb,** facial nerve bridge; **fo,** foramen ovale; **fr,** foramen rotundum; **fs,** facial sulcus; **fv,** fenestra vestibuli; **gf,** glenoid fossa; **gg,** geniculate ganglion; **gpn,** greater petrosal nerve; **hF,** hiatus Fallopii; **hff,** hypoglossal foramina; **i,** incus; **iam,** internal acoustic meatus; **ias,** intramembranous part of alisphenoid; **ijv,** internal jugular vein; **ips,** inferior petrosal sinus; **ju,** jugal; **lf,** lateral flange of petrosal; **lhv,** lateral head vein; **lp,** lamina parietalis; **lt,** lateral trough of petrosal; **lw,** lateral wall of epitympanic recess; **m,** malleus; **mcf,** part of petrosal in middle cranial fossa; **me,** mastoid exposure; **mp,** mastoid process; **oc,** otic capsule; **pa,** pila antotica; **pc,** prootic canal; **pcf,** prefacial commissure; **pcv,** vein of prootic canal; **pe,** petrosal; **pff,** primary facial foramen; **pgf,** postglenoid foramen; **pgp,** postglenoid process; **plf,** posterior lacerate foramen; **pm,** postmeatal process; **pp,** paroccipital process; **"pp,"** paroccipital process of exoccipital; **pr,** promontorium; **ps,** prootic sinus; **pt,** pterygoid; **ptf,** posttemporal foramen; **ptn,** posttemporal notch; **ri,** ramus inferior of stapedial artery; **rs,** ramus superior of stapedial artery; **rtpp,** rostral tympanic process of petrosal; **s,** stapes; **sa,** stapedial artery; **saf,** subarcuate fossa; **sag,** superior sagittal sinus; **sdv,** sulcus for diploetic vessels; **sev,** sphenoparietal emissary vein; **sf,** stapedius fossa; **sff,** secondary facial foramen; **sips,** sulcus for inferior petrosal sinus; **smn,** stylomastoid notch; **so,** supraoccipital; **spd,** sulcus for perilymphatic duct; **spf,** sphenoparietal fenestra; **sps,** sulcus for prootic sinus; **sq,** squamosal; **ss,** sigmoid sinus; **ssev,** sulcus for sphenoparietal emissary vein; **ssf,** subsquamosal foramen; **sss,** sulcus for sigmoid sinus; **tcf,** transverse canal foramen; **tf,** trigeminal fossa; **th,** tympanohyal; **ts,** transverse sinus; **ttf,** tensor tympani fossa; **vcn,** vestibulocochlear nerve; **vdm,** vena diploetica magna; **vv,** vertebral vein.

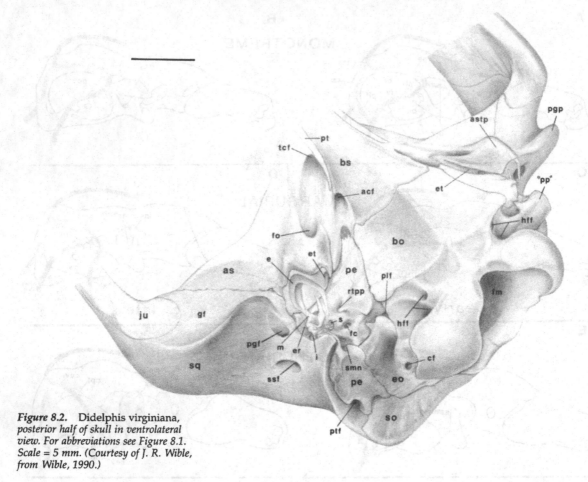

**Figure 8.2.** Didelphis virginiana, posterior half of skull in ventrolateral view. For abbreviations see Figure 8.1. Scale = 5 mm. (Courtesy of J. R. Wible, from Wible, 1990.)

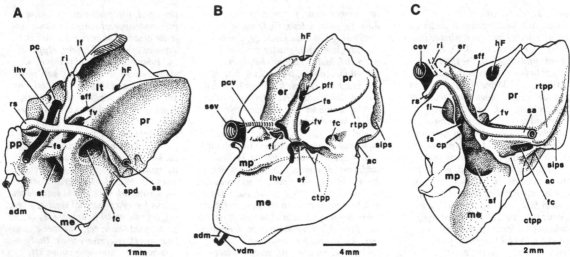

**Figure 8.3.** **A.** Eozostrodon (Triassic), **B.** Archimetatherian, Type A (FMNH No. PM53907, shown also in Figures 8.4–8.5; latest Cretaceous; Bug Creek), and **C.** Eutherian (AMNH No. 118643; latest Cretaceous; Bug Creek). Comparison of ventral (tympanic) surfaces of petrosals and their reconstructed vasculature. For abbreviations see Figure 8.1. (Courtesy of J. R. Wible, from Wible, 1990.)

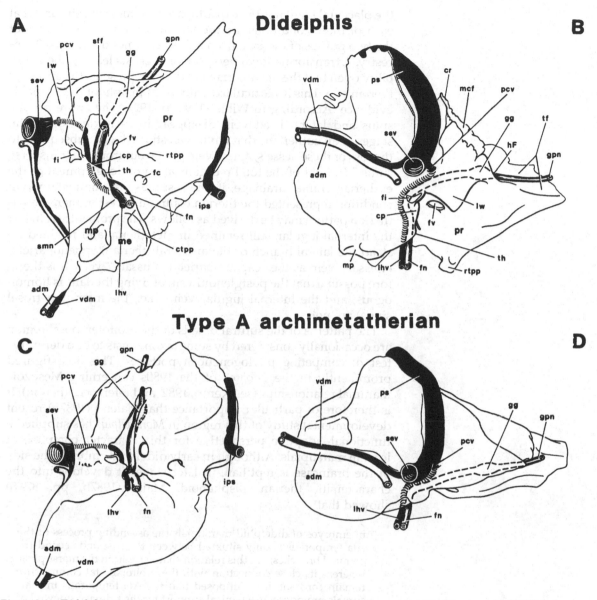

**Figure 8.4.** Didelphis virginiana (above) and archimetatherian, Type A (below). Comparison of ventral (tym-panic; C, D) and squamosal (A, B) sides with reconstructed vasculature and nerves. For abbreviations see Fig-ure 8.1. (Courtesy of J. R. Wible, from Wible, 1990.)

(1989) believes that the eutherian vessel, the capsuloparietal emissary vein, is not a strict homologue of the metatherian vein exiting the postglenoid foramen, and considers the two conditions equally derived from a postulated last common ancestor of the living therians. Yet there are potential problems with the notion that the metatherian jugular is the sphenoparietal emissary vein, whereas the placental jugular is the capsuloparietal emissary vein, and that each of these conditions are independently unique attributes of the two groups of therians from a different (monotreme-like?) ancestry. A transformational perspective suggests leaving the possibility open that the metatherian condition represents the ancestral therian one. This is particularly important in light of the available evidence. According to Wible (1990, p. 192) "The entire prootic sinus and lateral head vein disappear in early developmental stages of all eutherians that have been studied. . . . Subsequent to the loss of these vessels, a new vein grows out from the transverse sinus." In light of the total reorganization of development of the eutherian cranial drainage, if one assumes that the metatherian condition represented the therian common ancestor, then the eutherian pattern may be derived as follows. The protoeutherian lost the internal jugular, still retained in Metatheria, and retained instead the lateral branch of the ancestral (therian) sphenoparietal emissary vein as the "capsuloparietal emissary" vein. It is therefore possible that the postglenoid vein of living therians is homologous, and the internal jugular veins (i.e., the inferior petrosal sinus) are not.

The patterns of the sutural details of the orbitotemporal region are occasionally considered by some taxonomists to be a definitive test of competing phylogenetic hypotheses. This area figured prominently in the debates of the 1980s concerning Mesozoic mammal relationships (see Kemp, 1982, and references therein). It is therefore of particular importance that Maier's (1987b) recent developmental study of this region in *Monodelphis* has supplied a functional–adaptive perspective for this character complex, at least in marsupials. Although in early didelphid embryos the side of the braincase is reptilian, in later stages it develops into the charactersitic therian alisphenoid. Maier (1987b, pp. 87–8) showed that

> in embryos of didelphid marsupials the ascending process of the Ala temporalis is only situated between the first and second trigeminal branches; . . . this relation changes within the marsupials, whereas its close connection with the orbitoparietal commissure remain constant; it is supposed that the Ala temporalis in marsupials provides a mechanical support for the primary sidewall of the braincase, which tends to be stressed by the chewing muscles. It may therefore be considered as a structural adaptation to the precocious development of early suckling activity of young marsupials. . . . The Cavum epiptericum, which had become incorporated within the braincase of mammals is primarily occupied by the enlarged trigeminal ganglion. Only after birth do the cerebral hemispheres begin to expand and middle cranial fossae become moulded. The ascending process of the Ala temporalis then becomes strengthened by perichondral ossification and "Zuwachsknochen" (appositional bone), which spreads into the sphenoobturatory membrane; these ossifications are called the alisphenoid.

In Wible's (1990) recent study noted above he has marshaled arguments for the following taxonomic properties, specifically ancestral apomorphies, that characterize the last common ancestor of Metatheria and Eutheria. The chosen cranial null group by Wible, based on sound judgments of comparability, was the cranial conditions found in the Cretaceous eupantothere *Vincelestes*. In some instances the conditions in the latter were not known to him. Subsequently Rougier et al. (1992) did describe the basicranium of *Vincelestes*. The suggested cranial synapomorphies (from Wible, 1990, pp. 195–202) for a common ancestor of Metatheria and Eutheria are these : (1) absence of pila antotica; (2) an osseous wall separating the cavum epiptericum and cavum supracochleare (independently from *Tachyglossus*); (3) a "lateral trough of the petrosal that floors the geniculate ganglion but not the trigeminal ganglion" (p. 196); and (4) anterior lamina of the petrosal greatly reduced.

Attempts at an exclusively dental diagnosis of the Theria are difficult, yet by a biologically significant as well as pragmatically convenient definition the first therian, as I have suggested, was a species with a tribosphenic molar dentition. To expect no ambiguity at that stage of molar evolution would be very unrealistic, so debates surrounding future discoveries are likely to ensue. Employing the necessarily highly selected criteria of the fossil record, in exclusion of developmental and reproductive biology, carries with it the probably unavoidable danger of excluding one of the most fundamental and constraining aspects of evolutionary change: developmental evidence. Furthermore this evidence may eventually be causally linked to reproductive biology that clearly delineates the living groups of mammals. It is also clear that reproductive and developmental strategies have major effects on all other aspects of the biology of animals. There is as yet no convincing evidence that independent development of, for example, carnassials or selenodont teeth have constrained future evolutionary modifications in any way comparable to developmental–reproductive factors. For that reason, and for the distinct probability that future osteological discoveries will allow assessment of reproductive strategies, I informally diagnose the last common ancestor of the *living* Theria as the *ovoviviparous* ancestor of all fossil and living metatherians and placentals. Whether the stem tribotherians were such animals is an entirely moot point at present. There is as yet, of course, no currently conceivable association between tooth structure and reproductive biology. Postcranial evidence and its evolutionary analysis will help, however, as does Maier's (1993) analysis of cranial developmental and functional dynamics. Lucas (1990) explicitly warned, voicing the misgivings of many vertebrate paleontologists concerning the drawing of higher taxonomic boundaries, that employing "extinction criteria" for the diagnosis of higher categories when following the dictates of purely cladistic techniques runs the danger of ignoring certain aspects of monophyletic groups: their biologically most important evolutionary novelties that in fact may be causally responsible for their success. Lucas (1990, p. 33A) notes, for example, that: "Those who argue that biological significance is a subjective concept that cannot be used fail to recognize that all choices of characters rely on such concepts." This operational suggestion for classification, consequently, should result in higher taxon boundaries which have some causally, adaptively, compelling components.

As discussed in Chapter 3 and below, there is no agreement at present about the homologies of the tooth families in the protometatherian in contrast to the primitive condition in eutherians, and the homologies in these two groups in relation to the last common ancestor of the Theria. Ongoing work by W. P. Luckett (personal communication; 1993) suggests that the changes in the interpretation of the dental homologies proposed by Thomas (1887, 1888; also Archer, 1978) may not be warranted. The presence of seven postcanine tooth families in the living therians is of some significance in the eventual understanding of their homologies (Owen, 1868). Nevertheless, there is increasing evidence that a total of eight postcanine tooth families may be the primitive condition in the Theria (see especially Sigoneau-Russell, Dashzeveg, & Russell, 1992), although such a conclusion is far from being well established for the living Eutheria. I continue to maintain here that the metatherian dental formula and suppression of the deciduous anterior dentition, both of which I consider derived in the Theria, cannot be linked to the origins of the marsupial mode of postneonate development as has been suggested. Furthermore, judging by the condition displayed by the dentition of peramurids, dryolestids, and other eupantotheres, the gradual transition of the premolar–molar dentition of the earliest eutherians is the primitive condition in the Theria (*sensu stricto*). The sharp boundary between the metatherian **P3** and **M1** appears to be a decidedly apomorphous condition, as is the extreme reduction of diphyodonty, diagnosing the protometatherian along with other traits to be discussed.

Whether the hypothesis of Thomas (1887, 1888) that the eutherian molar formula represents the loss of the fourth molar family compared to the more primitive therian condition in metatherians, is correct or not is an issue yet to be understood more fully (see discussion in Luckett, 1993, and references therein). Luckett finds that, as in eutherians, primitive therian molars are unreplaced primary teeth of their own tooth families. An extensive discussion of this fundamental issue in therian phylogenetics can be found in Luckett (1993). As that review indicates, a number of unresolved issues remain concerning the homologies of the tooth families between the living metatherians and eutherians. The first relates to the already-noted question concerning the serial homology of the metatherian versus eutherian postcanine tooth families. Is the first molar of metatherians homologous to the ultimate or penultimate premolar of therian ancestry, given an ancestry that had five premolars? Although some of the issues raised by Clemens and Lillegraven (1986) about this problem are laid to rest by Luckett (1993), others remain unresolved. For example, based on groups of molariform teeth, Butler (1978) suggested that although not a single specimen shows complete cheek tooth numbers, there were four molars in the Pappotheriidae (*Pappotherium* and *Holoclemensia*; see Slaughter, 1968). The issue in case of the premolar numbers does not necessarily concern actual numbers of teeth (e.g., in living peramelids some specimens retain the deciduous last molar as the growing jaw accommodates it; Lynne, 1982; Luckett, 1993), but the inference based on them. The problem concerns the number of premolar families rather than tooth numbers, and this obviously primes one's perception of the nature of the evidence differently than would the mere search for tooth counts. This issue too, concerns raw data versus a taxonomic property. A recent report by Luckett and Hong (1989) supports the hypothesis that the P3 of various marsupial groups does indeed belong to the same tooth family as dP3, in spite of the fact that it may develop lingually (as in bandicoots, exemplified by *Perameles*) or anteriorly (as in dasyurids, exemplified by *Dasyurus*). The problems of the postcanine dentition homologies within Metatheria or the Eutheria are minor compared to the major difficulty such questions pose about the poorly known ancient clades of the Tribotheria. Until those are sorted out the tooth family equivalence between the roots of the living therians may not be understood.

The evidence for tooth replacement has been often tied to that of differential wear between the first molar and the ultimate premolar (i.e., presumably a replacement tooth); this evidence is empirically critical in testing the number of molars in eutherian fossils. This criterion, however, is not always useful to apply to living metatherians, at least in my experience. The solution of many of these problems relating to tribotherians, fossil metatherians, and eutherians simply requires new specimens and new perspectives. Nevertheless the possibility or likelihood must be kept in mind that the earliest eutherians, like *Prokennalestes* (see Kielan-Jaworowska & Dashzeveg, 1989; Sigoneau-Russell et al., 1992; Gheerbrant, 1992), do retain the original therian premolar–molar relationships. In spite of the (putative) reproductive therian plesiomorphy of the Metatheria, the dental formula and replacement pattern of the protometatherian (as seen in the living didelphids, *models* of ancient Metatheria in that respect) may well have been apomorphous compared to antecedent tribotherians and eutherians, as already noted in the last century. If the last molariform premolar of the early therian pattern of four premolar families was not replaced, as Owen suggested, then it would have become the first metatherian molar, a member of the first wave of teeth and a strict homologue of the eutherian **dP4** (or **dP5**). As it is, the first two premolars of living marsupials are unreplaced teeth of the first wave of tooth buds (Luckett, 1993), this attribute adding another apomorphy to the probably derived number of molars of metatherians. Although the presence of only seven postcanine *tooth families* in the first therian is perhaps more than possible (keeping in mind five *premolar numbers* in several intriguing fossils), this issue is still not settled. But even if eight or more postcanine families were present in the first therian, the homology of the first marsupial molar with the deciduous eutherian **dP4** (or **dP5**) constitutes a distinct and separate hypothesis that must be judged on its own merit. Given the complex (and as yet depauperate) nature of the dentitional (i.e., of whole dentitions), cranial, and postcranial evidence for early therian evolution, the possibility exists for a diagnosis of a holophyletic Metatheria based on both dental and some postcranial (carpal) traits (to be discussed later). One can also contemplate a *dentally* misdiagnosed Eutheria that included taxa more recently related to cladistic metatherians than to eutherians! The early members of such a "dental Eutheria" would have been reproductively (and physiologically) like the primitive living marsupials, but cladistically either metatherian or eutherian. As these early therians become better known postcranially, the concept of a paraphyletic Tribotheria might become even more widely attractive in order to delineate, holophyletically, not only the Metatheria but the early Eutheria as well. Holophyly of all taxa in any inclusive classificatory system that professes a commitment to reflect real phylogeny, and not a wedding of such taxa with Haeckelian palingenesis, is impossible.

The second major issue is the conformation of the molar topography, such as the relative size of various cusps; the absence, presence, and /or independent development of such landmarks as conules; stylar cusps with seemingly well-definable relative positions on the crown; and the relative width of talonid. The issue, for example, of an either primitive or derived hypertrophy of the paraconid in the Deltatheroida, the tribotherians *Aegialodon* and *Kielantherium* notwithstanding, is an example of a number of dilemmas concerning Cretaceous mammal phylogeny. Without a contextual analysis, can any and all topographic attributes be used as taxonomic properties? Even though the interpretation of the homologies of the crown topology of the first (i.e., tribosphenic) therians appears secure (Patterson, 1956), the transformation events from that tribosphenic base and their uniqueness or multiple occurrence pose an exceptionally daunting problem in early therian phylogenetics (see especially the discussions in Clemens & Lillagraven, 1986; Gheerbrant, 1992; Cifelli, 1993).

Although I consider the discussion of Clemens and Lillegraven (1986) and Cifelli (1993) about the evolution of tooth crown attributes very valuable, it will become apparent under the discussion of various taxa that I disagree with several of their proposed transformation series. Similarly, I find a number of serious difficulties with the definition, not diagnosis, provided recently by Marshall et al. (1990) for tribotherians. These differences are undoubtedly related to the phylogenetic nature of a diagnosis versus the general trait identification in a taxon that characterizes the definitional procedure. For example, in the earliest tribotherians the paracone was certainly not subequal to the metacone – as the former was considerably larger primitively. Specifically, it is almost impossible to speculate on at present when the various traits related to reproductive and developmental biology appeared in the early Cretaceous dental therians, the tribotherians.

The molariform upper tooth, *Comanchea*, described by Jacobs, Winkler, and Murry (1989) from the Texas Albian Early Cretaceous, as well as the excellent specimens of the dentition of the equally ancient *Prokennalestes* (Kielan-Jaworowska & Dashzeveg, 1989) raise important questions about the analysis, meaning, and description of rare and critical specimens like Cretaceous mammal teeth. I recognize that there is a strong and legitimate need to attempt to identify such important minutiae as the nature of linearity of the buccal margin or the relative size of the stylar cusps, to mention only a few recorded traits, as either primitive or derived. For example, *Comanchea* does not show a crest from the paracone to what Jacobs et al. (1989) identify as stylar cusp **B** (stylocone), and the cusp called **C** is twinned to the paracone, as a stylar cusp previously connected by a crest would be expected to be. The homology of cusp **B** is supported by an occlusal facet, and yet topographically cusp **C** could be conceived as the stylocone (cusp **B**). In other words there is possible doubt about the homologies of all the stylar cusps with later therians, and in particular, as noted by Jacobs et al., about the homology of the marsupial cusp **C** with that of *Comanchea*. As the recent analysis of the structurally primitive (dentally) eutherian *Bustylus* and relatives by Gheerbrant (1992) emphasizes, the earliest dental eutherians like *Prokennalestes* (these and relatives were, I believe, mistakenly referred to by Gheerbrant as "zalambdodont") had a well-developed series of stylar cusps, even a small (or twinned) mesostyle (cusp **C**). As widely suspected among students of early therians, then, the nature of early stylar cusp evolution, by itself, is not a clear indication of metatherian or eutherian taxonomic properties.

Epipubic bones, as discussed in Chapter 3, predate the therian ancestry, and the baculum of eutherians, as long suggested, may be an apomorphic modification of these. Klima's (1987, and references therein) paper, discussed in Chapter 3, documents and explains additional evidence, and is, therefore, important reading for any attempt to trace the homologies and transformations of the shoulder–breast complex in mammals and nonmammals.

As noted, this book does not present an adequate survey of carpal morphology in marsupials, except for a rather superficial glance (Figs. 3.2 and 8.16), for reasons already discussed. It appears obvious, however, that while the didelphids and caenolestids, and all other Metatheria known by the complete carpus, have incorporated the centrale probably within the scaphoid, the centrale remains distinct in most early eutherian lineages. The persistence of the centrale of most eutherians, then, appears to be more primitive than the known (living) metatherian ancestral state.

Marshall's (1979) ambitious review of mammalian, therian, metatherian, and eutherian characters, pertaining to the brain, head, reproduction, and dentition is an important, comprehensive, and useful compilation of some of the evidence pertinent to these groups. Understandably, much of the literature cited, however, was not critically reviewed, but major avoidable mistakes about bulla construction and postcranial morphology have been offered as facts (e.g., pp. 379, 393, 403). The characters given in the appendix for the Mammalia by Gauthier et al. (1988) strongly supplements Marshall's contribution. Nevertheless a causal and analytical assessment of what the noted putative mammalian characters might mean in functional–adaptive and historical terms, that is, their *testing* as to their usefulness as taxonomic properties, is a much desired future goal.

A note concerning the dental diversity of therians in the Cretaceous by Jacobs et al. (1989, p. 4994) sums up a judicious perspective on these poorly known animals:

> the morphological diversity of Cretaceous tribosphenic mammals [those with protocones and distal talonids] demonstrates that a twofold division into marsupials and placentals is simplistic and that other groups of comparable ... [ancestry], now extinct, evolved in an initial radiation. ...

Similar views are embodied in my support of the formal concept Tribotheria (and its numerous known and unknown lineages), also expressed in the extensive reviews by Clemens and Lillegraven (1986) and Cifelli (1993).

## Infraclass Metatheria Huxley, 1880

### "Diagnosis"

Protometatherian was probably similar reproductively, physiologically, cranially, and in some respects dentally to a putative ancestral therian (see above). It had: advanced therian reduction of diphyodonty as in Didelphidae (the therian pattern perhaps retained by stem eutherians) resulting in almost a single generation of erupted teeth; a dental formula of I1,2,3,4,5/4; C1; Pd1,d2, dP3 replaced by P3; M1,2,3,4; enamel aprismatic possibly with microtubules (primitive synapsid, or possibly without these tubules; C. Wood, personal communication); tribosphenic molars with retained, primitive therian stylar cusps, and with a relatively small protocone; one major lingual conule on both preprotocrista and postprotocrista (para- and metaconule); trigonid triangular and shearing, more developed than in earliest eutherians, but less so than in the derived deltatheroidans; paraconid perhaps better developed than metaconid (a putative primitive therian condition, as in the Aegialodontidae, but probably secondarily so in Deltatheroida), in contrast to earliest eutherian deemphasis (?) of paraconid; entoconid and hypoconulid large, the latter lingually displaced and somewhat twinned (untwinned in the secondarily reduced talonid of Deltatheroida), possibly close to the putative prevalent tribotherian condition (barely known); talonid perhaps not as wide relatively as in earliest eutherians and earliest ameridelphians; bony palate fenestrated; jugal posteriorly extended to border of glenoid fossa retained, as in other early (Cretaceous) mammals and cynodonts; incipient alisphenoid bulla; fenestra ovale round and fenestra rotunda transversely elongated and elliptical (the cranial and postcranial traits listed under the Ameridelphia were probably protometatherian characters also, although some of them may predate the Metatheria–Eutheria or even the Theria, i.e., the Tribotheria).

## Distribution

Early Cretaceous (Aptian–Albian) to Recent.

## Included Taxa

Cohorts Holarctidelphia, Ameridelphia, and Australidelphia.

## Discussion

The diagnosis is rather incomplete and uncertain, mainly because the metatherians without the as-yet not fully studied Asiadelphia (Trofimov & Szalay, 1993) and the deltatheroidans are far better known. In general, the glimmer of the critical Asian metatherian radiation is a looming hint that extreme caution is needed in any discussion of the origins of the Metatheria. The allocation of deltatheroidans (Kielan-Jaworowska & Nessov, 1990) and the recognition of *Asiatherium, nomen nudum* (Trofimov & Szalay, 1993), however, significantly reconfigure our understanding of the early evolution of the Metatheria. Consequently many of the traits listed under the diagnosis of the Ameridelphia, although possibly diagnostic of the Metatheria itself, will have to be reconsidered in the near future. This is certainly not a satisfactory state of affairs, but any attempt at greater precision might be unwarranted at present. The new Asiatic Late Cretaceous Udan Sayr asiadelphian *Asiatherium* is a metatherian represented by excellent cranial, dental, and postcranial material. Though undoubtedly a metatherian, this form is not an ameridelphian; its analysis will greatly clarify issues that can only be raised here. Because many marsupials, particularly didelphids, so generally retain the putative broad biological foundations that were probably present in the last common ancestor of the living therians, it is virtually impossible to determine how many meaningful differences can be cited between the last common ancestor of all known metatherians and their antecedent, ancestral lineage. Some of the characters I cite above as diagnostic for the Metatheria, or below, for the Ameridelphia, may well have been diagnostic for the last common ancestor of the Metatheria and Eutheria, or, to use Cifelli's uncondoned term the "higher" therians. I have repeated these characters in diagnoses, because until the somewhat derived, perhaps hypercarnivorous Delatatheroida and the seemingly very conservative Asiadelphia are better known, such issues will remain open.

As noted, the recognition and supporting arguments that the Deltatheroida are Metatheria by Kielan-Jaworowska and Nessov (1990) were of far-reaching significance for the understanding of the early evolution of the marsupials. Similarly the study of *Asiatherium* (Trofimov & Szalay, 1993 and in preparation) significantly adds to the data base and available hypotheses about early metatherian evolution. Although I disagree with the interpretation of the polarity of several deltatheroidan characters by Kielan-Jaworowska and Nessov (1990), and part company with the view of Kielan-Jaworowska (1992) that these metatherians go back independently to the Aegialodontia, I fully endorse the metatherian status of the Deltatheroida. Yet one of the most diligent and keen contributors to early metatherian phylogenetics, Cifelli (1993) disagrees with the proposed metatherian ties of deltatheroidans, but endorses the special relationship of the order with the Aegialodontia. Thus, given the views of two such astute students of Cretaceous mammals, the concept of the Metatheria (i.e., its diagnostic dental traits) will probably remain unsettled for awhile.

My reasons for disagreement with these workers can be found in both the morphology of the Deltatheroida, *Kielantherium*, and the interpretation of the similarities and differences. Although both of these taxa have relatively small talonids and large trigonids with a dominant paraconid on the lingual side, the remaining differences are significant for phylogenetics. The primitive deltatheroidan talonid (cf. *Sulestes*), like all other Metatheria and

Eutheria, has a distinct entoconid and well-developed talonid basin, features lacking in the aegialodontians (see also the recently described *Hapomylos*; Sigoneau-Russell, 1992). There is no unequivocal sign on the landmarks of lower molars (some wear facets notwithstanding), the small talonid in particular, that in the aegialodontidans there was a protocone developed to any significant degree, let alone with conules on its buccal wings. In deltatheroidans not only is the lingual side fully developed, but a carnassial notch on the postmetacrista of the upper molars, coupled with the hypertrophied prevallid (the protoconid–paraconid crest) strongly suggests that paraconid hypertrophy and talonid shrinkage are the result of carnassial specialization. The latter occurs repeatedly in both metatherian and eutherian carnivores. This interpretation is further supported by the construction of the talonid of the deltatheroidans. In these metatherians the cristid obliqua meets the distal wall of the talonid at the base of the protoconid. In aegialodontians the cristid obliqua meets the crestate distal margin of the metaconid, the postmetacristid, forming a continuous crest with it. The two conditions are significantly different. It is worth noting here that the cristid obliqua on the relatively wide talonid of the oldest known undoubted (dental) eutherian, *Prokennalestes* (mistakenly called "zalambdodont" by Gheerbrant, 1992), also meets the base of the metaconid. In fact the presence of a (pretribosphenic) postmetacristid in *Prokennalestes* suggests that the earliest known eutherians were somewhat more primitive in their talonid construction than known Metatheria. Nevertheless, the well-defined tribosphenic molar structure, including the well-developed protocone, conules, and talonid basin surrounded by the three well-defined cusps, remains a diagnostic hallmark of the last common ancestor of the Eutheria and Metatheria. The Aegialodontia thus may be considered to represent the (structural) "stem group" of the Tribotheria from which the Pappotherida originated. This does not negate the near certainty that the Peramura (as used by Kielan-Jaworowska, 1992) is the source of the Theria.

In light of the known holarctidelphians (deltatheroidans and the asiadelphian evidence to be described in detail by Trofimov & Szalay), the Metatheria cannot be diagnosed by the presence of hypertropphied stylar cusp **C** and **D**. The therian ancestor certainly had a parastyle (cusp **A**), stylocone (cusp **B**), a metastyle (cusp **E**) and probably (quantitatively variable) expressions of cusps **C–D**, and probably others as well (see earlier discussion). Phylogenetically eutherian taxa in fact also retain some (or all) of these cusps, even if **C** is only a single or twinned trace (see illustrations of *Prokennalestes* in Kielan-Jaworowska & Dashzeveg, 1989). The tribotherian (?) *Comanchea hilli* Jacobs et al. 1989, from the Texas Early Cretaceous (approx. 110 mybp), in spite its primitive therian attributes, displays a relatively enormous stylar cusp **C** (if it is that – see caveats discussed earlier), a feature of prominence in the marsupial taxonomic literature, but also modestly present in *Prokennalestes*. Yet as noted by Jacobs et al. (1989), because of its relatively much larger paracone than metacone, its poorly developed paraconule, and the extremely low protocone, *Comanchea* was unlikely to have been a metatherian phylogenetically. Though there are no lower teeth known for *Comanchea*, these authors note that the development of an increasingly larger metacone was functionally correlated with an increase in size of its occluding lower tooth counterpart, the hypoconid. They state unequivocally (Jacobs et. al, 1989, p. 4994) that "*Holoclemensia* is not a marsupial based on the degree of twinning [of entoconid and hypoconulid] in referred lower molars." This issue, however, is certainly not settled as yet.

The new dental material of Cretaceous deltatheroidans recently described by Kielan-Jaworowska and Nessov (1990) for the genus *Sulestes* has stylar cusps **A** and **B** and multiple lobulation of the buccal cingulum that cannot be homologized with the remaining stylar cusps of other metatherians. Nevertheless I am convinced, based particularly on the dentitional (formula) evidence these authors presented as well as their arguments, that the Deltatheroida, while a dentally somewhat derived hypercarnivorous group, is cladistically metatherian and should be included in the Metatheria. It is plausible that the ancestor of the clade Ameridelphia originated from a radiation closely related to the asiadelphians, a lineage of the Holarctidelphia that was the paraphyletic sister group of the ameridelphians. There are marsupials known from the Late Cretaceous of Asia, the above-noted *Asiatherium*, which, although they display entoconid–hypoconulid twinning and strong para- and metaconules, have no significant stylar cusps **C** and **D** (Trofimov & Szalay, 1993).

One of the most important diagnostic features of the first metatherians, which we can corroborate eventually, had to be their derived therian pattern of reduction of diphyodonty and the hypothesized extremely limited tooth replacement (only **dP3**) in the first metatherian, resulting perhaps in the sharp morphological break between the ultimate premolar and first molar. Based on this condition, I presume that the dental formula of the protometatherian was more advanced than what we encounter among the earliest eutherians (but see the discussion in Clemens & Lillegraven, 1986). Yet this taxonomic property is difficult to recognize without complete fossil dentitions. Both Deltatheroida and Asiadelphia show at least the dental formula shared with the protometatherian. In living didelphids the incisors, while not replaced, are preceded by deciduous buds which are resorbed (Luckett, personal communication, 1993).

I strongly suggest (see Chapter 3) that the reproductive biology of the lineage antecedent to placentals was "marsupial" (this is clearly stated, with some doubtful caveats, by Lillegraven et al., 1987). The female of that animal probably produced neonates developed approximately to the same degree as those in didelphids (a putative gestation period of about three weeks), and carried these attached to her nipples until weaning began. That such a pattern was also ancestral for the last common ancestor of living therians is suggested by Maier's study (1993). In other words, the uniquely derived reproductive biology of the last common ancestor of the concept of the living Theria espoused here was the same as what we can infer for the ancestry of the marsupial clade of the Cretaceous, Paleogene, and Neogene. I am reluctant to speculate on the poorly known Aegialodontia. Renfree (1991) makes some important observations on living therians and adds her reservations concerning my view (one which follows Lillegraven's and others). Nevertheless, marsupial mammals, at that coarse, yet descriptively and *phylogenetically* (but not taxically) meaningful level of resolution, were ancestral to the first placentals. This is a biologically well-corroborated hypothesis, and not the necessary taxonomic artificiality that accompanies systems of classification. Sometimes the latter are necessarily dependent on such distinctions as the minor differences discussed earlier in molar construction or in dental formulae, traits unlikely to be coupled with reproductive biology. Like most macrotaxonomic differences, these assume importance in retrospect, and they do not necessarily coincide with biologically more significant transformations. But for the two extant infraclasses of the Theria this marsupial background is, of course, of critical importance. It is integral to the evolutionary explanation of both metatherian and eutherian origins. If one intends classification to reflect some biologically meaningful historical realities, then perhaps it is important, on this level of resolution, to speak of eutherian origins from an early marsupial, albeit probably not a metatherian but some hitherto undiscovered (paraphyletic) tribotherian stock.

Among the Metatheria, particularly in the didelphid radiation and the descendent microbiotheriids, analysis of the auditory region helped in the recognition of the distinctness of *Dromiciops* and fossil *Microbiotherium* (Segall, 1969). Basicranial information was similarly useful to determine the derived nature of the Sparassocyninae (Reig & Simpson, 1972) and the probable primitive condition within the Didelphidae (Reig et. al., 1987; Maier, 1989a, b). Yet the ear region, or the entire skull for that matter, has not as yet yielded the transformational insights still needed for understanding the early differentiation of marsupials known by dental and tarsal material. Outstanding contributions to the basicranial osseous characters of marsupials, fossil and extant, may be found in Archer (1976c, pp. 302–17), and more recently in Wible (1990, 1991) and Rougier et al. (1992), who discussed this aspect of the earliest known ameridelphian radiation. While character transformations are not resolved in Archer's study, discussion of the traits is comprehensive and thoroughly comparative, and it remains one of the best sources for character analysis for future studies. Archer's account of sinuses and recesses (epitympanic recess, periotic hypotympanic sinus, mastoid epitympanic sinus, paroccipital hypotympanic sinus, squamosal epitympanic sinus, various accessory sinuses, postglenoid cavity, septa between sinuses) and of basicranial bones (alisphenoid, basisphenoid, basioccipital, ear ossicles, petrosal, paroccipital process, and squamosal) provide a rich storehouse for eventual functional–adaptive elucidation of the complex basicranial transformations. Much of this research has not yet been conceived and even considered beyond an anatomical distribution analysis. Wible (1990) has summarized some important aspects of the distribution of specifically ameridelphian character patterns as they are related to circulation, and petrosal and other cranial bony morphology, aspects also discussed under **Archimetatheria.** He concluded, significantly, based on twenty-four traits, that the fixation of the essentially identical circulatory patterns in virtually all living groups of marsupials signifies that this shared pattern predates the differentiation of the South American and Australasian stocks, and that the petrosals of the North American archaic radiation suggest a holophyletic group, independent of the Didelphidae and other extant marsupials. As I discuss below, my interpretation of the evidence, so lucidly presented by him, differs from his cladistic perspective on the evolution of the known patterns.

The basicranial information and, particularly the circulatory system as summarized by Wible (1984, 1990) and Rougier et al. (1992) and discussed under **Theria,** suggest some distinguishing features of both the holarctidelphians (Szalay & Muizon, in preparation) and the ameridelphians from eutherians. Therefore, some of these inferred protometatherian characters, such as the putative elimination of the stapedial arteries as judged by criteria of the absence of grooves from the surface of the promontorium, may be distinct from the therian stem group. The great difficulty with some of these characters, as noted by both Archer (1976c) and Archibald (1979), is that the absence of the appropriate grooves on fossil petrosals is not a totally satisfactory test of such hypotheses. Interestingly, didelphids retain the primitive therian bicrurate stapes. Additional features are even more difficult to perceive in the fossils. These features, cited by Wible (1984, 1990), in addition to those given in the diagnosis of the Ameridelphia below, are: (1) maxillary artery that supplies the upper and lower jaws, running beneath the skull base lateral to the mandibular nerve; (2) loss of a true occipital artery – the occiput is supplied by a branch of the vertebral artery; and (3) supraorbital vessels supplied through the ophthalmic branch of the internal carotid.

The ontogenetic study by Maier (1989a, b), functional-adaptively conceived, on the middle ear development of didelphid marsupials has tested a hypothesis concerning the developmental constraints that channel bulla development in marsupials. He specifically suggested that the relative importance of the alisphenoid, the side support of the pumpsucking skull (Maier, 1993), may mean that in this form it represents an ancient therian trait, and that the alisphenoid thus may be structurally predisposed to be the anchor of the the middle ear floor. The fossil record can supply information on either specific construction or the lack of this cranial area in the earliest therians and metatherians. Meanwhile it is possible that a small tympanic wing of the alisphenoid is an ancient metatherian (and possibly pre-metatherian) trait, in spite of the fact that some forms may completely lack this anterior support for a cartilaginous bulla.

The occasional presence of lunula-like ossification of a small patella in nonperamelid living marsupials, as well as its probable presence in borhyaenids (e.g., *Thylacosmilus*) strongly suggests that this feature was not lost in the first ameridelphians or else that the developmental potential inherited from a nonmetatherian ancestor is present in all of them. As I have noted, the loss (or more likely fusion) of the centrale in the carpus characterizing the known American lineages may be a diagnostic metatherian trait. In light of the virtual absence of any Cretaceous information however, this suggestion is conjectural.

The tarsal structure of the earliest Metatheria, the probable sister lineage of the Eutheria, and the first reproductively eutherian mammals may well have been extremely similar in their tarsal morphology. It is of course also questionable whether the first "dental eutherians" were at that point reproductively eutherian. There is probably no hope at present to correlate correctly the disparate systems and identify the ancestral eutherians by the tarsus or even by the remainder of the skeleton.

The papers of Archer (1984a, b) are perhaps the most synthetic ones on an overall perspective of marsupial systematics up to that time. More recently Aplin and Archer (1987) presented the most comprehensive discussion on marsupial classification anywhere in the literature. Marshall et al. (1990) subsequently discussed and offered characters for their taxonomic subdivision of the Metatheria. The groupings of these authors are discussed under the respective taxa.

One of the most significant contributions of Aplin and Archer (1987, p. xiii) to current attempts at marsupial phylogenetics is the clear recognition, as many others have understood before, for the need to weight a priori. These authors also state (as fully expressed and theoretically justified in Bock, 1977b) that construction of phylogenies and their taxonomic expressions require two distinct steps. Aplin and Archer aspire for a "syncretic" classification, yet curiously they also strive for a fully cladistic one (i.e., made up of exclusively holophyletic groupings). The adjective "syncretic" refers to their attempt to effect a reconciliation and union of opposing priniciples. Their search is for "phylogenetic concensus" and cladistic expression. Yet, as discussed in Chapter 2, an attempt at an entirely holophyletic classification never comes face to face with the reality (naturalness) and the necessity of paraphyla. Contrary to the stated belief of Aplin and Archer, the pattern of evolutionary relationships that link organisms are not *intrinsic properties* of the organisms. The genotype and its phenotypic expression are intrinsic properties of an organism, but homology statements or hypotheses of phyletic relationships tested against the properties of these organisms are not. Aplin and Archer (1987) attempt to break away, at least in stated principle, from a purely cladistic system of inference/classification. While their groupings are sometimes based on distance data to reflect monophyla (assuredly including paraphyla), often the only available information, their aspiration for a fully cladistic integration results in an unnecessary inflation of the various categorical levels used to express the same information, similarly to that of Marshall et al. (1990). My argument throughout is, and as I will contend in discussions of the specific taxa, that the number of ordinal and particularly familial groupings of marsupials recognized by such expert contributors to marsupial taxonomy is decidedly out of balance compared with the classification of the much more speciose and diversified eutherians. Metatherian phylogenetics can be often expressed with the categories of subfamily and tribe (or subtribe) instead of family or superfamily.

## Cohort Holarctidelphia, new

### "Diagnosis"

The protoholarctidelphian had a suite of apomorphous attributes that are probably more primitive than those of the protoameridelphian. It had: upper molars with stylar cusps that were not hypertrophied, but had parastylar and metastylar wings, delineated by a slight groove from the paracone and metacone, respectively; protoconid highest cusp on trigonid, the latter with subequal paraconid and metaconid (paraconid secondarily enlarged in Deltatheroida); talonid not narrowed as in Deltatheroida, but with tall and sharp twinned entoconid–hypoconulid; alisphenoid wing for cartilaginous bulla attachment and a wide C-shaped ectotympanic, probably retained from premetatherian condition; no arterial grooves on promontorium (Trofimov & Szalay, in preparation).

### Distribution

Cretaceous of the Northern Hemisphere.

*Included Taxa*

Orders Deltatheroida and Asiadelphia (Trofimov & Szalay, 1993, in preparation).

### Discussion

The earlier discussion of Metatheria is relevant to this still poorly known and not yet fully studied primarily Asiatic radiation. Kielan-Jaworowska and Nessov (1990, and references therein) present detailed factual evidence and a convincing discussion of the dentition of Deltatheroida. As noted, I support the metatherian status of these hypercarnivorous animals, and that of another new genus called *Asiatherium reshetovi nomen nudum* based on a unique skull and skeleton from Udan Sayr, Mongolia (Trofimov & Szalay, 1993) which represents another highly distinct clade, the Asiadelphia, of the Cretaceous Asiatic Metatheria. This as-yet undescribed cranial, dental, and postcranial material of a Mongolian Late Cretaceous metatherian, under study by Trofimov and Szalay, does not contradict the sister-group ties of the Asian holarctidelphians to the Ameridelphia. In fact the origins of the first didelphidan is perhaps best envisaged from a holarctidelphian, but not from a deltatheroidan which is a probably derived and holophyletic group within the Holarctidelphia. As I noted, I do not restrict the vernacular term "marsupial" to the nondeltatheroidan metatherians as do Kielan-Jaworowska and Nessov (1990). Because that term has a clear meaning in English, referring to animals with a particular type of reproductive strategy, its usefulness transcends the taxonomic meaning attached to "Marsupialia."

Kielan-Jaworowska and Nessov (1990) properly argue that the incipient alisphenoid bulla (in the deltatheroidan Gurlin Tsav skull which I had studied in some detail) and the dental formula of deltatheroidans strongly link them with traditionally recognized Metatheria. Some of the dental differences of the deltatheroidans from the Cretaceous Ameridelphia are best considered as being the more primitive character states of homologous features, whereas others such as the enlarged paraconid, highly narrowed talonid, and subsequently untwinned entoconid–hypoconulid are derived specializations of deltatheroidans along with their telling carnassial notch on the postmetacrista. But the relatively larger stylar shelf, and an incipient alisphenoid bulla are primitive metatherian features. *Asiatherium*, once described, will show a number of metatherian dental attributes probably primitive for the infraclass. Thus, none of the following three taxa within the Metatheria, the Holarctidelphia, the Didelphida, or the Gondwanadelphia are holophyletic – they are properly, phylogenetically, diagnosed monophyletic groups which will probably continue to hide antecedents.

The Deltatheroida as constituted by Kielan-Jaworowska and Nessov (1990), and as diagnosed here, is probably holophyletic. No deltatheroidan seems to have given rise to any other known higher taxon. Kielan-Jaworowska and Nessov (1990) have given a detailed comparison of the dental features with Aegialodontidae and the earliest eutherians, *Prokennalestes* and *Bobolestes*. Their suggestion of an early, possibly pre-Aptian split of the Metatheria and Eutheria appears reasonable (see Fig. 2.3).

## Order Deltatheroida Kielan-Jaworowska, 1982

### Diagnosis

Protodeltatheroidan had: hypertrophied postmetacrista with a well developed carnassial notch; hypertophied trigonid with an emphasis on the relative size of the paraconid; buccolingually reduced talonid retaining entoconid and contact of cristid obliqua with base of protoconid, unlike the talonid of aegialodontids.

### Distribution

Same as that of Holarctidelphia.

### Included taxa

Deltatheroididae and Deltatheridiidae (Deltatheridiinae and Sulestinae).

### Discussion

While their diversity is still poorly known, these metatherians along with the undescribed Asian asiadelphian *Asiatherium* from Udan Sayr may hold the key to the origin and early dispersal of the Metatheria, as discussed here and in Chapter 9, and particularly by Kielan-Jaworowska and Nessov (1990). Yet as I discuss under Metatheria, I disagree with the assessment of Kielan-Jaworowska (1992) concerning the affinities of the Deltatheroida. Neither do I consider her suggestion of deltatheroidan–sparassodontan affinities well founded (see discussion under Sparassodonta), nor do I believe that the Aegialodontia (*Agialodon* and *Kielantherium)* belong in the ancestry of the Metatheria, with the Deltatheroida retaining primitive similarities to these genera. The special similarities between the Deltatheroida and *Kielantherium*, and their holophyletic affinities, were also advocated by Cifelli (1993). Finally, Kielan-Jaworowska (1992) also suggests an independent eupantotherian ancestry for the Eutheria, a stem which leads back to the Peramura. It should also be noted again that these authors also disagree with each other's view on the metatherian ties of the Deltatheroida.

Butler and Kielan-Jaworowska (1973) were probably the first who recognized the metatherian affinities of these, closely followed by a note from Van Valen (1974) that gave support to this allocation. While not considering deltatheroidans metatherians, Kielan-Jaworowska (1975a), described excellent dental and facial skull material of *Deltatheridium* and has given a lasting and lucid account of the form–function relationships of the molars of that genus. Our current ignorance of postcranial structure in these animals will probably be relieved slightly in the near future. Nessov and Szalay (in preparation) are describing the mammalian postcranial remains from the Cretaceous of the Kizyl Kum desert, some of which undoubtedly belonged to deltatheroidans.

## Cohort Ameridelphia Szalay, 1982

### *"Diagnosis"*

Protoameridelphian had: tribosphenic molars with perhaps more robust and more mesiodistally expanded sharp protocone than in stem metatherians; matacone may or may not have been subequal (i.e., more advanced than in tribotherians, protoeutherians, and holarctidelphians); presence of at least three stylar cusps (note, as previously discussed, that current understanding of stylar cusp evolution is scanty); fully formed para- and metaconule retained from metatherian ancestry; buccal postcingulid; talonid wide in conjunction with a slightly hypertrophied protocone compared to primitive Holarctidelphia and protoeutherian; close twinning of entoconid and hypoconulid on lower molars retained from holarctidelphian ancestry; only dP3 replaced and adult dental formula I1,2,3,4,5/4; C1; Pd1,d2,3; M1,2,3,4 (probably identical to conditions of protometatherian); unique combination of traits in the basicranial circulation which may be synapomorphies shared with the living marsupials and could be detected in fossils (see discussions under Theria above, and below), such as the course of the internal carotid extrabullar medial to the tympanic cavity and the lost stapedial artery in the adults of *living* marsupials (but presence of adult stapedial artery in metatherian common ancestor is possible); retained bicrurate stapes from therian ancestry, even after loss of stapedial artery in adults; inflected angle of mandible, a retention, or a slightly hypertrophied condition from therian ancestry; incomplete, or cartilaginous connection of the lower arch of the atlas; retained patellar ossification (subsequently suppressed in didelphids and most australidelphians); shoulder–breast complex as described for the therian common ancestor; seven cervical vertebrae as in therian ancestry, and probably no more than twenty-six presacral vertebrae, with probably thirteen thoracic vertebrae; retained flabellum of fibula (dorsoposterior process); large peroneal process of the calcaneus (as in therian ancestry); astragalar canal more posterior and plantar than in oldest known eutherians; astragalar AFi facet primitively narrow, as the therian tricontact **UAJ** is retained; (putative) tricontact **UAJ** with weakly developed medial malleolus on tibia, and curved and extended **ATi** articular surface on the distal tibia; **UAJ** with the meniscus and lunula of most living metatherians, attached to the calcaneus, fibula, and tibia.

### Distribution

Early Cretaceous (Aptian–Albian) to Recent.

### Included taxa

Order Didelphida.

### Discussion

I know of no meaningful way to characterize the cranial attributes of the first ameridelphian other than simply to suggest that these traits were probably the "same" as those of the first metatherians. The first ameridelphian, in contrast to a holarctidelphian ancestry, had probably initiated or more likely continued the incipient metacone hypertrophy or paracone reduction in that ancestor, resulting in the subsequent lineages in a condition beyond subequality of these cusps. Most early ameridelphians stand in contrast to the few known holarctidelphians in that some of the latter (Deltatheroida) appear to retain a primitively (slightly) smaller metacone than paracone from the (putative) metatherian ancestry, have less robust protocones and narrower talonids, and are more primitive in the dental and other traits noted under Metatheria and Holarctidelphia.

The most difficult and challenging problems of ameridelphian evolution involve the Cretaceous and Paleogene differentiation of groups. The combined factors of (1) paucity of information (teeth only, with some significant exceptions noted above and below), (2) prevalence of primitive therian and metatherian traits, and (3) the greater probability of homoplasy because of the less altered ancestral morphology, while they barely allow identifications of samples as metatherians, often negate any confidence in the recognition of *homologously* shared derived features. It is no accident, therefore, that the greatest number of debates surrounding the early Ameridelphia involve the earliest history of Metatheria, and these undoubtedly will continue as new dental evidence becomes available. Highly competent students of therian dentitions have conflicting perspectives on the meaning of the slightest differences in molar morphology. The section on Dental Morphology in Chapter 3 is closely related to the discussion of the Ameridelphia.

Dedicated students of Cretaceous mammal evolution such as Butler, Cifelli, Clemens, Fox, Lillegraven, and Kielan-Jaworowska, among others, have grappled with these metatherian problems and sharpened our focus on them. Recently Eaton (1990) and Cifelli (1990a-f) presented evidence that supports either the presence of a full complement of stylar cusps in an *Alphadon*-like ancient form (Eaton, 1990), or the repeated independent "acquisition" (much more likely emphasis) of stylar cusp C in the Ameridelphia (Cifelli, 1993). As I noted above, any character that is present in an incipient form in several genera of non-metatherians has limited significance for taxon phylogeny. Cifelli (1989, p. 17a) has recently reiterated some of the difficulties discussed here and made the following important observations and interpretations about some Cretaceous North American ameridelphians.

> *Iqualadelphis* . . . is probably the most primitive of undoubted marsupials [i.e., of ameridelphians, in the present sense]; the distribution and variability in expression of cusp C among [ameridelphian] marsupials indicates that it may have appeared subsequent to cusp D, as suggested by Fox. New Turonian and early Campanian taxa are problematic in lacking any consistent stylar cusp except the stylocone, and would be excluded from the Marsupialia using currently accepted criteria. However, other presumed apomorphies of the group (enlarged metacone, anteroposteriorly expanded protocone, lingually placed paraconid and hypoconulid, labial postcingulid) are present in these taxa. The simplest hypothesis of relationships for these taxa is that they are primitive marsupials (or near marsupials, depending on definition), suggesting that the latter features appeared earlier in marsupial evolution than did characteristic apomorphies of the stylar shelf.

While *Alphadon* appears to be the oldest known taxon included in the Didelphida, *Iqualadelphis*, however, may be more primitive didelphidan in light of its lack of extensive stylar development according to Cifelli (see especially Fig. 8.4) and particularly in light of the Asian evidence. The Udan Sayr *Asiatherium* (Trofimov & Szalay, 1993) and deltatheroidans support Cifelli's (1989) suspicions. In fact the development of large stylar cusps may be the reason for a relatively wide stylar shelf. In spite of this complicating factor, the presence of *Alphadon*-like forms in beds dated Cenomanian (Eaton, 1990), 5 to 8 million years earlier than the Turonian pediomyine, must not be ignored. Another potential difficulty is the presence of an extensive stylar shelf not only in Albian Cretaceous tribotherian pappotheriids, but in *Prokennalestes* from Asia (Kielan-Jaworowska & Dashzeveg, 1989), and in the Holarctidelphia. *Prokennalestes* and *Bobolestes*, while most likely cladistic eutherians, have relatively reduced paraconids (compared to early metatherians), wide talonids and well-developed stylar shelves. In addition, as noted, like many of the related atribosphenic relatives of the therians, these eutherians lack the abrupt shape change between the ultimate premolar and first molar. *Prokennalestes* is dated by its describers, astonishingly, as Albian. My understanding of the dental evidence at this point, as it is critical in setting the initial stage for a historical–narrative explanation of the evolution of more completely known groups, is that the molar evidence by itself is often equivocal as a reliable test for any one detailed hypothesis of ameridelphian relationships and paleogeography (see also Cifelli, 1993).

In keeping with the methodological perspectives outlined in Chapter 2, I do not consider the presentation of either cladograms or trees especially meaningful prior to an attempted understanding of character transformations, which ultimately involves the testing of taxonomic properties against data. Without the support of well-corroborated taxonomic properties, phylogenetic hypotheses can be illusory or even absurd. Even though it remains a macrotaxonomic challenge, the paraphyletic suborder Archimentatheria discussed below, suggests to me that a major radiation of ameridelphian marsupials was already under way before the last common ancestor of the living marsupials were split off from a branch of this ancient assemblage. Thus, I am profoundly skeptical of the conclusions of Westerman and Edwards (1991, p. 123), based on DNA–DNA hybridization studies, that didelphids, *Dromiciops*, and the Australian marsupials (dasyurids and macropodids) show "a trichotomy of divergence . . . in the early to mid-Cretaceous." Similarly, the conclusions of Hershkovitz (1992, p. 207) that the "Cohort Microbiotheriomorphia with its particular residium of prototherian and metatherian–eutherian grade characters must have arisen earlier than the Didelphimorphia, possibly in middle Jurassic, either in South America or North America" must be judged in light of that study's morphological approach.

## Order Didelphida Szalay, 1982

### Diagnosis

See under **Ameridelphia,** above.

### Distribution

As for Ameridelphia.

### Included taxa

Suborders Archimetatheria, Sudameridelphia, Glirimetatheria, and Didelphimorphia.

### Discussion

Didelphida is a paraphyletic order within which the nexus of relationships is not well understood, although its alleged holophyletic status is occasionally firmly asserted. Some presently unspecifiable lineage of the Archimetatheria is probably the stem group for the Sudameridelphia, and two unknown lineages of the latter are the hypothesized source of the other two suborders. While much desired, I cannot obtain better understanding of this problem at present. While a didelphid didelphidan is undoubtedly the ancestor of the Australidelphia as I will demonstrate, nothing, but the psychologically reassuring adherence to an idealogy of pan-holophylism — a classificatory formalism — may be gained by including the American opossums with the primarily Australasian cohort.

Taken together, Wible's (1990) account of the petrosals from the North American Cretaceous and the tarsal evidence detailed in this volume point to the close relationship of the archaic assemblage and the roots of the Didelphidae. The dental and tarsal evidence leaves little doubt that this as yet unsorted close relationship extends to the origins of the South American radiations as well.

## Suborder Archimetatheria, new

### Diagnosis

See under **Ameridelphia**, above.

### Distribution

Early and Late Cretaceous of North America, and possibly Asia, and Paleogene of North America, Europe, Asia, South America, and Africa.

### Included taxa

Families Stagodontidae and Pediomyidae (including Pediomyinae, Alphadontinae, Glasbiinae, and Peradectinae).

### Discussion

I consider the Holoclemensiidae erected by Aplin and Archer (1987) to be best viewed as a synonym of the pappotheridan Pappotheriidae, and until other evidence becomes available, the group is more meaningfully classified in the Tribotheria. It is not at all certain that this minimal sample for a mammalian taxon, *Holoclemensia*, can be considered as either patristically or cladistically metatherian. The presence of a stylar cusp **C** is unconvincing as evidence of such ties. See particularly the discussion of Jacobs et al. (1989), Cifelli (1989, 1993), as well as previous discussions in this volume.

In my view, there are no grounds — phenetic, and certainly not phylogenetic — that would permit the recognition of more than two families (one with subfamilies) of didelphidans in the post-Cenomanian Cretaceous and the Paleogene of North America and Europe. The significance of the Paleogene Asiatic and the undoubted African metatherian evidence, which I consider probably derivative of the European peradectines, is discussed separately in Chapter 9; the archimetatherian evidence relevant here is treated under the earlier discussion of dental evidence and in the discussion to follow of the South American sparassodontans, paucituberculates, simpsonitherians, and didelphimorphians. I have discussed the important recent contributions of Wible (1990, 1991) and Rougier et al. (1992) to cranial morphology of the North American Cretaceous forms above, and I will remark on the additional contribution of Wible (1990) to this issue.

Wible's (1990) analysis suggests that the petrosal assemblage of the American Late Cretaceous (those of the stagodontid *Didelphodon* and the probably pediomyid petrosal types *A–D*) is clearly diagnosable (in the presently used sense) in contrast to the homologous morphology of the Didelphidae (see Figs. 8.2–8.6). It is of paramount importance to remember that unique clusters in phylogenetic analysis are axiomatically and properly viewed as synapomorphies in contrast to other such clusters. This is the inductive phase of analysis, which should be followed by deductive testing. Most of the time, however, it is not considered (i.e., examined in context) whether in recognizing two homologous clusters of synapomorphies one may be also examining evidence for a transformation series (a diachronic and linear modification of a homologous series), instead of two (assumed) divergent conditions. The latter is too often implied by the exclusive use of cladistic theory without phyletics to be a correct representation of real phylogenies. So an only cladistic, and nonphyletic, presentation of this important petrosal cluster implies it to be a holophyletic assemblage. I consider this, however, to be unlikely on the character level. While it may be argued that some of these traits represent autapomorphies, others are plausible representatives of the ancestral conditions from which the protodidelphidan ones evolved. The following list of petrosal traits, after Wible's careful study (pp. 195–202), *may* represent the apparently diagnostic traits of the archimetatherians (and therefore possibly also ameridelphians and even metatherians). It must be noted, however, that Wible considers some of these characters not verifiable in all of the petrosals he studied or as also present independently in living metatherians. These traits are (1) a prominent posterolaterally directed mastoid process (identified in stagodontids and petrosals *A* and *B*; see Figs. 8.4–8.6); (2) a low ridge on the promontorium in contrast to the raised one in living forms (Wible correctly cautions concerning the phyletic lability of this particular character); (3) a depression for origin of tensor tympani muscle on anterolateral part of promontorium; (4) an elongated sulcus for the greater petrosal

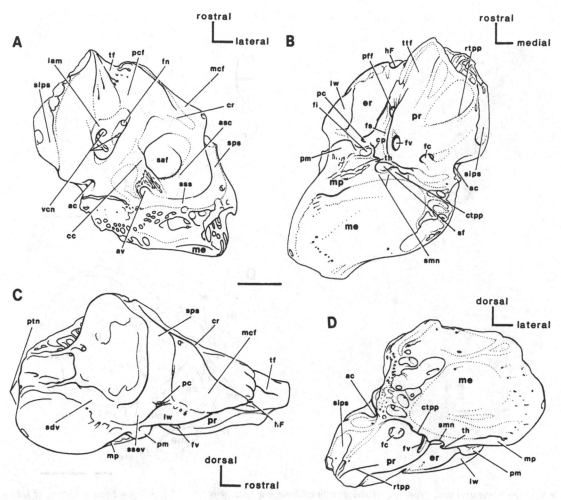

**Figure 8.5.** *Archimetatherian, Type A. Outline drawings of four sides of FMNH No. PM50937 shown in* cerebellar (A), tympanic (B), squamosal (C), and lambdoid (D) views. For abbreviations see Figure 8.1. Scale = 2 *mm. (Courtesy of J. R. Wible, from Wible, 1990.)*

nerve; and (5) a broad mastoid exposure, in contrast to the narrow didelphid one. Of all the traits analyzed, Wible considers only the "fossa incudis with a prominent lateral wall formed by a postmeatal process" (p. 199) as an unquestionable autapomorphy of the Creataceous clade. This condition stands in contrast to what he views as the primitive therian, eutherian, and metatherian condition of a fossa incudis, the lateral wall of which is formed by the squamosal. In an ongoing study (Szalay & Muizon) of Itaborí metatherian petrosals the putative autapomorphy of the fossa incudis claimed by Wible for the Creataceous marsupials he studied is being tested.

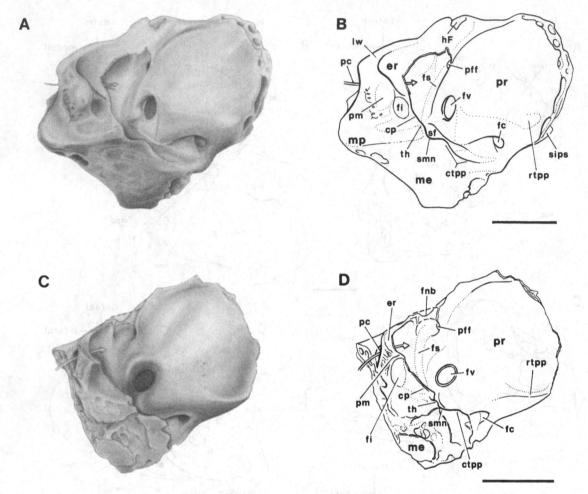

**Figure 8.6.** *Archimetatherians, Type B (AMNH No. 120124; above) and Type C (AMNH No. 120125; below).*

*Labeled outline drawings of specimens on right. For abbreviations see Figure*

*8.1. Scales = 2 mm. (Courtesy of J. R. Wible, from Wible, 1990).*

## Family Stagodontidae Marsh, 1889

*Diagnosis*

Protostagodontids had: "partially prismatic" (Wood & Clemens, 1992) enamel; relatively robust teeth with a hypertrophid paraconid and a reduced metaconid; upper molars without stylar cusp **C;** enlargement of premolars of younger samples (derived addition of protocones to **P2/** and **P3/** lingually) and robust mandibles suggesting some hard elements in the diet (e.g., malacophagy); promontorium characterized by having a caudal tympanic process of petrosal contacting it medial to the fenestra rotunda (fenestra cochlea); petrosal depression lateral to fossa incudis and postmeatal process; peroneal process on calcaneus proximally retracted, and a **CCJ** that is nearly an unmodified ovoid, suggesting a rotary mobility between the calcaneus and cuboid, and thus concordant with the hypothesis of an aquatic habitus.

## Distribution

Late Cretaceous, North America.

## Discussion

This modestly diverse, relatively abundant, probably aquatic and malacophagous group of marsupials of the Late Cretaceous of the Western Interior of North America were the largest known therians of their day. There is general consensus that they are closely related to the other Cretaceous ameridelphians, although the possibility of their more ancient metatherian ties are suggested by the study of Wood and Clemens (1992; for background see Gilkeson & Lester, 1989). Fox and Naylor's (1986) account is the most complete and analytical of the known lower jaw and dental evidence (see also Clemens, 1968a; Cifelli, 1987b). I know of no convincing arguments for real synapomorphies, beyond those that hold for the Ameridelphia or Metatheria, with other archimetatherian groups. I also agree with Kielan-Jaworowska and Nessov (1990) that the stagodontids have no special ties with deltatheroidans, primitive as the stagodontids may be in some respects within the Ameridelphia (Fox, 1987) or Metatheria. In a recent study of enamel structure in this group, Wood and Clemens (1992) consider the enamel structure of this family to be the most primitive among mammals with tribosphenic molars (studied so far). Their null group comparison with eupantothere enamel structure indicates close similarities to the latter. Although enamel patterns are difficult to evaluate, because of the extreme *developmental* and *adaptive* constraints imposed on this tissue (see especially Koenigswald et al., 1993), new data may still be potentially significant. Wood and Clemens (1992, p. 61a) state:

> Stagodontid enamel has strong seams and less dominant prism sheaths which are flat arcs rather than horse-shoe crescents of fully 180⁰ or more; this pattern has been called 'partially prismatic' in the pantothere. All latest Cretaceous marsupial genera except *Didelphodon* have fully prismatic enamel.

Tarsal morphology (Figs. 6.5 & 6.6) is significantly different from what I judge to be the more primitive metatherian pediomyine–alphadontine material (Figs. 6.1–6.14) from the same deposits, or from the late Cretaceous Lance calcaneal sample known to me. This derived metatherian morphology of the stagodontids is presumably based on extremely primitive metatherian features which are, I believe, as highly modified adaptively as is their premolar dentition. This assessment is based on the retracted peroneal process and the circular, unmodified ovoid articulation of the **CCJ.** The pedal evidence is entirely noncorroborative of ties to the Borhyaenidae (see Fig. 6.15), and the dental evidence is plagued by the same presently unanalyzable minutiae surrounding plesiomorphies and parallel adaptive solutions, a problem similar to the alleged didelphid affinities of other Creataceous and Paleogene dental phena. This is, I believe, a holophyletic family.

## Family Pediomyidae (Simpson, 1927)

### "*Diagnosis*"

The dental and tarsal attributes of the last common ancestor of this family are probably indistinguishable from what I consider to be the ancestral ameridelphian conditions discussed in this volume.

### *Distribution*

Early Late Cretaceous (Turonian) to Miocene, North America; Paleogene to Miocene of Europe, Paleogene of Asia, and Africa.

### *Included taxa*

Alphadontinae, Pediomyinae, Glasbiinae, and Peradectinae.

### *Discussion*

The ignorance displayed under "Diagnosis" clearly indicates the lack of meaningful assessment of an ancestral morph beyond the diagnostic didelphidan one. Nevertheless each of the recognized subfamilies can be diagnosed without any difficulty, and in the literature dealing with the relatively scant and only slightly diverse dental remains this has been done exceptionally well, with a few of those exceptions that attempt to allocate living dental phena to the pediomyids (like that of *Dromiciops*, for example). The synthetic contributions of Krishtalka and Stucky (1983a, b; 1984), Crochet (1978, 1980), and Fox (1987, and several references by that author therein), Clemens and Lillegraven (1986), Lillegraven and McKenna (1986), and most recently Cifelli (1990a,b,c) document and discuss the dental details of this and other relevant radiations. But the suite of cheek tooth attributes present in the last common ancestor of this family group, traits that separate this assemblage from other dentally defined *families* (not genera or subfamilies), is very difficult to establish. This group, at least dentally, is still the structurally unfocused and broadly delineated stem group of the concept of Ameridelphia based largely on dental, petrosal, and (barely) tarsal evidence. The cheek-tooth morphology of this ancestry was quite close not only to the various tribotherians discussed in Cifelli (1993, and references therein), but also to the known Holarctidelphia, and first eutherians (Otlestidae), although these can be differentiated, as this was discussed above (see also Fig. 8.7).

At present, in spite of the minutely argued points by Reig et al. (1987), Marshall (1987), Aplin and Archer (1987), and Marshall et al. (1990), there is no convincing evidence that some of the four subfamilies of this family recognized here have especially close phylogenetic ties to the South American marsupials. Yet Crochet's (1978) important synthesis (see also Koenigswald, 1970; Crochet 1977a,b; 1979a,b) suggests that a connection through some member of his concept of Peradectini (which, however, included not only peradectines but also both *Alphadon*, and *Bobbschaefferia*, the latter from Itaboraí, among others) may have been likely with the South American radiation. As I have argued, the possibility exists that a group of *Peradectes* (a structurally definable dental morph) may have had an independent history earlier than known at present. An African Peradectinae that dispersed northward is a possibility (see particularly Gheerbrant, 1990, 1992). Equally likely is an Eastern North American and Central North American presence for peradectines back into the Cretaceous that dispersed into the rest of the Americas during the Paleocene.

On dental grounds alone the ordering of the increasingly numerous taxa is not possible with any confidence for formal holophyletic taxonomic association. The known morphology of nonstagodontid marsupial tarsal remains, as well as their modest range of dental morphology, permits nothing more than the recognition of a probably paraphyletic family Pediomyidae. I believe that any other taxonomic arrangement now is merely a formalistic exegesis of principles and procedures that will often omit evidence, particularly the extremely important stratigraphic aspect of all the dental samples, from a phylogenetic synthesis. No widely or objectively acceptable taxonomic properties corroborate the special ties between the North American and European taxa of the Cretaceous and Paleogene and the better-known later groups. Yet the ongoing rapid expansion of dental and skeletal information on archimetatherians, along with an equal measure of attempts to understand the evolutionary morphology of dental and skeletal change, will sharpen the presently unclear picture of their history. Unequivocal diagnostic features will be hard to establish. Suggestions such as the reduction of cusps (paracone), evolution of dilambdodonty, continuous but difficult to polarize transitions between extremes of stylar shelf size, relative talonid dimensions, the relative size differences between various cusps, and the degree of separateness of the hypoconulid from the entoconid will have to be approached from a functionally integrated perspective, one that is yet to be attained. One will always need to be reminded that these traits like many others probably exhibited during their evolution a measurable – quantitative and gradual – continuum; their "qualitative" expressions are due only to the discontinuity of the existing record.

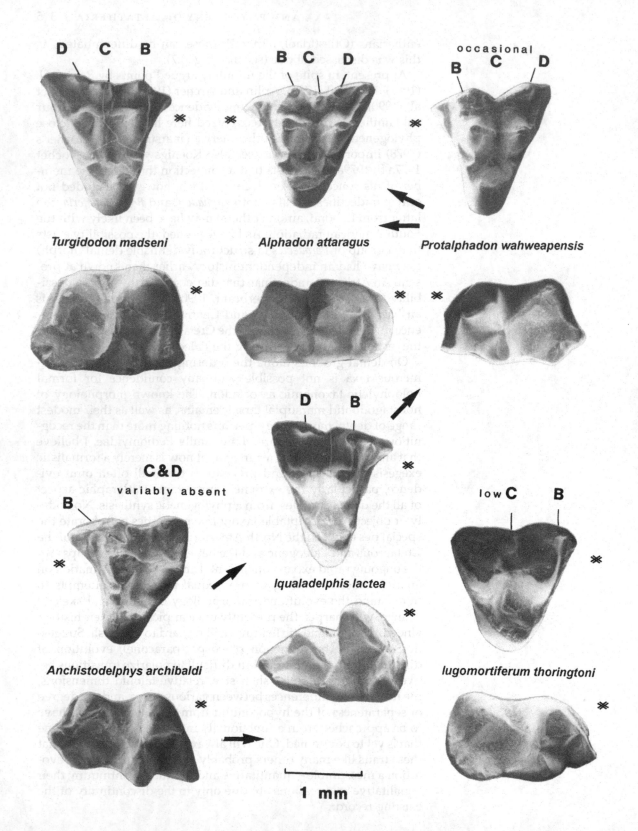

D C B

B C D

occasional
B C D

*Turgidodon madseni*

*Alphadon attaragus*

*Protalphadon wahweapensis*

C & D
variably absent

B

D B

low C B

*Anchistodelphys archibaldi*

*Iqualadelphis lactea*

*Iugomortiferum thoringtoni*

1 mm

In a tightly argued, important paper, while describing the Aqui-lan (early Campanian) Cretaceous genus *Iqualadelphis*, Fox (1988) presented some clearly enunciated hypotheses about early meta-therian evolution. Earlier than previously described species of *Pediomys* and *Alphadon* (*sensu lato*), this form is, according to Fox, structurally and actually ancestral to later Cretaceous amer-idelphians. The mesiodistally somewhat narrow and transversely wide upper molar of *Iqualadelphis* with well-developed occluding cusps has a relatively narrow stylar shelf buccal to the paracone but with a clearly identifiable stylocone (cusp **B**), a stylar cusp **D** more distally, but no cusp **C** in between. While Fox considers the Cretaceous taxa "marsupicarnivorans," he also asserts that the appearance of cusp **C** can be considered as a clear sign of a phy-letically unified Didelphidae. He does not, however, consider oth-er attributes reported in the literature – his taxic statements are essentially equivalents of his views of dental evolution. I believe as do others (e.g., Cifelli, 1993) that because various expressions of stylar cusp **C** obviously occur outside of the Metatheria, it is not unlikely that cusp **C** has appeared, became emphasized and deemphasized, and disappeared several times in the Theria (see also Cifelli, 1990a, b, c, 1993, for the same perspective as followed here on stylar cusp evolution). *Iqualadelphis* may well represent an excellent structural (and possibly actual) dental ancestor for the Pediomyinae, as suggested by Fox. In fact Fox's stand implicitly supports the view that such taxa, lacking stylar cusps, like the known representatives of Holarctidelphia, can be allocated to the Metatheria without any difficulty. The alphadontines of the ped-iomyid assemblage (the caveats expressed in this volume notwith-standing), however, may well be the stem-group for the heterogeneous radiation (possibly monophyletic, but not holophyletic) of the peradectines and the South American meta-therians. Of the latter the Didelphidae is only one clade which becomes well recognizable in the Paleogene on tarsal grounds. However, once the early European and African record signifi-cantly improves, the role of peradectines may be more mean-ingfully reflected in their recognition as a family. It is entirely possible that this group has deeper roots in Africa (or Asia) than in North America. This is a moot, but tantalizing, point at present.

Little may be said here of the two recognized species (Archi-bald, 1982) of the small frugivorous metatherian *Glasbius*. In this Late Cretaceous genus, teeth are low crowned, the stylar shelf is narrow, the stylar cusps are bulbous, the basal cingulids of lower teeth are well developed, and the last lower molar is reduced (Clemens, 1966). Its uniquely derived attributes prevent phy-logenetic association with any one of the archaic marsupial sub-groups. Marshall (1987) has compared *Glasbius* extensively with the Itoboraí marsupials, and Marshall et al. (1990) consider this genus allied to the South American Caroloameghiinae.

The pediomyine–alphadontine tarsal material from the Oldman Formation is far more primitive, given the criteria outlined for tarsal transformation, than either the North American Paleogene sample, or the most primitive material from Itboraí (compare Figs. 6.1–6.31). The **CaFi** facet is very large and the peroneal process is the least reduced of any marsupial samples known. In light of the tarsal information the differences between various dental samples of *Alphadon* and *Pediomys* are of such relatively minor importance that I do not believe they represent either ancient independent lineages or stem lineages referable to other groups. The allocation of *Pediomys, Aquiladelphis,* and *Glasbius* to the Microbiotheria by Reig et al. (1987) is, in my view, unwarranted. The puzzling dismissal of the special similarities of the microbiotherians to the Caluromyinae dentally and the failure to appreciate the nature of transformation of the tarsus of *Dromiciops* from a clearly derived didelphid metatherian antecedent condition mar both their descriptive procedure as well as their purely dental and largely phenetic approach to early marsupial phylogenetics. Similarly, the inclusion of a restricted concept of the Pediomyidae (*Aquiladelphis, Pediomys,* and *Iqualadelphis*) in the Microbiotheria by Marshall et al. (1990) is also categorically rejected.

The Tiffany, Wasatch, Bridger, and Tepee Trail tarsal materials are morphologically distinct and cluster outside of any other homologous sample. The poorly known Cretaceous groups (Fig. 6.1–6.7) stand in contrast, perhaps because of lack of known specimens, to the Paleogene ones (Figs. 6.8–6.14). But, as noted, this may be a real discontinuity due to the origin of the Peradectinae outside of North America. The traits recognized in these phena have been discussed in Chapter 6. The chronochange in the North American Eocene is traceable within the Peradectinae. The reduction of the **CaFi** facet in that lineage is well established by the early Eocene, and a slightly convex **UAJ** seen in the Bitter Creek and to a lesser degree in the Bridger samples is transformed into the caenolestid-like, highly angulated **ATim–ATil** facets, probably locally. This change most likely reflects the changes in a terrestrial lineage in which molar morphology remained conservative.

It is possible, therefore, that the Pediomyidae is a holophyletic group that gave rise to no others, but more likely it is a poorly known and decidedly paraphyletic group in which at least the structural (and probably actual) ancestry of the sudameridelphian radiation may be sought as the fossil record improves.

## Suborder Sudameridelphia, new

### Diagnosis

Protosudameridelphian had: relatively primitive ameridelphian molar structure, complete metatherian dental formula, and a tarsus (atragalus and calcaneus described under the Itboraí pedal evidence for *Itaboraí Metatherian Groups, IMGs* other than that of the didelphids, polydolopids, and the extremely divergent *IMG*

*II)* characterized by a tricontact **UAJ,** narrow **AFi** facets, somewhat reduced peroneal process, semicircular **CCJ,** very large **ampt,** and a retention of a ribbon-like taper of the **Su** facet proximally.

### Distribution

Early Paleocene (Tiupampian) to Pliocene of South America.

### Included taxa

Infraorders Itaboraiformes, Polydolopimorphia, and Sparassodonta.

### Discussion

It is my current judgment that this group, and the stem taxon Itaboraiformes included in it, are South American and paraphyletic groups. The tarsal evidence, still uncertainly associated with dental taxa, detailed in this volume, along with the great diversity of dental morphology known from the Itaboraí fissures, and more recently from the earlier Tiupampa local fauna (see Muizon, 1991), *suggests* that the entire South American radiation may be rooted in that continent. Such an ancestor would have been an archimetatherian which made its way there, i.e., not a vicariance event (see Chapter 9). Evidence to test such a suggestion made by several students will be long in accumulating, however. Nevertheless the tarsal evidence of the representatives of various itaboraiform family rank groups that I analyzed (even if exact matching with dental morphology was not possible), that of polydolopimorphians, and glirimetatherians, is suggestive of a monophyletic group. The group with the closest similarity, the *null-group,* for the majority of the tarsal evidence, if not the dental, is specifically an archimetatherian group, most likely the stem peradectines. This conclusion is similar to Crochet's analyses of the dental evidence. The direction of change for the differences are not clear (not "resolvable") given the present morphological and stratigraphic evidence. It is possible that an early itaboraiform gave rise to the Peradectinae, and a lineage of these made its way northward in the Paleocene-Eocene.

Didelphids are a tarsally and to some extent dentally highly derived group within the Didelphida, however, as I emphasize repeatedly. Furthermore, as I will demonstrate, I am inclined to view the origin of that South American family from an itaboraiform stem rather than another archimetatherian stock. It should be added that the North American Paleocene tarsal record for the archimetatherians is unknown, with the exception of the Tiffany calcaneus described above, leaving ample room for speculation. It may be that the connection to the origins of the sundry sudameridelphians (from an itaboraiform) occurred in the Early Paleocene in South America. The date of the Tiupampa fauna originally suggested as Cretaceous is now widely regarded as Early Paleocene (see Van Valen, 1988a, b; Ortiz Jaureguizar & Pascual, 1989; Pascual and Ortiz Jaureguizar, 1990; Muizon, 1991).

Different, as yet conjectured, primitive itaboraiforms may have given rise to the polydolopimorphians, sparassodonts, and glirimetatherians, and also to the didelphimorphians. Because this conjecture is so speculative, it is not shown in the phylogenetic hypothesis of Figure 2.3. Clearly it is not a very satisfactory arrangement that a group of families (Itaboraiformes), which while monophyletic (if the broadly sampled tarsal evidence is any indicator, or test, of this concept), should be the ill defined source for so many different higher categories. Yet I believe this to be a far more satisfactory arrangement than the alternative possibilities of virtually groundless sister-group arrangements (all implicitly claiming holophyly for sundry clusters) based either on some vaguely defined characters which may be homoplasies (tooth enlargement) or mere ameridelphian dental plesiomorphies. The tarsal traits may indicate the apomorphies of the first lineage of the Sudameridelphia. Phylogeny within Sudameridelphia is an area of great ignorance and an obvious future challenge for marsupial phylogenetics.

I consequently include a number of infraorders in this suborder because (1) they are demonstrably not specially related to the Didelphidae or the Microbiotheriidae, and (2) there is no reason to believe, based on other than primitive ameridelphian or convergent dental attributes, that they are specially related to any one of the known Archimetatheria of North America and Europe. It is possible, of course, that the sparassodontans, the polydolopimorphians, and the glirimetatherian caenolestids and simpsonitherians represent independent derivation from hitherto unknown archimetatherian stocks. Nevertheless the tarsal morphology of at least some representatives of these groups appears to be derivable from morphology that characterizes the more primitive itaboraiform condition. This condition entails the *combination* of ameridelphian attributes of sharply angled medial and lateral tibial facets on the astragalus, narrow fibular contact with the astragalus, a fully tricontact **UAJ** with large **CaFi** facet, a nearly triangular but ovoid-contoured **CCJ** contact, and an **AN** articulation that hints at habitual flexion–extension rather than abduction–adduction at this joint.

There is no reason to believe that the molar morphology of the first itaboraiform or of the first borhyaenid was more advanced than the most primitive morphs of a strictly defined Didelphidae. Of the taxa recently described by Marshall and Muizon (1988) and Muizon (1991), both the genera *Jaskhadelphys* (with well developed cusp **D** in addition to **B**) and *Allqokirus* (with well developed metacone hypertrophy and deemphasized entoconid) display primitive appearing transverse molars combined with the probably advanced (compared to the more ancient North American taxa) relatively mesiodistally shortened small protocones. Such traits are neither those that one would infer to have been present in the stem ameridelphians, nor are they significant similarities to deltatheroidans (contrary to the suggestion by Kielan-Jaworowska, 1992). These similarities of the Tiupampa genera, however, may be convergences to a more primitive therian state due to carnassial emphasis and deemphasis of the talonids (see particularly *Allqokirus*) in the kind of structurally ancestral sparassodontan that *Jaskhadelphys* may be.

Because no known occlusal constraints operate on the stylar cusps of the upper molars in marsupials, their relative importance in higher level phylogenetics may be questioned when such factors are absent. Lack of occlusal constraint, however, is not synonymous with adaptive unimportance. Undoubtedly this area of the stylar shelf plays a significant functional–adaptive role in mastication, and the change in that area in various groups of ameridelphians probably reflects this role. Nonetheless this lack of occlusal constraint does increase the possibility of some cameo appearance and disappearance of these structures (see also Cifelli, 1993).

Judging by the described and well-illustrated dental patterns (Marshall & Muizon, 1988, 1992; Muizon, 1991) of marsupials from the El Molino Formation, Tiupampa locality, Bolivia, this diversity is remarkable, and once studied the postcranial specimens will go a long way toward the reconciliation of apparently conflicting interpretations based on dental evidence on the one hand and tarsal evidence on the other. The two basic types of archaic upper molar morphs are represented, judging by the patterns of the crest connecting the paracone and metacone (i. e., the centrocrista and the crests mesial and distal to it being either U or W shaped in occlusal view). These adaptively critical patterns, together with the degree of reduction of the stylar shelf (as this feature is also affected by the nature of shear of the buccal crests) form the bases of taxonomic allocations. As we must expect at this stage of understanding, the only guide to classificatory allocation of several of these dentally relatively unmodified taxa is the available information concerning dentition. It is easy to base generic accounts on molar traits, but inferring higher taxonomic affinities, although routinely done, can be a questionable enterprise, as I have pointed out.

Given the general acceptance of the Early Paleocene age of the Tiupampa fauna, the essay by Case and Woodburne (1986) on the continued diversity of marsupials in South America from the Cretaceous into the Paleogene does not appear to be valid. They claimed that stability of the marsupial fauna in crossing the boundary between the Mesozoic and Cenozoic in South America is also supported by evidence of a stable fossil plant record on either side of this boundary, in sharp contrast to the apparent floral turnover in at least some Nearctic areas that suggests ecological instability in North America at that time. The condylarths and pantodonts (see especially Muizon & Marshall, 1992), however, strongly support an Early Paleocene age. As Van Valen (1988a,b) and Pascual and Ortiz Jaureguizar (1990) state, the various arguments advanced for a Late Cretaceous age can be used to support the Early Paleocene age of the fauna, their concept of the Tiupampian.

The following Tiupampa marsupial taxa have been reported by Marshall and Muizon (1988): *Peradectes, Khasia, Pucadelphys, Incadelphys, Mizquedelphys, Andinodelphys, Tiulordia, Jaskhadelphys, Roberthoffstetteria,* and *Allqokirus.* I exclude *Kollpania,* which I consider to be a deciduous tooth of a eutherian (see also Muizon, 1991). Further study of this well described and generously illustrated fauna by Marshall and Muizon will obviously play a decisive role in testing a number of hypotheses in ameridelphian marsupial relationships.

The marsupial fauna of the São José of Itaboraí fissures of Brazil represented for a long time the greatest diversity of the earliest substantial samples of marsupials on the South American continent. As Bonaparte (1990b) suggests, there may have been accumulation of diachronous genera; although the bulk of the taxa is decidedly Riochican, several of the animals may represent an earlier Paleocene time interaval. The Tiupampa assemblage, although less diverse, is probably earlier in age as noted. Table 6.3 is a list of the named taxa that I believe to be valid species and genera from Itaboraí. While the didelphids, borhyaenids, polydolopimorphians, and paucituberculatans are present in the fauna, I suggest that an adequate case for a monophyletic paraphylon Sudameridelphia can be made based on the tarsal evidence. The Tiupampa marsupials do not in any way invalidate this hypothesis. The first sudameridelphian may be the last common ancestor of the entire South American marsupial radiation, an explosive diversification that appears to be evident from the approximately sequential Tiupampa and Itaboraí faunas. With the exception of the more ancient but poorly known Peruvian Laguna Umayo ameridelphians described by Sigé (1972a, b), and the great diversity of the Bolivian Tiupampa forms, the Brazilian Itaboraí assemblage represents the largest sample of the continuation (the "near-beginning") of the archaic marsupial record in South America.

I believe that any discussion dealing with the macroevolutionary aspects of marsupials or any other group depends on the context and nature of all the evidence and the various hypotheses that are tested against it. When clearly apomorphous features or convincing vertical changes are available, then both phylogenetic discussions as well as macroevolutionary accounts of continental scope may be warranted. In the case of putative Cretaceous but probably Paleocene marsupials, however, the usual evidence has been limited to molar morphology, and more particularly upper teeth. In light of the extreme difficulties in sorting out the very few diagnostic attributes of many of these animals, therefore, most recent statements about which "family" or "subfamily" evolved, became extinct, or crossed boundaries must be met with extreme scepticism. For example, presence of a single tooth from the Tiupampa locality and the maxilla fragment of *Monodelphopsis* from Itaboraí form the bases for postulating the presence of Pediomyinae in both the Cretaceous and Paleogene of South America, according to Case and Woodburne (1986). Although such events may have been the case, the evidence as yet simply does not support such a hypothesis. The upper molars of *Monodelphopsis*, although they have a rectilinear centrocrista along with many other undoubted nonpediomyines, also have a relatively broad stylar shelf and an enlarged metacone that is distinctly nonpediomyine. If the latter feature is "didelphid," then how is it that the centrocrista is distinctly different from those of the majority of didelphids? Perhaps the dental concept of "didelphid," as the concept of dental "pediomyine," is still only a general structural one. The notion of "pediomyine," as generally used, embraces morphs with minor fluctuations and retentions. All of these must be clarified before the dental diagnoses of families of the ameridelphian Metatheria become genuinely reliable. Extreme care should be taken that certain diagnostic combinations on molars are not mere retentions (stylar shelf, one or the other type of centrocrista, etc.) and losses (e.g., of conules, or of the twinning of talonid cusps) coupled with easily convergent traces of independently derived traits. Such clusters can certainly be postulated to be a real grouping, yet tests for them are very weak; therefore, formally recognized taxa based on them are unreliable and remain easily contestable.

**Infraorder Itaboraiformes, new**

*Diagnosis*

As in **Sudameridelphia**.

*Distribution*

Early Paleocene (Tiupampian) through Eocene of South America.

*Included taxa*

Caroloameghiniidae with Eobrasilinae (including Eobrasilini, Derorhynchini, and Protodidelphini), Caroloameghiniinae, Mirandatheriinae, and the Monodelphopsinae.

## Discussion

The nature and relationships of the dental evidence has been discussed recently in some detail by Marshall (1987). As my classification indicates I regard these ties somewhat differently than he does. Under the Itaboraí pedal evidence I briefly note some of my specific disagreements with Marshall's (1987) important taxonomic contribution. I recognize no dental pediomyid or microbiotheriid at Itaboraí, and the tarsal evidence does not suggest their presence there, either. While the known teeth are diversified to reflect adaptive shifts related to diet, the morphological and size range of the tarsal evidence strongly suggests that we are witnessing the very beginning of a local (South American) radiation of the Metatheria. Much of the evidence from the Itaboraian Paleocene and the earlier Tiupampian is as yet unclear concerning the specific affinities of many itaboraiforms. The details of my views as they relate to the early South American dental evidence will be discussed elsewhere.

**Infraorder Polydolopimorphia (*sensu* Ameghino, 1897)**

## Diagnosis

Protopolydolopimorphian had: a combination of enlarged premolars (perhaps originally like those in the prepidolopids, and therefore *not* in a plagiaulacoid mode), and a reduced trigonid shear and bunodont crown pattern; a drastic reduction of the calcaneal peroneal process; a truncated **CCJ**, coupled with retention of the primitively elongated calcaneal **Su** facet, as well as the calcaneal **CaA** facet relatively larger than the **CaFi** facet.

## Distribution

Medial Paleocene (Itaboraian) through early Oligocene of South America and the Eocene of Antarctica.

## Included taxa

Prepidolopidae, Polydolopidae, and Bonapartheriidae.

## Discussion

Through his prolific studies Marshall (1980a, 1982a, b; 1987) has established, as far as I am able to judge, that there are no synapomorphies of any level of confidence, beyond ameridelphian ones, shared between the most primitive (or more advanced) polydolopimorphians, the genus *Prepidolops*, and the most primitive (or more advanced) caenolestids, the Paucituberculata. In fact the nonhomologous enlargement of the trenchant teeth in paucituberculatans (the M/1-P3/ pair) appears to be a shearing–

slicing, "carnassial-type" adaptation, quite distinct from the plagiaulacoid modifications of the premolars in polydolopimorphians. In light of the large canine of bonapartheriids, the enlargement of the anterior teeth is unlikely to be homologous with that condition of paucituberculatans. It is of course quite possible that the premolar and molar modifications of the Eocene Bonapartheriidae are independent of those of the Prepidolopidae, the stem family of the Polidolopimorphia. Any evidence for this possibility would reopen the question of polydolopimorphian and paucituberculatan affinities. But largely for the reasons outlined by Marshall (1987), I cannot see placing the polydolopimorphians together with paucituberculatans in the same suborder. The dental evidence for this infraorder has been clearly presented by Pascual (1980, 1981, 1983), Pascual and Bond (1981, 1986), and Pascual, Carlini, & de Santis (1986). Marshall et al. (1990) recognize the Reigiinae (based on *Reigia* Pascual, 1983) and Prepidolopinae as two subfamilies of the Prepidolopidae.

Marshall's (1987, pp. 143–7) analysis of the evolution of the known dental and cranial traits of polydolopimorphians is convincing to me, although his association of argyrolagid glirimetatherians with the former is less so. Although the exact number of upper and lower incisors of the last common ancestor of this group is still debatable, were this form known, it would probably be classified as a prepidolopid. Unlike glirimetatherians, which have an enlarged incisor, prepidolopids have an enlarged, procumbent, and gliriform canine, an apomorphy of the first polydolopimorphian. In prepidolopids the second and third premolars are already enlarged while the molars still retain some similarity to a more primitive itaboraiform pattern. On the upper molars – but not reflected on the lowers because these are not strictly occluding areas – stylar cusps **B** and **D** are hypertrophied and thus augment the bunodont function of the cheek teeth. The specialization of the second and third premolars into hypertrophied puncturing–slicing units is already well developed. The bonapartheriids carry some of these, what I believe to be genuine, trends (not due to the alleged stochastic vagaries of species selection) to a greater extent. The molars have become crushing surfaces with the cusps reduced to multituberculate-like dimensions, and the last two premolars are areas of hypertrophied crushing structures. The polydolopids all but eliminated **P2** in addition to **P1**; they came to emphasize the **P3** as a robustly based, strong, splitting–slicing tooth and further reduced the relief of the molars.

Of the dentally highly divergent but well-delineated families of marsupials included in this infraorder, only the Polydolopidae are open to conjecture concerning tarsal morphology. Well represented by numerous specimens, *IMG VII* is both in the size range of *Epidolops*, and calcaneally it is also one of the most modified from the common Itaboraí pattern. The extreme reduction of the peroneal process, the axis of rotation on the **CaA** facet, the large **CaFi** facet, and the truncated **CCJ**, along with the long but not deep calcaneal tuber all suggest a highly terrestrial animal. If the association of the astragali with the calcaneal sample is correct, then nothing prevents the tentative derivation of the Polydolopimorphia from an itaboraiform which probably lacked stylar cusp **C** and had a primitive itaboraiform tarsal structure. There is no evidence that would even hint at origins from Didelphidae, a point well made by Marshall (1987). Dental association of the other families included in the Polydolopimorphia with the Didelphidae, *sensu stricto,* is not probable in light of the difficulties posed by the ancestral didelphid dental attributes.

Both the long and shallow calcaneal tuber and the transverse axis of the **LAJ** indicate a nearly total commitment by these animals to a terrestrial habitat. The possibly contaxic astragali from Itaboraí strongly suggest derivation from an itaboraiform rather than a derived didelphid-like astragalar structure.

Flynn and Wyss (1990) and Flynn (1991) report the last early Oligocene co-occurrence of polydolopids with the newly arrived caviomorph rodents. The intriguing question, of course, is the possible causal relationship between the primarily terrestrial rodent diversification and polydolopid extinction in South America.

## Infraorder Sparassodonta Ameghino, 1894

### Diagnosis

Protosparassodontan, as in *Jaskhadelphys,* or a hypothetical hathlyacinine borhyaenid ancestor, had: (? secondarily) enlarged stylar shelf due to the demands of postmetacrista–protocristid shear and reduced stylar cusps; paracone slightly smaller than metacone; protocone mesiodistally reduced in light of trigonid reduction; sizeable epitympanic recess in basicranium; tarsal morphology similar to that commonly encountered in primitive itaboraiiforms.

### Distribution

Early Paleocene (Tiupampian) to Late Pliocene (Chapadlamalan) of South America.

### Included taxa

Borhyaenidae (including Hathlacyninae, Hondadelphinae, Prothylacyninae, Borhyaeninae, Proborhyaeninae) and Thylacosmilidae.

### Discussion

An excellent overview of the dental evidence for the Borhyaenidae is presented by Marshall (1978b), and the dental and some of the cranial evidence of the most primitive radiation of this group, the Hathlyacininae, is thoroughly documented and discussed by Marshall (1981a). The importance of this great radiation is signaled by Aplin and Archer (1987) in their almost "evolutionary" (Darwinian; syncretic and synthetic, for all intents and purposes) "rerecognition" of the group as an order. These carnivorous marsupials are perhaps roughly equivalent in diversity, size range, and intertaxonal morphological variation to the eutherian Creodonta. The large saber toothed genus *Thylacosmilus* has a unique mastoid bulla (Turnbull & Segall, 1974), but otherwise morphologically the group is quite similar in many ways to other archimetatherians and itaboraiforms, so that I prefer infraordinal ranking only. Reliable taxonomic properties have never been established that would corroborate a sparassodontan–thylacinid relationship (see Fig. 8.9)

In spite of arguments presented by Marshall et al. (1990), I strongly disagree with any special affinity between borhyaenids and stagodontids, as did Fox (1979c), as noted below. Much of the previous discussion on the affinities of this group has been a debate concerning structural transformation of the molar dentition. While valuable contributions (e.g., Marshall, 1978a, 1981b), these are character phylogeny accounts of the dentition, without taking into consideration other evidence bearing on group phylogeny. At any rate, contrary to Marshall's (1977a) suggestion, Fox (1979c) has convincingly pointed out that the stylar area reduction (given a putative *Iqualadelphis*-like ancestry) of pediomyines and borhyaenids was independent. Fox (1979c, p. 734) notes that, unlike borhyaenids, which coupled the reduction of stylar shelf and cusps and improved carnassial function

the chronostratigraphic succession of *Pediomys* fossils in the upper Cretaceous documents *reduction* of orthal (vertical) shear and *increase* in the horizontal surfaces associated with grinding and crushing. Evidence from fossils indicates that while facets from occlusal contact were produced in life, they mostly resulted in a truncation of cusps and crests on molars of pediomyids and not in a honing of these structures from wear along their sides. In sequences of *Pediomys* species from older to younger, the protocone became larger relative to the more labial parts of the crown, while in the Thylacinidae and Borhyaenidae, the protocone became relatively smaller and the more labial parts comparatively enormous, particularly along the shearing postmetacrista. . . . Moreover, the trends towards increase in protocone size and enhancement of grinding capacity occurred both in small . . . and . . . large *Pediomys*, . . . so [these] seem not a function of coronal size. . . . The lower molars in the Pediomyidae seem no more adapted for a carnassial function than do the uppers. In borhyaenids and thylacinids the posterolingual trigonid cusp, the metaconid, is reduced or lost, the paracristid is a trenchant shearing crest and the talonid is narrowed. But these trends are not in evidence, even incipiently, in known pediomyid species: in them the metaconid is the second largest trigonid cusp and is robustly developed, the cristid obliqua meets the posterior wall of the trigonid labially, the hypoconid is strong and the talonid basin wide, all of which features are associated with a broad protocone, having a grinding and crushing function.

Fox's analysis continues to be supported by new evidence from the Tiupampa marsupials and the consideration of metatherian dentitions.

Recently Marshall and Kielan-Jaworowska (1992), stated their conviction that the Aegialodontia was the ancestor of Metatheria to the exclusion of the Eutheria. This I commented on above under Metatheria. These workers have also expressed the view that the Deltatheroida and the Borhyaenoidea are specially related to one another. Both these groups are carnivorously adapted, as discussed in this volume under the Deltatheroida. The earliest taxa from the Tiupampa local fauna that may be considered relevant here are *Jaskhadelphys* and *Allqokirus*. Both of these genera have a well-developed metacone hypertrophy, a feature lacking in Deltatheroida. The South American taxa lack the well-defined sulci that delineate the mesiobuccal parastylar lobe and distobuccal metastylar lobe in deltatheroidans. Both of these specializations strongly suggest ameridelphian rather than deltatheroidan affinities for sparassodontans. The talonid in *Allqokirus* (the only one known of these two taxa), while considerably narrowed, still retains the entoconid, but without its "twinned" association with the hypoconulid. Interestingly, a well-developed postmetacristid appears to be present, forming a quasi-carnassial notch on the lingual side of the lower molar. All of these features suggest an adaptive convergence to Mesozoic conditions shown by Deltatheroida. Carnassial notches and paraconid hypertrophy are the hallmarks, independently, of various genera of didelphids (*Lutreolina, Didelphis, Chironectes*), of various dasyuromorphians, and of the eutherian Hyaenodonta, to mention only the most obvious. Mesonychids have reduced the talonid, an event obviously related to the alteration of the antecedent biological role of the tribosphenic talonid. I believe that the alleged dental taxonomic properties for a Deltatheroida–Borhyaenoidea clade, when considered in light of ecological morphology – the known tarsal patterns of borhyaenids and other ameridelphians – will prove to be convergences, results of the independent enlargement of the paraconid and reduction of the talonid.

It is my conviction that a number of metatherian genera known by a sample of tarsals from Itaboraí represent either the Borhyaenidae or descendants of the stem group that gave rise to borhyaenids. Only two species, represented by tarsal samples (calcanea), allow association with the Didelphidae, and as I have stated I consider the dental identification of didelphids in the Paleogene largely unreliable. The overwhelming majority of the Itaboraí specimens display borhyaenid-like morphology, with the noted exceptions of the samples making up *IMG II* (an entirely distinct family of marsupials), *IMG V* and *IMG XII* (didelphids), and *IMG VII* (probably polydolopids). The question, of course, immediately arises: What may be considered diagnostic borhyaenid tarsal features? *Sipalocyon*, a relatively late form, together with the large *Thylacosmilus*, displays the curious mixture of a greatly reduced peroneal process, a primitive retention of an extensive **CaFi** facet and narrow **AFi** facet, and a number of features which are best considered as *sui generis* transverse-stability-related adaptations, probably for terrestrial locomotion (Figs. 8.8 & 8.9). Thus, an extreme angle between the **ATim** facet and the **ATil** facet is such a feature – although it may represent a minor modification of ancestral metatherian retention, or convergence to what is seen in *younger* peradectines. Attempts to derive borhyaenid tarsal morphology from peradectines run into some difficulties. The latter, without known exception, have all but reduced the calcaneofibular contact. In the Itaboraí assemblage, however, there is a clear correlation between the presence on the calcanea of a large **CaFi** facet and a rather broad ("terrestrial") calcaneal tuber. Species lacking the calcaneofibular contact have a relatively deep and narrow tuber that indicates a strong plantar flexor musculature – and probably therefore arboreality (see Fig. 6.16). By this criterion of calcaneal tuber conformation, the didelphid samples of Itaboraí, given the **CCJ** configuration, are clearly arboreal, whereas the borhyaenids appear to be terrestrial.

## Suborder Glirimetatheria, new

### Etymology

For the gliriform adaptation of the central incisors.

### Diagnosis

Protoglirimetatherian had: incisors reduced to **4/4** and an incipient enlargement of the first pair of incisors independently from polydolopimorphians and australidelphian phalangeriforms; basically itaboraiiform pedal structure as seen in the majority of the Itaboraí taxa and in the living caenolestids; molar dentition that could be characterized as being reduced in height of trigonids and thus performed extensive horizontal rather than vertical shear, without any indication of the early bunodonty seen in the polydolopimorphians; hypocone which may be homologous to the metaconule of its ancestry.

### Included taxa

Infraorders Paucituberculata and Simpsonitheria.

*Figure 8.8.* *Comparison of left calcanea of two sparassodontans (A–D) and the dasyuromorphian* Thylacinus *(E, F) shown in dorsal (above) and plantar (below) views. For abbreviations see Table 1.1. Scales = 1 mm.*

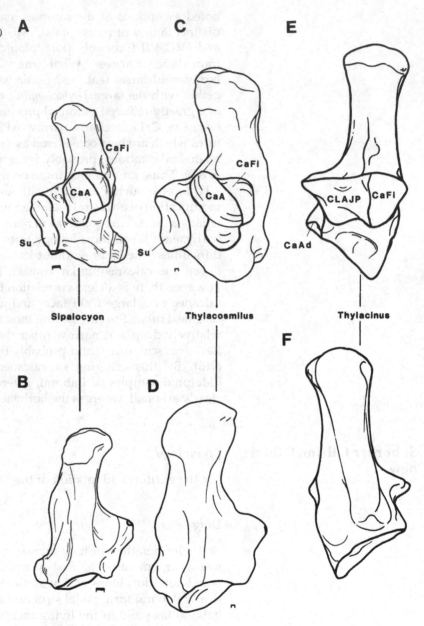

### Distribution

Medial Paleocene (Itaboraian) to Recent.

### Discussion

One of the subtle yet most compelling similarities between some (or all) caenolestids and argyrolagids is the emphasis of the first molars, particularly the trigonid of these, which is probably re-

**Figure 8.9.** *Comparison of left astragali of two sparassodontans (A–D) and the dasyuromorphian* Thylacinus *(E, F) shown in dorsal view on left and plantar view on right. For abbreviations see Table 1.1. Scales = 1 mm.*

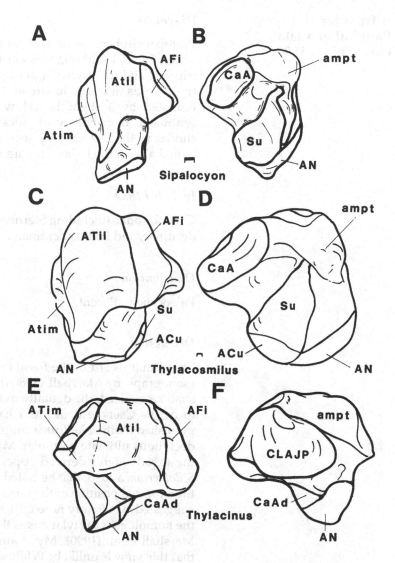

lated to the original emphasis between the caenolestid P3/ and M/1 for shearing. I am not convinced that the buccal upper molar cusps of caenolestids are homologues of stylar cusps **B** and **C**. Both the dentition and the postcranial apomorphies of the argyrolagids may be derived from a stem caenolestid-like condition. Because of the relatively primitive similarity of caenolestid tarsal morphology to the primitive sudameridelphian stock, derivation of the argyrolagid tarsal structure from such a condition is very likely. In either case the tarsal morphology of argyrolagids indicates an ancestry more primitive than a didelphid one. These primitive traits, such as a narrow **ATil** or a ribbonlike **Su** facet, are not compatible with the didelphid diagnostic traits and hypothesized subsequent constraints.

**Infraorder
Paucituberculata
(Ameghino, 1894)**

### Diagnosis

Protopaucituberculatan had: emphasized vertical shearing between **P3/M/1** (although its heritage indicates a loss of the high trigonid and transverse upper molars of the itaboraiiform ancestry); carpus in which lunate and magnum are in contact but are indented by a slight lateral wedge of the scaphoid (scapho-centrale?); morphology of calcaneus and astragalus extremely similar to the morphology seen in *IMG I* and *III*, apparently retained also in the living caenolestids.

### Included taxa

Caenolestidae (including Sternbergiinae, new, Caenolestinae, Abderitinae, and Palaeothentinae).

### Distribution

Paleocene to Recent.

### Discussion

The dental record of the fossil caenolestids has been treated in a monograph by Marshall (1980a). Of the three subfamilies Marshall recognized, the dentally most primitive one known appears to be the Caenolestinae. As I have stated, I consider the single published tooth of *Kollpania* enigmatic; it is most likely a eutherian deciduous ultimate premolar. Marshall (1980a) proposed that on the already highly derived upper molar of the Santacrucian genus *Stilotherium* a cusp, that he called "intermediate conule" located in the expected position of the metaconule is really the metacone. He suggested, as already noted, that the paracone and metacone are the homologues of stylar cusps **B** and **D,** a view that is still held by Marshall et al. (1990). My examination of *Stilotherium* suggests that this view is unlikely. While the expanded "hypocone" of that genus may be the homologue of the enlarged metaconule seen in *Sternbergia* from Itaboraí, the intermediate conule may be simply a neomorph, arising from a basal crest of the metacone, similar to the one seen in *Patene* (see also Bown and Fleagle, 1993). In spite of the suggestion of Marshall and Muizon (1988, p. 39) that the newly described genus *Kollpania* from Tiupampa represents an ideal structural ancestor for caenolestoids, I see no meaningful similarity of that "genus" to *Stilotherium* (see also Muizon, 1991).

The carpal pattern in the living taxa is enigmatic. If a centrale separated the lunate and the magnum, then a case could be made for a pattern that closely matches the primitive eutherian one. As it is, the unciform is very large compared to the proportions of the putative primitive eutherian one, and the caenolestid condition may be a terrestrial modification (see Fig. 8.16) of the condition seen in didelphids. Caenolestid and didelphid similarities are very difficult to evaluate without substantial knowledge of Cretaceous and Paleogene carpal patterns.

There is a pattern shared between the living caenolestids and the Borhyaenidae in the details of the tarsus. The fossil record of the hind foot is not especially good, but nevertheless this information has important bearing on these families. The **Mt1** is reduced in the caenolestids, as in many other groups committed primarily to a terrestrial habitat, and therefore any similarity is not phyletically significant. The relative length of the metatarsals to the phalanges is more comparable to that of dasyurids than to didelphids, reflecting the adaptive imperatives. The details of the foot bones and their articulation, however, display a pattern ancestral at a point which, in my view, predated the separation of the Didelphidae, Caenolestidae, and Borhyaenidae. In spite of the minor variation displayed by the three extant genera of caenolestids, the family-specific combination of features is present in all of these forms. In fact the similarities to some of the Itaboraí groups is such that the possibility of close ties to some caenolestids that are dentally unrecognizable should not be unexplored. As noted by Marshall (1980a, p. 129), the possibility that *Derorhynchus* from Itaboraí is a caenolestid cannot be dismissed. The smallest known tarsal sample from Itaboraí, *IMG I*, approximately "matching" in measurements the dental data of *Derorhynchus* (and of others), shares some of the primitive pattern with caenolestids – but with the notable differences I have already mentioned. The details of the caenolestid tarsus, closely relevant to their phylogenetics (and those of others), are discussed in Chapter 7.

Osgood's (1921) monograph on *Caenolestes* is the most comprehensive on any member of the family, fossil or extant. Yet as is often the case with descriptions that stem from widely divergent assumptions or perspectives, his treatment of the morphology is only marginally useful. It was strongly influenced by Osgood's belief of a peramelid–diprotodontian tie for the caenolestids. The osteology of the living caenolestids is in dire need of modern evolutionary analysis.

**Infraorder Simpsonitheria, new**

*Etymology*

Named in honor of G. G. Simpson, one of the most insightful and prolific students of evolutionary theory, biological systematics, and mammalian taxonomy in the twentieth century. Simpson's work has synthesized an enormous amount of information on mammal evolution in North and South America, and his perceptive monograph on the Argyrolagidae is a reflection of his exceptional and enduring powers as a phylogeneticist of mammalian macrotaxonomy.

### Diagnosis

Protosimpsonitherian was a small, ricochetal marsupial. It had: two functional, greatly elongated metatarsals (3 and 4); highly modified mortise–tenon **UAJ** and fused crus; apomorphic deemphasis of the **CaA** facet (known only in argyrolagids) and the hypertrophy of the **CaFi** facets, together with a "stepped" **CCJ**; short forelimb; skull with enormous palatal vacuities (probably for cooling); long and pointed rostrum extending well past the enlarged upper teeth; completely ossified alisphenoid bulla and mastoid region of the skull inflated; zygoma deep and downward arching; orbital region expanded and posteriorly isolated from the temporalis musculature; lower jaw deep with small coronoid process; two gliriform, enlarged anterior teeth on each side, both above and below, probably I1 and I2; cheek teeth reduced in number.

### Included taxa

Gashterniidae, Groeberiidae, Argyrolagidae, and Patagoniidae.

### Distribution

Late Paleocene (Riochican) to Pleistocene.

### Discussion

It has not been widely appreciated that the best known of these highly evolved, extremely unusual mammals, the Argyrolagidae, in spite of Simpson's (1970a,b) convincing efforts, are really metatherians. McKenna (1980b, p.58–9), for example, has gone to considerable lengths to express doubts about what, to him, were unconvincing synapomorphies of argyrolagids with marsupials. Nevertheless he considered these animals to be worthy of a level of taxonomic recognition higher than they were usually accorded. Reig (1981, p. 60) also doubted marsupial ties for argyrolagids and furthermore suggested eutherian anagalid affinities. Recently these doubts were expressed again by Kirsch, et al. (1991, p. 10468) who stated their views as: "groeberiids and argylolagids, if they are indeed marsupials and not peculiar endemic eutherians. . . ." I reject all these views along with the suggestion of Clemens et al. (1989, p. 537), who stated that the Argyrolagidae and Groeberiidae might be "neither eutherian or metatherian, but . . . another therian lineage of Gondwanan origin." The pedal and crural evidence described here, along with the robust analysis of

Simpson, leaves no doubt in my mind concerning the ameridelphian metatherian status of argyrolagids. Marshall et al. (1990)
also unequivocally support the metatherian ties of argyrolagids
and groeberiids. Though the marsupial status of argyrolagids is
not an issue anymore, the level at which one wishes to classify
such modified descendents of early glirimetatherians, as the families included here, is difficult to decide. Does a lagomorph-like
radiation of marsupials (of which we only know the barest sample) warrant ordinal recognition? Simpsonitherians are certainly
far more modified cranially, dentally, and tarsally, from an archimetatherian morphology than are the macropodids. Few other
groups of marsupials have undergone such extreme – and not
only dental, but cranial as well as postcranial – change as did these
animals. Because they have been poorly known until Simpson's
work, the magnitude of their evolutionary divergence has not
been widely appreciated. Pascual and associates (e.g., Pascual &
Carlini, 1987) and Flynn and Wyss (1990) are just beginning an
extensive uncovering of some aspects of this remarkable marsupial radiation.

The enlarged gliriform tooth in the mandible of the Divisaderan
(early Oligocene) *Groeberia* (Patterson, 1952; see also Koenigswald,
Martin, & Pfretzschner, 1993, p. 307) suggests affinities with argyrolagids, as noted by Marshall (1987; see also Simpson, 1970c,
1971). This relationship is far from having been established securely, although it is highly probable (see also Flynn & Wyss,
1990). The groeberiid cheek teeth, while already recognizably
hypsodont and prismatic as seen on the worn teeth of the type
specimen, are more primitive in general than those of the advanced argyrolagids. The most complete account of groeberiids,
that of Simpson (1970c), does not discuss the hypsodont cheek
teeth. The hypselodont, but perhaps more advanced, cheek dentition of argyrolagines (but not the more recently discovered pro-
argyrolagines), is discussed in detail by Simpson (1970a).

At present there can be little doubt at least, that the recently
described early Miocene (Colhuehuapian) family Patagoniidae
(Pascual & Carlini, 1987) also belongs in this infraordinal group.
In describing the small but highly diagnostic hypodigm made up
of lower teeth and mandibular elements, these authors note the
differences in dental formula (1.1.0.3) from the groeberiids, and
describe in detail the open-rooted enlarged incisor, the hypselodont, rectangular-shaped "rootless" (presumably open-rooted)
molars which are surrounded by enamel. As discussed by Pascual
and Carlini (1987), the family shares with the argyrolagids a deep
pterygoid fossa which is "limited ventrally by a flange." The implied lack of affinity of this family with groeberiids and argyrolagids by these authors, I believe, may not be correct.

The origin of the Simpsonitheria from pichipilin caenolestines is
not unlikely. If the simpsonitherians originated from the para-

*Figure 8.10.   Comparison of articulated left calcaneus and astragalus of IMG II (based on phena of ICS 3 and IAS 4; A) with articulated right calcaneus and astragalus of Argyrolagus (B). Heavy broken lines represent outlines of LAJ facets and astragalar AN facets. Integrated details of articular contacts are too dissimilar in these two animals to suggest close relationship.*

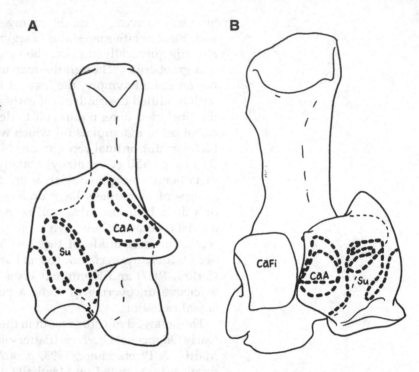

phyletic Caroloameghiniidae, independently from the caenolestids, then clearly the subordinal concept Glirimetatheria would need reexamination. I concur with Marshall's (1987) placement of the Gashterniidae in a monophyletic group with groeberiids and argyrolagids. The extremely poorly known Gashterniidae may represent the stem of this holophyletic assemblage.

An articular comparison of the proximal tarsus of the tentatively inferred terrestrial hopper and scamperer (?) of *IMG II* from the Paleocene and that of *Argyrolagus* (Fig. 8.10) indicates that the two extremely derived tarsal patterns are unrelated phylogenetically. While in *IMG II* the **CaFi** articulation is eliminated, it is extremely emphasized in the Miocene–Pliocene argyrolagids. Similarly, the **LAJ** modifications are equally divergent.

# Suborder Didelphimorphia (Gill, 1872)

## Diagnosis

Protodidelphimorphian (the protodidelphid) was dentally a relatively primitive ameridelphian, and it had: full complement of metatherian dentition, with molar morphology perhaps more similar to the ones found in caluromyines, i.e., a U-shaped, rather than the W-shaped, ectoloph seen in didelphines; only slight reduction of the paracone compared to the metacone; ear region probably with a small descending wing of the alisphenoid anchoring the cartilaginous ventral enclosure of the middle ear, as in protometatherian; carpal pattern that displays a relatively very large unciform, lack of an adult centrale, and a scaphoid (? scaphocentrale) with a distolateral process (which may be a captured centrale); highly modified sellar unciform–Mcp5 joint (Un-Mcp5J) resulting in a clasping hand; tarsal morphology with uniquely derived combination of an UAJ that lost the CaFi articulation and extended the AFi contact laterally; LAJ articulation with an astragalar Su facet proximally rounded rather than ribbonlike and tapered; reduced ampt; CCJ in which the proximally extended contact breaks up the original modified ovoid articulation into an articulation that has both a CaCud facet (the original therian and metatherian CaCu facet) and a didelphid CaCup facet (the latter condition prominently manifesting itself in a "pivotlike" projection of the cuboid underneath the calcaneus).

### Distribution

Middle Paleocene (Itaboraian) to Recent of South America, Plio-Pleistocene to Recent in North America.

### Included taxa

Didelphidae and Sparassocynidae.

### Discussion

The living opossums represent the most important, and in many ways still quite understudied, group of American marsupials. They have served in zoology and paleontology as the quintessential primitive marsupials, and both their didactic role and their ability to fascinate generations of students of mammalian evolution justifiably continue. But in phylogenetic analysis targeted at understanding evolutionary history, preoccupation with taxonomic statements about which animals are "primitive" and which are "advanced" can often do more harm than good to any rigorous attempt at deciphering this history. Axiomatically, only ancestors are completely primitive in relation to their descendants. Yes, opossums undoubtedly have many primitive metatherian characteristics, yet this cannot be made a rule for all of their character complexes, a point also stressed by Clemens (1968b, p. 13).

The didelphid fossil record is poor, and the undoubted dental remains go back to the Friasian middle Miocene. Although evidence from the tarsal record proves that Didelphidae are present at Itaboraí, their exact association with the name-bearing dental taxa is difficult (see above). Simpson (1970, p. 3), employing a concept of a dental (*sensu lato*) Didelphidae noted that "sampling of small fossil mammals is inadaquate in South America but also that much, perhaps most, of didelphid evolution was occurring outside the regions of known fossil fields." Interestingly, Simpson (1970) has astutely noted that several living didelphids are "structurally microbiotheriine," being undoubtedly aware of the suggestive similarity of the dentition of the Caluromyinae to *Dromiciops*.

Ecological morphology as it appears in the title of a section in Chapter 7 of this book may appear irrelevant to many taxonomists pursuing a narrowly defined taxonomic research program, but it is absolutely essential to appreciate the persistence of the character combinations in the American opossums in light of their ecological, particularly locomotor, diversity (for a very different approach to morphology and subsequent taxonomic perspective see Hershkovitz, 1992). In fact the understanding of adaptive diversity within the confines of certain phylogenetic constraints in a living group is far more important for the evaluation of character states in a phylogenetic approach to character evolution than is acknowledged by a method that is solely and "rigorously" oriented toward distribution analysis. A notable degree of morphological conservatism manifested in the dental and cranial morphology of the living opossums is obviously tied to ecology. This volume is not the proper forum for a full discussion of didelphid evolutionary dynamics, yet a few observations may be relevant. Much of evolutionary change is very likely the direct result of competitive interaction of sister species during their neo-sympatric phase (Bock, 1977a). Yet the highly omnivorous feeding habits of opossums have probably blunted the selective forces of competition usually due to sympatry. The wide range of their dietary regime may be the very evolutionary buffer that explains the relative stasis of their feeding mechanism. A widely omnivorous diet (although opossums are mistakenly considered "insectivorous" by many) also requires these arborealists to be competent arboreal "generalists" in order to pursue such a catholic diet, and given a specific set of phyletic constraints, this relative constancy of postcranial conservatism, at a different phyletic level than the feeding mechanism, also makes sense.

I have repeatedly noted my skepticism in this book about recognizing Didelphidae based on dental characters, in spite of the effort of outstanding taxonomists attempting to wrestle with this problem. Crochet's studies (1977a, b, 1978, 1979a, b, 1980), which have dealt in detail with the nature of "didelphoid" dental morphology of the fossil record, are examplary in their rigor and thoroughness in this regard. The incisor morphology of the living didelphids is difficult to compare with the Paleogene didelphidans because the paucity of such fossils. Detailed studies of the incisor morphoclines between taxa may prove to be highly rewarding, particularly if ecomorphological perspectives will focus such research. According to Takahashi (1974) it is the **I/1** that was lost from the lower series. I have in the past (Szalay, 1982) pointed out diagnostic differences of the incisors of *Dromiciops* from those of the common didelphid pattern. The significance of the incisor pattern of didelphids, however, will be difficult to decide until fossil didelphidans become far better known.

If the size increase of the molars from front to back exhibited by the Didelphinae and Borhyaenidae is a primitive itaboraiform condition, and it is certainly not peradectine, then the caluromyine condition is derived. Perhaps in light of evidence from the Udan Sayr *Asiatherium* (Trofimov & Szalay, 1993), the large ultimate molars are a primitive condition in the Metatheria, or possibly only in Ameridelphia. But it is possible that the shared didelphine and sparassodont conditions are shearing-related *independent* apomorphies, and the reduced ultimate lower molar of caluromyines represents the ancestral didelphid, and possibly didelphidan, pattern.

The diagnostic attributes of the tarsus of the first Didelphidae were presented in detail early in this volume (see also Figs. 8.11– 8.15), mainly because they are critical to the structure of many of the arguments about marsupial phylogenetics. The understanding of the taxon-specific attributes of the Didelphidae, like those of the Caenolestidae and Microbiotheriidae, has proved to be critical for the evaluation of all metatherian tarsal and selected carpal evidence. The highly diagnostic features are modified **UAJ**, characterized by the expansion of the **AFi** facet compared to the pediomyids and particularly the peradectines; the elimination of the **CaFi** facet; and the medial spread of the **ATim** facet (both of these possibly retained from a non-didelphid ancestry), along with the modification of the **CCJ**. Together these form an unequivocal set of features, amenable to testing against the fossil record that diagnose the protodidelphid. It is difficult to determine whether the relatively curved **UAJ** of the didelphids is a primitive archimetatherian or itaboraiform retention, or, more likely, a didelphimorphian change associated with improved arboreality, in contrast to the angular contact of the medial and lateral **ATi** articulations. I believe that it is an additional derived attribute of the didelphid tarsal complex.

The possibility that tarsally unquestioned Didelphidae, such as the caluromyines, retain a molar morphology probably more primitive than that seen in the didelphines should caution against the dental allocation of many Cretaceous and Paleogene teeth to the Didelphidae. The fact that only two tarsal morphs from the well sampled *Itaboraí Metatherian Groups*, those of *V* and *XII*, show undoubted didelphid attributes, in contrast to the various dental phena that have been routinely allocated to this family is an obvious incongruity which requires eventual explanation – a phy-

**Figure 8.11.** *Comparison of left calcanea in dorsal view of selected taxa of Didelphidae (Lutreolina, Caluromys derbianus, Philander opossum, Chironectes, Marmosa sp., and Metachirus). For abbreviations see Table 1.1. Scales = 1 mm.*

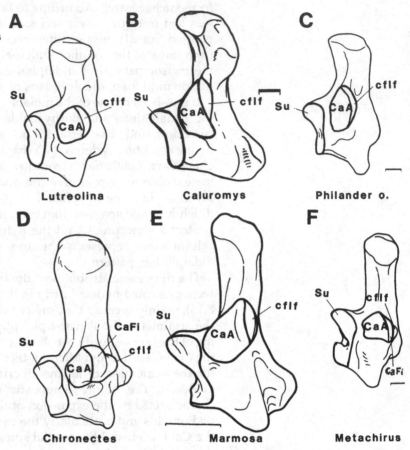

A **Lutreolina**
Su    cflf    CaA

B **Caluromys**
Su    CaA    cflf

C **Philander o.**
Su    CaA    cflf

D **Chironectes**
Su    CaA    CaFi    cflf

E **Marmosa**
Su    CaA    cflf

F **Metachirus**
Su    cflf    CaA    CaFi

logenetic explanation. Based on the independent criterion of calcaneal tuber proportions (see Fig. 6.16), the (arboreal) didelphids have deep tubers and therefore probably strong plantar flexing ability. It appears possible that these forms were decidedly more able arboreal pedal graspers than many of the other species we know from the tarsus at Itaboraí. These latter groups are the borhyaenids, the wastebasket (but more probably paraphyletic) caroloameghiniids, polydolopids (*IMG VII*), the possible caenolestid *Sternbergia* (to which, or to *Derorhynchus*, *IMG I* may belong), and the extraordinarily derived, probably hopping, tiny animals represented by *IMG II*.

### Caluromyinae (Figs. 7.2–7.5, 7.8, 7.9, & 8.11–8.13)

The genera *Caluromys, Caluromysiops,* and *Glironia* have been considered a holophyletic group by Kirsch (1977). The reasons for this association are based on the phenetic grounds of serology. In addition to serology, the relatively short facial skull, the morphology of the molars, specifically the near subequality of the paracone and

*Figure 8.12.* *Comparison of left calcanea, in plantar view, of selected taxa of Didelphidae (same as in Fig. 8.11). Note uniform presence of proximal calcaneocuboid facet, marked by X, irrespective of habitus of the taxa. Scales = 1 mm.*

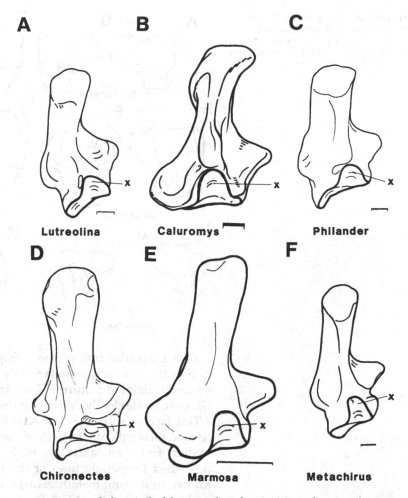

A     B     C

Lutreolina     Caluromys     Philander

D     E     F

Chironectes     Marmosa     Metachirus

metacone, and the probably correlated primitive absence (or possibly reduction) of carnassiality of the molars also cluster these taxa together. It is more difficult to determine, however, whether these attributes are primitive retentions from the ancestral didelphid or if these are diagnostic characters of a monophyletic (probably paraphyletic) subfamily Caluromyinae (see earlier discussion). This group is likely to be paraphyletic because the ancestor of the Gondwanadelphia may have been a caluromyine.

The relatively large peroneal process, while present in *Caluromys* and *Glironia* (I have not seen *Caluromysiops*), is almost certainly an archaic retention. More importantly, in spite of its shearing dentition, *Marmosa* also retains a relatively large peroneal process. This suggests primitive retention from the ancestor of the family. The living caluromyines are almost certainly fully arboreal. Details of this arboreal adaptation have not been studied, so only a few specific differences of these opossums from others can be evaluated.

**Figure 8.13.** *Comparison of left astragali, in dorsal and plantar views, of selected taxa of Didelphidae shown on Figures 8.11 and 8.12. For abbreviations see Table 1.1. Scales = 1 mm.*

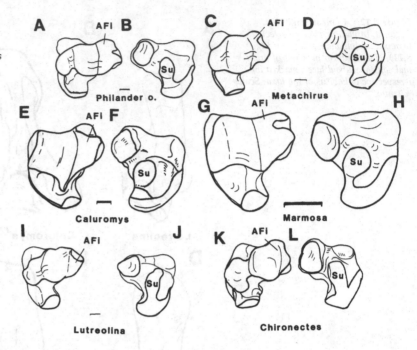

One particular trait of the astragalus appears to support the notion that the tarsal similarities uniting this group may be advanced didelphid features. The astragalus of both *Caluromys* and *Glironia*, while clearly displaying the relatively small and retracted **ATim** facet, also possess an **ATil** facet with a lateral tongue extended forward onto the neck of the astragalus. If didelphids originated from an ancestry with the astragalar morphology described for peradectines or the phenetically appropriate ones known in Itaboraí metatherians, then both the extreme retraction and reduction of the **ATim** facet and the extension of the lateral border of the **ATil** facet may be innovations and thus may signal the phyletic unity of the caluromyines. It may very well be, however, that the caluromyine pattern is the primitive didelphid pattern from which the other subgroups of the Didelphidae independently acquired this subtle difference.

***Didelphinae.*** *(Figs. 7.1, 7.6, 7.7, 7.10–7.12, 7.15, 8.2, 8.4, & 8.11–8.13)*

Perhaps the only genuinely qualitative and phylogenetically significant character complex of this assemblage will turn out to be the carnassified molar dentition (i.e., having developed advanced prevallid/postvallum shear). Though this is decidedly not a unique event among metatherians, it may nevertheless represent an apomorphy either of the ancestral didelphine (if this group is monophyletic within the Didelphidae) or of the ancestral

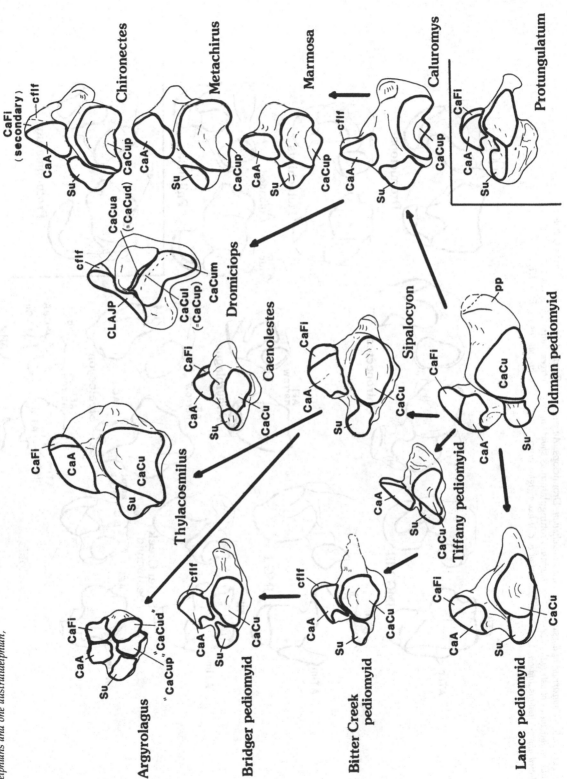

Figure 8.14. *Comparison of distal views of calcanea of selected amer-idelphians and one australidelphian,* Dromiciops. *An eutherian, Pro-tungulatum, is shown in partial box* *in lower right corner. For abbreviations see Table 1.1.*

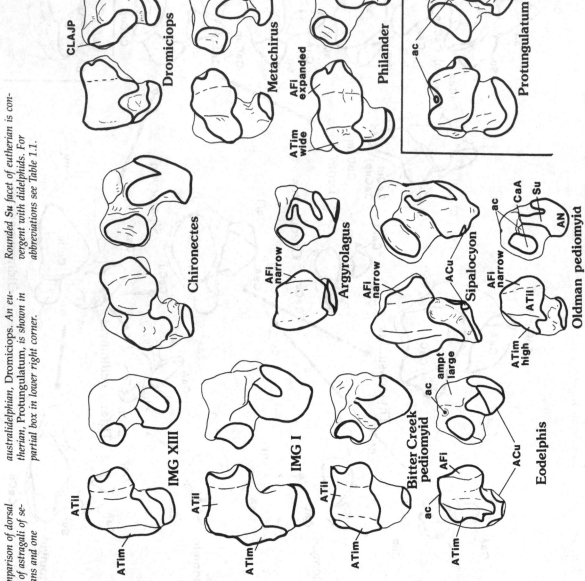

*Figure 8.15.* Comparison of dorsal and plantar views of astragali of selected ameridelphians and one australidelphian, Dromiciops. An eutherian, Protungulatum, is shown in partial box in lower right corner. Rounded **Su** facet of eutherian is convergent with didelphids. For abbreviations see Table 1.1.

didelphid. As I have noted, the probable hypertrophy of the prevallid shear, exhibited by an enlarged paraconid and high protoconid in a number of didelphids (see particularly *Lutreolina*) renders the evaluation of dentitions of deltatheroidans (and other groups) for taxonomy highly suspect without considering the ecological imperatives involved in dental evolution.

The reduced calcaneal peroneal process in this subfamily (but less so in *Marmosa, sensu stricto*) is not a particularly confidence-inspiring test of this concept. The reduction of the peroneal process is a common, and in its outcome a featureless, event; this clearly occurred many times independently in the Theria and at least two or three times in the Metatheria.

### Sparassocynidae

The attributes of the ancestor of this monogeneric family are necessarily those of the type genus. It occurs from the Late Miocene into the Pleistocene in Argentina. There are no unequivocal indications that this fossil, a dentally and cranially modified carnivorous marsupial genus, is more recently related to either some didelphid (such as *Lutreolina,* as suggested by Reig & Simpson, 1972), or some relatively primitive, hitherto undiscovered didelphimorphian group, or possibly (but not likely) even to some itaboraiforms. This animal is somewhat robust and badger-like in cranial conformation. The complete absence of stylar cusp **C,** the subsequent lack of the crista that connects the metacone to this cusp as seen in *Lutreolina,* the modified carnassiform trigonids with the metaconids incorporated into the base of the protoconids, and the drastically reduced talonids all would appear to prohibit an objective tie to the Didelphidae. While derivation from a didelphid is one of the possible explanations for the history of this family, the cranial morphology is also highly modified (see the thorough descriptive, functional, and comparative account in Reig & Simpson, 1972). The rostrum is short, and the neurocranium is greatly expanded posteriorly and laterally, involving primarily the squamosals. The zygomata are wide, the hypo- and epitympanic sinuses are relatively enormous, and the middle ear is fully enclosed by an alisphenoid bulla. Coupled with the enlarged ear region is the peculiar solid palate without vacuities, a condition in which, according to Reig and Simpson (1972) and confirmed by me, *Sparassocynus* resembles *Caluromys* and borhyaenids.

**Ameridelphia,** *incertae sedis*

### Necrolestidae

Opinions have varied as to the affinities of this enigmatic group represented by the genus *Necrolestes* from the Santacrucian Miocene of Argentina. Winge (1941, pp. 79–80) clearly believed that the genus was a marsupial:

*Necrolestes* ... must certainly have arisen from the primitive [borhyaenids] in which the number of upper incisors was reduced to four while the lower incisors were still four in number, and the cheekteeth had not yet acquired the carnivorous character, nor had the zygomatic arch become modified in the direction of the type found in carnivores. The genus is known from a fairly complete skull, although the base of the braincase is missing, as well as some of the bones in the forelimb. The dentition is adapted for food that requires little mastication and the head as well as the fore limbs were used as burrowing organs.

Simpson (1945) considered it as a possible "insectivore," and Patterson (1958), after reexamining the evidence, concluded that the genus had a 5/4 incisor formula and four molars, a slightly inflected angle, and a skull with the jugal extending all the way to the glenoid fossa in marsupial fashion. It is obvious today that the latter character is an ancient mammalian one indeed. Patterson suggested that it may be related to borhyaenoids. Both Archer (1984a) and Aplin and Archer (1987) questioned the marsupial affinities of this enigmatic burrower. Archer (1984a) has properly pointed out that the molar formula suggested by Patterson is entirely uncertain. I have examined the specimen and cannot offer any views against its marsupial status that may be backed by observations. Though large incisor number is not a diagnostic trait of Metatheria, the general nature of the Cretaceous and Paleogene fauna of South America, that is, the preponderance of many ameridelphian terrestrial lineages in the Paleogene in contrast to the known arboreal ameridelphian clades (the didelphids and microbiotherians), suggests to me, however, an ameridelphian source for this highly modified group. It is unlikely that this animal is a derivative of some hitherto unknown group of tribotherians, as there is no evidence for the presence of these archaic therians in South America. Its large number of incisors make it unlikely that it is eutherian. If it is not a eutherian, then in light of the known absence of Cretaceous therians in South America, it is probably a metatherian. Tarsal remains would conclusively resolve this puzzle.

## Cohort Australidelphia Szalay, 1982

### Diagnosis

Protoaustralidelphian had: complete metatherian dental formula; didelphid-like ear region without a fully ossified bulla, although probably with an alisphenoid wing and small inflated tympanic wing of the petrosal; complex tarsal modifications as in the living genus *Dromiciops* (detailed below); incisors mesiodistally expanded and with a slight talon on the lingual side of the lower incisors (the latter complex is a possible retention from an ancestry that would still be considered didelphid).

## Distribution

Probably Early Paleogene (but first unequivocal record in Miocene) to Recent in South America, probably Antarctica during the Early Paleogene, and probably the Early Paleogene to Recent in Australasia.

## Included taxa

Orders Gondwanadelphia and Syndactyla.

## Discussion

The tarsal attributes of the animals united in this holophyletic group leave no doubt in my mind that this character complex is uniquely derived and homologously shared. In fact this is the most complex, most rich in detail, and therefore most corroborated of all taxonomic properties that is extensively shared in all well-established taxa, the traces of which are present in even highly modified lineages. The importance of this character complex is potentially rivaled by the hitherto inadaquately studied carpal complexes and a few basicranial attributes. Reig et al. (1987) rejected this concept because, I believe, they misjudged the significance of the subtleties of the dental evidence and the account of the origins and transformations of the australidelphian tarsus. It is the analysis of the tarsus, dentition, and skull of the living *Dromiciops* (i.e., the morphological evidence) that has led to the hypothesis of Australidelphia as a holophyletic group of marsupials (Szalay, 1982). The work presented in this book further corroborates this hypothesis. Sharman (1982) has also reported cytological similarities between the chromosomes of *Dromiciops* and Australasian marsupials, and recent studies of Temple-Smith (1987) on sperm-structure evolution also support this hypothesis by showing a special similarity of the living microbiothere sperm morphology to that of possums.

The significance of details of the complex similarity of the tarsus of *Dromiciops* to the tarsus of the phalangeroids, and its structurally ancestral position to the derived attributes of the dasyuromorphians, syndactylans, and peramelinas is discussed later in this chapter. For some of the carpal evidence see Figs. 3.2 and 8.16. The origin of the distinctive australidelphian tarsal complex may be related to the considerable improvement of hyperinversion abilities of the protoaustralidelphian in contrast to its didelphid ancestor. This hypothesis should be tested against careful comparative accounts of small arboreal didelphids and *Dromiciops* as well as the small phalangers. The derivation of the microbiotherian tarsal complex from a didelphid one has become increasingly obvious during my rather long study. The sharing of the widened **AFi** facet and the transformation of the double-faceted **CCJ** of the Didelphidae into recognizably homologous, but altered, triple-faceted arrangements of the **CCJ** of the Australidelphia removes all of the doubts I expressed in 1982 concerning the origins of this cohort. It further makes microbiotherian ties to pediomyids, the hypothesis advocated by Reig et al. (1987) and Marshall et al. (1990), highly unlikely.

More recently Kirsch et al. (1991) endorsed the hypothesis that *Dromiciops* has more recent affinities with Australian marsupials than with didelphids, based on DNA–DNA hybridization studies. In their firm determination to view *Dromiciops* as the sister group of Diprotodontia, however, they have, as I pointed out in Chapter 3, failed, again, to understand the taxonomic property on the basis of which I made the original assignment of *Dromiciops* in 1982. Similarly, they set aside syndactyly and the clearly australidelphian aspects of the complex modifications of the pes and aspects of basicranial morphology in order to accommodate their DNA–DNA hybridization studies. This is surprising particularly in light of their open discussion of the complexities, nexus of assumptions, and possible artifacts involved in their "normalized percentage hybridization" (NPH) measurements. Another interesting DNA–DNA hybridization study, that of Westerman and Edwards (1991), supports a trichotomy of didelphids, *Dromiciops*, and Australian forms in the mid-Cretaceous.

## Order Gondwanadelphia, new

### Diagnosis

In addition to the attributes listed under the diagnosis of the Australidelphia, the protogondwanadelphian had: a carpal pattern in which the magnum and the well-developed lunate are separated from one another by the well-developed (distolateral) ulnar extension of the scaphoid (scaphocentrale?).

### Distribution

Probably the same as that of the Australidelphia.

### Included taxa

Suborders Microbiotheria and Dasyuromorphia.

### Discussion

The origin of this order is probably very close to the beginning of the Australidelphia itself. This event was probably in South America, from an animal which, prior to its acquisition of its apomorphies, any modern taxonomist would not hesitate to allocate to the Didelphidae, as diagnosed in this volume. The stem lineage had the full metatherian dental formula with molar morphology similar to the most primitive dasyuromorphians and a carpal (Fig. 8.16) and tarsal pattern identical to the one in the living *Dromiciops*. These shared carpal and tarsal patterns of the protogondwanadelphian were probably antecedent to Syndactyla.

Of the Australasian groups of marsupials only the Dasyuromorphia is usually not specifically allied with other australidelphian groups (except peramelids). Sarich (personal communication; 1993) suggests that immunological distance data link the microbiotheres more closely with the dasyurids than the latter with the peramelids. *Dromiciops* (Fig. 8.16) and dasyuromorphians share a carpal pattern. The Gondwanadelphia is a monophyletic group which included the antecedent to the first syndactylan.

## Suborder Microbiotheria Ameghino, 1889

### Diagnosis

Protomicrobiotherian had: composite bulla, formed by the tympanic wing of the alisphenoid and the inflated wings of the pars petrosa and pars mastoidea of the petrosal; ectotympanic enclosed within the bulla proper; incisors, unlike those of didelphids, not peglike but mesiodistally expanded, the lower ones with lingually slightly expanded bases; details of tarsal morphology in the **EMt1J, UAJ, LAJ,** and **CCJ** as in the australidelphian ancestry, discussed elsewhere in this volume.

### Distribution

Probably Early Paleogene of South America, but it is unequivocally known only from Miocene to Recent in the southern half of South America.

### Discussion

The living *Dromiciops* has in the smallest of details the inferred attributes of the tarsus that were probably present in the last common ancestor of the Dasyuromorphia and Syndactyla respectively. The characters noted are actually present in the phalangeroids and petauroids. The dasyuromorph tarsal pattern is readily derivable from this condition as well. This combination of gondwanadelphian (and microbiotherian) features are (1) the peroneal process reduced to a nubbin; (2) as in didelphids, the **ampt** is small; (3) the **EMt1J** is saddle-shaped, identical to that seen in the primitive condition of phalangeriforms, and it is derivable from the didelphid state (see Fig. 8.17); (4) the **UAJ** is slightly less angulated than that seen in didelphids, sharing with the latter the ancestral didelphid (but advanced ameridelphian) trait of an expanded **AFi** contact; (5) the **LAJ** articular facets are confluent, resulting in what has been dubbed the continuous lower ankle

*Figure 8.16.* Left: Dromiciops australis, *CNHM 127453.* Right: Caenolestes sp. *AMNH 64499. Both figures represent the distal end of left forearm, carpus, and metacarpus. Note the separation of the lunate and the magnum by the lateral process of the scaphoid (this process may represent the fused centrale, thus properly making this scaphoid the scaphocentrale) in Dromiciops, a condition more advanced than the didelphimorphian one in which the distolateral scaphoid process, while conspicuously present, is not wedged between the lunate and the magnum. In contrast note the primitive (?ameridelphian) contact of the lunate and magnum and the distolateral process of the scaphoid in the caenolestid (and in didelphids). Note the tightly interlocking upper wrist joint, along with other stability related locomotor adaptations in the terrestrial caenolestid compared to the primarily arboreal microbiotherian.*

The microbiotherian pattern is shared with dasyuromorphians, but differs from the putative protosyndactylan and protodiprotodontian ones. The gondwanadelphian pattern of microbiotherians and dasyuromorphians is more primitive than the protosyndactylan pattern shown by peramelids in which lunate reduction has not occurred but the distolateral process of the scaphoid is already lost. In the protodiprotodontian the lunate is greatly reduced, making space for a hypertrophied magnum and significantly, the scaphoid, as in its syndactylan ancestry, lacks the distolateral process. For abbreviations see Figure 3.2.

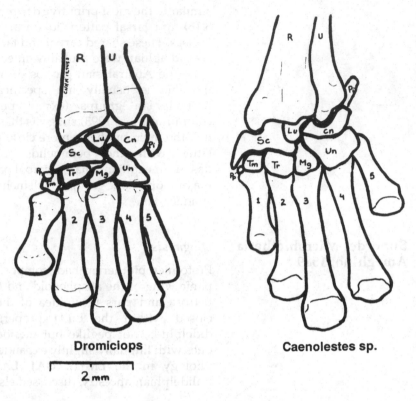

**Dromiciops**

**Caenolestes sp.**

2 mm

joint pattern (**CLAJP**), in contrast to the primitive separate lower ankle joint pattern (**SLAJP**) of nonaustralidelphian marsupials and other synapsids (Figs. 8.14 & 8.15); (6) the **CCJ** transformed from the double-faceted didelphid pattern of articulation to the highly modified triple-faceted one in which the didelphid **CaCud** facet equals the **CaCua** facet of australidelphians, the didelphid **CaCup** facet equals the **CaCul** facet of australidelphians, and the third facet of australidelphians, the **CaCum** facet, is a neomorph (Fig. 8.18); (7) in contrast to the didelphids and other ameridelphians the **UAJ** facets on the tibia and astragalus are smoothly continuous as in the putative ancestral syndactylan and phalangeriform.

The **CLAJP** is more particularly derivable, that is, more similar in derived attributes at each taxonomic level of comparison, from the condition found in didelphids, forms with an astragalar **Su** facet apomorphically rounded proximally, than from the more primitive condition found in Archimetatheria. In addition to the derived metatherian condition of the **UAJ** and **LAJ** of *Dromiciops,* the complex pattern of the **CCJ** is also more derivable from the advanced metatherian pattern of didelphids than from any archimetatherian condition. The didelphid **CCJ** can be divided into two angled facets, the distal and the proximal (Fig. 8.18). In the

*Figure 8.17.* *Comparison of the articular morphology of the right EMT1J in a representative didelphid (Didelphis: A, B) and two australidelphians from two different orders (Pseudocheirus, C, D; Dromiciops, E, F). Entocuneiforms are shown on the left and first metatarsals on the right. The morphological similarity and the great size discrepencies of articular morphology within either the sundry groups of ameridelphians or within the australidelphians preclude explanations of within-group similarities as being due to some nonhistorical, independent adaptive change, or ambiguous allometric trends. Dorsal (do) is toward the top of the page and medial (m) is to the middle of the figure. For abbreviations see Table 1.1. Scales = 1 mm.*

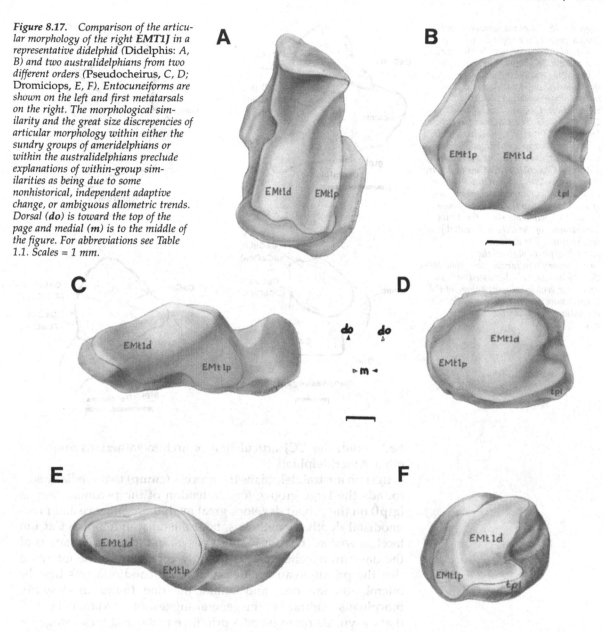

derivation of the australidelphian pattern a third articular area was added (the **CaCum** facet), while the slightly more distal **CaCul** facet (homologue of the most proximal one in didelphids, the **CaCup** facet) was retained approximately in the same proportion. The narrow australidelphian **CaCua** facet is the reduced, but persistent, original **CaCud** facet of didelphids, the last remnant of the unifaceted original synapsid, therian, and metatherian modi-

*Figure 8.18.* *Comparison of homologous aspects of the right cuboids in didelphid ameridelphians (Caluromys), and australidelphians (Dromiciops, Burramys, and Neophascogale). Note the tripartite subdivision of the australidelphian calcaneocuboid articulation,* **CaCua, CaCul,** *and* **CaCum** *facets. The* **CaCum** *facet is a "neomorph," if such a designation is permitted for a new subdivision of an ancestral articulation. The diagnostic didelphid* **CaCud** *and* **CaCup** *facets are homologous with the* **CaCua** *and* **CaCul** *facets of australidelphians, respectively. Unlike the ameridelphians, the australidelphians show the derived hypertrophy of the cuboidal medial process* (**cump**). *The derived australidelphian state of the dasyuromorphian tarsus, the cuboid obviously included, is discussed in detail under the ecological morphology of the Dasyuromorphia in Chapter 7. For abbreviations see Table 1.1. Scales = 1 mm.*

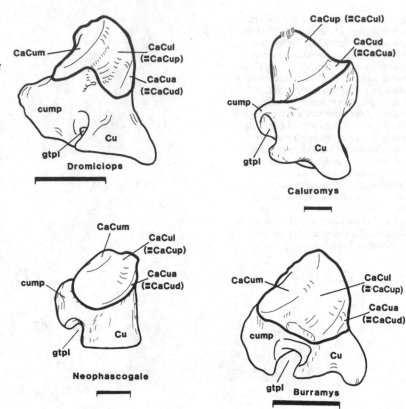

fied ovoid, the **CCJ** articulation of archimetatherians and most other Ameridelphia.

In stem australidelphians the process (**cump**) that medially surrounds the large groove for the tendon of the peroneus longus (**gtpl**) on the cuboid develops great medial prominence and proximodistal depth. Nonetheless, both the australidelphian **CaCum** facet as well as the robust **cump** persist in the modified tarsus of the dasyuromorphians. This strongly corroborates the inference that the primitive australidelphian tarsal condition was like the microbiotherian one, and unlike the one found in dasyuromorphians (contrary to the general suggestion of Marshall (1972) that dasyurids represented a primitive pedal and tarsal stage for Australian marsupials). The clearly derived australidelphian nature of the dasyuromorphian tarsal complex is discussed later in this chapter.

Marshall (1987) considered the dental similarities of the Itaboraí *Mirandatherium* so close to the (Casamayoran) Argentinian Eocene *Eomicrobiotherium* that he included both in the Microbiotheria. As noted under Itaboraí, there are no tarsals in that assemblage which would corroborate such ties. Marshall et al. (1990) include *Khasia* and *Monodelphopsis* in the Microbiotheriidae, and include their concept of Pediomyidae and the "Plesion" *Andinodelphys* in the Microbiotheria as well. The allocation of sundry Paleogene dental taxa to the Microbiotheria, with the exception of *Microbiotherium*, cannot be fully substantiated in light of the problems surrounding the concepts of what actually constituted the primitive didelphid molar structure.

## Suborder Dasyuromorphia (Gill, 1872)

### Diagnosis

Protodasyuromorphian had: petrous portion of periotic slightly inflated; incisor formula reduced to 4/3; relatively primitive metatherian dental morphology except for incisors that appear to have been channeled by the modifications described under the diagnosis of Australidelphia; talonids of molars reduced relative to the size of the trigonids when compared to the primitive didelphid and microbiotheriid dentitions; lack of magnum contact with lunate in the carpus, as in *Dromiciops*, retained from gondwanadelphian ancestry; transformed pedal structure comprised of an elongated and hallux-deemphasized terrestrially adapted foot; modified tarsals that reflect the selectional pressures exerted by a primarily nonarboreal substrate, leading to the abandonment of the obligate supination and inversion of the gondwanadelphian ancestral condition; modified **EMt1J** (probably consequence of hallucial reduction) from an australidelphian one; **CLAJP** retained, but the **LAJ** with additional new contact within it (**CaAd**); secondary structural and functional "simplification" of the **CCJ**, resulting in the near elimination of the angled appearance of the australidelphian condition (but with retained evidence for the ancestral gondwanadelphian triple-faceting of this joint).

### Included taxa

Dasyuridae, Myrmecobiidae, and Thylacinidae.

### Distribution

Paleogene to Recent of Australasia.

### Discussion

Virtually all previous efforts to understand the origins of dasyuromorphians have centered on dental attributes (Archer, 1976b, d), with attention also focused on the skull (Archer, 1976c). Excellent reviews of the literature can be found in the synthetic studies by Archer (1981; 1982a, b; 1984b; see also Wood-Jones, 1949). The issue of dental apomorphies of the first dasyurid, probably the stem dasyuromorphian, continues to be elusive with as yet unsurmountable difficulties for two reasons. The obvious problem is the lack of an adequate fossil record of teeth, a problem compounded by the relative uniformity of the molars within the Dasyuridae. Yet the recently described putatively Early Eocene therian mammal *Tingamarra* from Australia (Godthelp et al., 1992) is interesting for a number of reasons in addition to its antiquity for that continent. More particularly its relatively very large trigonid and somewhat small talonid, with a hypoconulid perhaps unusually large in eutherians, are somewhat suggestive of the stem specialization of the Dasyuromorphia. The second problem is the great phenetic proximity of molar structure of dasyurids to that found in some didelphids and other dentally relatively unmodified metatherians. Consequently, those espousing the concept of the "Marsupicarnivora" have expressed a belief that the relative phenetic uniformity of dental features of these groups also mirrors the history of the groups themselves. But clearly, if various "marsupicarnivorans" may be shown to be more recently related to other groups not originally included in the concept of the Marsupicarnivora, then that taxon is of little use (e.g., see concept of Marsupicarnivora applied in Hunt and Tedford, 1993, Figs. 5.4). Thus, if for example, the sparassodonts prove to be more recently related to the glirimetatherians or the polydolopimorphians, then the concept of Marsupicarnivora would express virtually nothing of either phylogeny or adaptively significant boundaries. Dasyuromorphians are more recently related to both the microbiotherians and syndactylans than to any other group, and these together as the Australidelphia are holophyletic.

At first glance it appears that the carpal pattern of dasyuromorphians, in which there is no contact of the magnum with the lunate (Fig. 3.2), is a primitive therian holdover. The ulnar extension of the scaphoid is evidence against that (see also *Dromiciops*,Fig. 8.16). The evidence for the derived terrestriality of the stem of this marsupial group apppears overwhelming. In my view therefore the apparent-similarity of this partial pattern to the putative terrestrial eutherian ancestry (with its magnum and lunate of approximately equal size and with a centrale intervening between them) is not homologous. In fact the distolateral, ulnarly oriented process of the scaphoid is reminiscent of the didelphimorphian and putative gondwanadelphian condition. The gondwanadelphian pattern is distinct from the paramelid pattern which, like stem syndactylans, lacks the ulnarly oriented distolateral process of the scaphoid.

The single most important factor in understanding the transformational significance of the various states on the feet of the families included in this group is hallucial reduction (see Figs. 4.1 & 7.32). This state is unquestionably derived; as the clawless hallux is either very small or lost, compared to the didelphid, microbiotheriid, or phalangeroid grasping adaptations. It is equally important that certain tarsal specializations accompany the dasyurid

*Figure 8.19.* *Comparison of left calcanea in dorsal view of selected taxa of dasyurids (Phascolosorex, Dasycercus, Phascogale, Antechinomys, Smithopsis, and Murexia). For abbreviations see Table 1.1. Scales = 1 mm.*

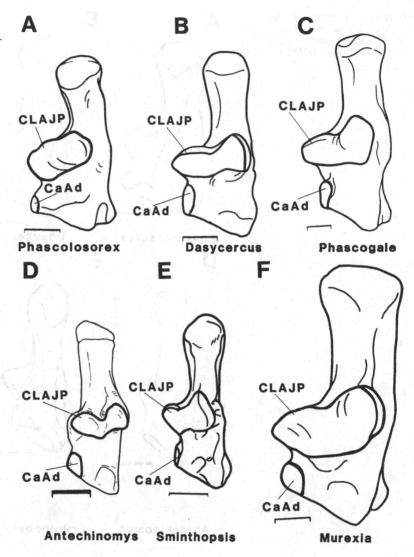

hallucial reduction and the (proposed) ancestral dasyurid terrestriality. There is little question that the members of the stem lineages of the Myrmecobiidae and Thylacinidae, were they known, probably could not be differentiated from the Dasyuridae proper, as constituted now (see Figs. 8.19–8.21). So the Dasyuridae will remain paraphyletic, some dictates notwithstanding, whether one likes it or not. Consequently it is *causally* logical, in a nonmonotonic sense, that the persistence in the family of particular attributes on the bones related to grasping and habitual inversion mirror an ancestral condition, allowing derivation from (or possibly, but unlikely, to) these conditions. Again, as in other macrotaxonomic evalutations, ecological morphology and

*Figure 8.20.  Comparison of left calcanea in plantar view of the selected taxa of dasyurids shown in Figure 8.19. For abbreviations see Table 1.1. Scales = 1 mm.*

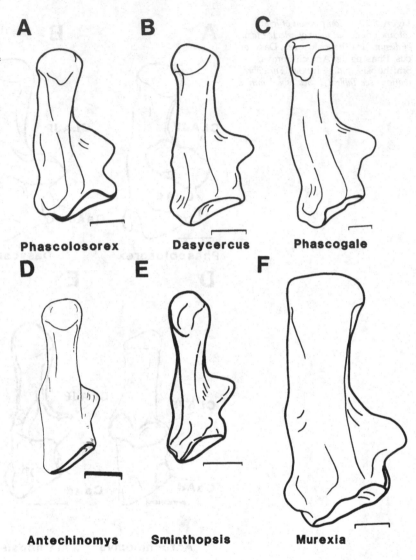

**A** Phascolosorex
**B** Dasycercus
**C** Phascogale
**D** Antechinomys
**E** Sminthopsis
**F** Murexia

functional analysis in an evolutionary framework is necessary, albeit such efforts are far less simple than clustering techniques.

Given the functional complex of the whole foot, the vertical comparisons compellingly point to the transformation of the morphocline discussed earlier under Microbiotheria and in Chapter 7 under Dasyuromorphia, the reverse of the transformation postulated by Marshall (1972). Marshall (1972, p. 55) has proposed a clearly stated explanation for foot evolution in Australian marsupials, essentially the same as Winge's (1941) account. That hypothesis, I believe, was dominated entirely by a taxic outgroup approach based on a dental perspective rather than an independent analysis of the superficially gleaned character complex itself. Marshall stated that

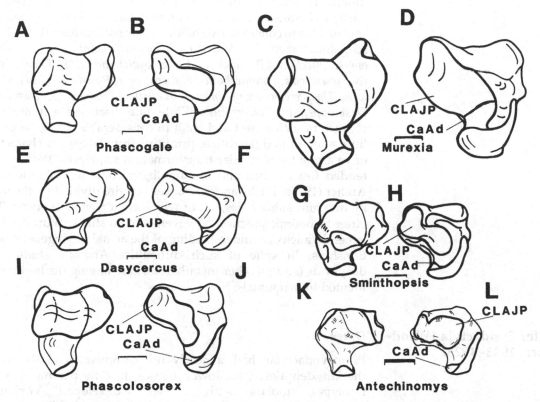

***Figure 8.21.*** *Comparison of left as-tragali in dorsal and plantar views of* *the selected taxa of dasyurids shown in Figures 8.19 and 8.20. For abbrevia-* *tions see Table 1.1. Scales = 1 mm.*

The acquisition of syndactyly in dasyurids would result in a foot structure identical to a phalangerid foot, save for relative differences in digit size. It is thus proposed that the foot structure as seen in dasyurids may have been the condition seen in the ancestors of all Australian marsupials.

An analysis of the details of tarsal morphology, one that apparently was not carried out by Marshall, does not support that hypothesis; in fact it strongly supports a different explanation. In the account of the crus and tarsus of dasyurids in Chapter 7 under the pedal evidence, the comparative aspects are examined. In addition to the discussion of transformation series, evidence is presented there, that the myrmecobiids and thylacinids, as long recognized, are modified dasyurids.

The internal phylogeny of the Dasyuridae has been the focus of some of the most intense molecular and genetic studies (Baverstock, 1984; Baverstock et al., 1982; Kirsch et al., 1990b) as well as dental, and to some degree cranial, analyses. It is an extremely difficult and challenging group to understand in light of its persistently primitive dental and cranial traits. A serious lack of functional–adaptive studies of the skeletal system, however, is partly to blame for this state of affairs. Tate (1947) considered the Dasyuridae to consist of two holophyletic subfamilies: (1) the Phascogalinae (including *Murexia*, *Thylacinus*, *Sminthopsis*, *Antechinomys*, *Antechinus*, *Planigale*, and *Phascogale*) and (2) the Dasyurinae (*Neophascogale*, *Parantechinus*, *Phascolosorex*, *Pseudantechinus*, *Myoictis*, *Dasycercus*, *Dasyuroides*, *Satanellus*, *Dasyurinus*, *Dasyurus*, *Dasyurops*, and *Sarcophilus*). While Tate discussed a number of characters such as feet and teeth in considerable detail, he made little attempt to differentiate primitive from advanced characters or to argue for particular transformation sequences. With an extended fossil record and immunological data available to him, Archer (1977b, 1982b, and references therein) divided up the family into nine subfamilies, out of which eight were monotypic. The latter taxonomic practice, however, is often still an indicator of a general paucity of understanding of the actual phylogenetic relationships. In spite of such difficulties Archer's studies on dasyurids (and on other metatherians) are among the best documented in marsupials.

## Order Syndactyla (Wood-Jones, 1923–1925)

### Diagnosis

Protosyndactylan had: a nearly full complement of the metatherian dentition (last lower incisor lost in peramelid morphotype), molars with a well developed W-shaped (dilambdodont) ectoloph; petromastoid that probably expanded onto occiput; a basicranium not pneumatized but probably with an internal overlapping by the squamosal process of the roof of the middle ear (Fig. 8.22); carpus without an ulnar extension of the scaphoid; tarsus virtually identical to that in the Microbiotheria, the pes however displaying syndactyly of the second and third cheiridia.

### Included taxa

Semiorders Peramelina and Diprotodontia.

### Distribution

Paleogene to Recent of Australasia.

### Discussion

The exact relationships within the order are poorly known largely because the geological record continues to hide the ancestors and early representatives of the known syndactylans, the stem lineages for the numerous diprotodontian, the notoryctid, and per-

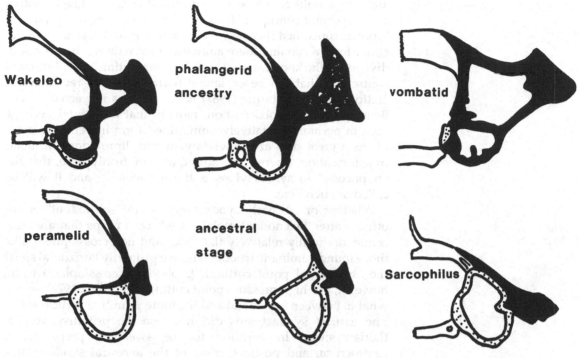

**Figure 8.22.** *Schematized cross-sections of posterior halves of one side of skulls, of selected australidelphians to illustrate the pattern of distribution of squamosal (solid black) and alisphenoid (stippled) components. Peramelids already show the beginnings of the intracranial extension of the squamosal, so characteristic of the diprotodontians. (Redrawn from Murray et al., 1987.)*

amelid radiations. Nonetheless there appears to be not a shred of evidence that the dasyuromorphians are members of this group or that they lost syndactyly (*contra* Marshall et al., 1990).

The various taxonomic concepts tacitly tied to theories of dental evolution, while extremely important for understanding the evolutionary transformation of the teeth themselves, have sometimes had a confusing influence on attempts to understand evolution of *taxonomic groups*. The greater availability of a dental record tended to slow the construction of nondental taxonomic properties of whole organisms. Many "hard morphologists" and paleontologists, certainly including myself, still need to be reminded that the teeth of the taxa we study are not the animals themselves. While they are entities that change (or do not) through time independent of other aspects of organisms, they are only one of the many mosaics that constitute the whole organism. Teeth can rarely be, by themselves, a reliable mirror of history of various lineages. The postulated radiation of the ancestral syndactylan species and their diverse descendants comprised far more than the changes in dental attributes with which they are often equated. Thus, the notion that "parameloids" (i.e., teeth like those found in some peramelids) would be ancestral to diprotodontians, while an astute and important concept as far as the understanding of dental evolution is concerned (Bensley, 1903; Archer, 1976b), has no meaning either for envisioning these animals, or for refining the taxic subdivision of the known samples, or for formulating an evolutionary continuum that can be explained narratively. In a recent study, Tedford and Woodburne (1987) have offered a newer version of Bensley's original observation, namely that the dental heritage seen in perhaps a relatively unmodified form in peramelids explains a great deal about syndactylan and diprotodontian molar transformation. There is no doubt, as they firmly state, that the "hypocone" in syndactylans is the metaconule, and it will be called as such here.

Whether or not early syndactylans were "selenodont" is another matter. Selenodont molars (teeth with cusps that are crescentic, occlusally relatively flattened, and non-crosslophed), like those found ruminant artiodactyls, use primarily horizontal shear (i.e., horizontal point cutting). Lophodont (crosslophed) teeth, however, use high pressure point cutting (Seligsohn, 1977), somewhat in between horizontal and the more primitive vertical shear. The earliest syndactylans did not lose the primitive vertical therian shear. In addition to the sweeping preparacrista, centrocrista, and postmetacrista of the ancestral syndactylans, these forms possessed a tooth structure with clear lophids on the lower molars and lophs running lingually from the apices of the paracone and metacone. In spite of the sweeping buccal crests on the upper teeth in primitive peramelids, the molars are prismatic and have crosslophs, and the lower molar talonids and trigonids form crosslophids. Similarly the wynyardiids, ilariids, and early phascolarctids have prismatic, lophodont teeth on which vertical shear dominates. In selenodont artiodactyls, although the teeth are high crowned, the occlusal surfaces have minimal relief, due to the slightly differential height of the enamel and exposed dentine. Particularly in the well-documented ilariids (Tedford & Woodburne, 1987) the high relief of the lophs and lophids leaves little doubt that the correct descriptive term, as for macropodid teeth, is lophodont, in spite of the ancestral undulations of the buccal crests. Only the pseudocheirin molars approach a genuinely selenodont occlusal pattern, although they also resemble some equid molar configurations.

The most appropriate dental null group, the one most similar to the inferred most primitive syndactylans, is the relatively primitive didelphimorphian dentition, which is most similar to the peramelid condition. For the foot structure in syndactylans, however, the null group is what we see in the living Microbiotheria, in spite of the fact that Dasyuromorphia is often considered to be the sister group of the Syndactyla.

# Semiorder Peramelina (Gray, 1825)

## Diagnosis

Peramelina is presently a concept probably diagnosable in the manner detailed under the diagnosis of the Syndactyla. It is not clear to what extent, if at all, an incipiently or partly bilaminar construction of the middle ear capsule may have been present in the protoperamelinan, and to what degree its foot structure was different from a primitive diprotodontian condition. The combination of a deep navicular and reduced mesocuneiform shared by notoryctids and peramelids might also diagnose the Peramelina.

## Included taxa

Notoryctidea and Peramelidea.

## Distribution

Probably Paleogene to Recent (first record from the Miocene).

## Discussion

An entire, complex radiation of diverse syndactylans is probably as yet unknown. Bensley, with so much vision, postulated the concept and name of "Properameloidea" as an ancestral group of the syndactylans. The last common ancestor of this hypothetical paraphyletic group, the putative stem of all the syndactylans, was almost certainly arboreal. Two lineages of this group were probably antecedent to the Peramelina and to the Diprotodontia, respectively.

### Notoryctidae (Figs. 7.41–7.42)

The small, solitary, and superbly fossorial living genus *Notoryctes* is best considered a single species, a synspecies, populations of which are sometimes recognized as two taxonomic species. These two are significantly allopatric, and probably merely the recently disjunct distribution of a single biological species. This animal represents a long standing historical problem in marsupial phylogenetics. The tarsal evidence removes some of the confusion not only in relation to other living groups, but most definitely concerning the fossil record of mammals and other marsupials. These animals are certainly not "Mesozoic mammals" as it has been once suggested. Their australidelphian affinities cannot be doubted given the pattern of articular morphology in the tarsus. Evolutionary analysis of the entire skeleton, in the outstanding tradition begun by Bensley (1903), and continued by Winge (1923, 1941; but written between 1887–1918) and Gregory (1910; 1951), however, is very necessary to understand this animal more fully.

I interpret the dental formula (which can be variable, with fewer teeth) as I1,2,3/1; C1; Pd1,3; M1,2,3,4. The living species has zalambdodont molars with a "fused" paracone and metacone (or reduced metacone), and lower molars that lost the talonid but retain the prominent trigonid. It has a digging manus, shoveling pes, greatly modified cranial structure, and numerous highly apomorphic surface features. Aspects of the postcranium of *Notoryctes* are discussed in Chapter 7. Its clavicles are thin and vestigial, the praeclavia are minute cartilaginous remnants, the broad wings of the manubrium are fused with the sternal portion of the first rib, and the vertebral portion of this rib is also firmly fused with the latter. In the carpus the relatively huge scaphoid appears to have incorporated the trapezoid, and the unciform, in correspondence with the hypertrophied fourth metacarpal. The magnum is not easily visible as it is tucked in under the flexad (plantar) side of the "scaphoid". The large distal and laterally oriented process of the "scaphoid," to which the lunate is nested proximally, is reminiscent of the primitive didelphimorphian and gondwanadelphian condition. The **UAJ** is uniquely mortise–tenon independent of the similar argyrolagid condition, with a nearly equal contribution to the mortise by both the tibia and the fibula, with no **CaFi** contact.

The great reduction of the epipubic bones is an apomorphic condition, and in general the numerous skeletal modifications causally related to a burrowing life are all unique attributes of this genus. The columelliform stapes of *Notoryctes* has been noted by many (Doran, 1877; Fleischer, 1973; Novacek & Wyss, 1986). Fleischer's functional–adaptive explanation for the loss of the bicrurate, or stirruplike, primitive therian (*sensu stricto*) condition is entirely convincing, in spite of the carefully argued account, from cladistic theory, of Novacek and Wyss (1986) on mammalian stapedial transformation. The latter based their polarity determinations on cladistic outgroup comparisons, explicitly setting aside the convincing functional–adaptive arguments.

Although notoryctids and peramelids in fact do share a rodlike stapes, a synapomorphy, I consider such a modified condition from an ancestral bicrurate stapes in two fossorial groups an extremely poor character. Similarly, the unquestionably fossorial modifications of the forefoot in protoperamelids may or may not have been shared in a common ancestry with notoryctids. The carpal and tarsal patterns of the two groups are certainly divergent. There are, however, a number of attributes that appear to corroborate the hypothesis of syndactylan and possibly more specific peramelinan affinities. The **UAJ** of *Notoryctes* is derived in the Australidelphia, but differently than the **UAJ** seen in the bandicoots, which have the "eutherian" mortise–tenon modification. The notoryctid mortise is more "marsupial" in having a greater participation by the fibula. In fact this notoryctid condition may represent the one antecedent to the more derived peramelid one in which both the proximal and distal extremities of the fibula are reduced to an eutherian-like conformation (Fig. 4.5). Added to this possible transformational association is the peculiar separation of the hypertrophied **Mt4** from **Mt1–3,** and the lateral and plantar, "tucked under" conformation of the fifth ray of the foot which is greatly reduced externally (but not as much osteologically). The reduction of **Mt5** is compensated by the enormously hypertrophied proximal lateral process of that bone. This process is probably a buttress against the peroneal region of the calcaneus; it is functionally convergent with the fossorial monotremes, in which the peroneal processes is ligamentously tied to the lateral proximal section of **Mt5.** The lateral ray of the foot in notoryctids suggests an ancestral constraint in which which **Mt5** was reduced.

Unlike the reduced **Mt1** of the protoperamelid, there is reason to believe that the notoryctid ancestry had a large first pedal ray – a decidedly more primitive australidelphian condition than that displayed by bandicoots (or dasyurids) – unless this condition is a reversal as is possibly the phascolarctid one. The skin webbing uniting **Mt1–3,** but not as much **Mt4,** may represent the medial extension of the original syndactylous association in an ancestor that has opted for terrestrial locomotion, and in that process emphasized **Mt4.** As I discuss under both the Peramelidae and Macropodidae and as it was noted by Winge (1923, 1941), Gregory (1951), and Marshall (1972) for these groups, the syndactyly of ancestry undoubtedly chaneled the emphasis onto **Mt4** as opposed to **Mt3.** Given these factors, the unenlarged incisor morphology strongly suggests primitive, nondiprotodont syndactylan ties, but perhaps not specifically with peramelids as early argued by Bensley (1903). The great depth of the navicular and contrasting smallness of the mesocuneiform, probably associated with the mechanically relatively unloaded syndactylous rays in contrast to the compressively loaded adjacent **Mt4,** is a peramelid-like feature of *Notoryctes.* Yet there is no reason to suspect that **Mt1–3** are not more or less equally loaded while shoveling dirt. This, therefore, may be an important indication of the ancestral constraints acquired in the first peramelinan lineage, terrestrial animals perhaps in several ways similar to primitive peramelids. As I have pointed out, the cuboid and ectocuneiform fuse in *Notoryctes.*

I believe that the large, well-developed paraconid on the trigonid of *Notoryctes* makes it likely that the ancestry of this family was close to the ancestral condition of the syndactylan molars.

*Peramelidae (Figs. 7.43–7.49 and 8.23–8.25)*
The protoperamelid probably had: chorioallantoic placenta; almost full metatherian complement of teeth (**I1,2,3,4,5/1,2,3; C1; Pd1,d2,3; M1,2,3,4);** molars with well-developed W-shaped ecto-

**Figure 8.23.** *Comparison of left calcanea in dorsal view of representative taxa of Peramelidae (Echymipera doryana, Peroryctes longicauda, Microperoryctes, Perameles nasua, Macrotis, and Isoodon macrourus). For abbreviations see Table 1.1. Scales = 1 mm.*

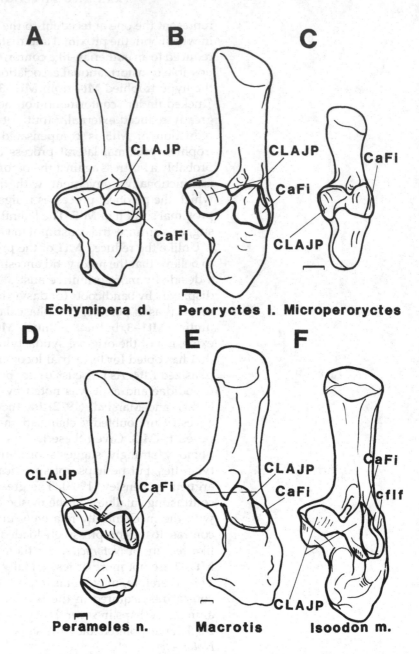

A. Echymipera d.  B. Peroryctes l.  C. Microperoryctes

D. Perameles n.  E. Macrotis  F. Isoodon m.

loph; basicranium with relatively large alisphenoid wing and with smaller petrosal wing; highly derived "jointed" neonate shoulder–breast arch in which the bracing of the shoulder and breast region is accomplished through the loading of the humeral head; adult clavicles absent; spinous processes of thoracic vertebrae exceptionally long and inclined posteriorly; spinous processes of lumbar vertebrae exceptionally robust and long and

*Figure 8.24.* *Comparison of left calcanea in plantar view of the representative peramelid taxa shown in Figure 8.23. Scales = 1 mm.*

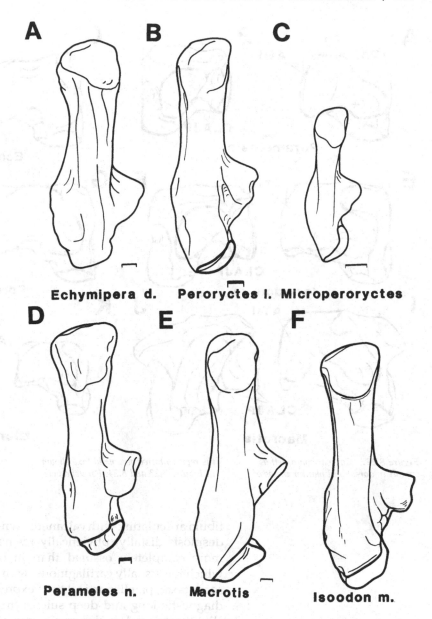

inclined forward; mamillary processes of vertebrae lengthened; lumbar vertebrae with elongated transverse processes; pronation–supination reduced in forearm, and ulna distally reduced in importance; scaphoid with reduced ulnarly oriented process as in diprotodontians, and large lunate in contact with the magnum; hind limb longer than fore limb; middle three digits well developed on fossorial manus and lateral ones deemphasized; reduced fibula distally forming the lateral side of the UAJ mortise and proximally withdrawn from contact with the femur;

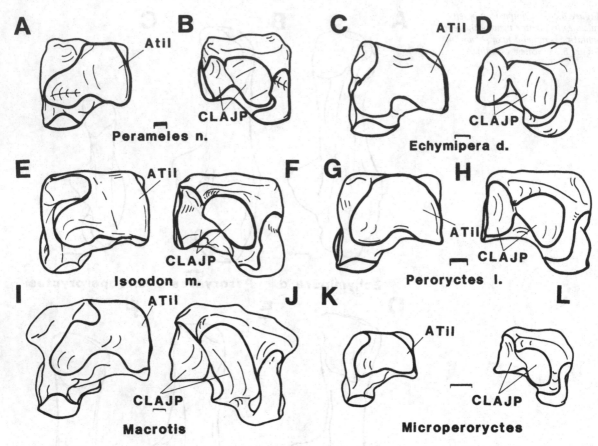

**Figure 8.25.** *Comparison of left astragali in dorsal and plantar views of the representative peramelid taxa shown in Figures 8.23 and 8.24. For abbrevia-tions see Table 1.1. Scales = 1 mm.*

fibula articulating with calcaneus; synovial **TFJ** modified into syndesmosis distally; phyletically reemphasized patella larger and more completely ossified than in other living marsupials (in which it is usually cartilaginous, with small ossification centers in some taxa); patellar articulation expressed on the distal femur in a diagnostic long and deep sulcus; first digit of syndactylous pes slightly reduced in size and nongrasping and lined up with the remaining rays; navicular deep, and mesocuneiform quite small and possibly related to an enlarged condition of the **Mt4** (probably present in a group antecedent to the notoryctids and peramelids); **Mt4** more hypertrophied than the also robust **Mt5**; **Mt4** proximally in contact with both the cuboid and the ectocuneiform (the latter having shifted its distal articular contact from **Mt3** to **Mt4**, retaining only small and nonweight bearing contact with the third metatarsal). Bandicoots have a posteriorly opening pouch (see also Abbie, 1937).

As the diagnosis plainly suggests, bandicoots, in spite of their retention of the near ancestral ameridelphian incisor numbers ("polyprotodont"), represent one of the more highly modified holophyletic clades of australidelphian marsupials, both osteologically and reproductively. Their life history strategies have been reviewed by Lee and Cockburn (1985). To even assume, however, that the Peramelidae could remain holophyletic (e.g., Archer & Kirsch, 1977; Kirsch, 1977) with the recognition of the Thylacomyidae is unrealistic. But such issues aside, *Thylacomys* is another kind of bandicoot. Immunological distance measures with their phenetic results that lack a true relative time dimension, are powerful clustering procedures, but are not complete substitutes for tested taxonomic properties such as transformation series and uniquely shared morphological attributes. The monotypic subfamily Thylacomyinae is not distinct enough from other peramelids in its morphological attributes – to judge by some eutherian standards – to warrant family recognition. It is, I believe, difficult if not impossible to support such strong statements about ranking as that of Kirsch et al. (1990a, p. 434) that "Limitied [DNA–DNA] hybridizations with *Macrotis lagotis* suggest that its current position as representative of an entirely distinct family is correct."

The intra-peramelid relationship of the bandicoots has been pursued through DNA–DNA hyridization studies by Kirsch et al. (1990a). I consider it remarkable that in their summary phylogenetic phenogram the outgroup is *Vombatus*. In light of their belief expressed in the same paper (p. 435) concerning the unacceptability of Syndactyla based on immunological evidence, it appears that their choice for an outgroup, in order to be consistent with their stated views, should have been the Dasyuridae. Would the summary phenogram, in that case, have presented the same branching sequence?

Unfortunate, ladder-based terminology and an accompanying curiously "single time plane" perspective of preevolutionary zoology continue to plague some descriptions of peramelids. They are not "intermediate" type of animals between other "polyprotodonts" and syndactylous diprotodonts. Furthermore, as a family taxon, they do not represent the evolutionary base of the primitive diprotodontian radiation, as I have already emphasized. Mosaic evolution clearly accounts for the fact that the earliest syndactylans, in retrospect, could not (if we are even slightly careful with all our studied characters) be considered peramelids! Yet such implied or explicit statements perpetuate a tenaciously clinging traditional perspective in mammalian evolutionary studies, one which is overdue for abandonment.

There are no fully satisfactory explanations why the peramelids have attained the unique marsupial attributes of their fetal membrane pattern and placentation, the discontinuous and joint-braced neonate shoulder-breast apparatus, or their postcranial morphology. This quandary remains in spite of the fact that their foreleg and hind leg adaptations can be explained by the demands of digging and bounding–running terrestriality, based on the constraints of a syndactylous ancestry. The full explanation of the origin of these animals, however, may supply us with the best available partial living *model* for the study of origins of the Eutheria (but its history should not be considered a substitute historical–narrative explanation).

The chorioallantoic placenta is likely to be related to the intra-uterine phase of the peramelid life cycle. Tyndale-Biscoe (1973, p. 68) noted in relation to the exceptionally short period of gestation of peramelids that

> In this context the very intimate attachment of the bandicoot placenta may have conferred advantages in efficient transport, but may also have necessitated a shortening of gestation to avoid allograft rejection. Thus bandicoots have not been able to exploit one adaptation for lack of another, and the result is the birth of young unmistakably marsupial in their immaturity.

Yet Padykula and Taylor (1977, p. 310) offer some serious arguments on the alternate significance of the peramelid chorioallantoic placenta. They state that

> It is doubtful that trophoblastic invasion in marsupials serves only as anchorage of fetal to maternal tissue. Our cytological evidence indicates a more efficient avenue for functional interaction between fetal and maternal substances is provided through this hybrid modification of the placental barrier that brings the two bloodstreams into closer proximity. It should be noted, however, that thinness of a placental barrier may not necessarily correlate with efficiency of transfer. . . . The fact remains, however, that in the two extremes of placental involvement among marsupials, in peramelids and *Didelphis* (both with the same gestation length), the peramelid young are at birth twice the weight and are morphologically more differentiated. . . . For example, in newborn *Perameles nasuta*, tastebuds, partially formed Mullerian ducts, and ear pinnae are recognizable, whereas in newborn *Didelphis virginiana* they are not. The head start for peramelids at birth is maintained in the developmental history of the pouch young and the time at which independence is attained. . . . The bandicoots . . . develop with greater rapidity than is yet known for any other marsupial. . . . Thus, once intrauterine hatching from the shell membrane takes place, the peramelid strategy appears to be directed toward acceleration of pre- and post-natal development which in this regard outstrips any other marsupial yet studied. It may be that the chorioallantoic placenta acts as the initial facilitator of such rapid development.

It appears that the departure in the development of the shoulder–breast apparatus is probably causally related to the post–neonate or post-pouch young phase of locomotor behavior.

This connection is likely to be so in spite of the fact that the reduction of the anlage is initiated before birth. Despite Klima's (1987) suggestion, the reasons for this pattern are not merely another way of solving the problem of the emerging neonate, but are probably, I believe, a solution to the problems of peramelid postneonate life. In that phase of ontogeny the skeleton first had to accommodate the demands of the crawling neonate, but later (in the adult phase) it had to serve a moderately cursorial quadruped. This reduction is probably a parallel in a general way to the condition that must have occurred very early in eutherian phylogeny when the demands created by neonates crawling onto nipples were eliminated (see Chapter 3). Clearly, the (proximal) causal factors may have been distinct, although the possibility of similar demands for adaptive locomotor solutions in the origin of the first eutherians, as they occurred in peramelids, should not be overlooked.

The carpal pattern is interestingly derived but difficult to evaluate. The enlarged squared-off lunate, in contact with the magnum, makes sense as the centrally wedged proximal load-bearing element of a highly terrestrial wrist. In fact primitive eutherians probably displayed this general pattern, with the retained presence of the centrale added to it. As in other syndactylans (but not like the extremely altered notoryctids), the scaphoid lacks primitive distolateral extension, this being the reason for reacquired contact of the lunate and the magnum from a gondwanadelphian.

It is obvious that the functional peculiarities of bandicoot foot mechanics required by their locomotor adaptations had to be resolved around the syndactyly of the second and third rays. This development was clearly established by Winge (1923, 1941, pp. 83 & 92). The same point was also repeated later by Marshall (1972). Therefore, the hypothesis (as well as the rather nonevolutionary perceptions of morphological transformations) that this syndactyly is *homoplastic* with other syndactylans, as merely stated without any arguments by Kirsch (1977), has been rejected (see Goodrich, 1935; Szalay, 1982; Hall, 1987). Such a posteriori reasoning is the direct consequence of a weighting scheme that gives more emphasis to both clustered morphological similarities (clearly metatherian and australidelphian plesiomorphies) and to *clustered distance* data (Kirsch, 1982); it is not a conceptually independent attempt to evaluate syndactyly in its own phylogenetic and functional context.

Reig et al. (1987, p. 78), skeptical about previously unutilized but relevant morphoclines, have also misunderstood the corroborated hypothesis of transformation of tarsal and other traits, including the significance of the "pinched" LAJ of bandicoots and kangaroos. Their objection to the Australidelphia hypothesis in which their own view of the peramelid tarsal evidence played a role, therefore, lacks the necessary arguments to explain the evidence differently than I (Szalay, 1982) did. Under the Dasyuromorphia I have discussed the issue relevant to the peramelids, namely a view of Marshall (1972, p. 55) that dasyurid pedal morphology is ancestral to the syndactylan one. Although this hypothesis is clearly stated, it is, I believe, based on dental evidence and not on character analysis of the complex in question. It cannot be maintained anymore.

Tate and Archbold (1937) made a strong case for the general peramelid primitiveness of the New Guinea genera, a general point with great merit. In their phylogenetic tree of the family (p. 349) they gave a now outdated diagnosis of the protoperamelid (and therefore in the sense used here, of the family): "triangular upper molars; didelphoid lower molars; five upper incisors; fossorial feet; cursorial gait; moderate length of tail; moderately developed ears; didelphoid bulla; coarse but not unduly spinous pelage; pouch opening backwards." Tate's later (1948) work, while also dated now, is a most comprehensive and explicit morphological effort to describe the interrelationships of peramelid genera. It is difficult, however, to match the various character combinations dispersed through his text with the nodes of his phylogram. Yet Tate's phylogeny is a major hypothesis of bandicoot relationships based on a large number of morphological characters that should be further tested against a more complete understanding of character clines (see also Kirsch et al., 1990a, for DNA–DNA hybridization analysis of bandicoot intergeneric relationships). Tate (1948, p. 317) believed that syndactyly arose independently in bandicoots and he considered such primitive metatherian traits as polyprotodonty as being evidence for dasyurid–peramelid ties. The internal phylogeny of the Peramelidae, hypothesized in detail at first by Tate (1948) based on external and osteological traits, was followed up later by Kirsch (1977) in his immunodiffusion studies. While Kirsch has emphasized that *Macrotis* is "distinct" from other bandicoots, a fact also amply shown before by morphology, the ability of his various clusters to reflect taxon phylogeny seems extremely doubtful to me. My reluctance is partly based on Kirsch's (1977) detailed and lucid description and discussion of the numerous caveats concerning the immunological methodology in his monograph, in spite of which he consistently grouped peramelids with dasyurids (contraindicated in later results obtained by Sarich, 1993). Kirsch considered convergent such morphological features as syndactyly, a view based, I believe, on his unstated belief that comparative serology is a true and objective reflection of group phylogenetic ties, and thus that morphological assessment of features should be judged accordingly.

Tate's (1948) seminal study, on the other hand, in spite of his mistaken use of primitive features and his rejection of the monophyly of syndactyly, laid the taxonomic foundation for a reexamination of the cranium and external morphology of these animals. It is not at all clear, however, whether the enlargement of the metaconule (to form a peramelid "hypocone") and changes in bullar shape and size occurred in the family only once or several times independently. These questions should be tackled by detailed functional–adaptive studies before any internal phylogeny of the bandicoots can receive meaningful morphological testing.

## Semiorder Diprotodontia (Owen, 1866)

### Diagnosis

Protodiprotodontian had: incisor number probably I3/3, with the enlargement of a lower incisor (I/1?), and the less pronounced enlargement of the first upper one; last premolar with shearing emphasis (retained in the dentally more primitive descendants); an M/1 that retained a low trigonid, probably because of influence of P3/3 form–function; M/2–4 with four cusps emphasized, the protoconids, metaconids, hypoconids, and entoconids, all of which together were incipiently and functionally near-bilophodont; upper molars with "hypocone," probably derived from metaconule, also incipiently lophodont but retaining the crescentic (quasi-selenodont) sweep of the preparacone and postmetacone crests; basicranium probably with a relatively small alisphenoid wing but without a completely ossified bulla (bandicoot-like); carpus in which the scaphoid (without an ulnar extension) is large but the lunate is present, albeit reduced to a small bone, compared to the more primitive syndactylan (and ameridelphian) condition (where the lunate is relatively large); tarsal morphology as described under Australidelphia and Phalangeriformes. The living forms have a superficial thymus gland; they are known to share the *fasciculus aberrans*, a connection between the cerebral hemispheres, homoplastic with the eutherian solution of the *corpus callosum*.

### Included taxa

Suborders Phalangeriformes, Vombatiformes, and Yalkaparidontia.

### Distribution

Probably Paleogene to Recent of Australasia (first record in Miocene).

### Discussion

An important background to the understanding of the diprotodontian radiations is the assessment of the paleoecology of Antarctica and Australasia during the early Tertiary. In a synthetic analysis Case (1989b) has reviewed relevant climatic, geographical, and ecological factors and has proposed several important hypotheses. He envisages the Eocene as already having distinct possum (phalangeriform as the group is designated here), macropodoid, and vombatiform representatives, a view with which I concur. According to him the podocarp- to *Nothofagus* forests were altered into a greater diversity of forest habitats favored the diversification of the original syndactylan possums. As the forest habitats began to open up in the Oligocene, providing a lusher angiosperm-dominated habitat, the radiation of the vombatiforms got under way, and by the time the Miocene grasslands began to appear, the macropodoids, particularly the macropodines, started their great radiation.

It appears to me, and there is probably a concensus regarding this, that the protodiprotodontian ancestor coped far better with most plant-derived foods than the non-diprotodontian syndactylan antecedent to it, the latter a species that probably had primitive peramelid-like cheek teeth. The diprotodontian forward-projecting incisors may have been originally evolved for exudate access, and the cheek teeth for mixed insect and plant foods. Were we to find the skeleton of a primitive diprotodontian, after careful study it would be recognized as a possum-like (not bandicoot-like) animal, an arboreal, prehensile-tailed form with no appreciable ground-related adaptations.

For a long time now the outlines of the major groups of the Diprotodontia have been well established. Questions concerning holo- and paraphyly, however, have no simple answers, whether these are sought through morphology or molecular studies. The most recent authoritative discussion of the taxonomic history of this group is by Aplin and Archer (1987), but the character analysis provided by Murray et al. (1987) is the most closely argued source for understanding cranial character transformations based on causal mechanisms. I will review some of the critical issues as there is clearly disagreement over the significance of cranial data corroborated by several competent studies. The tarsal information adds another factor to be considered in making a reasoned judgment about the phylogeny of this radiation.

According to Aplin (1987, pp. 379–380) "six major morphological 'types' [of basicrania of diprotodontians] can be distinguished. These are a '*Phascolarctos*' type; 2, a vombatimorphian type; 3, a 'central' phalangeridan type; 4, a macropodoid type; 5, a '*Tarsipes*' type; and 6, an acrobatid type." These are, noted Aplin, variably distributed between the three major infraordinal-level taxa recognized by Aplin and Archer (1987) as follows: "Phascolarctimorphia (type 1 only); Vombatomorphia (type 2); Phalangerida (types 3–6)." In order to attach phylogenetic significance to these generally defined "types" of Aplin, I see a real need for a reexamination of morphogenetic pattern formation and functional–adaptive demands.

Using other lines of evidence, Harding (1987) reviewed spermatozoan microstructure, and Springer and Kirsch (1989; 1991) have presented a review of single-copy DNA hybridization studies. The latter used a number of numerical corrections, and with the aid of pairwise tree construction algorithms they outlined a diprotodontian phylogeny. In spite of Kirsch's past misgivings about the concept of Syndactyla, the study used *Peroryctes* as an outgroup. This is particularly curious because Kirsch et al. (1990a, p. 435) state that "many authors have regarded the bandicoots' possession of a syndactyl pes as a synapomorphy with Australian diprotodont marsupials (e.g., the "Syndactyla" of Szalay, 1982), but serology suggests a special relationship with dasyuroids. . . ." The choice of one's taxic outgroup appears to influence one's taxic conclusions, as it has been repeatedly stated (e.g., Szalay & Bock, 1991). These results of Springer and Kirsch (1991) confirm the existence of three major groups: the vombatiforms, the kangaroos, and the phalangeriforms.

In a developmental pattern-oriented phylogenetic study of the newly described primitive thaylacoleonid *Wakeleo*, Murray, et al. (1987) have presented a character analysis of cranial attributes and offered tested taxonomic properties that have not been previously adequately appreciated. They have opened up for examination various issues of ontogenetic and phylogenetic dynamics. In addition, an assessment by Springer and Woodburne (1989) of the distribution of some basicranial features of most diprotodontians, with emphasis on the phalangeriforms (present sense), has been also significant for the understanding of basicranial diversity. The latter study summarizes the presence and clarifies the distribution of some characters, but it uses a method of polarity analysis that is based exclusively on taxic outgroup comparison, the analytical results of which I question. Some of the basic information presented by them (e.g., on *Dromiciops*) is open to different interpretations (see Figs. 3.1 & 8.22).

Springer and Woodburne (1989) discussed diprotodontian phylogeny following their survey of basicranial patterns. They have, however, not attempted to construct taxonomic properties of this complex and developmentally, functionally, and adaptively influenced area of the skull. In one notable instance, that of the phascolarctids, the functional–adaptively highly altered squamosal–alisphenoid complex (one that is extremely pneumatized), the ventrally deep bulla, and the drastically reduced ectotympanic were considered by them as a plesiomorphic state compared to the condition of the vombatid ear region. I believe, however, it is more likely that the phascolarctid condition, although reminiscent of that aspect of the ancestral condition where the bulla is also alisphenoid derived, may be secondary. One should entertain probabilities rooted both in developmental mechanisms and adaptations not only when considering the origins of the huge bulla of these animals, but in all cases where taxonomic properties should be tested against all of the evidence. When such a relatively large component of the basal part of the skull was to be constructed ontogenetically, then rapid growth of the alisphenoid during ontogeny would prevent squamosal takeover of the area, a condition which was probably present in the antecedent to the koala. In other words, the derivation of the basicranium in koalas may be a consequence of the pneumatized hypertrophy of the bones noted. Added to this is the still poorly understood functional dynamics of the angle to middle ear floor connections in metatherians and the demands of such constraints. The koala bulla, then, may be a *derived* vombatiform trait, and not a retention of the primitive marsupial condition inherited from the stem vombatiform. The large postglenoid process is probably a primitive condition inherited from a diprotodontian ancestry.

It was stated by Murray et al. (1987, p. 460) that only tentative hypotheses can be formulated about diprotodontian relationships because so much developmental information is lacking; this perspective sounds realistic. To this caveat must be added a warning concerning the equal paucity of comparative, functional, and ecomorphology analyses, as I emphasized in the discussion of cranial morphology in Chapter 3. The following tentative narrative explanations are based on what little is understood in light of sundry mechanisms. The stem diprotodontians were probably arboreal possum-like animals with a molar dentition that was perhaps not far removed from the primitive bandicoot crown pattern. These animals had a basicranium with a ventromedially bilaminar chamber for the middle ear cavity, and a partial bony bulla, formed presumably by the alisphenoid, with the enlarged ectotympanic half-tube partly enclosed in the bulla. The putative Miocene wynyardiid genus *Muramura* named by Pledge (1987b) holds the promise to understand some important issues of diprotodontian phylogeny. So far postcranial attempts to reconstruct the root of this diversity are restricted to the carpal and tarsal evidence, which suggests possum-like animals, although perhaps with vombatiform affinities (see also Munson, 1992).

The origins of the suborder Vombatiformes, if it is monophyletic (see the caveats that follow), is from a species which evolved from early phalangeriforms (perhaps a lineage of a hitherto unknown family) as diagnosed here, an early clade which did not have the complete bony floor provided by the alisphenoid for the bulla. We would have great difficulty recognizing such a form as a precursor to the vombatiforms because of the numerous primitive phalangeriform traits it would have had. Similarly, the combined osteological evidence, and taken in light of a line of arguments provided by Murray et al. (1987), make it quite likely that both the semisuborder Vombatomorphia and the semisuborder Diprotodontiformes (Wynyardioidea, Thylacoleonoidea, and Diprotodontoidea) were derived from an animal that would be also in many ways possum-like, but perhaps already of vombatiform ancestry. Phalangeriform-like precursors to the vombatiforms and diprotodontiforms are yet to be discovered. The ancestors could be very easily misidentified as Phalangeriformes, were we to discover them early enough in geological time without the known diagnostic parts of the vombatiforms.

While the recently discovered Yalkaparidontia is known only from its enlarged anterior teeth, phalangeriform-like skull, and uniquely zalambdodont molars, it also further complicates the possible ties among the phalangeriform groups. The basicranial similarity of *Yalkaparidon* to peramelids (Archer et al., 1988) suggests that it is an early offshoot of the stem diprotodontians. Nonetheless it is apparent to me that any effort to apply the principle of taxic pan-holophylism to the classification of the diprotodontian subgroups is unwarranted at present, without the necessary functional–adaptive research in various areas against which one would want to test such proposals.

Judging from the coadapted morphology of the incisors and cheek teeth in burramyines, petaurines, and phalangerids, there appears to be no convincing dental evidence that would corroborate Kirsch's (1977, p. 31) view that the protodiprotodontian condition of incisor enlargement is an adaptation for carnivorous stabbing. This view has its origins in observations of caenolestid predatory behavior. The cheek dentition of the caenolestids of course represents an adaptive shift toward a highly specific quadrate molar condition – a pattern in general rarely associated with exclusively carnivorous or insectivorous feeding. Smith's (1980) study of the dentally, cranially, and postcranially primitive petaurid *Gymnobelideus*, however, does suggest that incisor hypertrophy in the ancestor of the Diprotodontia may be related to both gum harvesting and associated insectivory, and not to carnivorous stabbing.

## Suborder Phalangeriformes (Szalay, 1982)

### Diagnosis

Protophalangeriform had: full diprotodontian dental formula; molars nearly triangular, with retained stylar cusps; masseteric foramen; overlap of squamosal by alisphenoid component of basicranium, more extensive than in syndactylan ancestry; bilaminar alisphenoid and squamosal tympanic wing probably (?) opened posteriorly; pneumatized spongy bone in squamosal and perhaps in sphenoids as well (this character may be size related, and therefore of little use in this diagnosis); reduced primitive diprotodontian presence of the lunate; primitive syndactylan pedal, carpal, and tarsal morphology.

### Included taxa

Superfamilies Petauroidea (including Petauridae and Tarsipedidae), Phalangeroidea (including Phalangeridae), and Macropodoidea (including Hypsiprymnodontidae and Macropodidae).

### Distribution

Paleogene to Recent (first record in Oligocene).

### Discussion

The Phalangeriformes is paraphyletic because of my inclusion within it of the putative stem groups of the other taxa of the entire semiorder Diprotodontia. The stem group of the Diprotodontia is currently best envisaged as the protophalangeriform. Lineages derived from this stem group, as hypothetically constituted, were probably ancestral to the other two suborders as well. The overwhelming number of the lineages in the Phalangeriformes were and are arboreal in their substrate preference, although species obviously vary in positional and locomotor behaviors in addition to other critical attributes. The last common ancestor, as well as most of the living species (except for the gliders to varying degrees), retained the fully prehensile tail of the first australidelphians. The internal relationships of these lineages are some of the most problematic ones among marsupials (see contributions in Archer, 1987, especially Flannery et al., 1987a, b). While Aplin and Archer (1987) recognize eleven families in five superfamilies in their Phalangerida, I consider the splitting of the family level an excessive arrangement. The use of subfamilies, tribes, and even subtribes is far more appropriate to express the clusters of morphological diversity, even if our phylogenetic understanding of these animals is still rudimentary. If the rodents were split proportionately to their morphological divergence and in a manner commensurate with that suggested by these and other authors for the phalengeriforms, we would erect several orders for the known rodents and a far greater number of families than we have today, serving no particular purpose. It is important to remember the recent study of Norris (1993) concerning the variation in the make up of the bulla, discussed in Chapter 3. Distinct, population-specific form–patterns of basicranial morphology are documented by him – in the same species.

The evidence for the monophyletic nature of the Phalangeriformes and its subdivisions, as I diagnose these, is morphological, and not serological. The serological and DNA–DNA hybridization information clusters the taxa based on distance, and not on characters, posing problems for any attempt to interdigitate fossil groups with living ones. The stem group of the Phalangeriformes is diagnosable, without any doubt as to its monophyly. Yet, as Aplin and Archer (1987, p. XlIX) so appropriately remarked, "anatomical knowledge of even the most basic kind is glaringly incomplete for a number of major groups . . . ." To this I must add that in addition to this highly desirable information yet to come, their use for testing new taxonomic properties will be even more significant for phalangeriform phylogenetics.

The basicranial synapomorphy of the Phalangeriformes (their Phalangerida), according to Aplin (1987, p. 388), is based on

> completion of the bony auditory bulla by proliferation of an alisphenoid tympanic process . . . and a caudal tympanic process of the petrosal . . . ; development of a deeply invasive squamosal epitympanic sinus . . . ; pneumatization of the tegmen tympani resulting in dorsal displacement and lateral expansion of this element; establishment of firm osseous contact between the posteromedial surface of the ectotympanic and the posterior surface of the promontory; formation of a tubular outer ear canal which incorporates the postglenoid process of the squamosal; and placement of the 'postglenoid' foramen anterior of the actual postglenoid process.

Such a combination, presumably, would not have been the ancestral one to other Diprotodontia. Such a characterization of the basicranium for the present concept of the Phalangeriformes, however, is too restrictive.

Nevertheless, I suspect that only both developmental and functional–adaptive studies probing the various postneonate stages will be able to determine whether the tarsipedid, macropodoid, and acrobatin conditions of the basicranium, as well as those of other diprotodontians, are independently derived from a primitive phalangeriform with an incompletely ossified bulla, or whether other transformations should be considered equally probable. The interaction of the alisphenoid, petrosal, and ectotympanic, with the exception of the latter, involves the same elements one finds in the relatively primitive didelphid ameridelphian radiation; these are the developmentally and functional–adaptively determined variables involved in the formation of the auditory region. Added to this is the occasional role of the squamosal in bulla formation (see Norris, 1993), and the caveats it signifies about bullar characters.

Understanding such history, based on a maze of conflicting distribution patterns of seemingly randomly chosen characters, without the application of a causal-mechanism-oriented analysis, appears to be hopeless. For example, one considerable complicating factor is the direction of absolute and relative size change in phylogeny and the various adaptive correlates of these changes. Size reduction most often entails a hypertrophy of the middle ear cavity to accommodate changing demands of hearing. This can lead to differential utilization of the roofing and flooring elements of the bones surrounding the tympanic cavity. Fluctuations in the size and number of the epitympanic sinuses, and other cavities, become formidable complicating factors in understanding trans-formational series, while at the same time they add to the diversity that can help bridge both gaps and give direction to one's under-standing of the change. The pneumatization of the basicranial area, a widespread phenomenon in mammals, adds further com-plexity. So it is not surprising that the difficult analyses of transfor-mations of the states of basicranial characters in marsupials, particularly within the diverse diprotodontians, are yet to be car-ried out. Defining any one homogeneous group by what at-tributes it has should not interfere with the analytical assessment of the transformations responsible for these patterns. Characters are transformed in phyla, and therefore it is inappropriate to "define" these statically.

The pedal and tarsal morphology of known phalangeriforms is relatively undifferentiated from an ancestral australidelphian con-dition (Figs. 8.26–8.31). In spite of the diagnostic australidelphian and phalangeriform similarities, however, species-specific distinc-tion of the individual foot bones is relatively easy. The syndac-tylous condition of the foot notwithstanding I could not successfully delineate significant phalangeriform features (i.e., not minor ones, which might easily be homoplastic) on the tarsus that could distinguish the last common ancestor of this taxon from the microbiotherian *Dromiciops*.

The superfamilies in the semiorder Diprotodontia are derivable from an animal that was an arboreal "possum." It seems to me that a paleontologist would unhesitatingly identify its various re-mains, were they found independently, as a phalangeriform postcranially, but perhaps as a peramelinan basicranially, and ei-ther peramelinan or vombatiform (not vombatomorphian) den-tally. Consequently, difficulties will continue to exist as to the particular cladistic ties of these separate but commonly rooted radiations. Clearly, however, at the point of origin of *all* taxa pa-tristic affinities are issues that must also be addressed and solved if one is to practice phylogenetics in an evolutionarily meaningful way.

It can be said of at least two of the superfamilies included in the Phalangeriformes that the probability that they are paraphyletic is great, and so these are certainly acceptable monophyletic taxa. Furthermore several of the families themselves may be also para-phyletic. Either of two distinct lineages of phalangeroids might have been ancestral to a thylacoleonid or stem wynyardiid (note that the Vombatiformes, then, would not be a monophyletic tax-

**Figure 8.26.** *Comparison of left calcanea, in dorsal view, of selected taxa of microbiotheriids (Dromiciops) and burramyines (Acrobates, Distoechurus, Eudromicia, Dromiciella, and Cercartetus) showing the apomorphous pattern of the australidelphian ancestry. For abbreviations see Table 1.1. Scales = 1 mm.*

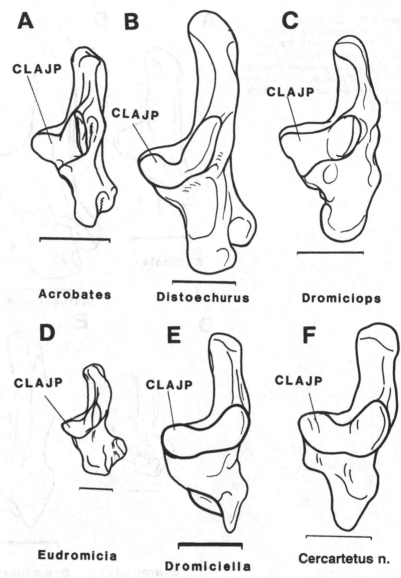

on), on one hand, and to a stem macropodoid on the other hand. The nature of the evidence is such that the nexus of phyletic ties between burramyins, acrobatins, tarsipedids, and petaurines, and a number of the fossil subfamilies known from teeth alone, may well demand accommodation in paraphyla.

The peculiar petrous expansion into the bulla in *Acrobates* and *Tarsipes*, for example, should not be taken at face value (e.g., Springer & Woodburne, 1989) without considering the various functional–adaptive factors that might have delayed alisphenoid posterior expansion. Although these two small possums may well be sister lineages, the appropriate studies are needed for convinc-

**Figure 8.27.** *Comparison of left calcanea in plantar view of selected taxa of microbiotheriids and burramyines shown in Figure 8.26. The highly characteristic lateral calcaneal flange (bearing the **CaCul** facet) is shown as y. For abbreviations see Table 1.1. Scales = 1 mm.*

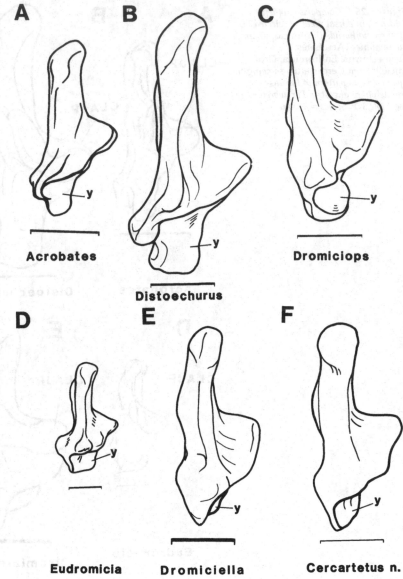

ing corroboration. Baverstock et al. (1987, 1990) have recently examined the albumin immunologic relationships of phalangeriforms, and Springer et al. (1990) conducted studies of DNA–DNA hybridization within the phalangerids. In the latter contribution the authors entertain the idea (p. 308) that "If we accept the DNA tree, we may also ask which characters exhibit the most homoplasy and are consequently less reliable indicators of phylogeny." This is, of course a blatant attempt to "interpret" morphological evidence in light of the "historical truth" provided by molecular "trees." Why "less reliable" ex cathedra? What deter-

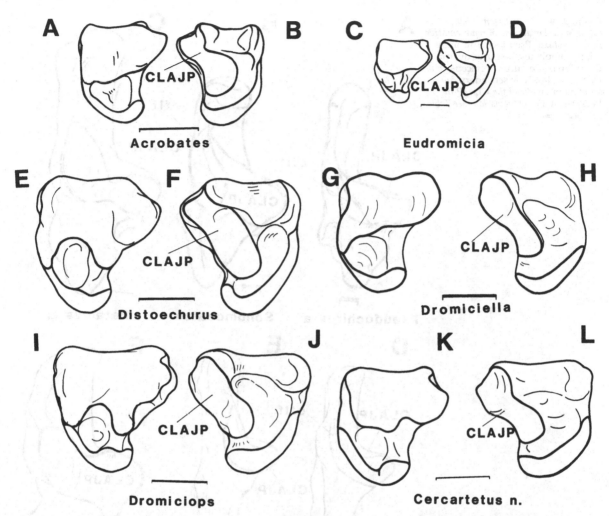

***Figure 8.28.*** *Comparison of left as-tragali in dorsal and plantar views of* *selected taxa of micropiotheriids and burramyines shown in Figures 8.26* *and 8.27. For abbreviations see Table 1.1. Scales = 1 mm.*

mines without some biologically sound reasons what should be the most reliable taxonomic properties in organisms? Springer and Kirsch (1991) maintain that within the Phalangeriformes (which they consider holophyletic) the phalangerids and bur-ramyids are one holophyletic group, whereas the "acrobatids, pe-taurids, and pseudocheirids" are another.

*Petauridae*

I know of no compelling morphological and phylogenetic reasons why the Petaurinae and the Burramyinae should not be included in one monophyletic family. The Petaurinae, like the burramyines discussed below, may be paraphyletic also. The three tribes in

*Figure 8.29.* *Comparison of left calcanea in dorsal view of representative petaurine taxa (Pseudocheirus archeri, Schoinobates, Petaurus australis, Dactylonax, Petaurus breviceps, and Dactylopsila). Note secondary constriction in middle of the CLAJP in Petaurus a. For abbreviations see Table 1.1. Scales = 1 mm.*

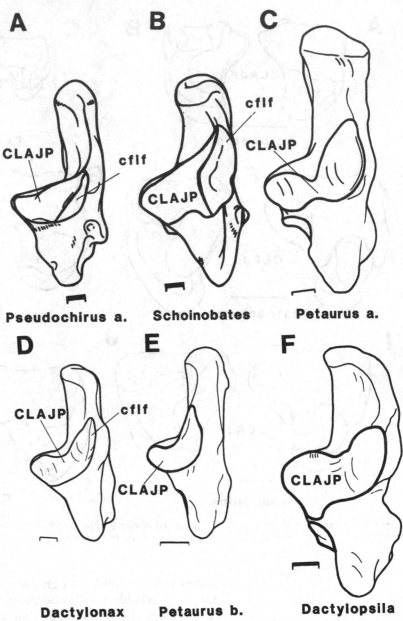

which I arrange them, the Petaurini, Dactylopsilini, and the Pseudocheirini, are extremely conservative postcranially and to some extent cranially as well (see Turnbull, Rich, & Lundelius, 1987a, b, c, for Plio-Pleistocene diversity; see Woodburne et al., 1987b for the Pseudocheirini). While the Petaurini and Dactylopsilini may be more similar to one another in some respects (e.g, teeth; or chromosomes – for the latter see Murray et al., 1990) than either is to the Pseudocheirini, the meaning of these similarities is not clear.

**Figure 8.30.** *Comparison of left calcanea in plantar view of representative petaurine taxa shown on Figure 8.29. For abbreviations see Table 1.1. Scales = 1 mm.*

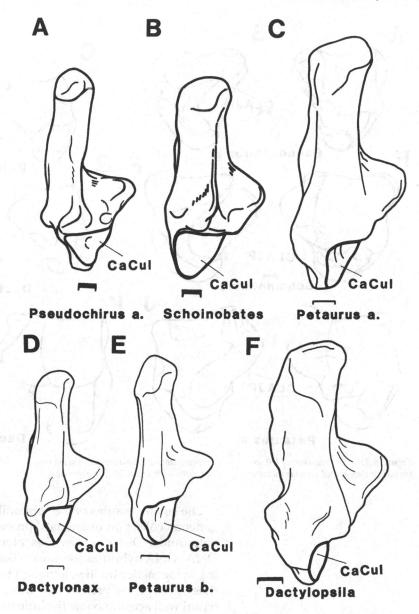

A  B  C

CaCul  CaCul  CaCul

Pseudochirus a.  Schoinobates  Petaurus a.

D  E  F

CaCul  CaCul  CaCul

Dactylonax  Petaurus b.  Dactylopsila

The highly folivorous adaptations of the ringtail possum group may have masked their specific cladistic ties with an early lineage of the Petauridae. I cannot envisage the Pseudocheirini to be more diverse than a host of monophyletic placental tribes – these phalangeriforms are not diverse, although moderately speciose (approximately thirteen species). The gliding *Schoinobates* has long been recognized to have special ties to *Hemibelideus lemuroides* based on an incipient patagium on the latter (see Chapter 7). Baverstock et al. (1990) have recently assessed the albumin immunologic relationships of the pseudocheirins.

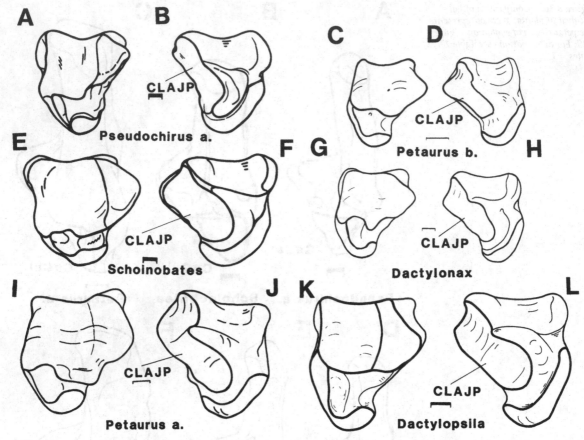

**A** **B**

CLAJP

Pseudochirus a.

**C** **D**

CLAJP

Petaurus b.

**E** **F** **G** **H**

CLAJP

Schoinobates

CLAJP

Dactylonax

**I** **J** **K** **L**

CLAJP

Petaurus a.

CLAJP

Dactylopsila

***Figure 8.31.*** *Comparison of left as-tragali in dorsal and plantar views of* *representative petaurine taxa shown on Figures 8.29 and 8.30. For abbrevia-* *tions see Table 1.1. Scales = 1 mm.*

The small possums of the subfamily Burramyinae are a hetero-geneous collection of animals, probably a paraphyletic group of petauroids. Distance measures alone, such as immunology or DNA–DNA hybridization, are unlikely to shed light on the specif-ic phylogenetic affinities, although they can further help to cluster these groups. At present two tribes, the Burramyini and Acro-batini well accommodate the little we know of their diversity and phylogeny. The problems of *Burramys*, as far as its cranial and dental attributes are concerned, have been discussed in some detail by Ride (1956). In considering the affinities of *Burramys* – then presumed to be extinct – based on detailed considerations of what was available for study, Ride concluded (p. 424) that "it is extremely unlikely that *Burramys* has any macropod affinities whatever." Although in general Ride's appraisal of the cranial and dental evidence is acceptable, I should point out that this small phalangeroid, in addition to its **P3** and posterior palatal similarity to *Hypsiprymnodon* also has an enlarged **CaCua** facet, which is a

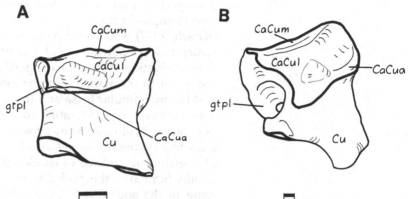

**Figure 8.32.** *Comparison of right cuboids of the macropodoid Hypsiprymnodon (A) and the phalangerid Phalanger (B). Some living phalangerids show cuboid and dental modifications similar to the primitive macropodoid conditions. For abbreviations see Table 1.1. Scales = 1 mm.*

diagnostic macropodoid attribute (see Fig. 8.32). This "stepped" configuration of the "angled" australidelphian **CCJ,** however, is related to the mechanics of stability in the terrestrial macropodoids. Since the mountain pygmy possum is reported to be substantially terrestrial in addition to its excellent shrub climbing ability (Calaby, Dimpel, & McTaggart, 1971), this feature is probably a homoplastic similarity between *Burramys* and kangaroos.

The acrobatin brace of genera, *Acrobates* and *Distoechurus*, show "an alisphenoid contribution to the anteromedial margin of the glenoid fossa; a separate foramen, located well lateral to the f. ovale, for passage of the masseteric nerve; and a unique ectotympanic "disc" located at the medial end of the outer ear canal" (Aplin, 1987, p. 386). I consider these differences, in light of the dental evidence, as synapomorphous derivations and modifications from a petauroid ancestry rather than independent specializations from a phalangeriform stem.

### Tarsipedidae

*Tarsipes* is the sole genus known of this group, and its family status is largely based on its uniquely different combination of derived characters compared to other small phalangeriforms; the tiny honey possum continues to be a phylogenetic enigma. The skull and its feeding mechanism (see discussion above) is highly and uniquely derived, a singularly important factor to consider when evaluating basicranial ontogeny, the feeding mechanism, or the morphology of the teeth themselves. Either the functional or ecological morphology of these features should not be decoupled from a yet to be conducted analysis of their homologies, a point so clearly made recently by Zeller (1989), while using primarily developmental criteria (embryology) in an adaptational perspective to understand character relationships of monotremes.

The diprotodont dentition of *Tarsipes* is vestigial, the foot is unquestionably syndactylan, the tarsals are somewhat enigmatic pri-

marily in their secondary modification from a primitive syndactylan pattern, and the morphology of the spermatozoa is inconclusive (Temple-Smith, 1987). Comparative serology of whole sera
(Kirsch, 1977) is so vague as to be of little use. I suspect that the
basicranial specializations of *Tarsipes* do not signal an independent derivation from a phalangeriform ancestry in exclusion of
the small Burramyinae. Elimination of the postglenoid process
and the mandibular fossa and the modest outer ear canal are expressions of the considerable reduction of stresses rather than independent evolution from protophalangeriforms. Its sliver-like
jaws and zygomatic arches clearly reflect the lack of significance
of chewing in contrast to its elongated tongue-oriented feeding
habits. Similarly, the probable replacement of the epitympanic
wing of the squamosal "by a substantial petrosal epitympanic
wing; the presence of a large [rostral tympanic process of pe-
trosal]; the presence of a small epitympanic sinus of squamosal;
and the marked inflation of both the anterior tympanic cavity and
the sinus tympani" (Aplin, 1987, p. 386) are best perceived, until
documented otherwise, as transformations from some burra-
myine.

Though *Acrobates* and *Distoechurus* (see especially Hughes et al.,
1987) may be members of a clade most recently related to the
tarsipedids (Aplin & Archer, 1987; Springer & Woodburne, 1989),
I include the latter in the Burramyinae, within the probably paraphyletic Acrobatini, because the uniqueness of *Tarsipes* makes that
hypothesis still inconclusive, in my view, for the reasons I have
noted.

*Phalangeridae*
George (1987) and Flannery et al. (1987b) present updated taxonomic accounts of the subfamily Phalangerinae (this subfamily,
as I use it here, is equivalent to the Phalangeridae of Flannery et
al., 1987b; see also Flannery & Archer, 1987c). I follow their subdivision into the more cranially and dentally plesiomorphic
Ailuropini and the Phalangerini. I believe that a real danger of
splitting and supraspecific taxonomic inflation may follow an increase in understanding of phalanger distribution and the superficial pelage and biochemical differences of widespread
populations. The meaning of the allopatric populations of somewhat divergent phena of the cuscuses should be, I believe, explored in possum systematics in general, but for the phalangers in
particular (see Groves, 1987a, b; Norris, 1992). Attempts would be
very desirable to balance the routine application of a taxonomic
species notion (see Szalay & Bock, 1991) with investigated cases,
and therefore appropriate evolutionary models, of polytypic *single species* of marsupials with proven reproductive continuity.

I have reduced the traditionally ranked family Phalangeridae to
a subfamily status because I believe that their morphological (os-

teological) diversity is far more modest than equivalent placental families. The entirely fossil subfamilies of the phalangerids are known only from dental and jaw material. The materials and their phylogenetic significance are presented and discussed in excellent accounts by Woodburne, Pledge, and Archer (1987) for the Miralininae, by Woodburne and Clemens (1986) and by Woodburne (1987) for the Ectopodontinae, and by Archer et al. (1987) for the Pilkipildrinae. It is likely that with such relatively limited material available, the exact cladistic ties of these groups – not their teeth – cannot as yet be definitively established. Even with the inclusion of the fossil groups in the family, some phalangerines may have been ancestral to the other subfamilies, and probably also to the macropodoids.

### Macropodoidea

The protomacropodoid was primarily distinguishable from an ancestral phalangeroid in its adaptive complex related to bounding–running and possibly hopping locomotion, and in the extensively herbivorous chewing musculature and cheek teeth. It had: dental formula which included **dP2**, a tooth probably as plagiaulacoid as the **P3; P3** replaced not only its deciduous homologue but also **dP2** during ontogeny; masseteric fossa that developed into a masseteric canal for insertion of deep masseter deep within dentary (masseter canal of dentary was not yet confluent with dental canal); P3 probably plagiaulacoid, probably retained in this stem from unknown phalangeroid ancestry; hypertrophied greater trochanter raised far above the femoral head, far beyond the condition seen in peramelids, and prominent scar for adductor magnus muscle on the back of the femur; patellar trochlea on the femur undeveloped compared to peramelids; emphasis of the fourth pedal ray established, while the hallux was retained; mortise–tenon **UAJ; CCJ** stepped, with the derived extension of the **CaCua** facet.

On the bases of dental, cranial, and postcranial attributes (Figs. 8.33–8.41) I cannot see an alternative to the hypothesis that views the origins of this radiation from some ancient phalangerine in the strict sense. Although I discussed the pedal evidence in some detail in Chapter 7, Aplin (1987, p. 385) noted the following differences of the kangaroos from other phalangeriforms:

> the incomplete ventral and lateral enclosure of the [epitympanic sinus of the squamosal]; the antero-medial displacement of the [caudal tympanic process of petrosal] to closely approximate the posterolateral face of the promontory; the more rounded stapedial footplate; the lack of inflation of the post-tympanic process of the squamosal; the location of the tensor tympani on the anterolateral face of the promontory; and the lack of intimate union between the anterior limb of the ectotympanic and the postglenoid process.

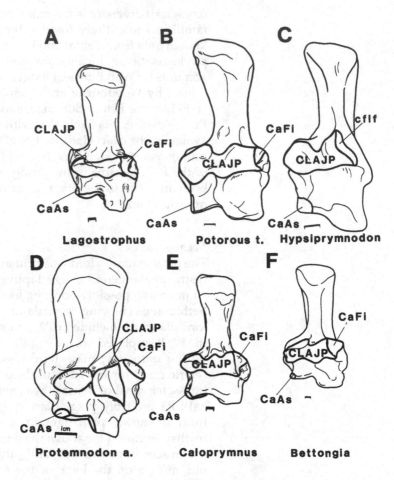

*Figure 8.33. Comparison of left calcanea in dorsal view of representative hypsiprymnodontid and macropodid taxa (Hypsiprymnodon, Potorous tridactylus, Lagostrophus, Protemnodon anak, Caloprymnus, and Bettongia). For abbreviations see Table 1.1. Scales = 1 mm unless indicated otherwise.*

These traits are clearly derivable from a phalangerine.

Flannery (1987) has contributed a balanced analysis of macropodoid character evolution and putative relationships. He notes (pp. 743–4) that

> It is interesting that almost all of the macropodoid synapomorphies observable in the skeleton and dentition seem to center on just two anatomical regions: the anterior cheektooth row and its associated musculature; and the hindlimb. . . . Perhaps this observation indicates that the ancestor of the macropodoid differed from its possum-like relatives in adaptation to a terrestrial lifestyle and in increased emphasis on the strength of premolar shear, possibly in response to utilization of some terrestrial food source. Once certain hindlimb and dental adaptations were gained, they were never completely eradicated in descendant forms, even when the adaptational pressures that led to their development was removed or reversed.

In response to the baffling distribution of characters that are becoming obvious from an ever richer fossil record, and as an anti-

**Figure 8.34.** *Comparison of left calcanea in plantar view of the representative hypsiprymnodontid and macropodid taxa shown on Figure 8.33. Scales = 1 mm unless indicated otherwise.*

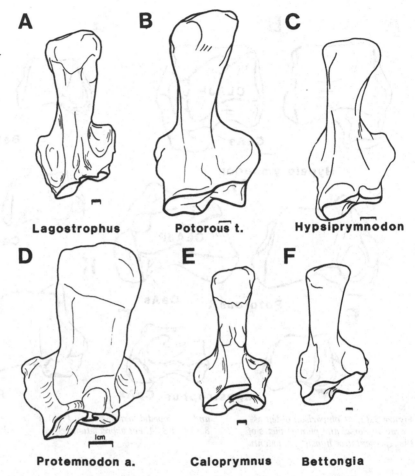

**A** Lagostrophus

**B** Potorous t.

**C** Hypsiprymnodon

**D** Protemnodon a.

**E** Caloprymnus

**F** Bettongia

dote to his essentially cladistic outgroup oriented commitments to the analysis of character polarities, he has noted (p. 744) that "the acquisition of synapomorphies in the stem macropodoid may have pre-adapted descendant species to the development of further traits in parallel." Flannery considered and discussed what to him were the two plausibilities, namely that macropodoids were either the sister group of the Phalanageridae, *sensu stricto*, or that the kangaroos were the sister group of all other phalangeriforms. While either hypothesis may appear to be supported by some similarities, the issues of which characters evolved in what manner, from what antecedent condition, how and why, will form eventually the bases for the most tested taxonomic properties of the stem kangaroos.

In a subsequent and more detailed analysis, Flannery (1989) has presented a detailed analysis of both fossil and living kangaroos unsurpassed to date, apparently relying on cladistic and parsimony-based computer procedures, but with a generous

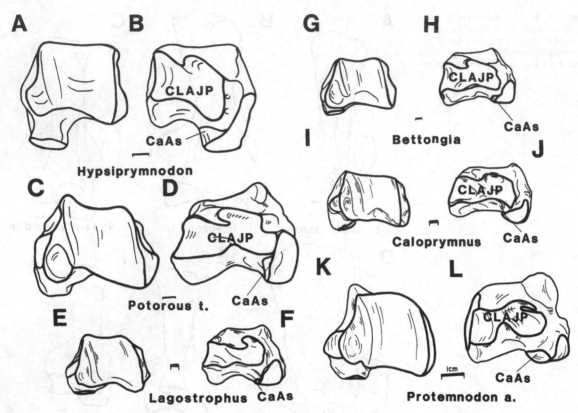

*Figure 8.35.* *Comparison of left astragali in dorsal and plantar views of the representative hypsiprymnodontid* *and macropodid taxa shown on Figures 8.33 and 8.34. For abbreviations* *see Table 1.1. Scales = 1 mm unless indicated otherwise.*

sprinkling of astute and eclectic phyletic assessment of characters rooted in his extensive knowledge of these animals. He has also presented a phylogenetic tree with a narrative account within it. The literature and taxonomic history are presented in extremely useful detail, and a number of hypotheses are clearly enunciated. Flannery's final statement does, however, cast some doubt on the validity of some of the binary coding of the characters and on some of the characters themselves, as well as on the subsequent and largely axiomatic results that tend to follow from parsimony-based algorithms. He notes (p. 43) that

> In overview, the most striking aspect of the evolutionary history of kangaroos has been the remarkable, indeed all pervasive, convergence that characterizes the lineage. . . . Finally it is worth meditating on the fact that without their unsurpassed fossil record, any attempt to unravel the relationships of modern genera based upon parsimony would result in incorrect phylogenies, precisely because of this widespread convergence.

**Figure 8.36.** *Comparison of left calcanea in dorsal view of representative macropodine macropodid taxa* (Lagorchestes leporides, Onychogalea fraenata, Setonyx, Prionotemnus rufus, Petrogale brachyotis, *and* Peradorcas). *Note the* obviously secondary *constriction of the* ***CLAJP****, along with the further fragmentation of the original sustentacular facet. The lower ankle joint is virtually immobile in kangaroos. For abbreviations see Table 1.1. Scales = 1 mm unless indicated otherwise.*

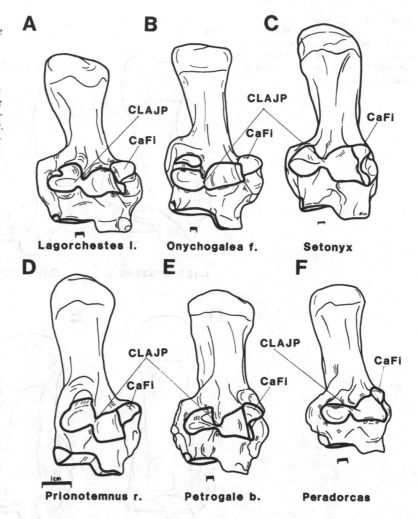

One does wonder that if convergence is so pervasive, was this assessed from the algorithms or from the analysis of the characters? For example, I do not judge the attainment of pedal characters of potoroines and macropodines convergent, as Flannery does, and I consider the two groups taken together to make up a holophyletic Macropodidae. The hypsiprymnodontids are a relatively homogeneous and paraphyletic assemblage of kangaroos, and if their various representatives were to be given cladistically dictated ranks, one would have a plethora of families without any meaningful (and certainly not very reliable) phylogenetic content. Several subfamily level groupings that Flannery endorses I reduce in rank in the classification in Chapter 2 of this volume. In the Hypsiprimnodontidae I include the Hypsiprymnodontinae and Propleopinae, and in the Macropodidae I include the Potoroinae

**Figure 8.37.** *Comparison of left calcanea in plantar view of the representative macropodine macropodid taxa shown in Figure 8.36. For abbreviations see Table 1.1. Scales = 1 mm unless indicated otherwise.*

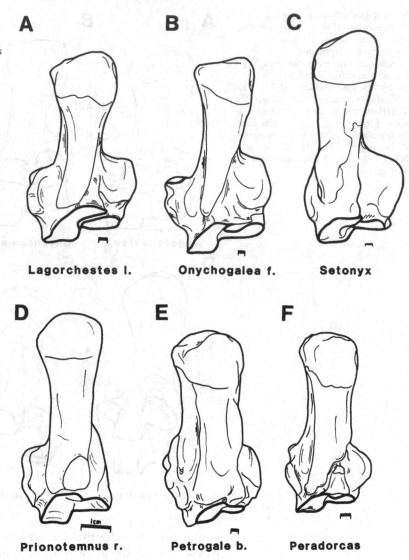

**A** Lagorchestes l.  **B** Onychogalea f.  **C** Setonyx

**D** Prionotemnus r.  **E** Petrogale b.  **F** Peradorcas

(Potoroini and Bulungamayini) and the Macropodinae (Balbarini, Macropodini, Sthenurini, and Dendrolagini).

In my present view concerning the Phalangeriformes, it is obvious that the complex research and a far better fossil record are required to answer questions about the evolution of key traits for the entire skeleton. It seems unproductive, therefore, to try to express in "precise" cladograms the still-unresolved uncertainties about the phylogenetic relationships of the various monophyletic groups within the Phalangeriformes (for additional views see, Archer & Flannery, 1985).

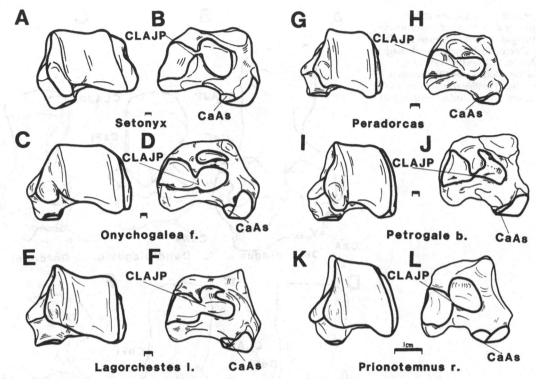

**A** **B**
CLAJP
CaAs
Setonyx

**C** **D**
CLAJP
CaAs
Onychogalea f.

**E** **F**
CLAJP
CaAs
Lagorchestes l.

**G** **H**
CLAJP
CaAs
Peradorcas

**I** **J**
CLAJP
CaAs
Petrogale b.

**K** **L**
CLAJP
CaAs
Prionotemnus r.

*Figure 8.38.* *Comparison of left astragali in dorsal and plantar views of* representative macropodine macropodid *taxa shown in Figures 8.36–8.37. For* abbreviations see Table 1.1. Scales = 1 *mm unless indicated otherwise.*

## Suborder Vombatiformes Woodburne, 1984

### Diagnosis

I have relatively little confidence in the validity of this group. The protovombatiform *may have been* at least partly terrestrial with a retained grasping hallux, a slightly hypertrophied **Mt5**, and a bilaminar bulla. It may have had: quasi-selenodont molars with well-separated mesial and distal moieties, and with well developed stylar cusps and lingual cuspids, resembling *Ilaria*. It may have had the following apomorphies from which the two included semisuborders derived (perhaps in parallel) their shared advanced similarities: relatively robust limbs; the scaphoid (the ancestral scaphocentrale, not the scapholunar) enlarged and the lunate completely eliminated, and consequently a cuneiform slightly more proximal than the scaphoid; slightly trochleated astragalar **ATil** facet.

### Included taxa

Semisuborders Vombatomorphia and Diprotodontiformes.

*Figure 8.39. Comparison of left calcanea in dorsal view of representative dendrolagin and sthenurin macropodid taxa (Dorcopsis hageni, Dendrolagus lumholtzi, D. goodfellowi, D. matschiei, Bohra, and Sthenurus occidentalis). For abbreviations see Table 1.1. Scales = 1 mm unless indicated otherwise.*

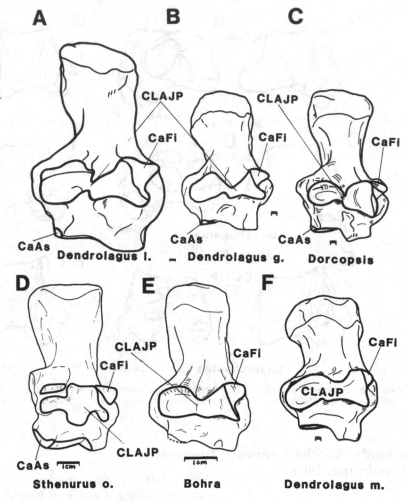

## Distribution

Oligocene to Recent.

## Discussion

The most complete recent discussion of this group is that of Munson (1992). While Marshall et al. (1990) have provided a valuable tabulation of traits they consider synapomorphies for the group and those included within it, Munson has discussed, in addition to both cranial and dental characters, the postcranial features as well. Munson's contribution is particularly valuable in that she clearly differentiates between hypotheses of apomorphies and homoplasies, facilitating future work on the various character complexes. Yet, almost all of the traits carefully noted by her as potential synapomorphies, either for the tenuous concept of the

*Figure 8.40.* *Comparison of left calcanea in plantar view of the representative dendrolagin and sthenurin macropodid taxa shown in Figure 8.39. For abbreviations see Table 1.1. Scales = 1 mm unless indicated otherwise.*

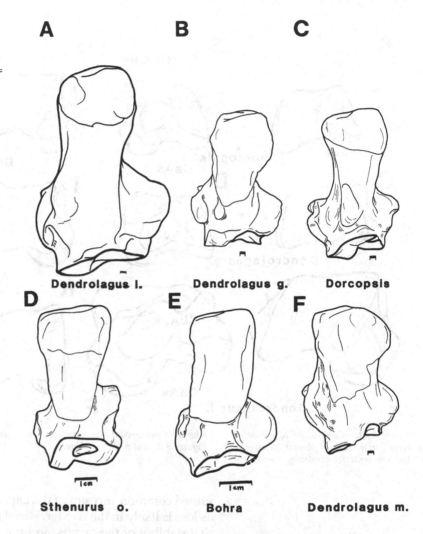

**A** Dendrolagus l.

**B** Dendrolagus g.

**C** Dorcopsis

**D** Sthenurus o.

**E** Bohra

**F** Dendrolagus m.

Vombatiformes or the tenable concept of the Vombatomorphia, appear to be quantitatively continuous minor ones, and therefore could prove to be convergences in two or three independently phalangeriform-derived terrestrial lineages.

The contention of Springer and Woodburne (1989) and Marshall et al. (1990) that the the alisphenoid bulla of phascolarctids represents the primitive diprotodontian condition appears to be unlikely, as I have mentioned. That ear region is one of the most apomorphous basicranial conditions in the Vombatiformes, along with its peculiar herbivorous jaw mechanics without a significantly inflected mandibular angle.

The possible reality of the Vombatiformes, for me, rests on a few relatively weak characters, that may not stand up to future tests probing their reliability as taxonomic properties. The lunate is already reduced in the living phalangeriforms and also in the pre-

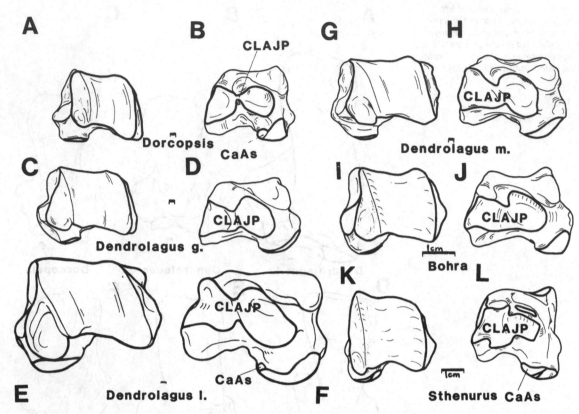

**Figure 8.41.** Comparison of left astragali in dorsal and plantar views of the representative dendrolagin and sthenurin macropodid taxa shown on Figures 8.39 and 8.40. For abbreviations see Table 1.1. Scales = 1 mm unless indicated otherwise.

sumed common ancestor of the Diprotodontia. So the potential for its loss is likely in the two terrestrial groups; in these the compressive stability of the carpus, and prevention of injury from such a small carpal element that may have concentrated point loads might have been at a premium. As noted elsewhere, a distinctive sulcus of the astragalar **ATil** facet occurs in other terrestrial groups as well – this character also might be convergent and not homologously derived.

## Semisuborder Vombatomorphia (Aplin & Archer, 1987)

### Diagnosis

Protovombatomorphian was a terrestrial animal with a retained grasping hallux. It had: basicranium construction that was bilaminar either advanced beyond or similar to the partially bilaminar condition of phalangerines, probably with a prominent tympanic process of the alisphenoid under the middle ear and exposed ventrally; firmly fused mandibular symphysis; molar morphology which consisted of occluding W-shaped ectoloph with ectolophid, and lophodont protocone and metacone with lophodont metaconid and entoconid; stepped, ancestral diprotodontian tarsal articulations.

## Included taxa

Superfamilies Phascolarctoidea and Vombatoidea (including Ilariidae and Vombatidae).

## Distribution

Miocene to Recent.

## Discussion

The living wombats and the koala, both tailless and with backwardly opening pouches, are the only representatives today of a once major Australasian radiation of the Vombatomorphia (see Archer, 1976a, e, 1977a, 1984c; Dawson, 1980; Haight & Nelson, 1987; Harding, 1987; Harding et al., 1987; Pledge, 1987b, c; Woodburne et al., 1987c). I emended this concept, from which I exclude the diprotodontiforms and include in it the phascolarctids. In addition to the terrestrial Peramelina, which had its origins from a syndactylan arboreal ancestor, the known Vombatomorphia (Figs. 8.42—8.46) also originated as a terrestrial, or semiterrestrial, radiation from either a common vombatiform ancestor or some putative member which could be classified either as an early phalangeriform or possibly a diprotodontiform wynyardiid. I may add here that the origin of ground squirrels from tree squirrels was probably a somewhat analogous transformation to that of the origins of the first vombatimorphian, the size discrepancies notwithstanding.

By all accounts (see Tedford & Woodburne, 1987), *Ilaria* is firmly allied with the vombatids according to Munson's (1992) detailed analysis of the postcranial evidence. Among the probable synapomorphies of the Vombatoidea, Munson believed the following to be convincing: scapula with a concave medial surface; smoothly concave distal facet of the magnum; trapezoid facet of second metacarpal is directed posteriorly; the expanded proximal end of the second metacarpal is concave; the facet for magnum of the third metacarpal is convex; and the facet for the second metacarpal is continuous with the facet for the magnum. A number of other traits listed by Munson, while shared apomorphies of the two groups, are not convincing synapomorphies.

In spite of their grasping hind feet (Fig. 7.74) there are enough details known of the surface biology of koalas, of the stomach and sperm structure (Hughes, 1965; Temple-Smith, 1987), and of the

*Figure 8.42. Comparison of left calcanea in dorsal view of representative taxa of vombatids (Vombatus, Lasiorhinus), phascolarctids (Phascolarctos), and phalangerids (Trichosurus vulpecula, Phalanger vestitus, and P. gymnotis). For abbreviations see Table 1.1. Scales = 1 mm.*

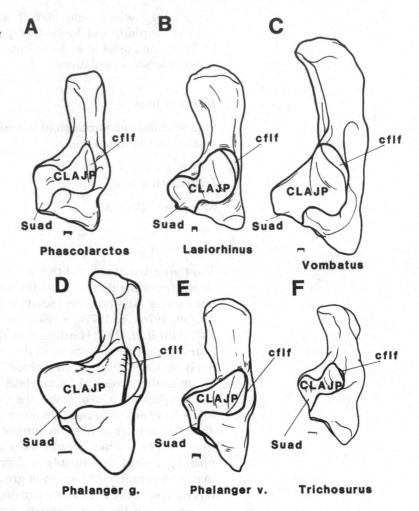

**A**    cflf    CLAJP    Suad    **Phascolarctos**

**B**    cflf    CLAJP    Suad    **Lasiorhinus**

**C**    cflf    CLAJP    Suad    **Vombatus**

**D**    cflf    CLAJP    Suad    **Phalanger g.**

**E**    cflf    CLAJP    Suad    **Phalanger v.**

**F**    cflf    CLAJP    Suad    **Trichosurus**

tarsus that strongly corroborate the hypothesis that these slow and clumsy arborealists are the *living* sister group of the wombats, and that they are the descendants of a primarily terrestrial ancestry (see Pledge, 1987a, for references on fossils). The extreme specializations of the ear region in both wombats and koalas, as well as the secondary vombatomorphian arboreality of koalas, make it very difficult to judge their most recent affinity. The two living groups, however, are extremely different in their dental morphology, the wombats having evolved the highly derived, rootless, hypselodont, ever-growing dentition adapted to grazing.

The basicranium is also widely different in the two living groups of vombatiforms. The articulation of the glenoid and the structure of the articular condyle of vombatids is reminiscent of the condition seen in pigs, which also lack a postglenoid process and have a transversely wide and very narrow arching contact between the mandible and the skull. Vombatoids have an articular

**Figure 8.43.** *Comparison of left calcanea in plantar view of representative taxa shown in Figure 8.42. Scales = 1 mm.*

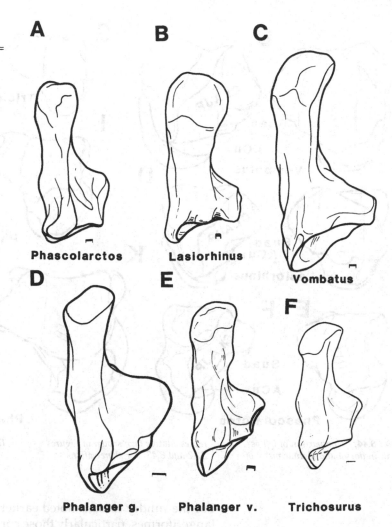

**A** Phascolarctos

**B** Lasiorhinus

**C** Vombatus

**D** Phalanger g.

**E** Phalanger v.

**F** Trichosurus

condyle that is highly derived, mediolaterally elongated, and arc shaped, and the glenoid eminence is correspondingly concave and narrow without any posterior bony buttressing. The medial downward extension of this eminence may represent the homologue of a postglenoid process. Aplin (1987, p. 389) believes that the concept Vombatomorphia (Aplin and Archer, 1987), a group in which they include the Diprotodontiformes of this volume, but from which they exclude the Phascolarctidae, can be characterized as follows:

> development of a squamosal tympanic process as the primary bullar element; presence of an entoglenoid extension of the articular eminence; and elimination of the postglenoid emissary vein with associated rerouting of dural drainage into the internal jugular vein. . . . [;] squamosal epitympanic sinus of rather variable relations and likely multiple origins. . . .

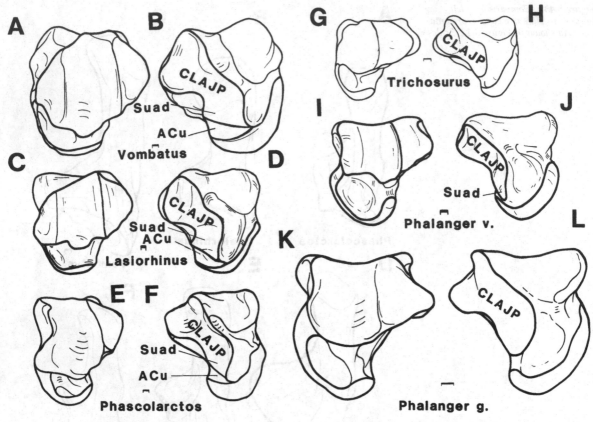

**Figure 8.44.** *Comparison of left astragali in dorsal and plantar views of* *representative taxa shown in Figures 8.42 and 8.43. For abbreviations see* Table 1.1. Scales = 1 mm.

For the sundry reasons noted earlier in the discussion of the Phalangeriformes, particularly those concerning the hypertrophy and pneumatization of the bulla, I do not accept this view or that of Springer and Woodburne (1989). The basicranial morphology of the koala is far more derived in many respects than the mere indication of the degree of alisphenoid posterior extension would imply. The latter may well be an ontogenetically mediated consequence for the very large bulla of koalas. The relative contribution of the alisphenoid (see Fig. 3.1) is only one isolated mosaic in the complex issue of phyletic transformation that must be considered within the context of ontogeny, the functional demands of the skull, and the adaptive bioroles performed by the feeding mechanism. If such a large bulla is to be constructed during ontogeny, then the hypertrophy of any one of the bones contributing to it (and the alisphenoid is the most common bullar element in marsupials) would, because of its speeded-up extension, exclude other elements from the adult bulla that were present in the

**Figure 8.45.** *Comparison of left calcanea in dorsal view of representative diprotodontiforms (Ngapakaldia, pal-orchestid,* Zygomaturus, *and* Thylacoleo) *with a phalangerid (Pha-langer vestitus) and a vombatid (Lasiorhinus). For abbreviations see Table 1.1. Scales = 1 mm unless indicated otherwise.*

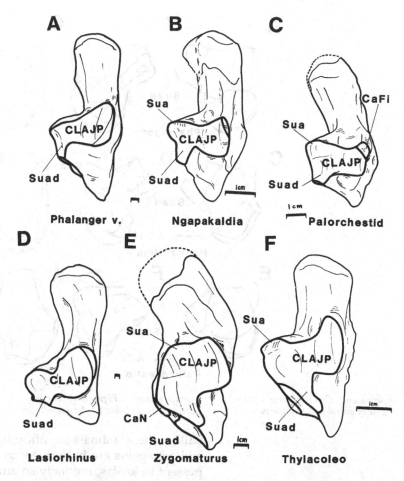

antecedent evolutionary stages. My disagreement with their interpretation of the cranial evidence is by no means the only one contending view, as the highly analytical contribution of Murray et al. (1987) suggests.

Wombat traits (see Flannery & Pledge, 1987, and Pledge, 1987b, for references on fossils) should be viewed in the context of their terrestriality in order to construct corroborated taxonomic properties. Murray et al. (1987, p. 460) consider the following cranial attributes to be unique to these animals:

1, extreme lateral extension of open epitympanic sinuses into squamosal; 2, loss of postglenoid process; 3, narrow mastoid strip bounded laterally by invasion of squamosal onto occiput; 4, reduced antemolar dentition and open rooted premolars and molars; and 5, reduced or absent lacrimal foramina.

Postcranial taxonomic properties are also of considerable significance. In addition to greatly reducing, but not losing, the grasping

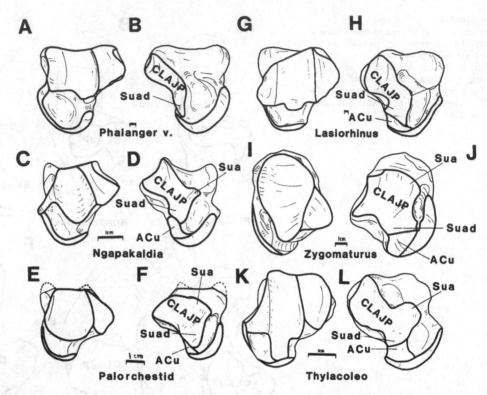

**Figure 8.46.** *Comparison of left astragali in dorsal and plantar views of groups shown on Figure 8.45. For abbreviations see Table 1.1. Scales = 1 mm unless indicated otherwise.*

hallux, these animals significantly altered the **CCJ**. The cuboid and astragalus are in derived articular contact, a condition also present in koalas, uniquely so among arboreal syndactylans. As expected in terrestrial mammals, the wombats secondarily increase the angle between the **ATim** and **ATil** facets in the **UAJ**. A number of other traits discussed under the analysis of pedal traits in Chapter 7 and under **Australidelphia** and **Phalangeriformes** above in this chapter are pertinent to this section.

The peculiarity of the relatively uninflected mandibular angle of *Phascolarctos* noted above is due to the strong developmental constraint (which is phylogenetic at its point of origin) imposed by the original relationship of the angle with the bulla in marsupials (see Maier, 1989a,b, for a detailed account of the primitive marsupial condition and its ontogenetic history). The hypertrophy of the pneumatized koala bulla, which may be responsible for the small gape of these animals, for exact developmental–functional or biorole reasons as yet unknown, has forced the angle laterally and downward in phascolarctid evolution – in effect eradicating the inflection of the adult angle of the mandible (a fact also noticed by Abbie, 1939), so pervasively retained in other marsupials.

# Semisuborder Diprotodontiformes, new

## Diagnosis

Protodiprotodontiform had: either primitive syndactylan or wynyardiid-like upper molars and a pedal morphology closely similar to that of some living phalangerids, but with an apomorphous medial extension of the **Su** facet (a **Sua** facet) on the astragalus and calcaneus. The vestigial presence or complete absence of a reduced lunate (or of any other small bone) in this ancestry cannot be established as yet because of the nature of the fossil evidence.

## Included taxa

Superfamilies Wynyardioidea, Thylacoleonoidea, and Diprotodontoidea (including Palorchestidae and Diprotodontidae).

## Distribution

Oligocene to Recent.

## Discussion

There are no living representatives of this once literally weighty radiation of Australasian marsupials. Yet the wynyardiids, palorchestids, diprotodontids, and thylacoleonids appear to be monophyletic to the exclusion of other vombatiforms. In a character analysis oriented discussion of the relationships of cranial anatomy of most known diprotodontians Murray et al. (1987) aimed to ascertain thylacoleonid ties. The list of characters they give for the common ancestry of the Thylacoleonidae, as they note (p. 460), probably includes characters which may have been present in the ancestral vombatiform or phalangeriform. In future accounts of the skull, the possible relevance of diprotodontoids to either vombatiforms or thylacoleonids will be undoubtedly also entertained. Murray (1990a) has recently contributed a well-tested hypothesis of transformation of diprotodontoid molar structure from a wynyardiid-like condition.

The study of the remains of the Wynyardiidae is essential for the understanding of the diprotodontiform radiation. The detailed analysis of the cranium and postcranium of the tantalizing genus *Muramura*, recently named by Pledge (1987b), will undoubtedly place all comparisons of diprotodontians in a fresh perspective. Until it is studied, the various views surrounding the significance of this family remain anchored to the poorly known *Wynyardia* itself. Haight and Murray (1981) noted the close (but clearly primitive diprotodontian) similarity of the external brain morphology, as seen on an endocast, of *Wynyardia* to such phalangerids as *Trichosurus*. Murray's (1990a,b) recent transformational analyses of molar structure strongly suggest that the wynyardiids may be close, and primitive, diprotodontoid relatives.

The attribute from the tarsus that suggests the reality of the Diprotodontiformes is the conformation of the **LAJ**. A unique extension of the sustentacular articulation among syndactylans, the **Sua** facet, is present in the diversely sized (and adapted) diprotodontoids, suggesting that this feature is homologous in the families. In addition to a well-developed **ACu** contact, the medial and distal extensions of the sustentaculum, the **Sua** and **Sud** facets respectively, closely link thylacoleonids with the best known of the primitive palorchestids, *Ngapakaldia*, and all other members of that family, as well as with the diprotodontids (Figs. 7.80–7.84, 8.45, & 8.46).

The Diprotodontoidea probably stem from an ancestry with fully lophodont cheek dentition, as long recognized (see Archer, 1984b; Marshall et al., 1990; Murray, 1990a,b). So it is possible that the incipient lophodonty of stem vombatomorphians may represent the stem condition for the diprotodontiform cheek teeth as suggested by Marshall et al. (1990) and Munson (1992). In spite of their molar similarities, however, there appears to be a consistent difference in the known calcaneal morphology of palorchestids and diprotodontids. The smaller and generally less modified palorchestids develop a **CaFi** facet that is absent in known diprotodontids. The latter probably display the advanced diprotodontiform condition in this regard. The discussion of the pedal traits in Chapter 7 is closely relevant to this section.

Given the constraints of syndactyly and of large body size, the diprotodontids have evolved one of the most bizarre feet among mammals (Fig. 7.74). Yet in spite of the large size of these spectacular beasts and the attendant modifications of the tarsus, it appears to me that the close special similarity of the tarsals of *Ngapakaldia* and *Thylacoleo* (unique **LAJ** attribute of a **Sua** facet) signal monophyly. The derivation of thylacoleonid molars, however, is not understood as yet. For a review of the ongoing appraisal of inter- and intra-thylacoleonid affinities, Archer and Dawson (1982), Archer and Rich (1982), Murray et al. (1987), and Murray and Megirian (1990) should be consulted.

## Suborder Yalkaparidontia (Archer et al., 1988)

*Diagnosis*

This diagnosis is slightly emended after Archer et al. (1988). Only known yalkaparidontian genus, *Yalkaparidontia*, has: **I1,2,3/1; C1/0; Pd1,d2,3; M1,2,3** dental formula; extremely zalambdodont cheek teeth, the lowers lacking traces of a talonid; enlarged and procumbent most anterior upper and lower incisors evergrowing; absence of squamosal epitympanic sinus; ossified components of bulla from alisphenoid and petrosal; foramina for transverse canal posterior to carotid foramen; foramen pseudovale; tiny postglenoid process.

## Distribution

Middle Miocene.

## Included taxa

Yalkapridontidae.

## Discussion

In the original recent description of this new kind of exciting Australian mammal, Archer et al. (1988) give a thorough analysis of the relationships, dental function, and possible feeding habits of this cranially phalangeriform-like animal. This similarity to phalangeriforms is evident at least in general skull conformation and the anterior dental pattern. In light of the developmental dynamics of the compound bullae of marsupials, it is conceivable to me that the basicranial morphology of *Yalkaparidon* represents a near primitive condition for that suborder. Unlike the describers, I consider the most recent ties of this ancient group to be with diprotodontians and not with peramelids. It is best classified here as a suborder of the semiorder Diprotodontia rather than as an order.

# 9

# Paleobiogeography and metatherian evolution

> *Nowhere in biology is the misleading and arbitrary use of 'parsimony' so disabling as in vicariance and panbiogeography, with their determined exclusion of virtually all biological data on vagility, reproductive strategy, niche-width, and so on.*
>
> Kirsch (1990, p. 162)

> *I have attempted to sketch how these oppositions (selection v. orthogenesis, ecology v. history) might be resolved if dichotomous views of causation are abandoned for a more complex view emphasizing interacting mutually contingent networks of causes. In practice achieving such a synthesis will require a breakdown of the conceptual and methodological boundaries between biogeography, evolutionary biology, systematics, ecology and geology.*
>
> Gray (1989, p. 804)

Few constraints limit and guide the nature of macro- and megaevolutionary patterns of terrestrial organisms as decisively as the nature of distribution, isolation, and connectedness of various land masses. Such important geographic factors are partly behind climatic and subsequently the environmental change that drives faunal extinctions and new ecological opportunities through ecological release (Van Valen, 1978), in other words opportunity-based adaptive evolution. These are some of the reasons why few subjects have engendered as much interest in a wide variety of subdisciplines of natural history as the geographic enigma of marsupial distribution and evolution through time and space. Given the background of the Cretaceous spread of the angiosperms (Brenner, 1976) and the faunal upheavals at the beginning and end of the Eocene in the northern hemisphere (Stucky, 1990; Krause & Maas, 1990, etc.) the occurrence of marsupials in space and time has been one of the knottiest of puzzles in mammalian paleobiogeography. The occurrence of these animals apparently by happenstance, now on all continents, has stimulated many attempts to make sense of their history or use their record for or against specific limiting paradigms of biogeography. Geologists (e.g., Jardine & McKenzie, 1972), paleontologists (a long and venerable tradition of marsupial studies in space and time, recently summarized by Marshall, 1980a), mammalogists (e.g., Fooden, 1972, 1973), and taxonomic theorists (e.g., C. Patterson, 1983), beginning with Darwin, have concerned themselves with marsupial phylogeny and global geography. Plate tectonic theory, however, has revolutionized and excitingly complicated or simplified (depending on the data available) the understanding of evolutionary history. It is appropriate, therefore, to cite in this section a passage that provides the geohistorical framework of today as a "mobilist" as opposed to a "stabilist" position for biogeographical analysis. As Jardine and McKenzie (1972, p. 20) stated:

> Since Darwin and Wallace, biologists have emphasized the crucial importance of geographical isolation in speciation and subsequent adaptation and differentiation of species. Many land animals and plants cannot cross an ocean, which therefore acts as a barrier to their dispersal. Thus continents act as boats whose passengers are organisms and during the period in which they are passengers evolution can take place. A simple rearrangement of landmasses will affect the evolution and dispersal of organisms. When two landmasses separate, the passengers of each become isolated from potential competitors which evolve on the other, so that they have a better chance to adapt and differentiate themselves to exploit their changing environment. When a landmass approaches and comes into contact with another the passengers of each can disembark, compete and differentiate in the territory of the other.

It has been emphasized throughout this volume that it is singularly pretentious to assume that the rigorous practice of any one taxonomic technique based on taxonomic rather than tested evolutionary assumptions can be considered foundational for not only phylogenetic history but also for (vicariance) biogeography. Contributions to the fields of panbiogeography, primarily concerned with taxonomic patterns and cladistic or vicariance biogeography, with its main concern for relationship of areas, are often nearly impossible to distinguish. It is of little relevance to this discussion that the two schools are at theoretical loggerheads with one another. Both approaches use atemporal "tracts" of Croizat (either "generalized" or "standard") and inasmuch as they consider vicariance as a deductive mode of explanatory panacea, they represent not only an incomplete but also a flawed perspective. Vicariance *may be* part of the H-N E for marsupial distribution, but dispersal of these animals appears to be a dominant theme.

In the following account of the geographic dispersal, evolution, and isolation of groups of marsupials I will avoid the term "scenario," as hypothesis is the proper designation, one which is tested by specific, available evidence. Explanations of past and present geographic distributions are H-N Es like any other explanatory account of biological or geological history (Bock, 1981). This, of course, is not story "telling," but testing. In the parlance of the cladistic techniques the term "scenario," as discussed in Chapter 2, is something to be derived from the "objective pattern" (i.e., based on distribution analysis alone). It is a term that has become heavily agenda laden in order to give "primary" scientific merit to cladistic analysis over all the other methodologically equally evolutionary (phylogenetic) endeavors related to natural history. This hierarchically ordained primacy of the evidence, both in phylogenetic and geographical studies, mistakenly described by some as rigorous (when in reality it is merely axiomatized, and in practice, often faulty and inflexible assumptions predominate, driving this formalism to often untenable and unreasonable conclusions), adheres to the belief that patterns can be established independent of a mechanism- and process-based understanding. For different reasons, the notion of a "model" for specific H-N Es is incorrect and has been discussed in detail elsewhere in the literature.

Extreme practitioners of cladistic techniques have maintained and attempted to proselytize that fossils should not be treated differently from Recent organisms (but see also Schaeffer et al., 1972). Yet obviously the new morphological evidence and data fossils provide in space and time are of the essence for both phylogenetics and subsequently in paleogeography, the formalist rhetoric of a few extremists notwithstanding. Paleontology is at the core, along with the neontological information, of not only the analysis of the evidence for phylogenetics but also for biogeography.

In a curious display of mixing evidence, coupled with an overlay of methodological circularity, marsupials have been used to create an "area cladogram," and thus demonstrate the virtue of formalized "vicariance" biogeography and the unimportance of the fossil record (C. Patterson, 1983), an approach that purports to do without centers of origins or dispersal (the sympatry of tracks, which are in fact postulated, being a sort of evidence for the latter). The special properties of taxa on which Patterson's account was based appear to be primitive metatherian and ameridelphian features. It does not matter, given our relatively poor level of understanding, that one means, in this case, either "phyletically" or morphologically primitive; the time of branching off for both the European and South American metatherian derivatives was probably latest Cretaceous to earliest Paleocene, and the degree of relative morphological primitiveness of these ancestries within the Metatheria (whichever way it is diagnosed) was probably comparable. The incomplete geographical and phenetically ordered morphological information available was simply taken as primary, in spite of the alleged use of purely cladistic procedures. The purportedly objective data base for that cladistic effort was in fact a nonsense phylogeny of marsupials.

The merits of vicariance biogeography continue to be hailed perhaps excessively in spite of the widespread rejection of Croizat's (1958, 1964, 1982) approach to historical biogeography.

Yet examples where an author relies on phylogenetically equivocal evidence and equally problematic interpretations in the primary literature, such as the hypothesis of marsupial biogeography presented by C. Patterson, are not uncommonly used to bolster narrowly conceived approaches to paleobiogeography. Another tack was taken by Nelson and Platnick (1981) in their tome on cladistics and vicariance biogeography. These authors derive biogeographic theory not from the attempted total understanding of cause and mechanisms steeped phylogeny and a similarly explanation based geophysical and paleoecological accounts that drive paleogeographic reconstructions, but from parsimoniously conceived cladistic schemes that are pruned of the uncomfortable reality of contracting and expanding paraphyletic stem groups. The extent of speciation and dispersal abilities of these paraphyletic groups often differ markedly. This real world is ignored. There are no examples given in Nelson and Platnick (1981) of actual records of lineages of organisms in a time–space continuum that may be used as tests or illustrations of their formulations of vicariance biogeography. Their exclusive use of such simple models rather than real examples underscores the inapplicability of such philosophy to actual H-N Es. As Briggs (1984) and others so aptly conveyed, the origins and dispersal of particular kinds of organisms, together with independently and often variously caused geographical subdivisions of biotas (vicariance events) and geography, are the facts of evolutionary history. Such a synthetic and pluralistic perspective as that expressed by Briggs (1984) or Gray (1989) is unlikely to become "obsolescent," or "eliminated," or "digested" by increasingly narrowly focused techniques heralding the false dawn of new understanding. In any H-N E, but particularly in such a synthetic area of natural history as paleobiogeography, all the knowable details that the analysis of the geological and biotic evidence can offer are extremely relevant. And even, as Simpson (1953, p. 61) plainly stated, beyond the facts of geological information, "The numerous different theories of paleogeography are susceptible to crucial testing by the facts of biogeography."

## Paleogeography

A review of Mesozoic physical geography (Hamilton, 1983) and climates, with a summary of the mammal faunas, is provided by Lillegraven et al. (1979b), as known up to that time (see also other contributions in Lillegraven et al., 1979a). Although our understanding is still unfocused, the origin and latitudinal spread of angiosperm floras form a critical background for the Laurasian radiation of tribosphenic mammals (see Hughes, 1976; Wing & Tiffney, 1977, and references therein). The separation of Laurasia and Gondwana throughout the Cretaceous by oceanic barriers has been well established (Fig. 9.1), and the recent paleontological advances in the Cretaceous of South America (Bonaparte et al., 1986; Bonaparte, 1987, 1990; Bonaparte & Pascual, 1987) provide strong evidence of endemism in the terrestrial fauna not only in the Mesozoic, but also early in the Paleocene (Muizon, 1991; Marshall & Muizon, 1992). The absence of tribosphenic mammals in the Los Alamitos fauna, which is Campanian in age (Bonaparte, 1990b), appears to support the hypothesis that there were no tribosphenic mammals in the southern continents prior to the Maastrichtian in the Late Cretaceous. This Patagonian fauna stands in stark contrast to the early Paleocene Tiupampa assemblage from Bolivia (Marshall and Muizon, 1988). The extremely fragmentary (?)tribosphenic mammal from the Bagua Late Cretaceous of North Peru (Mourier et al., 1986) suggests the presence of Theria in South America, and hopefully this record will be elaborated on in the near future.

It is evident from the discussions in the literature by Lillegraven (1974), Lillegraven et al. (1979b), and Kielan-Jaworowska (1982), or the various rapidly evolving paleogeographic reconstructions of the world, however, that agreement is still far away about the extent and actual physical or ecological connection during the Late Cretaceous of the various land masses (Asiamerica, Euramerica; the nature of the archipelago of Europe; Africa) or faunal provinces (western North American and Asian within the Asiamerican one). Critical geophysical and paleontological evidence is wanting, particularly for most of the Cretaceous of the more southern, moister, latitudes of Asia, for the Paleogene of Africa, and for the Cretaceous of the European archipelago. What makes matters particularly difficult for both the phylogeny and distribution of mammals in the Cretaceous is that the angiosperm floras do not become spectacularly dominant in the middle and northern latitudes of the Northern Hemisphere until the early Late Cretaceous (e.g., Brenner, 1976). If the rise of the tribosphenics, the therians, was tracking the radiation of the angiosperms, then the earliest and most southern of these radiations from the Early Cretaceous will remain difficult to recover due to the restrictions placed on paleontology by vegetational zonations of today.

Many careful, relatively recent accounts of marsupial paleobiogeography (see in particular Lillegraven, 1974; Marshall, 1980a; Kielan-Jaworowska, 1982; Clemens et al., 1989, etc.) have already explored the various general probabilities related to *their diagnoses of what a metatherian is* and the plate tectonic evidence. I will briefly review below some of the numerous components of these hypotheses and will attempt to place the phylogenetic conclusion I have reached in this study into a historically updated paleogeographic framework. The map outlines used in Figure 9.1 are redrawn and modified from Briggs (1987), but as noted, such reconstructions are steadily evolving due to the refinement of

**LATEST CRETACEOUS-PALEOCENE**

**EOCENE**

(only Metatheria and Monotremata are shown)

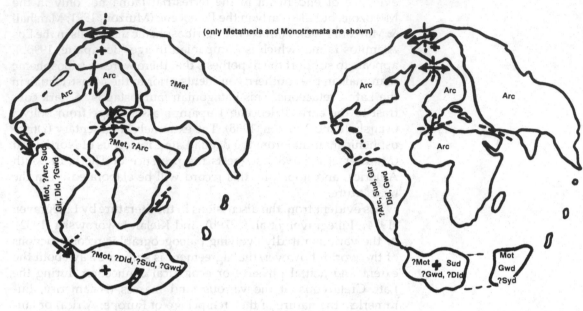

**MID-CRETACEOUS**

LAURASIA: tribosphenic and atribosphenic mammals

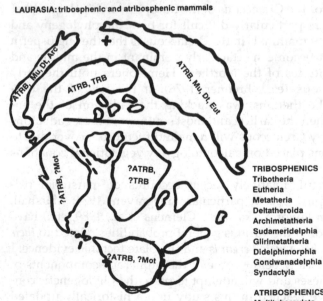

GONDWANA: atribosphenic mammals

**LATE CRETACEOUS**

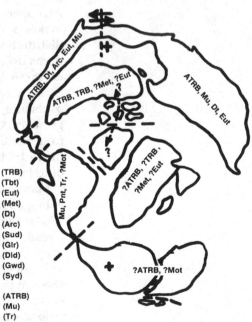

| TRIBOSPHENICS | (TRB) |
| Tribotheria | (Tbt) |
| Eutheria | (Eut) |
| Metatheria | (Met) |
| Deltatheroida | (Dt) |
| Archimetatheria | (Arc) |
| Sudameridelphia | (Sud) |
| Glirimetatheria | (Glr) |
| Didelphimorphia | (Did) |
| Gondwanadelphia | (Gwd) |
| Syndactyla | (Syd) |

| ATRIBOSPHENICS | (ATRB) |
| Multituberculata | (Mu) |
| Triconodonta | (Tr) |
| Pantotheria | (Pnt) |
| Monotremata | (Mot) |

available evidence. These distribution maps show occurrence (or lack) of mammals that can be positively identified as being nontherian (atribosphenics), merely therian (tribosphenics; with the attendant uncertainty as to affinity within the Theria), tribotherian, metatherians of various sorts, or eutherian (to contrast and show the therian faunal context of metatherian distribution in the Cretaceous). This summarizes the currently known and interpreted evidence – the pattern of fossils and their taxonomic allocation, which is critically dependent on rigorous testing and understanding of their taxonomic properties. As the evidence, the fossils, becomes more abundant and diverse, interpretation of the taxonomic properties and the phylogenetic relationships of these taxa will improve. Both the increased abundance and diversity, as well as the hypotheses of relationships tested against these, will in turn further test the H-N Es of paleobiogeography.

Although the pertinent geological literature is rapidly growing, taken alone, without paleobiogeographical information, it cannot adequately explain the history of metatherian distribution. Briggs (1987) presented an excellent general account of the relationships of geographical and faunal realms to plate tectonics. McKenna's (1973) contribution to conceptual analysis in "mobilist" biogeography was an important application and extension of concepts developed in a "stabilist" perspective by W. D. Matthew and

**Figure 9.1.** *Paleogeographic relationships of landmasses, hypotheses, and known records of distributions, probabilities of dispersal, and possible vicariance events of metatherian groups. The continental relationships of the four time segments depicted, redrawn and modified from Briggs (1987; Lambert equal-area projections), represent probably only crude approximations which will continue to be improved by both geophysical and paleontological data. The two lower maps show the "mid"- and Late Cretaceous and the known, and assumed or probable (indicated by "?"), distribution of therian (tribosphenic mammals only) and nontherian (atribosphenic) mammals. The two upper maps show the latest Cretaceous-Paleocene and the Eocene. Exact time (while extremely important for both evolutionary and paleobiogeographical accounts) is not available for the critical dispersal events of deltatheroidan and probably other holarctidelphians from Asia to North America or vice versa. In light of Cretaceous deltatheroidan and asiadelphian metatherian presence in Asia, the origin of Archimetatheria in North America*

*cannot be unequivocally assumed. The Holarctic origin of the Theria, Metatheria, and Eutheria is perhaps more probable than competing hypotheses at present. The possibility, unlikely as it appears now, of African origins of metatherians, and their dispersal at widely different times to give rise to the stocks in Asia, North America, and South America cannot be excluded. Late Cretaceous vicariance as an explanation for Euroamerica-Westafrica distribution of marsupials (also assumed by some for Europe and Africa in the Cretaceous) is possible, but not considered likely. Vicariance as an explanation for the distribution of American marsupials, in light of the South American Cretaceous mammal faunas, is improbable. Archimetatherians probably dispersed from Central America or Southern North America in the latest Cretaceous. Marsupials dispersed into Antarctica and Australia from South America probably in the early Paleogene. Origin and dispersal of the peradectines northward in the early Paleogene in North America, thence to Europe and Asia cannot be ruled out.*

*The map for the temporally long Eocene is more of an average after the earliest Eocene than a precise summary. The widely recognized North American and European faunal exchange (a corridor) in the latest Paleocene–earliest Eocene (but not later between Europe and North America), for example, is shown by the double headed arrows, as are the Late Cretaceous, and latest Paleocene-Eocene exchanges indicated between Asia and North America. Solid lines represent hypothesized barriers, whereas broken lines suggest either ecological or geographical filters, or possibilities for sweepstakes (chance) dispersal. Arrows indicate dispersal. Definitions of the abbreviations in parentheses are as shown on the bottom of the figure. The informal designations as tribosphenic as opposed to atribosphenic, while useful, also indicate uncertainty regarding the more precise taxonomic allocation of these groups or a postulate of such groups in a given area and time interval. Asiadelphia is not shown; the group was recognized after this figure has gone to press.*

G. G. Simpson. This extension grew directly from the understanding of the mechanisms responsible for plate tectonics. While plate tectonic theory is not universal in its appeal (see papers in Chatterjee & Hotton, 1992), it is significant that the newly established geological "rules of correspondence" for plate tectonics set the framework for an increasingly refined understanding of paleogeography. This is equivalent in significance to the realization that the functional–adaptive understanding of features of organisms, fundamentally mechanisms-dependent, should begin to set the limits and point the way for their usefulness for the construction of real phylogenies. In the case of both paleogeography and phylogenetics it is *causal analysis* based on the application of mechanisms (both causal and consequence laws) which make the activities properly scientific – i.e., both processual and historical.

The broadly conceived (i.e., still poorly known) paleogeography depicted in Figure 9.1 summarizes, for the purposes of this brief discussion, the available paleogeographic contingency throughout the Cretaceous and Paleogene. It appears that North and South America were isolated during the period from the late Jurassic to nearly the end of the Cretaceous. As Lillegraven (1974, p. 278) stated:

> During most of the earliest Cretaceous marine barriers, both oceanic and epicontinental, developed to the point that land vertebrate dispersal between virtually any of the present continental masses would have been unlikely. Exceptions probably were: (a) South America-Africa, and (b) South America-Antactica-Australia. The resulting prolonged isolation of land vertebrate assemblages on the various continents, each with its own ecological peculiarities, would have been conducive to the development of pronounced continental endemism with different "experimental" morphological and physiological adaptations developing in the different areas. The restriction of land vertebrate intercontinental exchange became even stronger through the early Late Cretaceous with rapidly widening oceanic basins (e.g. final separation of South America and Africa) and the attendant increase in extent of epicontinental seas. During the last part of the Late Cretaceous, however, opportunities for intercontinental exchange increased, especially between North America and Asia and possibly between North and South America.

Yet there continues to be a range of serious uncertainties of information about the time of separation of these landmasses. Added to the problem of geological evidence so critical in paleogeographic reconstruction is the ever-present probability of either filter route dispersal across ecological barriers or of chance (sweepstakes) dispersal between slowly separating or approaching landmasses, some "bridged" by island chains (Simpson, 1953, 1965). These continuing doubts, uncertainties, and other discrepancies related to geological information make the morphological and geographical components of paleontological information particularly valuable, independent of the geological evidence. McGowran (1973), Lillegraven (1974), P. V. Rich (1975), Raven and Axelrod (1975), Hallam (1981), Zinsmeister (1982), Rich and Rich (1983), and later Woodburne and Zinsmeister (1984a, b) have examined in some detail different aspects of the Late Cretaceous–Paleogene paleogeography in general, as well as geologic history of the southern circum-Pacific. Briggs (1987) examines in detail the relationships between biogeographical evidence and plate tectonics for all of the continental landmasses. More recently Gheerbrant (1990) contributed to the understanding of the Cretaceous–Paleogene biogeography of Africa and the northern landmasses of that time period. Zinsmeister (1982), and Woodburne and Zinsmeister (1984a, b) in their detailed historical–narratives deal with the critical issue of the final breakup of Gondwana, with obvious bearing on marsupial distribution on the southern continents. In their analysis and summary of the literature, Woodburne and Zinsmeister (1984b) postulate the following: (1) Late Cretaceous continuity from South America to Australia, without the artifactual breaks between crustal blocks, suggesting continuous overland connections; (2) final cessation of dispersal probabilities for land animals between South America and Antarctica toward the end of Eocene; (3) final separation of Australia from Antarctica occurring approximately 38 MYBP; and (4) latest Cretaceous complete separation of New Zealand from Greater Antarctica.

**The metatherians**

The diagnosis of what constitutes a taxon, in this case the Metatheria (and others "within" it or "above" it), has profound consequences on one's reconstruction of the phylogeny and biogeography of that group. The diagnoses and discussions of the Metatheria, Holarctidelphia, Ameridelphia, and Australidelphia presented in the preceding chapter are necessary background for the attempted synthesis in this chapter. The review of the distribution of the known Metatheria should be also prefaced by some remarks about the known, probable, or even possible distribution of other tribosphenic and atribosphenic mammals. The known historical relationship of the continents prior to the late Jurassic and during most of the Jurassic leaves little doubt that atrisbosphenic mammals had a world-wide distribution. What is difficult to understand both phylogenetically and geographically is the origin and radiation of the Aegialodontidae and the more precisely tribosphenic groups (the pappotheridan tribotherians, metatherians, and eutherians with well developed protocones, conules, and three cusped talonid) during the Early Cretaceous. This problem is partly the result of their poorly sampled diversity and cranioskeletal morphology, but also a consequence of our incomplete understanding of the geography of the lands and the role of ecological factors. It is well to remember that the Cretaceous of Eastern North America, Africa, Europe, and Asia may hold additional hidden surprises for any account of the origins and early diversification of metatherians that one may favor at present.

### Cretaceous Asian metatherians

While the African history of the Metatheria is virtually unknown (see below), the Asian events were undoubtedly long and complex. The recent suggestion that the Late Cretaceous, primarily Asiatic (but with recorded North American presence) Deltatheroida is Metatheria (fide Kielan-Jaworowska & Nessov, 1990) appears to be well corroborated by the derived therian dental formula and assumed replacement pattern. This classification holds in spite of the fact that these metatherians appear to have evolved carnivory-related molar adaptations from a more primitive protometatherian condition that make them appear more primitive than their analysis suggests (see above). It is highly unlikely that these animals are the sister group of the South American Sparassodonta (see above, contra Kielan-Jaworowska, 1992). An exeptional specimen of the genus *Asiatherium* (*nomen nudum*, Trofimov & Szalay, 1993 and in preparation), represented by skull, dentition, and a nearly complete skeleton from the latest Cretaceous (Campanian) Mongolian Udan Sayr locality, a temporal "equivalent" of the Barun Goyot Formation. This animal with a marsupial dental formula, characteristically twinned entoconid and hypoconulid, and diagnostic metatherian aspects of its basicranial morphology represents a lineage quite distinct from the Deltatheroida. Thus the presence of the Metatheria in Asia as a taxon with some diversity distinct from what is known in North America appears well established. Judged from Fox's (1974) description of *Deltatheroides*-like mammals from the Late Cretaceous of North America, deltatheroidans may have entered that continent along with the first appearance of the eutherians, which probably differentiated in the Early Cretaceous of Asia (Albian, see Kielan-Jaworowska & Dashzeveg, 1989). In light of the fact that much of the North American fossil record for that time is restricted to the Western Interior, all that may be said with confidence is that the sundry earliest groups of Metatheria may well have been widespread in Holarctica. While the placentals were derived from animals that I would consider "marsupials" in a vernacular sense, based on an admittedly highly inferential judgment about the stage of their reproductive biology and developmental anatomy (see Chapter 3), they appear to have retained the more primitive cheek tooth morphology, inferred tooth replacement pattern and dental formula in contrast to the metatherians (see Chapter 8).

*American metatherians*

While the presence of eutherians and the holarctidelphian asi-adelphians and deltatheroidans in Asia prior to their appearance in North America appears to be well substantiated (Kielan-Jaworowska & Nessov, 1990; Trofimov & Szalay, 1993), the known record (if not the time of origin) of the Ameridelphia in North America is so far uncontradicted. While this origin is early in the Cretaceous as certainly supported by the evidence of *Alphadon*-like morphs reported by Eaton and Cifelli (1988), Eaton (1990) and Cifelli (1990a), it does not answer the intriguing question about the provenance and the last common ancestor of the two ancient metatherian cohorts (see particularly Cifelli, 1990a,b,c,f, 1993), the Holarctidelphia and Ameridelphia. The presence of the holarctidelphian Deltatheroida is likely in North America in the Late Cretaceous (cf. Fox, 1974). As noted in Chapter 8, Kielan-Jaworowska's (1992) suggestion that the Metatheria and Eutheria evolved independently from an aegilodontian stage does not appear probable to me.

The issue of time of entry of metatherians into North America, if they are an immigrant group there, is a thorny one; it cannot be answered based on the known Cretaceous record. Assuming that the source for the Sudameridelphia was North American emigrants, there appears to be a concensus that if this did not happen in the earliest Paleocene, then it probably happened in the latest Cretaceous. It seems to me that the great South American metatherian dental diversity in the Paleocene is best explained by an explosive adaptive radiation of a colonizing species or species cluster. Sigé (1972a,b) described the metatherian *"Alphadon" austrinum*, later considered *"Peradectes,"* together with the eutherian *Perutherium*, from the presumed Cretaceous deposits of Laguna Umayo in Peru. These were potentially the earliest of the therian immigrants to South America. The fragmentary Late Cretaceous Bagua mammal reported by Mourier et al. (1986) is momentarily irrelevant to this discussion because it is undiagnosable. The generously and extremely well documented mammals from the Tiupampa locality of the El Molino Formation of Bolivia described by Marshall and Muizon (1988, 1992) and Muizon (1991) were originally viewed as late Cretaceous by these researchers. This age determination, however, has been rejected by Pascual (1986), Bonaparte and Pascual (1987), Van Valen (1988a,b), and Pascual and Jaureguizar (1990), and now both Marshall and Muizon (1992; also Muizon, 1991) concur with a Paleocene age. All of these authors consider the Tiupampian Land Mammal Age Early Paleocene, followed in time by the Middle Paleocene Itaboraian. [For discussions of the significance of the hypothetical (but now recognized to be unlikely) distributions of early therians in South America, prior to the discovery of the Tiupampa and Los Alamitos vertebrates, see Clemens (1968b; 1979), Hoffstetter (1971; 1975), and Tedford (1974)] Nevertheless, in spite of the age of the Tiupampa fauna, Muizon (personal communication) maintains based on new evidence that therians were present in the latest Cretaceous of South America, a fact that I accept here. As I see it at present, the entry into South America of metatherians of North American archimetatherian stock was the source for the explosive adaptive radiation of the marsupials in the "splendid isolation" of South America, to paraphrase G. G. Simpson.

## Paleogene European, Asian, and African metatherians

The post-Cretaceous history of metatherians in the Old World appears to be consistent with the hypothesis that these forms are archimetatherian derivatives, descendants of a probably peradectine stock which dispersed during early faunal interchanges with North America. While this appears to be a relatively well-corroborated hypothesis, many complexities must be also contemplated. The paleogeography of the European land areas during the Cretaceous and Paleogene was complex and their exact dimensions remain elusive; the different geological moments of time meant differences in connectedness with and isolation from the neighboring continents, and combinations of these factors were exacerbated by the ecology of the many islands which probably made up the archipelago of Europe (Szalay & McKenna, 1971; McKenna, 1980a; Szalay & Li, 1986; Gheerbrant, 1990, 1992; Krause & Maas, 1990; Storch, 1990; and references therein.)

Peradectines, the genus *Peradectes*, while previously reported as early as the Middle Paleocene (Montian) of Belgium (Crochet & Sigé, 1983) are now restricted to the Eocene (Gheerbrandt, 1992). Metatherians, however, appear to have been always only a modest part of the faunal diversity of Europe, all the way into the Middle Miocene, without any significant structural dental departure from the peradectine ancestry (see Crochet 1977a,b, 1979a,b, 1981, and especially Crochet's 1978 monographic account; see also Cray, 1973; Hooker, 1986). It also appears that perhaps this original European stock has dispersed both into Asia and Africa and has undergone a very modest multiplication of lineages but no significant diversification (i.e., phylogenetic departures from the known skeletal adaptive pattern of its ancestral stock), as far as such a statement can be made given the still poor Paleogene fossil record particularly of Africa (Simons & Bown, 1984; Crochet et al., 1992; Gabunia et al., 1990; Ducroq et al., 1992). Although understanding multiplicity of lineages of a group depends on a good fossil record, expressions of diversification may show up in a few (two or more) lineages.

For the time being, the combined record of archimetatherian marsupials in the Paleogene of Europe, Africa, and Asia merely suggests that these highly fragmented areas of the world during the Paleogene, and possibly even in the Cretaceous (see Gheerbrant, 1990; 1992) received their ameridelphian (or so identified) metatherian faunas from somewhere else (Simpson, 1978; Benton, 1985). None of the material so far described is very extensive, as most of the elements recognized are a few molar teeth. Gabunia et al. (1984) described a small didelphine-like peradectine(?), a *Herpetotherium*-like undoubted marsupial, from the Early Oligocene of Kazakhstan. This form was later (Gabunia et al., 1990) given the name *Asiadidelphis. Peradectes* itself, called *Siamoperadectes,* was recently reported from Thailand (Ducrocq et al., 1992), adding credence to the hypothesis that the Asian Paleogene stock of metatherians, like the European one, is ameridelphian. Needless to say that the enigma is very real, posed by absence of the Cretaceous Asian marsupials from known Paleogene mammal faunas of Asia.

Nevertheless, the possibility of an African-derived, non-ameridelphian "Peradectidae" (rather than the presently considered pediomyid peradectines), that made its way into Europe, North America, and Asia during the early part of the Paleogene cannot be ruled out. Equally possible, or perhaps even more likely, is the Cretaceous–Paleocene presence of peradectines in Eastern North America or Central America.

The African distribution of metatherians is likely to result in hitherto unanticipated taxonomic surprises as the record of the Paleogene improves there. Gheerbrant's (1990) analysis of the biogeography of some mammals in the Paleogene suggests that a series of trans-Tethyan or even trans-Turgai Straits dispersal events will not only supply the eventual explanation of the origin of several placental groups, but also the puzzle of African metatherians. In 1983 the first specimen of a putative marsupial (later described as *Geratherium*) was reported from Africa by Mahboubi et al. (1983), and another one of an undoubted metatherian was reported by Bown and Simons (1984). The latter, a left lower jaw with two molars, premolars, and canine, was considered to be *Peratherium* from the Oligocene of Egypt by Bown and Simons (1984), a species that was subsequently raised to the generic rank of *Qatranitherium* by Crochet et al. (1992), with new material referred to it from Oman. Crochet (1984) described the previously reported *Geratherium,* based on an upper molar, from the Early Eocene El Kohol fauna of Algeria. In regards to *Geratherium,* Van Valen (1988a) speculated on a host of possible eutherian relationships; tupaiid, dilambdodont palaeoryctid, and mixodectid affinities were entertained by him. *Geratherium* may or may not be a metatherian. Another genus described as a marsupial, the third one known up to that time from Africa, by Crochet (1986), if it is a metatherian, *Kasserinotherium,* from the lower Eocene of Tunisia, was based on two upper molars.

## Antarctic and Australian metatherians

Given the size of the Antarctic landmass, its inhospitable surroundings for yielding a Cretaceous–Paleogene fossil record, paleontologists will probably continue with guesses based on the South American and Australian record. Nevertheless, it was against such odds that Woodburne and Zinsmeister (1984a) described the pre-Divisaderan Late Eocene sudameridelphian polydolopid genus *Antarctodolops* from beds approximately 40 MYBP from Seymour Island, a former Antarctic peninsula.

As it was emphasized in several reviews (Archer, 1984a,b; Case, 1989b; etc.), nothing was known of the therian, metatherian, and particularly the australidelphian radiation in Australia itself before the Miocene (see also especially Tedford et al., 1975; Tedford et al., 1977; Tedford, 1974, 1985). Previous estimates of the beginnings of the diversity have been put between the Middle Eocene to Oligocene (Clemens, 1977; Case, 1989b). Yet a recent discovery of a fauna near Murgon, southeastern Queensland (Godthelp et al., 1992) is reported to be Early Eocene in age, but considered more likely to be Oligocene according to J. Case (personal communction). In addition to reporting on bats, birds, frogs, trionychid turtles, and madtsoiid snakes, and teeth of a variety of dilambdodont "plesiomorphic insectivorous marsupials," these authors record a new genus (*Tingamarra*) that they consider to be an eutherian based on a single lower molar. As I noted above under the **Dasyuromorphia**, *Tingamarra*, based on a minimum sample for a mammalian taxon, with its sizable hypoconulid does not appear to be eutherian. It needs further corroboration from additional specimens before its eutherian status can be accepted.

## A historical–narrative explanation of metatherian paleobiogeography

The most tested, corroborated, historical–narrative explanation of metatherian biogeography is the one that accounts best for the known distribution and most corroborated phylogeny. Like all other aspects of the status quo of any historical science, it is the most probable hypothesis (or set of hypotheses) that may be favored, given the explicitly analyzed total of geological and biological contingencies. The geological understanding of Cretaceous paleogeography and the breakup of Laurasia and then Gondwana, together with the still extremely incomplete existing record of organisms, gives a framework of relative dates for the distribution of metatherians. Until it is shown to be otherwise, I am assuming that such isolation of the northern and southern continents, and then later the isolation of individual southern landmasses, were the major initial and boundary conditions necessary for reconstructing not only metatherian but tetrapod paleobiogeography in general (but see especially Gheerbrant's 1990 and 1992 thoughtful analyses).

The spatiotemporal origins of metatherians pose a fundamental problem, given the spotty Cretaceous record. Few have put this general issue in better perspective than Clemens (1977, p. 57) when he stated that

> Rarely if ever have the time and place of origin of an order or class of vertebrates been determined with great precision or reliability. In part this stems from the nature of the event being sought. It is the differentiation of a species of a group of closely related lineages that might not have stood out as remarkably different from contemporaneous species but whose descendants form a group worthy of taxonomic recognition. Inadequacy of the available fossil record – both the frequency of stratigraphic (chronological) gaps separating collections, and incomplete sampling of contempraneous faunas documenting the range of environmental variation – contribute to the lack of precision.

Analyses of the dental record of Cretaceous therians (e.g., Clemens & Lillegraven, 1986; Cifelli, 1990a,b,c,d,e,f, 1993; Kielan-Jaworowska, 1992; Gheerbrant, 1992; and references therein) illustrate the difficulties of tracing tribotherian, metatherian, and eutherian lineages as the fossil record improves. Nevertheless great strides have been made both in the various arenas of discoveries as well as in the understanding of the affinities of previously enigmatic tribosphenic mammals.

Given the level of understanding of the evolution of Holarctidelphia, Ameridelphia, and Australidelphia, particularly the allocation of the deltatheroidans (by Kielan-Jaworowska & Nessov, 1990) and the asiadelphians (by Trofimov & Szalay, 1993, and in preparation) to the Metatheria, makes it necessary to rethink the origin and biogeographic pattern of the whole infraclass – pushing limits of reasonable probabilities, rather than Popperian exercises, as far as this is possible. Over a decade ago Marshall (1980b, p. 347) summarized the views of a number of workers, and noted that

> Vicariance and geographic isolation on a major scale were responsible for the marsupial–placental dichotomy. The Early Cretaceous was a time of exceptional restriction of intercontinental migration of terrestrial faunas due to marine barriers, and probably was a time of development of "island continent" centers of endemism (Lillegraven, 1974; Tedford, 1974). Some workers believe that marsupials and placentals are too alike . . . to have developed simultaneously in the same region.

In light of the co-occurrence of Eutheria and holarctidelphian metatherians in the Cretaceous of Asia, such a view (see also Hoffstetter, 1970, 1972, 1975; Keast, 1977; etc.) advocating geographic isolation between the northern continents cannot be maintained anymore. Shortly after Marshall's synthesis, Kielan-Jaworowska (1982) strongly and at that time convincingly, argued that in an evolutionary theater of the Northern Hemisphere, during the

Neocomian Early Cretaceous, continental geography dictated diversification from a widely distributed (world-wide, according to Kielan-Jaworowska) tribosphenic ancestry (some taxon perhaps similar to the primitive, and putatively fully tribosphenic genus *Aegialodon*). She suggested that the Eutheria was Asiatic (as put foreward by Beliajeva et al., 1974), whereas the Metatheria (different from the concept used here) was western North American (as suggested by Clemens, 1977, 1979; see also Clemens et al., 1989) or South American in origin. She also suggested that the Pappotheriidae was eastern North American in origin.

As the fossil record improves, one fact becomes increasingly clear about these issues: We probably know a lot less than what appears to be missing from the record, that is, what we should know, before we can make an assessment with reasonable probability about early therian phylogeny and paleogeography. Even if the Deltatheroida would prove to be not the cladistic metatherian ally of Ameridelphia, an alternative which I do not believe, the Mongolian asiadelphian *Asiatherium* (*nomen nudum*) is without doubt a Late Cretaceous metatherian. The lusher forest faunas of southern Asia during the Late Cretaceous may well have been even more densely populated with metatherians than was the North American Western Interior. The preponderance of (largely terrestrial) eutherians in Mongolia, for example, in contrast to the two known groups of holarctidelphian metatherians may well be nothing else but a reflection of latitudinal zonation. In fact the metatherian *Asiatherium*, like the eutherians of the Cretaceous of Mongolia (Nessov & Kielan-Jaworowska, 1991), also appears to have been terrestrial, similar in habitus to *Monodelphis* (Trofimov & Szalay, in preparation)

Though the approach of linking differentiation of stem groups of various therian higher taxa to landmass fragmentation is undoubtedly a proper one when evidence so suggests, the absence of a Cretaceous record from Africa, Eastern North America, and the various southern reaches of the immense landmass of Asia during the Cretaceous calls for caution. The precious and erratic dental record of the tribotherians and metatherians of the North American Cretaceous appears to offer solutions for the origin of various groups. One should, nevertheless, recall the abundance of phylogenetic hypotheses and implied schemes for their Cretaceous and Paleogene biogeography prior to recent times when it gradually and dramatically became apparent to paleomammalogists that the evolution of the eutherians could not be understood without their complex spatial and temporal distribution (with repeated dispersal events) and phylogenetic dynamics both in northern and southern Asia in addition to that of North America.

Most likely the North American history of the Metatheria only appears relatively simple – given our ignorance for most areas, with a few exceptions besides the Western Interior for the entire

Cretaceous and Paelogene. Eaton's (1990, and references therein), Cifelli's (1993, and references therein), and Fox's (1987, and references therein) work in the Early and Late Cretaceous holds great promise for some revealing looks into both tribotherian and metatherian history. The families Stagodontidae and Pediomyidae (see Chapter 8) are morphologically relatively conservative, and there is no evidence for the variety shown by the Paleogene sudameridelphians. Their diversity and relative abundance drastically dwindled at the end of the Cretaceous, while close relatives of the North American peradectines (possible immigrants) clung to existence in Europe and Asia into the Miocene and mid Oligocene (unpublished evidence from Kazakhstan), respectively. Peradectines, probably of European (or possibly South American) origin, reached Africa as well in the Paleogene as detailed above. The archimetatherians of North America were the likely source for not only for the Paleogene metatherians of the Old World, but also for the Sudameridelphia of South America as well, given all the serious caveats I have discussed.

But there is no general agreement on the provenance of Old World marsupials in the Paleogene (see especially Gheerbrant, 1990, 1992). In regards to the unusual *Geratherium*, Mahboubi et al. (1986, p. 44) suggested that

> the last common ancestor of the genera *Geratherium* and *Alphadon* must have had a distribution which included at least Africa and South America before the definitive opening of the South Atlantic in the late Albian. *Geratherium* is thus of Gondwanian origin (Jaeger & Martin, 1984). All other recent paleontological data also lead to assign such a cradle to the superorder Marsupialia. . . .

This assessment, however, is a deduction based entirely on vicariance assumptions, one which is certainly unsupported by the composition of the Late Cretaceous Los Alamitos mammal fauna from Patagonia (see below). If dispersal occurred, then the reasoning seemingly based on morphology is faulty, as dispersal and phylogeny together could account for the known distribution. Origins of *Geratherium* from an *Alphadon*-like ancestry would not by itself support Gondwanan origins. The presence of dentally extremely similar species of lipotyphlans, *Cimolestes* and *Palaeoryctes*, in both Morocco and western North America in the latest Cretaceous and early Paleocene, as postulated by Gheerbrant (1990, 1992), may explain all the marsupials, even a form like *Geratherium*, as lineages that dispersed into Africa, probably via Europe and across the Tethys. The presence of the Peradectinae in Africa certainly more strongly supports dispersal into Africa than any northward dispersal of hitherto unknown Mesozoic stocks of African metatherians. And in fact Crochet et al. (1992, p. 539) remained silent on the enigmatic *Geratherium* when they stated that "With the exception of *Geratherium*, all the known

Afro-Arabian marsupials are endemic to this continent at least as far back as the Paleocene–Eocene boundary, the period during which they probably dispersed southwards into Africa from the northern landmasses." Gheerbrant's (1990, p. 114) suggestion appears likely to me that a form like the North African and Arabian *Qatranitherium* would be an independent dispersal into Africa during the late Eocene to early Oligocene, distinct from the earlier peradectine dispersal event. None of these records, not even the problematic *Geratherium* or *Kasserinotherium*, given their morphology, suggests the presence of an ancient metatherian fauna in Africa.

There is a beginning of an understanding of the South American faunal dynamics, and consequently the story of the metatherians, as a result of a Patagonian mammal fauna. The evidence published by Bonaparte (1986, 1987, 1990a) of a vertebrate fauna from the Los Alamitos Formation of the Campanian (at least 80 MYBP) Late Cretaceous of northern Patagonia supplies a very important datum for both eutherian and metatherian biogeography. There appear to be no therians in South America at that time! The study of Krause and Bonaparte (1990) has established that multituberculates, instead of tribosphenics such as the alleged edentates, were present in the Campanian Cretaceous of South America (but see Bonaparte, 1990a, b). While small atribosphenic mammals (tricondontids, symmetrodonts, dryolestids, mesungulatids, and multituberculates) are found in the large and diverse mammal fauna, the absence of any tribosphenic, either eutherian or metatherian therian, is highly significant. The enigmatic Cretaceous Bagua mammal of Peru notwithstanding, this atribosphenic mammal fauna from Patagonia (as well as the Hauterivian–Barremian Early Cretaceous La Amarga specimen, which is also atribosphenic) strongly suggests the relatively late arrival (possible latest Cretaceous) of therians into South America (Bonaparte, 1986; Bonaparte & Pascual, 1987; Van Valen, 1988a, b; Muizon, 1991). Bonaparte and Pascual (1987, p. 361) have unequivocally stated that the composition of the Los Alamitos fauna of Patagonia is "due to the long Late Jurassic-Cretaceous isolation of the Laurasian and Gondwanian supercontinents." In light of the significant Los Alamitos discoveries and analyses, Van Valen (1988a,b) recently reiterated the hypothesis perceptively advanced by G. G. Simpson (1950) and B. Patterson and Pascual (1968), that therians came late to South America from North America, perhaps in the latest Cretaceous or earliest Paleocene. Given such evidence, these assessments represent the most corroborated H-N E. The Los Alamitos faunal evidence appears to negate Tedford's (1974) influential suggestion that the taxon Metatheria originated during the Cretaceous in a southern continent and only eventually reached North America. The same may be said of Woodburne and Zinsmeister's (1984b, p. 941) factually stated postulate that marsupials were present in South America 80 MYBP, or of the statement by Clemens et al. (1989, p. 535), probably a lapsus calami, that "Mesozoic mammalian faunas of the Southern Hemisphere were not composed only of monotremes and marsupials. . . ."

Bonaparte (1986, 1987) considered the currently available evidence as favoring the Laurasian origin of the Theria, as previously stated by Kielan-Jaworowska (1982), a conclusion that cannot be fully supported anymore. The status of Mesozoic therian evolution in Africa is still entirely unsettled. The more ancient history of metatherians, however, appears to involve Laurasia rather than Gondwana. The currently understood phylogeny of the Metatheria, as presented in Chapter 8, considered in light of plate tectonics, suggests the places of origin for a number of groups (tentatively), dispersal events (unequivocally), and vicariance events (equivocally) related to our knowledge of groups. Given the modification necessary in light of Cretaceous Asiatic Metatheria, and in spite of the still valid statement that "we simply do not know where [proto-] marsupials evolved" (Simpson, 1978, p. 23), the general hypothesis I favor has been advocated in one form or another by Jardine and McKenzie (1972), Lillegraven (1974), Clemens (1977), Cifelli (1989), Marshall et al. (1990), and others, and described in Marshall's (1980b) review as the "Mobilist Model 1" hypothesis, one of nine different hypotheses of marsupial paleobiogeography that considered plate tectonics as a major contingecy for explanation.

Six years before his death at the end of a career of unique distinction in evolutionary biology and vertebrate paleontology, Simpson (1978, pp. 22–23), fully embraced the mobilist perspective in paleobiogeography. He also stated emphatically that the position of Australia in the Paleogene made it impossible for marsupials to reach that landmass from Asia, a well corroborated hypothesis no one seriously disputes today (Simpson, 1961b). This obviously leaves the Latest Cretaceous, or more likely post-Cretaceous time of the Antarctic realm for the source of the Australasian marsupial fauna. The presence of Metatheria at least as early as the Cenomanian in the Late Cretaceous of North America and its synchronous absence from South America strongly suggests, at present, the conclusion that the North American stock spread into South America. Bonaparte (1990a, p. 63) sums up the singular importance of the Campanian Los Alamitos from northeastern Patagonia as follows:

[The] Los Alamitos fauna demonstrates strong compositional differences from other Late Cretaceous assemblages of mammals. A long period of isolation between Gondwana and Laurasia is most likely responsible for such distinct faunal differences. The dryolestoid and symmetrodont adaptive radiation recorded . . . is interpreted as a possible result of the absence of tribosphenic mammals. It is suggested that the Australian monotremes may be a local product of the dryolestoid radiation. . . . The Cretaceous mammals of South America and Australia suggest long-term isolation of Gondwana and Laurasia, perhaps from the Late Jurassic until the Late Cretaceous. Such separation may have resulted in the dominance of tribosphenic mammals in Laurasia, and derived non-tribosphenic mammals in South America (and possibly in Gondwana in general).

While Bonaparte's generalizations for all of Gondwana appear tempting, the events concerning Africa are largely unknown.

The microbiotherians represent evolved South American members of the Gondwanadelphia, a group which I believe was the source of the entire known great Australasian radiation of marsupials (contra Reig et al., 1987; Kirsch et al., 1991; Westerman & Edwards, 1991; Hershkovitz, 1992). While South American endemics penetrated Antarctica (Woodburne & Zinsmeister, 1984a,b), we know nothing else of the nature of the pre- and post-Cretaceous mammal fauna of that landmass. The Polydolopidae, until then, have been known as an endemic South American group. The significance of that find beyond the mere record of the polydolopids lies in the bridging of the South American and Antarctic fossil record of mammals. If polydolopids radiated in this way, then clearly a number of other representatives of the remaining lineages in the Paleocene and Eocene also might have dispersed along the same route. The inferred direction of change in basicranial, carpal, and particularly the tarsal morphoclines of the Metatheria, as I see it, supplies strong corroboration for this previously proposed hypothesis by several students.

This view is not shared by Kirsch et al. (1991, pp. 10468–9), who consider the Microbiotheria to be a sister group of the Diprotodontia (see my responses in Chapters 3, 7, and 8 under **Microbiotheria**). Because they maintain such a view of phylogeny, Kirsch et al. see the Microbiotheria in South America as a late dispersal event from Antarctica–Australia, or vicariance. In light of the nature of the morphological evidence discussed in this volume, particularly the derivation of the tarsus based taxonomic properties of the Australidelphia (see discussions under *Dasyuromorphia*), and the cranial and pedal attributes of Syndactyla, I find their hypothesis unlikely.

An exteremely important datum, whatever the level of its eventual accuracy, is the latest Cretaceous separation of New Zealand from Antarctica. If that date is indeed Late Cretaceous, then the absence of mammals from New Zealand should mean an absence of mammals from Greater Antarctica in the Late Cretaceous, as noted by Woodburne and Zinsmeister (1984b, p. 939). The occurrence of marsupials in the Australasian region, however, is not the "obvious" answer to the question of their origin, the rather tenuous arguments for an Australian marsupial homeland notwithstanding. The original existence of all or some of these groups in Antarctica during the Paleocene–Eocene is a distinct possibility. The presence of the gondwanadelphian Microbiotheria in South America, known unequivocally in the southern half only, may represent an Antarctic derivation of that lineage of the Australidelphia, which nevertheless originated from South American Didelphidae. In fact the seasonally cool marine climates of an early Tertiary Antarctica may have been indirectly responsible for the transformation of a didelphid lineage that dispersed into Greater Antarctica and evolved into the stem gondwanadelphians. Given our nearly complete absence of knowledge of the Paleogene terrestrial faunas of the Antarctic–Australian theater of evolution, it is not clear to me whether dispersal into these regions by marsupials was part of an ecologically filtered exchange, or whether it represented sweepstakes dispersal across water. Two recent papers have examined both the geological and ecological circumstances of origins of marsupials (Case, 1989b) and the origin and further differentiation of the Australo–Pacific land mammal fauna (Flannery, 1988).

Case (1989b) presented an interesting new and highly synthetic hypothesis bridging several subdisciplines of geology and systematics, concerning the effect of high latitude heterochroneity on the origin of the Australian marsupial radiation. According to his view (pp. 224–5):

> Antarctica's position during the Maastrichtian to Early Eocene within the high latitude Weddellian Biogeographic Province played an important role in retarding the radiation of the Australian marsupials relative to the extensive marsupial radiations in the mid-to low altitude regions of South America in the same interval. The high latitude province served as a holding tank for the Weddellian marsupials due to low diversity of habitats resulting from the widespread podocarp/*Nothofagus* closed forests with non-angiosperm understories. Extensive familial level cladogenesis probably did not occur among the Australian marsupials until the mid- to late Eocene when the continent had sufficiently separated from Antarctica and generated a more diverse habitat structure. The Weddellian marsupial fauna remaining on the Antarctic continent subsequent to the separation of Australia from Antarctica became extinct sometime after the beginning of the Oligocene due to the deteriorating climatic conditions and the development of the polar ice cap.

Corroboration for this interpretation, however, will be difficult to attain paleontologically. Substantiation of his hypothesis, as Case correctly stressed in his paper, would mean a conformation of a 20- to 25-million-year hiatus between the onset of the earliest South American and Australian adaptive radiations. I suspect that the radiation of the Australian marsupial fauna began immediately after the first colonizers landed. The first gondwanadelphians that bridged the temporal and geographical gaps between the South American and Greater Antarctic landmasses would have represented the initial modest diversity of the australidelphians from which, beginning in the Paleocene or Eocene after the delay perhaps caused by ecological constraints, the major groups of Australian metatherians that make up the Syndactyla began their diversification.

# References

Abbie, A. A. 1937. Some general observations of the Marsupialia, with especial reference to the position of the Peramelidae and Caenolestidae. *J. Anat.* **71:** 424–36.

Abbie, A. A. 1939. A masticatory adaptation peculiar to some diprotodont marsupials. *Proc. Zool. Soc. London Ser. B* **109:** 261–79.

Alberch, P. 1980. Ontogenesis and morphological diversification. *Amer. Zool.* **20:** 653–67.

Alberch, P. 1982. Developmental constraints in evolutionary processes. In: Boner, J. T. (ed.), *Evolution and development.* Berlin: Springer-Verlag, pp. 313—32.

Ameghino, F. 1887. Enumeracion sistematica de las especies de mamiferos fosiles coleccionados por Carlos Ameghino en los terranos eocenos de la Patagonia austral y depositados en el Museo La Plata. *Bol. Mus. de la Plata* **1:** 1–26.

Ameghino, F. 1889. Contribucion al conocimiento de los mamiferos fosiles de la Republica Argentina, obra escrito bajo los auspicios de la Academia Nacional de Ciencias de la Republica Argentina para presentarla a la Exposicion Universal de Paris de 1889, *Actas Acad. Cienc. Cordoba* **6:** 1–1027.

Ameghino, F. 1891. Los plaiaulacidos argentidos, y sus relaciones zoologicas, geologicas, y geograficas. *Rev. Arg. Hist. Nat.* **1:** 38–44.

Ameghino, F. 1894. Enumération synoptique des espèce de mamifères fossiles des formations éocènes de Patagonie. *Bol. Acad. Cienc. Cordoba* **13:** 259–452; *Ameghino Obras* **10:** 594–863.

Ameghino, F. 1897. Mammifères crétacés de l'argentine (Deuxième contribution à la Connaissance de la faune mammalogique des couches a *Pyrotherium*). *Bol. Inst. Geog. Argent.* **18:** 406–521.

Ameghino, F. 1901. Notices preliminaires sur des ongulés nouveaux des terrains crètacé de Patagonie. *Bol. Acad. Nac. Cienc. Cordoba* **16:** 349–426; *Ameghinos Obras* **13:** 92–203.

Ameghino, F. 1904. Nouves especies de mamiferos cretaceos y terciarios de la Republica Argentina. *An. Soc. Cien. Argent.* **56:** 193–208 (1903); **57:** 162–175, 327–341 (1904); **58:** 35–71, 182–192, 225–291 (1904). Reissued as separate dated 1904, pp. 1–142.

Anderson, S. 1982. *Monodelphis kunsi. Mammalian Species* **190:** 1–3.

Aplin, K. P. 1987. Basicranial anatomy of the early Miocene diprotodontian *Wynyardia bassania* (Marsupialia: Wynynardiidae) and its implications for wynyardiid phylogeny and classification. *In:* M. Archer (ed.), *Possums and opossums: studies in evolution,* vol. 1. Sydney: Surrey Beatty & Sons, pp. 269–91.

Aplin, K. P., & Archer, M. 1987. Recent advances in Marsupial systematics with a new syncratic classification In: M. Archer (ed.) *Possums and Opossums: Studies in Evolution,* vol. 1. Sydney: Surrey Beatty & Sons, pp. xv–lxxii.

Archer, M. 1976a. Koalas (phascolarctids) and their significance in marsupial evolution. *In:* T. J. Bergin (ed.), *The koala.* Proceedings of the Taronga Symposium on Koala Biology, Management, and Medicine, Sydney, March 11, 12, 1976, pp. 20–28.

Archer, M. 1976b. The dasyurid dentition and its relationships to that of didelphids, thylacinids, borhyaenids (Marsupicarnivora) and peramelids (Peramelina: Marsupialia). *Aust. J. Zool., Suppl. Ser.* **39:** 1–34.

Archer, M. 1976c. The basicranial region of marsupicarnivores (Marsupialia), interrelationships of carnivorous marsupials, and affinities of the insectivorous marsupial peramelids. *Zool. J. Linn. Soc.* **59:** 217–322.

Archer, M. 1976d. A revision of the marsupial genus *Planigale* Troughton (Dasyuridae). *Mem. Queensland Mus.* **17:** 341—65.

Archer, M. 1976e. Phascolarctid origins and the potential of the selenodont molar in the evolution of diprotodont marsupials. *Mem. Queensland Mus.* **17:** 367–71.

Archer, M. 1977a. *Koobor notabilis* (De Vis), an unusual koala from the Pliocene Chinchilla Sand. *Mem. Queensland Mus.* **18:** 31–5.

Archer, M. 1977b. Revision of the dasyurid marsupial genus *Antechinomys Krefft*. *Mem. Queensland Mus*. **18**: 17–29.

Archer, M. 1977c. Origins and subfamilial relationships of *Diprotodon* (Diprotodontidae, Marsupialia). *Mem. Queensland Mus*. **18**: 37–9.

Archer, M. 1978. The nature of the molar-premolar boundary in marsupials and a reinterpretation of the homology of marsupial cheekteeth. *Mem. Queensland Mus*. **18**: 157–64.

Archer, M. 1981. A review of the origins and radiations of Australian mammals *In:* A. Keast (ed.), *Ecological Biogeography of Australia*. The Hague: Junk Press, pp. 1437—88.

Archer, M. ed., 1982a. *Carnivorous Marsupials*, vols. 1 & 2. Sydney: Royal Zoological Society of New South Wales.

Archer, M. 1982b. Review of the dasyurid (Marsupialia) fossil record, integration of data bearing on phylogenetic interpretation and suprageneric classification. *In:* M. Archer (ed.) *Carnivorous Marsupials*, vol. 2. Sydney: Royal Zoological Society of New South Wales, pp. 397–443.

Archer, M. 1982c. A review of Miocene thylacinids (Thylacinidae, Marsupialia), the phylogenetic position of the Thylacinidae and the problem of apriorisms in character analysis. *In:* M. Archer (ed.), *Carnivorous Marsupials*, vol. 2. Sydney: Royal Zoological Society of New South Wales, pp. 445–76.

Archer, M. 1984a. Origins and early radiations of marsupials. *In:* M. Archer & G. Clayton (eds.) *Vertebrate Zoogeography and Evolution in Australasia - Animals in Space and Time*. Carlisle, Western Australia: Hesperian Press, pp. 585—625.

Archer, M. 1984b. The Australian marsupial radiation *In:* M. Archer & G. Clayton (eds.), *Vertebrate Zoogeography and Evolution in Australasia – Animals in Space and Time*. Carlisle, Hesperian Press: Western Australia, pp. 633–808.

Archer, M. 1984c. On the importance of being a koala. *In:* M. Archer & G. Clayton (eds.), *Vertebrate Zoogeography and Evolution in Australasia – Animals in Space and Time*. Carlisle: Hesperian Press, Western Australia, pp. 809–15.

Archer, M., ed. 1987. *Possums and Opossums: Studies in Evolution*, vols. 1 & 2. Sydney: Surrey Beatty & Sons.

Archer, M. & Clayton, G., eds. 1984. *Vertebrate Zoogeography and Evolution in Australasia – Animals in Space and Time*. Carlisle, Western Australia: Hesperian Press.

Archer, M., & Dawson, L., 1982. Revision of marsupial lions of the genus *Thylacoleo* Gervais (Thylacoleonidae, Marsupialia) and thylacoleonid evolution in the late Cainozoic *In:* M. Archer (ed.), *Carnivorous Marsupials*, vol. 2.

Sydney: Royal Zoological Society of New South Wales, pp. 477–94.

Archer, M., & Flannery, T. F. 1985. Revision of the extinct rat kangaroos (Potoroidea: Marsupialia) with description of a new Miocene genus and species and a new Pleistocene species of *Propleopus*. *J. Paleontology*. **59**: 1331–49.

Archer, M., Flannery, T. F., Ritchie, A., & Molnar, R. E. 1985. First Mesozoic mammal from Australia – an early Cretaceous monotreme. *Nature* **318**: 363–6.

Archer, M., Hand, S., & Godthelp, H. 1988. A new order of Tertiary zalambdodont marsupials. *Science* **239**: 1528–31.

Archer, M., & Kirsch, J. A. W. 1977. The case for the Thylacomyidae and Myrmecobiidae, Gill, 1872, or why are marsupial families so extended? *Proc. Linn. Soc. New South Wales* **102**: 19–25.

Archer, M., & Rich, T. H. 1982. Results of the Ray E. Lemley Expeditions. *Wakaleo alcootaensis* n.sp. (Thylacoleonidae, Marsupialia), a new marsupial lion from the Miocene of the Northern Territory with a consideration of early radiation in the family. *In* M. Archer (ed.), *Carnivorous Marsupials*, vol. 2. Sydney: Royal Zoological Society of New South Wales, pp. 495–502.

Archer, M., Tedford, R. H., & Rich, T. H. 1987. The Pilkipildridae, a new family and four new species of ?petauroid possums (Marsupialia: Phalangerida) from the Australian Miocene *In* M. Archer (ed.), *Possums and Opossums: Studies in Evolution*, vol. 2. Chipping Norton, NSW: Surrey Beatty & Sons Pty Limited, pp. 607–27.

Archer, M., Murray, P., Hand, S., & Godthelp, H. 1993. Reconsideration of monotreme relationships based on the skull and dentition of the Miocene *Obdurodon dicksoni*. *In* F. S. Szalay, M. J. Novacek, & M. C. McKenna (eds.) *Mammalian Phylogeny. Mesozoic Differentiation, Multituberculates, Monotremes, Early Therians, and Marsupials*. New York: Springer-Verlag, pp. 75–94.

Archibald, J. D. 1979. Oldest known eutherian stapes and a marsupial petrosal bone from the Late Cretaceous of North America. *Nature* **281**: 669–70.

Archibald, J. D. 1982. A study of Mammalia and geology across the Cretaceous-Tertiary boundary in Garfield County, Montana. *Univ. California Geol. Sci.* **122**: 1–286.

Ashlock, P. D. 1971. Monophyly and associated terms. *Syst. Zool.* **20**: 63–9.

Ashlock, P. D. 1972. Monophyly again. *Syst. Zool.* **21**: 430–8.

Ashlock, P. D. 1979. An evolutionary systematist's view of classification. *Syst. Zool.* **28**: 441–50.

Atramentowicz, M. 1982. Influence du milieu sur l'activité et la reproduction de *Caluromys philander*. *Rev. Ecol. (Terre et Vie)* **36**: 373–95.

Ax, P. 1987. *The Phylogenetic System: the Systematization of Organisms on the Basis of their Phylogenesis*. Chichester. John Wiley & Sons.

Barnett, C. H. 1955. Some factors influencing angulation of the neck of the mammalian talus. *J. Anat.* **89:** 225–30.

Barnett, C. H., Davies, D. V., & MacConaill, M. A. 1961. *Joints, their Structure and Mechanics*. London: Longman's.

Barnett, C. H., & Napier, J. R. 1953. The form and mobility of the fibula in metatherian mammals. *J. Anat.* **86:** 1–9.

Baverstock, P. R. 1984. The molecular relationships of Australasian possums and gliders. *In* A. Smith & I. Hume (eds.), *Possums and Gliders*, Sydney: Surrey Beatty & Sons, pp. 1–8.

Baverstock, P. R., Archer, M., Adams, M., & Richardson, B. J. 1982. 52 Genetic relationships among 32 species of Australian dasyurid marsupials. *In* M. Acher (ed.) *Carnivorous Marsupials*, Sydney: Royal Zoological Society of New South Wales, pp. 641–50.

Baverstock, P. R., Birrell, J., & Krieg, M. 1987. Albumin immunologic relationships among Australian opossums: a progress report *In* M. Archer (ed.) *Possums and Opossums: Studies in Evolution*, vol. 1. Sydney: Surrey Beatty & Sons, Pty Limited, pp. 229–23.

Baverstock, P. R., Krieg, M., Birrell, J., & McKay, G. M. 1990. Albumin immunologic relationships of Australian marsupials. II. Pseudocheiridae. *Australian J. Zool.* **38:** 519–26.

Beliajeva, E. I., Trofimov, B. A., & Reshetov, B. J. 1974. Osnovnye etapy evolyutsii memlekopitayushchikh v pozdnem mezozoe-paleogene Tsentral'noi Azii *In* N. N. Kramarenko (ed.), *Fauna i Biostratigrafiya Mezozoya i Kainozoya Mongolii*. *Tr. Sovm. Sov.-Mong. Paleont, Eksp.* **1:** 19–45.

Bensley, B. A. 1903. On the evolution of the Australian Marsupialia; with remarks on the relationships of the marsupials in general. *Trans. Linnean Soc. London, Series 2* **9:** 83–217.

Benton, M. J. 1985. First marsupial fossil from Asia. *Nature* **318:** 313.

Biewener, A. A. 1989. Mammalian terrestrial locomotion and size. *BioScience* **39:** 776–83.

Blackburn, D. G., Hayssen, V., & Murphy, C.J. 1989. The origins of lactation and the evolution of milk: a review with new hypotheses. *Mammal Rev.* **19:** 1–26.

Bock, W. J. 1967. The use of adaptive characters in avian classification. *Proceedings of the XIV International Ornithological Congress*, pp. 61–74.

Bock, W. J. 1973. Philosophical foundations of classical evolutionary classification. *Syst. Zool.* **22:** 375–92.

Bock, W. J. 1977a. Adaptation and the comparative method. *In* M. K. Hecht, P. C. Goody, & B. M. Hecht (eds.), *Major Patterns in Vertebrate Evolution*, NATO. Advanced Institute Series A. New York: Plenum Press, pp. 57–82.

Bock, W. J. 1977b. Foundations and methods of evolutionary classification. *In* M. K. Hecht, P. C. goody, & B. M. Hecht (eds.), *Major Patterns in Vertebrate Evolution* NATO. Advanced Study Institute Series A. New York: Plenum Press, pp. 851–95.

Bock, W. J. 1979. A synthetic explanation of macroevolutionary change – a reductionistic approach. *Bull. Carnegie Mus.* **13:** 20–69.

Bock, W. J. 1981. Functional-adaptive analysis in evolutionary classification. *Amer. Zool.* **21:** 5–20.

Bock, W. J. 1985. The arboreal theory for the origin of birds. *In* M. K. Hecht, J. H. Ostrom, G. Viohl, & P. Wellnhofer (eds.), *The Beginnings of Birds*, Eichstatt: Jura Museum, pp. 199–207.

Bock, W. J. 1986. Species concepts, speciation, and macroevolution. *In* K. Iwatsuki, P. H. Raven, & W. J. Bock (eds.), *Modern Aspects of Species*, Tokyo: University of Tokyo Press, pp, 31–57.

Bock, W. J. 1988. The nature of explanations in morphology. *Amer. Zool.* **28:** 205–15.

Bock, W. J. 1989a. Comments on "Populations and their place in evolutionary biology." *In* M. K. Hecht (ed.), *Evolutionary Biology at the Crossroads*. Flushing, New York: Queens College Press, pp. 53–8.

Bock, W.J. 1989b. The homology concept: its philosophical foundation and practical methodology. *Zool. Beitr. N.F.* **32:** 327–53.

Bock, W. J. 1990. From biologische anatomie to ecomorphology. *Netherland J. Zool.* **40:** 254–77.

Bock, W. J. 1991. Explanations in konstruction-morphologie and evolutionary morphology. *In* D. Schmidt, D. Kittler, & K. Vogel (eds.). *Constructional Morphology and Biomechanics: Concepts and Implications*, Heidelberg: Springer-Verlag.

Bock, W. J. & von Whalert, G. 1965. Adaptation and the form–function complex. *Evolution* **19:** 269–99.

Böker, H. 1927. Die Entstehung der Wirbeltiertypen und der Ursprung der Extremitaten. *Z. Morph. Anthr.* **26:** 1–58.

Bonaparte, C. L. J. L. 1838. Synopsis vertebratorum systematis. *Nuovi Ann. Sci. Nat., Bologna.* **2:** 105–33.

Bonaparte, C. L. J. L. 1850. *Conspectus Systematis*. Batavia: E. J. Brill.

Bonaparte, J. F. 1984. Late Cretaceous faunal interchange of terrestrial vertebrates between the Americas. *In* W. F. Reif & F. Wesphal (eds.), *Third Symposium of Terrestrial Ecosystems*, Tübingen: Attempo Verlag, pp. 19–24.

Bonaparte, J. F. 1986. History of the terrestrial Cretaceous vertebrates of Gondwana. *Actas IV Congr. Arg. Paleont. y Bioestrat.* **2**: 63–95.

Bonaparte, J.F. 1987. The Late Cretaceous fauna of Los Alamitos, Patagonia, Argentina. *Paleontologia* **3**: 103–78.

Bonaparte, J. F. 1990a. New Late Cretaceous mammals from the Los Alamitos Formation, northern Patagonia. *Nat. Geog. Res.* **6**: 63–93.

Bonaparte, J. F. 1990b. Book review of *Dinosaur Tracks and Traces*. *Historical Biology* **4**: 139–50.

Bonaparte, J. F., Franchi, M. R., Powell, J. E., & Sepulveda, E. G. 1984. La Formacion Los Alamitos (Campaniano-Maastrichtiano) del sudeste de Rio Negro, con descripcion de *Kristosaurus australis* n. sp. (Hadrosauridae). Significado paleogeografico de los vertebrados. *Rev. Ass. Geol. Argentina* **39**: 284–99.

Bonaparte, J. F., & Pascual, R. 1987. Los mamiferos (Eotheria, Allotheria y Theria) de la Formacion Los Alamitos, Campaniano de Patagonia, Argentina. *IV Congreso Latinoamericano de Paleontologia, Bolivia*, **1**: 361–72.

Bown, T. M. & Fleagle, J. G. 1993. Systemmatics, biostratigraphy, and dental evolution of the Palaeothentidae, later Oligocene to early-middle Miocene (Deseadan-Santacrucian) caenolestoid marsupials of south America. *J. Paleontology* **67**(supplement to No. 2): 1–76.

Bown, T. M., & Rose, K. D. 1979. *Mimoperadectes*, a new marsupial, and *Worlandia*, a new dermopteran, from the lower part of the Willwood Formation (Early Eocene), Bighorn Basin, Wyoming. *Contr. Mus. Paleont.* **25**: 89–104.

Bown, T. M., Simons, E. L. 1984. First record of marsupials (Metatheria: Polyprotodonta) from the Oligocene in Africa. *Nature* **308**: 447–9.

Brenner, G. J., 1976. Middle Cretaceous floral provinces and early migrations of angiosperms. *In* C. B. Beck (ed.), *Origin and Early Evolution of Angiosperms*. New York: Columbia University Press, pp. 23–47.

Briggs, J. C. 1984. *Centers of Origin in Biogeography*. *Biogeographical Monographs No. 1*, Leeds: University of Leeds.

Briggs, J. C. 1987. *Biogeography and Plate Tectonics*. Amsterdam: Elsevier Science Publishers B. V.

Broom, R. 1898. On the affinities and habits of *Thylacoleo*. *Proc. Linn. Soc. N. S. W.* **23**: 57–74.

Brown, J. C. & Yalden, D. W. 1973. The description of mammals. 2. Limbs and locomotion of terrestrial mammals. *Mammal Review* **3**: 107–34.

Burnett, G. T. 1830. Illustrations of the Quadrupeda, or quadrupeds, being the arrangement of the true four-footed beasts indicated in outline. *Quart. J. Sci. Lit. Art.* **28**: 336–53.

Butler, P. M. 1978. A new interpretation of the mammalian teeth of tribosphenic pattern from the Albian of Texas. Breviora: Museum of Comparative Zoology **446**: 1–27.

Butler, P. M. & Kielan-Jaworowska, Z. 1973. Is *Deltatheridium* a marsupial? *Nature* **245**: 105–6.

Cabrera, A. 1919. *Genera Mammalium. Monotremata. Marsupialia*. Madrid: Museo Nacional de Ciencias Naturales.

Cairn, E. J. & Grant, R. 1890. Report of a collecting trip to northeastern Queensland during April to September, 1889. *Rec. Australian Mus.* **1**: 27–31.

Calaby, J. H. 1960. Observations on the banded anteater *Myrmecobius f. fasciatus* Waterhouse (Marsupialia), with particular reference to its food habits. *Rep. Proc. Zool. Soc. London* **135**: 183–207.

Calaby, J. H., Dimpel, H., & Cowan, I. M. 1971. The mountain pigmy possum, *Burramys parvus* Broom (Marsupialia), Kosciusko National Park, New South Wales, *New South Wales, Division of Wildlife Research Technical Paper No. 23.*

Carter, J. T. 1920. Microscopic structure of the enamel of the teeth of two sparassodonts, *Cladosictis* and *Pharsophorus*, as evidence of their marsupial character: together with a note on the value on the pattern of the enamel as a test of affinity. *J. Anat.* **54**:189–95.

Cartmill, M. 1975. Strepsirhine basicranial structure and the affinities of the Cheirogaleidae, *In* W. P. Luckett & F. S. Szalay (eds.), *Phylogeny of the Primates: a Multidisciplinary Approach*, New York: Plenum Press, pp. 313–54.

Cartmill, M. 1981. Hypotheses testing and phylogenetic reconstruction. *Z. Zool. Syst. Evolut. Forsch.* **19**: 73–96.

Case, J. A. 1984. A new genus of Potoroinae (Marsupialia: Macropodidae) from the Miocene Ngapakaldi local fauna, south Australia, and a definition of the Potoroinae. *J. Paleont.* **58**: 1074–86.

Case, J. A. 1989a. Cranial isometry in carnivorous marsupials and phyletic relationships of the dog like thylacines. *J. Vert. Paleont.* **9**(supplement): 16A.

Case, J. A., ed. 1989b. Antarctica: the effect of high latitude heterochroneity on the origin of the Australian marsupials. *In* J. A. Crame (ed.), *Origins and Evolution of the Antarctic Biota, Geological Society Special Publication* **47**: 217–26.

Case, J. A., & Woodburne, M. O. 1986. South American marsupials: A successful crossing of the Cretaceous-Tertiary boundary. *Palaios* **1**: 413–16.

Charig, A. J. 1982. Systematics in biology: a fundamental comparison of some major schools of thought. *In* K. A. Joysey & A. E. Friday (eds.), *Problems of Phylogenetic Reconstruction. Systematics Associations Special Volume* **21**, London: Academic Press, pp. 363–440.

Charles-Dominique, P. 1983. Ecology and social adaptations in didelphid marsupials: comparison with eutherians of similar ecology. *Soc. Publ. Amer. Soc. Mamm.* **7**: 395–422.

Charles-Dominique, P., Atramentowicz, M., Charles-Dominique, M., Gerard, H., Hladik, H., Hladik, C. M., & Prevost, M. F. 1981. Les mammifères frugivores arbicoles nocturnes d'une fôret guyanaise: interrelations plantes-animaux. *Rev. Ecol. (Terre et Vie)* **35**: 343–435.

Chattejee, S. & Hotton, N. III. 1993. *New Concepts in Global Tectonics.* Lubbock: Texas Tech University.

Cifelli, R. L. 1983. Eutherian tarsals from the late Paleocene of Brazil. *Amer. Mus. Novit.* **2761**: 1–31.

Cifelli, R. L., 1987a, Therian mammals from the Late Cretaceous of the Kaiparowits Region, Utah. *J. Vert. Paleont.* **7**(supplement): 14A.

Cifelli, R. L. 1987b. Marsupial from the earliest Late Cretaceous of western US. *Nature* **325**: 520–2.

Cifelli, R. L., 1989. The origin and early evolution of marsupials. *J. Vert. Paleont.* **9**(supplement): 17A.

Cifelli, R. L. 1990a. Cretaceous mammals of Southern Utah. I. Marsupials from the Kaiparowits Formation (Judithian). *J. Vert. Paleont.* **10**: 295–319.

Cifelli, R. L. 1990b. Cretaceous mammals of Southern Utah. II. Marsupials and marsupial-like mammals from the Wahweap Formation (Early Campanian). *J. Vert. Paleont.* **10**: 320–31.

Cifelli, R. L. 1990c. Cretaceous mammals of Southern Utah. III. Therian mammals from the Turonian (Early Late Cretaceous). *J. Vert. Paleont.* **10**: 332–45.

Cifelli, R. L. 1990d. Cretaceous mammals of Southern Utah. IV. Eutherian mammals from the Wahweap (Aquilan) and Kaiparowits (Judithian) Formations. *J. Vert. Paleont.* **10**: 346–60.

Cifelli, R. L. 1l90e. Therian mammals from the Late Cretaceous of South Texas. *J. Vert. Paleont.* **10**(supplement): 18A.

Cifelli, R. L. 1990f. A primitive higher mammal from the Late Cretaceous of Southern Utah. *J. Mamm.* **71**: 343–50.

Cifelli, R. L. 1993. Theria of Metatherian-Eutherian grade and the origin of marsupials. *In* F. S. Szalay, M. J. Novacek, & M. C. McKenna (eds.), *Mammalian Phylogeny. Mesozoic Differentiation, Multituberculates, Monotremes, Early Therians, and Marsupials,* New York: Springer-Verlag, pp. 208–15.

Cifelli, R. L., & Eaton, J. G., 1987. Marsupial from the earliest Late Cretaceous of Western US. *Nature* **325**: 520–2.

Clark, W. E. Le Gros. 1936. The problem of the claw in primates. *Proc. Zool. Soc. London* **1936**: 1–25.

Clemens, W. A. 1966. Fossil mammals of the Type Lance Formation Wyoming. Part II. Marsupialia, *Univ. California Publ. Geol. Sci.* **62**: 1–122.

Clemens, W. A. 1968a. A mandible of *Didelphodon vorax* (Marsupialia: Mammalia). *Contrib. in Sci.* **133**: 1–11.

Clemens, W. A. 1968b. Origins and early evolution of marsupials. *Evolution* **22**: 1–18.

Clemens, W. A. 1970. Mesozoic mammalian evolution. *Ann. Rev. Ecol. System.* **1**: 357–90.

Clemens, W. A. 1977. Phylogeny of the marsupials. *In* B. Stonehouse & D. Gilmore (eds.), *The Biology of Marsupials,* New York: Macmillan Press Ltd., pp. 51–68.

Clemens, W. A. 1979. Marsupialia. *In* J. A. Lillegraven, Z. Kielan-Jaworowska, & W. A. Clemens (eds.). *Mesozoic Mammals: the first Two Thirds of Mammalian History.* Berkeley: University of California Press, pp. 192–220..

Clemens, W. A., & Lillegraven, J. A. 1986. New Late Cretaceous North American advanced therian mammals that fit neither the marsupial nor the eutherian molds. *Contrib. Geol. Special Paper* **3**: 55–85.

Clemens, W. A., Marshall, L. G. 1976. American and European Marsupialia. *In* F. Westphal (ed.), *Fossilum Catalogus, I: Animalia,* Gravenhage: Dr. W. Junk B.V. pp. 3–31.

Clemens, W. A., Richardson, B. J., & Baverstock, P. R. 1989. Biogeography and phylogeny of the Metatheria. *In Fauna of Australia, volume 1B Mammalia,* Walton, D. W., and Dyne, G. R., eds.), Canberra: Australian Government Publishing Service, pp. 527–48.

Cockburn, A. 1989. Adaptive patterns in marsupial reproduction. *Trends Ecol. Evol.* **4**: 126–30.

Collett, R. 1887. On a collection of mammals from central and northern Queensland. *Zool. Jahrb.* **2**: 829–940.

Collins, L. R. 1973. Monotremes and marsupials, a reference for zoological institutions, Smithsonian Publication **4888**. Washington, D.C.: Smithsonian Institution Press, 323 pp.

Cox, C. B. 1973. Systematics and plate tectonics in the spread of marsupials. *In* N. F. Hughes (ed.), *Organisms and Continents Through Time,* Paleontological Association of London, pp. 113–19.

Cracraft, J. 1981. The use of functional and adaptive criteria in phyogenetic systematics. *Amer. Zool.* **21**: 21–36.

Cray, P. E. 1973. Marsupialia, Insectivora, Primates, Creodonta, and Carnivora from the Headon Beds (Upper Eocene) of southern England, *Bull. Brit. Mus. (Nat. Hist.) Geol.* **23**: 1–102.

Crochet, J.-Y. 1977a. Les didelphides Paleogènes Holarctiques: historique et tendances evolutives. *Geobios, Mem. Special* **1**: 127–34.

Crochet, J.-Y. 1977b Les Didelphidae (Marsupicarnivora, Marsupialia) Holarctiques tertiaires, *C. R. Acad. Sci. Paris* **284**(Series D): 357–60.

Crochet, J.-Y. 1978. *Les Marsupiaux du Tertiare d'Europe.* Ph.D. thesis, Académie de Montpellier, Université des Sciences et Techniques du Languedoc, vols. 1 & 2.

Crochet, J.-Y. 1979a. Diversité systematique des Didelphidae (Marsupialia) Européens tertiaires. *Geobios* **12**: 365–78.

Crochet, J.-Y. 1979b. Données nouvelles sur l'histoire paleogeographique des Didelphidae (Marsupialia). *C. R. Acad. Sci. Paris* **288**(Series D): 1457–60.

Crochet, J.-Y. 1980. L'occlusion dentaire chez *Paradectes, Amphiperatherium* et *Peratherium,* Marsupiaux du Tertiaire d'Europe. Paleovertebrata, Montpellier: Mm. Jubil. R. Lavocat, pp. 79–80.

Crochet, J.-Y. 1984. *Garatherium mahboubii* nov. gen., nov. sp., marsupial de l'Eocène Inferieur d'El Kohol (Sud-Oranais, Algérie). *Annales Paleontologie (Vert.-Invert.)* **70**: 275–94.

Crochet, J.-Y. 1986. *Kasserinotherium tunesiense* nov. gen., nov. sp., third marsupial from Africa (lower Eocene), Tunisia. *C. R. Acad. Sci. Paris* **302**(Series II,14): 923–926.

Crochet, J.-Y., Hartenberger, J. L., Rage, J. C., Remy, J. A., Sigé, B. Sudre, J., & Vianey-Liaud, M. 1981. Les nouvelles faunes de vertebres antérieures a la <<Grande Coupure>> découvertes dans les phosphorites du Quercy. *Bull. Mus. Nat. Hist. Paris* 4 Ser, 3, section C(3): 245–66.

Crochet, J.-Y., & Sige, B. 1983. Les mammifères Montiens de Hainin (Paleocene moyen de Belgique). Part III. Marsupiaux. Palaeovertebrata **13**: 51–64.

Crochet, J-I., Thomas, H., Sen, S., Roger, J., Gheerbrandt, E., & Al-Sulaimani, Z. 1992. Découverte d'un peradectide (Marsupialia) dans l'Oligocene inférieur du Sultanat d'Oman: nouvelle données sur la paleobiogeographie des marsupiaux de la plaque arabo-africaine *C. R. Acad. Sci. Paris* **314**: 539–45.

Croizat, L. 1958. *Panbiogeography.* Caracas: published by the author.

Croizat, L. 1964. *Space, Time, and Form, the Biological Synthesis.* Caracas: published by the author.

Croizat, L. 1982. Vicariance, vicariism, panbiogeography, "vicariance biogeography," etc. A clarification. *Syst. Zool.* **31**: 291–304.

Currey, J. 1984. *The Mechanical Adaptations of Bones.* Princeton: Princeton University Press.

Dagosto, M. 1985. The distal tibia of primates with special reference to the Omomyidae, *Intl. J. Primat.* **6**: 45–75.

Dagosto, M. 1986. *The Joints of the Tarsus in the Strepsirhine Primates.* Ph.D. Thesis, City University of New York.

Darwin, C. 1859. *On the Origin of Species by Means of Natural Selection, or Preservation of Favored Races in the Struggle for Life.* London: John Murray.

Darwin, F. 1887. *The Life and Letters of Charles Darwin, Including an Autobiographical Chapter.* London: Murray.

Davis, D. D. 1964. The giant panda: a morphological study of evolutionary mechanisms. *Fieldiana: Zoological Memoirs* **3**: 1–339.

Dawson, L. 1980. The status of the taxa of extinct giant wombats (Vombatidae: Marsupialia), and a consideration of vombatid phylogeny. *Australian Mammal.* **4**: 65–79.

Dawson, T. J. 1977. Kangaroos. *Sci. American* **237**: 78–89.

De Queiroz, K. 1988. Systematics and the Darwinian revolution. *Philos. Science* **55**: 238–59.

Decker, R. L., & Szalay, F. S. 1974. Origin and function of the pes in the Eocene Adapidae (Lemuriformes, Primates). *In* F. A. Jenkins, Jr. (ed.), *Primate Locomotion.* New York: Academic Press, pp. 261–91.

Dollo, L. 1899. Les ancêtres des Marsupiaux étaient-ils arboricoles? *Trav. Stat. Zool. Wimerrius* **7**: 188–203.

Dollo, L. 1900. Le pied du *Diprotodon* et l'origine arboricole des marsupiaux. *Bull. Biol. France Belgique* **33**: 275–80.

Dollo L. 1906. Le pied de l'*Amphiproviverra* et l'origine arboricole des marsupiaux. *Bull. Soc. Belge Geol.* **20**: 166–8.

Donoghue, M. J., Doyle, J. A., Gauthier, J., Kluge, A. J., & Rowe, T., 1989. The importance of fossils in phylogeny reconstruction. *Ann. Rev. Ecol. Syst.* **20**: 431–60.

Doran, A. H. G. 1877. Morphology of the mammalian ossicula auditus. *Trans. Linn. Soc. London* **1**: 371–497.

Ducrocq, S., Buffetaut, E., Buffetaut-Tong, H., Jaeger, J-J., Jongkanjanasoontorn, Y., & Suteethorn, V. 1992. First fossil marsupial from south Asia. *J. Vert. Paleo.* **12**: 395–9.

Eaton, J. G. 1987. The Campanian-Maastrichtian boundary in the western interior of North America. *Newsl. Statigr.* **18**: 31–9.

Eaton, J. G. 1990. Therian mammals of the Cenomanian (Late Cretaceous) Dakota Formation, southwestern Utah. *J. Vert. Paleont.* **10**(supplement): 21A.

Eaton, J. G., & Cifelli, R. L. 1988. Preliminary report on Late Cretaceous mammals of the Kaiparowits Plateau, southern Utah. *Contrib. Geol.* **26**: 45–55.

Eldredge, N. 1985. *Unfinished Synthesis,* New York: Oxford University Press.

Eldredge, N., & Cracraft, J. 1980. *Phylogenetic Patterns and the Evolutionary Process: Method and Theory in Comparative Biology.* New York: Columbia University Press.

Eldredge, N., & Novacek, M. J. 1985. Systematics and paleobiology. *Paleobiol.* **11**: 65–74.

Eldredge, N., & Salthe, S. N. 1984. Hierarchy and evolution. *Oxford Surv. Evol. Biol.* **1**: 184–208.

Elftman, H. O. 1929. Functional adaptations of the pelvis in marsupials. *Bull. Amer. Mus. Nat. Hist.* **58**: 189–232.

Emery, C. 1897. Beitrage zur Entwicklungsgeschichte und Morphologie des Hand-Fusskelets der Marsupialier. *Semons Zool. Forsch.* **2**: 371–400.

Emery, C. 1901. Hand- und Fusskelet von *Echidna hystrix*. Semons Zool. Forsch. **3**: 663–76.

Erwin, D. H. 1990. Variations on a theme. *Paleobiol.* **16**: 96–101.

Farris, J. S. 1974. Formal definitions of paraphyly and polyphyly. *Syst. Zool.* **23**: 548–54.

Filan, S. L. 1990. Myology of the head and neck of the bandicoot (Marsupialia: Peramelemorphia). *Aust. J. Zool.* **38**: 617–34.

Finch, M. E. 1982. Discovery and interpretation of *Thylacoleo carnifex* (Thylacoleonidae, Marsupialia). *In* M. Archer (ed.), *Carnivorous Marsupials*, vol. 2. Sydney: Royal Zoological Society of New South Wales, pp. 537–61.

Finch, M. E., & Freedman, L. 1988. Functional morphology of the limbs of *Thylacoleo carnifex* Owen (Thylacoleonidae: Marsupialia). *Aust. J. Zool.* **36**: 251–72.

Flannery, T. F. 1982. Hindlimb structure and evolution in the kangaroos (Marsupialia: Macropodoidea). *In* M. Archer & G. Clayton (eds.), *Vertebrate Zoogeography and Evolution in Australasia*. Perth, Carlisle, Western Australia: Hesperian Press, pp. 508–24.

Flannery, T. F. 1983a. A unique trunked giant. *In* S. Quirk & M. Archer (eds.), *Prehistoric Animals of Australia*. Sydney: Australian Museum.

Flannery, T. F. 1983b. Review of the subfamily Sthenurinae (Marsupialia) and the relationships of the species of *Troposodon* and *Lagostrophus*. *Aust. Mammal.* **6**: 15–28.

Flannery, T. F. 1984. Kangaroos: 15 million years of Australian bounders. *In* M. Archer & G. Clayton (eds.), *Vertebrate Zoogeography and Evolution in Australasia – Animals in Space and Time*. Carlisle, New South Wales: Hesperian Press, pp. 817–35.

Flannery, T. F. 1987. The relationships of the Macropodidae (Marsupialia) and the polarity of some morphological features within the Phalangeriformes. *In* M. Archer (ed.), *Possums and Opossums: Studies in Evolution*, vol. 2. Sydney: Surrey Beatty & Sons, pp. 741–7.

Flannery, T. F. 1988. Origins of the Australo-Pacific land mammal fauna. *Aust. Zool. Rev.* **1**: 15–24.

Flannery, T. F. 1989. Phylogeny of the Macropodoidea; a study in convergence. *In* G. Grigg, P. Jarman, & I. Hume (eds.). *Kangaroos, Wallabies, and Rat-kangaroos*. Sydney: Surrey Beatty & Sons Pty Limited, pp. 1–46.

Flannery T. F., & Archer, M. 1987a. *Hypsiprymnodon bartholomaii* (Marsupialia: Potoroidae), a new species from the Miocene Dwornamor local fauna and a reassessment of the phylogenetic position of *H. moschatus*. *In* M. Archer (ed.), *Possums and Opossums: Studies in Evolution*. vol. 2. Sydney: Surrey Beatty & Sons, pp. 749–58.

Flannery, T. F., & Archer, M. 1987b. *Bettongia moyesi*, a new and plesiomorphic kangaroo (Marsupialia: Potoroidae) from Miocene sediments of northwestern Queensland. *In* M. Archer (ed.), *Possums and opossums: studies in evolution*, vol. 2. Sydney: Surrey Beatty & Sons, pp. 759–69.

Flannery, T. F., & Archer, M., 1987c. *Strigocuscus reidi* and *Trichosurus dicksoni*, two new fossil phalangerids (Marsupialia: Phalangeridae) from the Miocene of northwestern Queensland. *In* M. Archer (ed.), *Possums and opossums: studies in evolution*, vol. 2. Sydney: Surrey Beatty & Sons, pp. 527–36.

Flannery, T. F., Archer, M., & Maynes, G. 1987. The phylogenetic relationships of living phalangerids (Phalangeroidea: Marsupialia) with a suggested new taxonomy. *In* M. Archer (ed.), *Possums and Opossums: Studies in Evolution*, vol. 2. Sydney: Surrey Beatty & Sons, pp. 477–506.

Flannery, T. F., Archer, M., & Plane, M. 1983. Miocene kangaroos (Macropodidae: Marsupialia) from three localities in northern Australia, with a description of two new subfamilies. *Bull. Bur. Miner. Res. Aust. Geol. Geophys.* **7**: 287–302.

Flannery, T. F., & Pledge, N. S. 1987. Specimens of *Warendja wakefieldi* (Vombatidae: Marsupialia) from the Pleistocene of South Australia. *In* M. Archer (ed.), *Possums and Opossums: Studies in Evolution*. vol. 1. Sydney: Surrey Beatty & Sons, pp. 365–8.

Flannery, T. F., & Szalay, F. S. 1982. *Gohra paulae*, a new giant fossil tree-kangaroo (Marsupialia: Macropodidae) from New South Wales, Australia. *Aust. Mammal.* **5**: 261–5.

Flannery, T. F., Turnbull, W. D., Rich, T. H. V., & Lundelius, E. L. 1987. The phalangerids (Marsupialia: Phalangeridae) of the early Pliocene Hamilton local fauna, southwestern Victoria. *In* M. Archer (ed.), *Possums and Opossums: Studies in Evolution*, vol. 2. Sydney: Surrey Beatty & Sons, pp. 537–46.

Fleagle, J. G. 1977. Locomotor behavior and skeletal anatomy of sympatric Malaysian leaf monkeys *Presbytis obscura* and *Presbytis melalophus*. *Ybk. Phys. Anthro.* **20**: 440–53.

Fleischer, G. 1973. Studien am Skelett des Gehororganes der Säugetiere, einschliesslich des Menschen. *Säugetierkundl. Mitteilungen (München)* **21:** 131–239.

Fleischer, G. 1978. *Evolutionary Principles of the Mammalian Middle Ear.* Berlin: Springer-Verlag.

Flynn, J. J. 1991. Pre-Deseadan, post-Mustersan mammals from Central Chile: and update. *J. Vert Paleont.* **11:** 29A.

Flynn, J. J., & Wyss, A. R. 1990. New Early Oligocene marsupials from the Andean Cordillera, Chile. *J. Vert. Paleont.* **9:** 22A.

Fooden, J. 1972. Breakup of Pangaea and isolation of relict mammals in Australia, South America, and Madagascar. *Science* **175:** 894–8.

Fooden, J. 1973. Rifting and drift of Australia and the migration of mammals. *Science* **180:** 759–62.

Forey, P. L. 1982. Neontological analysis versus paleontological stories. *In* K. A. Joysey & A. E. Friday (eds.), *Problems of Phylogenetic Reconstruction. Systematics Association Special Papers* 21, London: Academic Press, 119–57.

Fox, R. C. 1971. Marsupial mammals from the early Campanian Milk River Formation, Alberta, Canada. *Zool. J. Lin. Soc.* **50**(supplement 1): 145–64.

Fox, R. C. 1974. *Deltatheroides*-like mammals from the Upper Cretaceous of North America. *Nature* **249:** 392.

Fox, R. C. 1975. Molar structure and function in the early Cretaceous mammal *Pappotherium*: evolutionary implications for Mesozoic Theria. *Canad. J. Earth Sci.* **12:** 412–42.

Fox, R. C. 1976. Additions to the mammalian local fauna from the Upper Milk River Formation (Upper Cretaceous), Alberta. *Canad. J. Earth Sci.* **13:** 1105–18.

Fox, R. C. 1979a. Mammals from the Upper Cretaceous Oldman Formation, Alberta. I. *Alphadon* Simpson (Marsupialia). *Canad. J. Earth Sci.* **16:** 91–102.

Fox, R. C. 1979b. Mammals from the Upper Cretaceous Oldman Formation, Alberta. II. *Pediomys* Marsh (Marsupialia). *Canad. J. Earth Sci.* **16:** 103–13.

Fox, R. C. 1979c. Ancestry of the "dog-like" marsupials. *J. Paleont.* **53:** 733–5.

Fox, R. C. 1981. Mammals from the Upper Cretaceous Oldman Formation, Alberta. V. *Eodelphis* Matthew, and the evolution of the Stagodontidae (Marsupialia). *Canad. J. Earth Sci.* **18:** 350–65.

Fox, R. C. 1983. Notes on the North American Tertiary marsupials *Herpetotherium* and *Peradectes*. *Canad. J. Earth Sci.* **20:** 1565–78.

Fox, R. C. 1987. Palaeontology and the early evolution of marsupials. *In* M. Archer (eds.), *Possums and Opossums: Studies in Evolution*, vol. 1. Sydney: Surrey Beatty & Sons, pp. 161–9.

Fox, R. C. 1988a. Patterns of mammalian evolution towards the end of the Cretaceous, Saskatchewan, Canada. *In* P. M. Currie & E. H. Koster, *Fourth Symposium on Mesozoic Terrestrial Ecosystems, Short Papers:* pp. 96–100.

Fox, R. C. 1988b. An ancestral marsupial and its implications for early marsupial evolution. *In* P. M. Currie & E. H. Koster (eds.), *Fourth symposium on Mesozoic terrestrial ecosystems, Short Papers: pp. 101–5.*

Fox, R. C., & Naylor, B. G. 1986. A new species of *Didelphodon* Marsh (Marsupialia) from the Upper Cretaceous of Alberta, Canada: paleobiology and phylogeny. *N. Jb. Geol. Palaeot. Abh.* **172:** 357–80.

Francisco, B. H. R., & Souza Cunha, F. L. de 1978. Geologia e estratigrafia da Bacia de São José, municipio de Itaboraí, RJ. *An. Acad. Brasil Cienc.* **50:** 381–416.

Friday, A. 1987. Models of evolutionary change and the estimation of evolutionary trees, *Oxford Sur. Evol. Biol.* **4:** 61–88.

Gabunia, L. K., Shevyreva, & Gabunia, V. D. 1984. On the presence of fossil marsupials in Asia. *Bull. Acad. Sci. Georgian SSR* **116:** 169–71.

Gabunia, L. K., Schevyreva, N. S., & Gabunia, V. J. 1990. A new opossum (Didelphidae, Marsupialia, Metatheria, Mammalia) from the lowest Oligocene of Zaisan Depression (Eastern Kazakstan). *Palaeontologycheski Journal* **1:** 101–9.

Ganslosser, U. 1977. Observations on the behavior of Doria-Tree-Kangaroos and Grizzled-gray Tree-Kangaroos in Zoological Gardens *Zool. Anz.* **198:** 393–412.

Ganslosser, U. 1980. Muskelansatze, Knochenstarke, Hand- und Fusswurzelskelet einiger Baumkanguruharten (*Dendrolagus* Müller, 1838). *Zool. Anz.* **205:** 68–78.

Gardner, A. L. 1973. The systematics of the genus *Didelphis* (Marsupialia: Didelphidae) in North and Middle America. *Specl. Pub. Mus. Texas Tech. Univ.* **4:** 1–81.

Gardner, A. L. 1982. Virginia opossum. *In* J. A. Chapman & G. A. Feldhamer (eds.). *Wild Mammals of North America.* Baltimore: John Hopkins University Press, pp. 3–36.

Gauthier, J., Kluge, A. G., & Rowe, T. 1988. Amniote phylogeny and the importance of fossils. *Cladistics* **4:** 105–209.

Gebo, D. L. 1985. The nature of the primate grasping foot. *Am. J. Phys. Anthrop.* **67:** 269–78.

Gebo, D. L. 1986. *The Anatomy of the Prosimian Foot and its Application to the Primate Fossil Record.* Ph. D. Thesis, Duke University.

Gebo, D. L. 1987. Functional anatomy of the tarsier foot. *Am. J. Phys. Anthrop.* **73:** 9–31.

Gelderen, C. van. 1924. Die Morphologie der Sinus

durae matris. 2. Die vergleichenden Ontogenie der neurokraniellen Venen der Vögel und Säugetiere. *Z. Anat. Entwickelungsgeschichte* **74**: 432–508.

Gemmell, R. T., Johnston, G., & Barnes, A. 1984. The uniformity of growth within the litter of the marsupial *Isoodon macrourus*. *Growth* **48**: 221–33.

George, G. 1987. Characterization of the living species of Cuscus (Marsupialia: Phalangeridae). *In* M. Archer (ed.) *Possums and Opossums: Studies in Evolution*. Vol. 2. Sydney: Surrey Beatty & Sons, pp. 507–26.

Gervais, P. & Verreaux, J. 1842. On a new genus of marsupial animals (*Tarsipes rostratus*), *Proc. Zool. Soc. London* **1842**: 1–5.

Gheerbrant, E. 1990. On the early biogeographical history of the African placentals. *Hist. Biol.* **4**:107–16.

Gheerbrant, E. 1992. *Bustylus* (Eutheria, Adapisoriculidae) and the absence of ascertained marsupials in the Paleocene of Europe. *Terra Nova* **3**: 586–92.

Gheerbrant, E., & Russell, D. E. 1989. Presence of the genus *Afrodon* (Mammalia, Lipotyphla, Adapisoriculidae) in Europe; new data for the problem of trans-Tethyan relations between Africa and Europe around the K/T boundary. *Palaeogeogr. Palaeclimatol. Palaeoecol.* **76**: 1–15.

Ghiselin, M. T. 1991. Classical and molecular phylogenetics. *Boll. Zool.* **58**: 289–94.

Gilkeson, C. F., & Lester, K. S. 1989. Ultrastructural variation in enamel of Australian marsupials. *Scanning Microscopy* **3**: 177–91.

Gill, T. 1872. Arrangement of the families of mammals with analytical tables. *Smithsonian Misc. Coll.* **11**: 1–98

Gingerich, P. D. 1979. Stratophenetic approach to phylogeny reconstruction in vertebrate paleontology. *In* J. Cracraft, & N. Eldredge (eds.), *Phylogenetic Analysis and Paleontology* New York: Columbia University Press, pp. 41–79.

Glaesmer, E. 1908. Untersuchung über die Flexorengruppe am Unterschenkel und Fuss der Säugetiere. *Morphologisch. Jahrb.* **38**: 36–90.

Glaesmer, E. 1910. Die Beugemuskeln am Unterschenkel und Fuss bei den Marsupialia, Insectivora, Edentata, Prosimiae und Simiae. *Morphologisch. Jahrb.* **41**: 149–336.

Glauert, L. 1926. A list of Western Australian fossils, Supplement 1. *Bull. Geol. Surv. West. Australia* **88**: 36–72.

Godinot, M. 1985. Evolutionary implications of morphological changes in Paleogene primates. *Spec. Pap. Paleontol.* **33**: 39–47.

Godthelp, H., Archer, M., Cifelli, R., Hand, S. J., & Gilkeson, C. F. 1992. Earliest known Australian Tertiary mammal fauna. *Nature* **356**: 514–16.

Goin, F. J., Carlini, A. A., & Pascual R. 1986. Un probable marsupial del Cretacico Tardio del Norte de Patagonia, Argentina. *IV Congr. Arg. Paleont. y Biostr., Simposio Evolucion de los vertebrados cenozoicos de America del Sur* **2**: 43–7.

Goin, F. J., & Montalvo, C. 1988. Revision sistematica y reconocimiento de una nueva especie del genero *Thylatheridium* Reig (Marsupialia, Didelphidae). *Ameghiniana* **25**: 161–7.

Goldfuss, G. A. 1820. *Handbuch der Zoologie, 2 vols.* Nurnberg: J. J. Schrag.

Goldschmid, A., & Kotrschal, K. 1989. Ecomorphology: development and concepts. *In* H. Splechtna & H. Hilgers (eds.), *Trends in Vertebrate Morphology.* Stuttgart: Gustav Fischer Verlag, pp. 501–11.

Goodrich, E.S. 1935. Syndactyly in marsupials. *Rep. Proc. Zool. Soc.* **Part I**: 175–8.

Gordon, G. 1974. Movements and activity of the short-nosed bandicoot *Isodoon macrourus* Gould (Marsupialia). *Mammalia* **38**: 405–31.

Gordon, K. R. 1989. Adaptive nature of skeletal design. *BioScience* **39**: 784–90.

Gould, S. J. 1985. A clock of evolution. *Nat. Hist.* **85(4)**: 12–25.

Gould, S. J. 1986. Evolution and the triumph of homology, or why history matters. *Amer. Sci.* **74**: 60–9.

Gould, S. J. 1990. Enigmas of the small shellies. *Nat. Hist.* **90(10)**: 6–17.

Gould, S. J. 1991. Eight (or fewer) little piggies. *Nat. Hist.* **91(1)**: 22–9.

Gow, C. E. 1985. Apomorphies of the Mammalia. *S.-African Jour. Sci.* **81**: 558–60.

Gow, C. E. 1991. Vascular system associated with the sidewall of the braincase and the prootic canals of cynodonts, including mammals. *S.-Afr. Tydskr. Dierk.* **26**: 140–4.

Grand, T. I. 1967. The functional anatomy of the ankle and foot of the slow loris (*Nycticebus coucang*). *Amer. J. Phys. Anthropol.* **26**: 207–18.

Grand, T. I. 1983. Body weight in relationship to tissue composition, segmental distribution of mass, and motor function. III. The Didelphidae of French Guyana. *Australian J. Zool.* **31**: 299–312.

Grant, T. R. 1989. Ornithorhynchidae. *In* D. W. Walton & G. R. Dyne (eds.), *Fauna of Australia, Mammalia*, vol. 1B. Canberra: Australian Government Publishing Service, pp. 436–50.

Grant, T. R. & Temple-Smith, P. D. 1987. Observations on torpor in the small marsupial *Dromiciops australis* (Marsupiala: Microbiotheriidae) from Southern Chile. *In* M. Archer (ed.), *Possums and Opossums: Studies in Evolution*, vol. 1. Sydney: Surrey Beatty & Sons, pp. 273–7.

Gray, J. E. 1821. On the natural arrangement of vertebrose animals. *London Med. Reposit.* **15**: 296–310.

Gray, J. E. 1825. Outline of an attempt at the disposition of the Mammalia into tribes and families with a list of the genera apparently appertaining to each tribe. *Ann. Philos. n. s.* **10**: 336–44 (vol. 26 of the whole series).

Gray, R. 1989. Oppositions in panbiogeography – can the conflicts between selection, constraint, ecology, and history be resolved? *New Zealand J. Zool.* **16**: 787–806.

Green, M., & Martin, J. E. 1976. *Peratherium* (Marsupalia: Didelphidae) from the Oligocene and Miocene of South Dakota. *In: Essays on Paleontology in Honour of Loris Shano Russell.* Ontario Museum, Life Sciences Miscellaneous Publications.

Gregory, W. K. 1910. The orders of mammals. *Bull. Amer. Mus. Nat. Hist.* **27**: 3–524.

Gregory, W. K. 1922. On the "habitus" and "heritage" of *Caenolestes.* *J. Mammal.* **3**: 106–14.

Gregory, W. K. 1951. *Evolution Emerging: a Survey of Changing Patterns from Primeval Life to Man.* vol. 1 & 2, New York: Macmillan Press.

Gregory, W. K. & Simpson, G. G. 1926. Cretaceous mammal skulls from Mongolia. *Amer. Mus. Novitates* **225**: 1–20.

Groves, C. P. 1982. The systematics of tree kangaroos (*Dendrolagus;* Marsupialia, Macropodidae). *Australian Mammal.* **5**: 157–86.

Groves, C. P. 1987a. On the highland cuscuses (Marsupialia: Phalangeridae) of New Guinea. *In* M. Archer (ed.), *Possums and Opossums: Studies in Evolution,* vol. 2, Sydney: Surrey Beatty & Sons, pp. 559–67.

Groves, C. P. 1987b. On the cuscuses (Marsupialia: Phalangeridae) of the *Phalanger orientalis* group from Indonesian territory. *In* M. Archer (ed.), *Possums and Opossums: Studies in Evolution,* vol. 2. Sydney: Surrey Beatty & Sons, pp. 569–79.

Guillette, L. J., & Hotton, N., 1986. The evolution of mammalian reproductive characteristics in therapsid reptiles. *In* N. Hotton III. et al. (eds.), *The Ecology and Biology of Mammal-like Reptiles.* Washington, D.C.: Smithsonisn Institution Press, pp. 239–50.

Haeckel, E. 1866. *Generelle Morphologie der Organismen,* vol. 2. Berlin: Georg Reimer.

Haight, J. R., & Murray, P. F. 1981. The cranial endocast of the early Miocene marsupial, *Wynyardia bassiana:* an assessment of taxonomic relationships based upon comparisons with recent forms. *Brain, Behav. Evol.* **19**: 17–36.

Haight, J. R., & Nelson, J. E. 1987. A brain that doesn't fit its skull: a comparative study of the brain and endocranium of the koala, *Phascolarctos cinereus* (Marsupialia: Phascolarctidae) *In* M. Archer (ed.), *Possums and Opossums: Studies in Evolution,* vol. 1, Sydney: Surrey Beatty & Sons, pp. 331–52.

Haines. R. W. 1958. Arboreal and terrestrial ancestry of placental mammals. *Quart. Rev. Biol.* **33**: 1–23.

Hall, L. S., & Hughes, R. L. 1987. An evolutionary perspective of structural adaptations for environmental perceptions and utilization by the neonatal marsupials *Trichosurus vulpecula* (Phalangeridae) and *Didelphis virginiana* (Didelphidae). *In* M. Archer (ed.), *Possums and opossums: studies in evolution,* vol. 1. Sydney: Surrey Beatty & Sons, pp. 251–71.

Hall, L. S. 1987. Syndactyly in marsupials – problems and prophecies. *In* M. Archer (ed.), *Possums and Opossums: Studies in Evolution,* vol. 1. Sydney: Surrey Beatty & Sons, pp. 245–55.

Hall-Craggs, E. C. B. 1966. Rotational movements in the foot of *Galago senegalensis.* *Anat. Rec.* **154**: 287–94.

Hallam, A. 1981. Biogeographic relations between the northern and southern continents during the Mesozoic and Cenozoic. *Aufsatze* **70**: 583–95.

Hamilton, W. 1983. Cretaceous and Cenozoic history of the northern continents. Ann. Missouri Bot. Gard. **70**: 440–58.

Harding, H. R. 1987. Interrelationships of the families of the Diprodonta – a view based on spematozoan microstructure. *In* M. Archer (ed.), *Possums and Opossums: Studies in Evolution,* vol. 1. Sydney: Surrey Beatty & Sons, pp. 195–216.

Harding, H. R., Carrick, F. N., & Shorey, C. D. 1987. The affinities of the koala *Phascolarctus cinereus* (Marsupialia: Phascolarctidae) on the basis of sperm ultrastructure and development. *In* M. Archer (ed.), *Possums and Opossums: Studies in Evolution,* vol. 1. Sydney: Surrey Beatty & Sons, pp. 353–64.

Hartenberger, J. et al. 1982. Mammals and stratigraphy: The Paleogene of Europe. *Paleovertebrata* **23**: 1–77

Hayman, D. L., Kirsch, J. A. W., Martin, P. G., & Waller, P. F., 1971. Chromosomal and serological studies of the Caenolestidae and their implications for marsupial evolution. *Nature* **213**: 194–5.

Hecht, M. K., & Edwards, J. L. 1976. The determination of parallel or monophyletic relationships: the proteid salamanders – a test case. *Amer. Natur.* **110**: 653–7.

Hennig, W. 1950. *Grundzuge einer Theorie der phylogenetischen Systematik.* Berlin: Deutsche Zentralverlag.

Hennig, W. 1965. Phylogenetic systematics. *Annual Rev. Ent.* **10**: 97–116.

Hennig, W. 1966. *Phylogenetic systematics.* Urbana: University of Illinois Press.

Hershkovitz, P. 1992. Ankle bones: The Chilean opossum *Dromiciops gliroides* Thomas, and marsupial phylogeny. *Bonner Zoologische Beiträge* **43:** 181–213.

Hildebrand, M. 1985. Digging in quadrupeds. *In* M. Hildebrand et al. (eds.), *Functional Vertebrate Morphology.* Cambridge: Belknap Press of Harvard University Press, pp. 89–109.

Hildebrand, M., Bramble, D. M., Liem, K. F., & Wake, D. B. (eds.) 1985. *Functional Vertebrate Morphology.* Cambridge: Belknap Press of Harvard University Press.

Hill, J. P. 1895. Preliminary note on the occurrence of a placental connection in *Perameles obesula*, and on the foetal membranes of certain macropods. *Proc. Linnean Soc. New South Wales.* **10:** 578.

Hill, J. P., & Hill, W. C. O. 1955. The growth stages of the pouch young of the native cat (*Dasyurus viverrinus*), together with observations on the anatomy of the new-born young. *Trans. Zool. Soc. London* **28:** 349–453.

Hofer, H., 1952, Uber das gegenwartige Bild der evolution der Beuteltiere. *Zool. Jahrb. Abt. Anat. Antog.* **72:** 365–437.

Hoffstetter, M. R. 1970. L'histoire biogeographique des Marsupiaux et la dichotomie Marsupiaux-Placentaires. *C.R. Acad. Sci. (Paris)* **271(D):** 388–91.

Hoffstetter, M. R., 1971. Le peuplement mammalien de l'Amerique du Sud, role des continents austraux comme centres d'origine, de diversification et de dispersion pour certains groupes mammaliens. *An. Acad. Brasil. Cienc.* **43:** 125–44.

Hoffstetter, M. R. 1972. Données et hypothèses concernant l'origine et l'histoire biogéographique des marsupiaux. C.R. Acad. Sci. (Paris) **274(D):** 2635–8.

Hoffstetter, M. R. 1975. Les marsupiaux et l'histoire des mammifères: aspects phylogéniques et chronologiques. *Colloque Inren, C.N.R.S., Evolution des Vertébrés, Paris 1973* **218:** 591–610.

Hooker, J. J. 1986. Mammals from the Bartonian (middle/late Eocene) of the Hampshire Basin, southern England. *Bull. Br. Mus. nat. Hist. (Geol)* **39:** 191–478.

Howell, A. B. 1944. *Speed in Animals.* Chicago: University of Chicago Press.

Hughes, N. F. 1976. Cretaceous paleobotanic problems. *In* C. B. Beck (ed.), *Origin and Early Evolution of Angiosperms.* New York: Columbia University Press, pp. 11–22.

Hughes, R. L. 1965. Comparative morphology of spermatozoa from five marsupial families. *Rep. Aust. J. Zool.* **13:** 533–43.

Hughes, R. L., & Hall, L. S. 1988. Structural adaptations of the newborn marsupial. *In* C. H. Tyndale-Biscoe & P. A. Janssens (eds.). *Developing Marsupial. Models for Biomedical Research.* Berlin: Springer-Verlag, pp. 8–27.

Hughes, R. L., Hall, L. S., Aplin, K. P., & Archer, M. 1987. Organogenesis and fetal membranes in the New Guinea Pen-Tailed possum, *Distoechurus pennatus* (Acrobatidae: Marsupialia) *In* M. Archer (ed.), *Possums and Opossums: Studies in Evolution,* vol. 2. Sydney: Surrey Beatty & Sons, pp. 715–24.

Hunsaker D. II 1977a. *The Biology of Marsupials.* New York: Academic Press.

Hunsaker D. II 1977b. Ecology of New World marsupials. *In* D. Hunsaker II (ed.). *The Biology of Marsupials.* New York: Academic Press, pp. 95–156.

Hunt, R. M. Jr. & Tedford, R. A. 1993. Phylogenetic relationships within the aeluroid Carnivora and implications of their temporal and geographic distribution *In* F. S. Szalay, M. J. Novacek, & M. C. McKenna, (eds.), Mammalian Phylogeny: Placentals. New York: Springer-Verlag, pp. 53–73.

Huxley, T. H. 1880. On the application of the laws of evolution to the arrangement of the Vertebrata and more particularly of the Mammalia. *Proc. Zool. Soc. London* **1880:** 649–62.

Izor, R. J., & Pine, R. H., 1987. Notes on the black shouldered opossum, *Caluromys irrupta. Fieldiana Zool. n. s.* **39:** 117–24.

Jacobs, L. L., Winkler, D. A., & Murry, P. A. 1989. Modern mammal origins: evolutionary grades in the Early Cretaceous of North America. *Proc. Natl. Acad. Sci.* **86:** 4992–5.

Jaeger, J. J., & Martin, M. 1984. African marsupials – vicariance or dispersion? *Nature* **312:** 379.

Janis, C. M., 1990. Why kangaroos (Marsupialia: Macropodidae) are not as hypsodont as ungulates (Eutheria)? *Australian Mammal.* **13:** 49–53.

Jardine, N., & McKenzie, D. 1972. Continental drift and the dispersal and evolution of organisms. *Nature* **235:** 20–24.

Jenkins, F. A., Jr. 1970a. Cynodont postcranial anatomy of the prototherian level of mammalian organization. *Evolution* **24:** 230–52.

Jenkins, F. A., Jr. 1970b. Limb movements in a monotreme (*Tachyglossus aculeatus*): a cineradiographic analysis. *Science* **168:** 1473–5.

Jenkins, F. A., Jr. 1971. Limb posture and locomotion in the Virginia opossum (*Didelphis marsupialis*) and in other non-cursorial mammals. *J. Zool. Lond.* **165:** 303–15.

Jenkins, F. A., Jr. 1973. The functional anatomy and evolution of the mammalian humero-ulnar articulation. *Am. J. Anat.* **137:** 281–98.

Jenkins, F. A., Jr. 1974. The movement of the shoulder in claviculate and aclaviculate mammals. *J. Morph.* **144:** 71–84.

Jenkins, F. A., Jr. 1990. Monotremes and the biology of Mesozoic mammals. *Netherlands J. Zool.* **40:** 5–31.

Jenkins, F. A., Jr. & Krause, D. W. 1983. Adaptations for climbing in North American multituberculates (Mammalia). *Science* **220:** 712–15.

Jenkins, F. A., Jr., & McClearn, D. 1984. Mechanisms of hind foot reversal in climbing mammals. *J. Morph.* **182:** 197–219.

Jenkins, F. A., Jr.. & Parrington, F. A. 1976. The postcranial skeleton of the Triassic mammals *Eozostrodon, Megazostrodon, and Erythrotherium Phil. Trans. Roy. Soc. London* **B273:** 387–431.

Jenkins, F. A., Jr., & Schaff, C. R. 1988. The Early Cretaceous mammal *Gobiconodon* (Mammalia, Triconodonta) from the Cloverly Formation in Montana. *J. Vert. Paleont.* **8:** 1–24.

Jenkins F. A., Jr., Weijs, W. A. 1979. The functional anatomy of the shoulder in the Virginia opossum *(Didelphis virginiana). J. Zool. London* **188:** 379–410.

Jimenez, J., & Rageot, R., 1979, Notas sobre la biologia del monito del monte *(Dromiciops australis* Philippi, 1893). *An. Mus. Hist. Nat. Valparaiso* **12:** 83–8.

Johnson, P. M., & Strahan, R. 1982. A further description of the Musky rat-kangaroo, *Hypsiprymnodon moschatus* Ramsay, 1876 (Marsupialia, Potoroidea), with notes on its biology. *Australian Zool.* **21:** 27–46.

Johnson, J. I., Jr. 1977. Central nervous system of marsupials. *In* D. Hunsaker II (ed.), *The Biology of Marsupials.* New York: Academic Press, pp. 157–278.

Johnson-Murray, J. L. 1987. The comparative myology of the gliding membranes of *Acrobates, Petauroides,* and *Petaurus* contrasted with the cutaneous myology of *Hemibelideus and Pseudocheirus* (Marsupialia: Phalangeridae) and with selected gliding Rodentia (Sciuridae and Anomaluridae). *Aust. J. Zool.* **35:** 101–13.

Keast, A. 1977. Historical biogeography of the marsupials. *In* B. Stonehouse & D. Gilmore (eds.), *The Biology of Marsupials.* New York: Macmillan Press Ltd., pp. 69–95.

Kelt, D. A., & Martinez, D. R. 1989. Notes on the distribution and ecology of two marsupials endemic to the valdividian forests of southern South America. *J. Mamm.* **70:** 220–24.

Kemp, T. S. 1982. *Mammal-like Reptiles and the Origin of Mammals.* London: Academic Press Inc.

Kemp, T. S. 1985. Models of diversity and phylogenetic reconstruction. *Oxford Surv. Evol. Biol.* **2:** 136–58.

Kermack, D. M., & Kermack, K. A. 1984. *The Evolution of Mammalian Characters.* Washington, D. C.: Kapitan Szabo Publishers.

Kermack, K. A., Lees, P. M., & Mussett, F. 1965. *Aegialodon dawsoni,* a new tuberculosectorial

tooth from the lower Wealden. *Roy. Soc. (B) Proc.* **162:** 535–54.

Kielan-Jaworowska, Z. 1975a. Evolution of the therian mammals in the Late Cretaceous of Asia. Part I. Deltatheridiidae. *Paleontologia Polonica* **33:** 103–32.

Kielan-Jaworowska, Z. 1975b. Possible occurrence of marsupial bones in Cretaceous eutherian mammals. *Nature* **255:** 698–9.

Kielan-Jaworowska, Z. 1979. Pelvic structure and nature of reproduction in Multituberculata. *Nature* **277:** 402–3.

Kielan-Jaworowska, Z. 1982. Marsupial-placental dichotomy and paleogeography of Cretaceous Theria. *Proceedings of the First International Meeting on Paleontology,* Venice, 2–4 June 1981, pp. 367–83.

Kielan-Jaworowska, Z. 1992. Interrelationships of Mesozoic mammals. *Historical Biology* **6:** 185–202.

Kielan-Jaworowska, Z., Crompton, A. W., & Jenkins, F. A., Jr. 1987. The origin of egg-laying mammals. *Nature* **326:** 871–3.

Kielan-Jaworowska, Z., & Dashzeveg, D. 1989. Eutherian mammals from the Early Cretaceous of Mongolia. *Zoologica Scripta* **18:** 347–55.

Kielan-Jaworowska, Z., & Nessov, L. A. 1990. On the metatherian nature of the Deltatheroida, a sister group of the Marsupialia. *Lethaia* **23:** 1–10.

Kirsch, J. A. W. 1977. The comparative serology of Marsupialia, and a classification of marsupials. *Aust. J. Zool.* **52**(supplement): 1–152.

Kirsch, J. A. W. 1982. The builder and the bricks: notes toward a philosophy of characters. *In* M. Archer (ed.), *Carnivorous Marsupials,* vol. 2. Sydney: Royal Zoological Society of New South Wales, pp. 587–94.

Kirsch, J. A. W. 1984. Marsupial origins: taxonomic and biological considerations. *In* M. Archer & G. Clayton (eds.), *Vertebrate Zoogeography and Evolution in Australasia – Animals in Space and Time.* Carlisle, New South Wales: Hesperian Press, pp. 627–31.

Kirsch, J. A. W. 1990. Review of *Science as a Process. Am. Sci.* **78:** 161–2.

Kirsch, J. A. W., Dickerman, A. W., Reig, O. A., & Springer, M. S. 1991. DNA hybridization evidence for the Australasian affinity of the American marsupial *Dromiciops australis. Proc. Natl. Acad. Sci.* **88:** 10465–9.

Kirsch, J. A. W., Krajewski, C., Springer, M. S., & Archer, M. 1990b. DNA-DNA hybridisation studies of carnivorous marsupials. II. Relationships among dasyurids (Marsupialia: Dasyuridae). *Aust. J. Zool.* **38:** 673–96.

Kirsch, J. A. W., Springer, M. S., Krajewski, C., Archer, M., Aplin, K., & Dickerman, A. W. 1990a.

DNA/DNA hybridization studies of the carnivorous marsupials. I: The intergeneric relationships of bandicoots (Marsupialia: Perameloidea). *J. Mol. Evol.* **30:** 434–48.

Klima, M. 1987. *Early Development of the Shoulder Girdle and Sterum in Marsupials. Advances in Anatomy Embryology and Cell Biology* **109**. Berlin: Springer Verlag.

Koenigswald, W. von. 1970. *Peratherium* (Marsupialia) im Ober-Oligozän und Miozän von Europa. *Bayerische Acad. der Wissens.* **144:** 1–79.

Koenigswald, W. von, Martin, Th., & Pfretzschner, H. U. 1993. Phylogenetic interpretation of enamel structures in mammalian teeth: possibilities and problems. *In* F. S. Szalay, M. J. Novacek & M. C. McKenna (eds.), *Mammalian phylogeny. Placentals.* New York: Springer-Verlag, pp. 303–14.

Krause, D. W., & Bonaparte, J. F. 1990. The Gondwanatheria, a new suborder of Multituberculata from South America. *J. Vert. Paleont.* **9**(supplement): 31A.

Krause, D. W., & Jenkins, F. A. Jr. 1983. The postcranial skeleton of North American multituberculates. *Harvard Univ. Mus. Comp. Zool. Bull.* **150:** 199–246.

Krause, D. W. & Maas, M. C. 1990. The biogeographic origins of Late Paleocene-Early Eocene mammalian immigrants to the Western Interior of North America. *Geol. Soc. Am. Spec. Paper* **243:** 71–105.

Krishtalka, L., & Stucky, R. K. 1983a. Revision of the Wind River Faunas, early Eocene of Central Wyoming. Marsupialia. *Ann. Carnegie Mus.* **52:** 205–27.

Krishtalka, L., & Stucky, R. K., 1983b. Paleocene and Eocene marsupials of North America. *Ann. Carnegie Mus.* **52:** 229–63.

Krishtalka, L., & Stucky, R. K., 1984, Middle Eocene marsupials from northeastern Utah and the mammalian fauna from Powder Wash. *Ann. Carnegie Mus.* **53:** 31–45.

Kühne, W. G. 1956. *The Liassic Therapsid Oligokyphus.* London: British Museum (Natural History).

Landsmeer, J. M. F. 1979. The extensor assembly in two species of opossum, *Philander opossum* and *Didelphis marsupialis.* J. Morph. **161:** 337–46.

Langdon, J. H. 1986. Functional morphology of the Miocene hominoid foot. *Contrib. Primatol.* **22:** 1–225.

Lauder, G. V. 1990. Functional morphology and systematics: studying functional patterns in a historical context. *Annu. Rev. Ecol. Syst.* **21:** 317–40.

Laws, R. A., & Fastovsky, D. E. 1987. Characters, stratigraphy, and "depopperate" logic: an essay on phylogenetic reconstruction. *PaleoBios* **44:** 1–9.

Lee, A. K., & Carrick, F. N. 1989. Phascolarctidae. *In*

D. W. Walton & G. R. Dyne (eds.), *Mammalia,* Canberra: Australian Government Publishing Service, pp. 740–54.

Lee, A. K., & Cockburn, A. 1985. *Evolutionary Ecology of Marsupials.* Cambridge: Cambridge University Press.

Levinton, J. 1988. *Genetics, Paleontology, and Macroevolution.* Cambridge: Cambridge University Press.

Lewis, O. J. 1962a. The phylogeny of the crural and pedal flexor musculature. *Proc. Zool. Soc. London* **138:** 77–109.

Lewis, O. J. 1962b. The comparative morphology of *M. flexor accessorius* and the associated long flexor tendons. *J. Anat. London* **96:** 321–33.

Lewis, O. J. 1963. The monotreme cruro-pedal flexor musculature. *J. Anat. London* **97:** 55–63.

Lewis, O. J. 1964a. The homologies of the mammalian tarsal bones. *J. Anat. London* **98:** 195–208.

Lewis, O. J. 1964b. The tibialis posterior in the primate foot. *J. Anat. London* **98:** 209–18.

Lewis, O. J. 1964c. The evolution of the long flexor muscles of the leg and foot. *Intl. Rev. Genl. Expl. Zool.* **1:** 165–85.

Lewis, O. J. 1972. The evolution of the hallucial tarsometatarsal joint in the Anthropoidea. *Amer. J. Phys. Anthr.* **37:** 13–34.

Lewis, O. J. 1980a. The joints of the evolving foot. I. The ankle joint. *J. Anat.* **130:** 527–43.

Lewis, O. J. 1980b. The joints of the evolving foot. II. The intrinsic joints. *J. Anat.* **130:** 833–57.

Lewis, O. J. 1980c. The joints of the evolving foot. III. The fossil evidence. *J. Anat.* **131:** 275–98.

Lewis, O. J. 1983. The evolutionary emergence and refinement of the mammalian pattern of foot architecture. *J. Anat.* **137:** 21–45.

Lewis, O. J. 1989. *Functional Morphology of the Evolving Hand and Foot.* Oxford: Oxford University Press.

Lillegraven, J. A. 1969. Latest Cretaceous mammals of upper part of Edmonton Formation of Alberta, Canada, and review of marsupial-placental dichotomy in mammalian evolution, *Univ. Kansas Paleont. Contrib. Vert.* **12:** 1–122.

Lillegraven, J. A. 1974. Biogeographical considerations of the marsupial-placental dichotomy, *Ann. Rev. Ecol. Sysmt.* **5:** 263–83.

Lillegraven, J. A. 1975. Biological considerations of the marsupial-placental dichotomy. *Evol.* **29:** 707–22.

Lillegraven, J. A. 1976. Didelphids (Marsupialia) and *Uintasorex* (?Primates) from later Eocene sediments of San Diego County, California. *Trans., San Diego Soc. Nat. Hist.* **18:** 85–112.

Lillegraven, J. A., 1979, Reproduction in Mesozoic mammals. *In* J. A. Lillegraven, Z. Kielan-Jaworowsak, & W. A. Clemens (eds.), *Mesozoic Mammals. The First Two-thirds of Mammalian History.*

Berkeley: University of California Press, pp. 26–276.

Lillegraven, J. A. 1985. Use of the term "trophoblast" for tissues in therian mammals. *J. Morph.* **183:** 293–9.

Lillegraven, J. A., Kielan-Jaworowska, Z, & Clemens, W. A. 1979a. *Mesozoic Mammals. The First Two-thirds of Mammalian History.* Berkeley: University of California Press.

Lillegraven, J. A., Kraus, M. J., & Bown, T. M. 1979b. Paleogeography of the world of the Mesozoic *In* J. A. Lillegraven, Z. Kielan-Jaworowska, & W. A. Clemens (eds.), *Mesozoic Mammals. The First Two-thirds of Mammalian History.* Berkeley: University of California Press, pp. 277–308.

Lillegraven, J. A., & McKenna, M. C., 1986, Fossil mammals from the "Mesaverde" Formation (Late Cretaceous, Judithian) of the Bighorn and Wind River Basins, Wyoming, with definitions of Late Cretaceous North American land-mammal "ages." *Amer. Mus. Nov.* **2840:** 1–68.

Lillegraven, J. A., Thompson, S. D., McNab, B. K., & Patton, J. L. 1987. The origin of eutherian mammals. *Bio. J. Linnean Soc.* **32:** 281–336.

Lombard, E. R., 1991. Experiment and comprehending the evolution of function. *Amer. Zool.* **31:** 743–56.

Lucas, S. G. 1990. The extinction criterion and the definition of the Class Mammalia. *J. Vert. Paleo.* **9:** 33A

Luckett, W. P. 1988. Early development and homology of the dental lamina and tooth buds in eutherian, metatherian, and prototherian mammals. *Anat. Rec.* **220:** 60A.

Luckett, W. P. 1989. Developmental evidence for dental homologies in the marsupial family Dasyuridae. *Anat. Rec.* **223:** 70A.

Luckett, W. P. 1991. An ontogenetic assessment of dental homologies in therian mammals. *In* F. S. Szalay, M. J. Novacek, & M. C. McKenna (eds.), *Mammalian Phylogeny. Mesozoic Differentiation, Multituberculates, Monotremes, Early Therians, and Marsupials.* New York: Springer Verlag, pp. 182–204.

Luckett, W. P., Bangma, G., & Hong, N. 1991. Ontogenetic evidence for premolar homologies in marsupials. *Anat. Rec.* **229:** 55A–56A.

Luckett, W. P. & Hong, N. 1989. Ontogenetic evidence for tooth replacement in marsupials (Mammalia). *Cell. Differ. Dev.* **27**(suppl.): s45.

Lydekker, R. 1887. *Catalogue of the Fossil Mammalia in the British Museum (Natural History) Cromwell Road, S. W. Part 5. Containing the Group Tillodontia, the Orders Sirenia, Cetacea, Edentata, Marsupialia, Monotremata, and Supplement.* London: Trustees of the British Museum (Natural History).

Lynne, A. G. 1982. Observations on skull growth and eruption of teeth in the marsupial bandicoot *Perameles nasuta* (Marsupialia: Peramelidae). *Aust. Mammal.* **5:** 113–26.

MacConaill, M. A. 1946a. Studies in the mechanics of synovial joints. I. Fundamental principles and diadochal movements. *Irish J. Med. Sci.* **6:** 190–9.

MacConaill, M. A. 1946b. Studies in the mechanics of synovial joints. II. Displacements on articular surfaces and the significance of saddle joints. *Irish J. Med. Sci.* **6:** 223–35.

MacConaill, M. A. 1946c. Studies in the mechanics of synovial joints. III. Hinge joints and the nature of intra-articular displacements. *Irish J. Med. Sci.* **6:** 620–6.

MacConaill, M. A. 1948. The movements of bones and joints. I. Fundamental principles with particular reference to rotational movement. *J. Bone Jt. Surg.* **30(B):** 322–6.

MacConaill, M. A. 1953a. The movements of bones and joints. V. The significance of shape. *J. Bone Jt. Surg.* **35(B):** 290–7.

MacConaill, M. A. 1953b. Close packed position of joints and its practical bearing. *J. Bone Jt. Surg.* **35(B):** 486.

MacFadden, B. J., Campbell, K. D. Jr., Cifelli, R. L., Siles, O., Johnson, N. M., Naeser, C. W., & Zeitler, P. K. 1985. Magnetic polarity stratigraphy and mammalian fauna of the Deseadan (Late Oligocene-Early Miocene) Salla Beds of northern Bolivia. *J. Geol.* **93:** 223–50.

Maclean, N. 1976. *A River Runs Through it and Other Stories.* Chicago: University of Chicago Press.

MacPhee, R. D. E. 1981. Auditory regions of primates and eutherian insectivores: morphology, ontogeny and character analysis. *Contrib. Primatol.* **18:** 1–282.

Mahboubi, M., Ameur, R., Crochet, J. Y., & Jaeger, J. J. 1983. First discovery of a marsupial in Africa. *C. R. Acad. Sci (Paris)* **297(II):** 691–4.

Mahboubi, M., Ameur, R., Crochet, J. – Y., & Jaeger, J.-J. 1986. El Kohol (Saharan Atlas, Algeria): a new Eocene mammal locality in northwestern Africa. Stratigraphical, phylogenetic and paleobiogeographical data. *Palaeont. Abt.* **A 192(1–3):** 15–49.

Maier, W. 1987a. The ontogenetic development of the orbitotemporal region in the skull of *Monodelphis domestica* (Didelphidae: Marsupialia), and the problem of the mammalian alisphenoid. *In* H.-J. Kuhn & U. Zeller (eds.), *Morphogenesis of the Mammalian Skull,* Hamburg: Verlag Paul Harvey, pp. 71–90.

Maier, W. 1987b. The angular process in *Monodelphis domestica* (Didelphidae: Marsupialia) and its relationships to the middle ear: an ontogenetic and evolutionary study. *Gegenbaurs Morphol. Jahrb. (Leipzig)* **133:** 123–61.

Maier, W. 1989a. Morphologische Untersuchungen am Mittelhor der Marsupialia. *Z. Zool. Syst. Evolut.-forsch.* **27:** 149–68.

Maier, W. 1989b. Ala temporalis and alisphenoid in therian mammals. *In* H. Splechtna and H. Hilgers, (eds.), *Trends in Vertebrate Morphology: Proceedings of the 2nd International Symposium on Vertebrate Morphology,* New York: Gustav Fischer Verlag, pp. 398–400.

Maier, W. 1991. Cranial morphology of the therian common ancestor – as suggested by the adaptations of neonate marsupials. *In* F. S. Szalay, M. J. Novacek, & M. C. McKenna (eds.). *Mammalian Phylogeny. Mesozoic Differentiation, Multituberculates, Monotremes, Early Therians, and Marsupials.* New York: Springer-Verlag, pp. 165–181.

Mann, G. 1955. Monito del Monte, *Dromiciops australis philippi. Invest. Zool. Chil.* **2:** 159–166.

Mann, G. 1978. Los pequenos mamiferos de Chile. *Sayana Zoologia* **40:** 1–342.

Marsh, O. C. 1889. Discovery of Cretaceous Mammalia. *Am. J. Sci. (ser. 3)* **38:** 81–92; **38:** 177–80.

Marshall, L. G. 1972. Evolution of the peramelid tarsus. *Rep. Roy. Soc. Vict.* **5:** 51–60.

Marshall, L. G. 1977a. Cladistic analysis of borhyaenoid, dasyuroid, didelphoid, and thylacinid (Marsupialia: Mammalia) affinity. *Syst. Zool.* **26:** 410–25.

Marshall, L. G. 1977b. *Lestodelphys halli. Mammalian Species* **81:** 1–3.

Marshall, L. G. 1978a. *Lutreolina crassicaudata. Mammalian Species* **91:** 1–4.

Marshall, L. G. 1978b. Evolution of the Borhyaenidae, extinct South American predaceous marsupials. *Univ. Calif. Publ. Geol. Sci.* **177:** 1–89.

Marshall, L. G. 1978c. *Dromiciops australis. Mammalian Species* **99:** 1–5.

Marshall, L. G. 1978d. *Glironia venusta. Mammalian Species* **107:** 1–3.

Marshall, L. G. 1978e. *Chironectes minimus. Mammalian Species* **109:** 1–6.

Marshall, L. G. 1979. Evolution of metatherian and eutherian (mammalian) characters: a review based on cladistic methodology. *Zool. J. Linn. Soc.* **66:** 369–410.

Marshall, L. G. 1980a. Systematics of the South American marsupial family Caenolestidae, *Fieldiana: Geol.* **5:** 1–145.

Marshall., L. G. 1980b. Marsupial paleobiogeography *In* L. L. Jacobs (ed.), *Aspects of Vertebrate History.* Flagstaff: Museum of Northern Arizona Press, pp. 345–86.

Marshall, L. G. 1981a. Review of the Hathylacyninae, an extinct subfamily of South American "dog-like" marsupials. *Fieldiana: Geol.* **7:** 1–120.

Marshall, L. G. 1981b. Systematics of the South American marsupial family Microbiotheriidae. *Fieldiana: Geol.* **10:** 1–75.

Marshall, L. G. 1981c. The families and genera of Marsupialia. *Fieldiana: Geol.* **8:** 1–65.

Marshall, L. G. 1982a. Evolution of South American Marsupialia. *In* M. M. Mares & H. H. Genoways (eds.), *Mammalian Biology in South America,* vol. 6, Pittsburgh: University of Pittsburgh Press, pp. 251–72.

Marshall, L. G. 1982b. A new genus of Caroloameghiniinae (Marsupialia: Didelphoidae: Didelphidae) from the Paleocene of Brazil. *J. Mammal.* **63(4):** 709–16.

Marshall, L. G. 1987. Systematics of Itaborian (Middle Paleocene) age "opossum-like" marsupials from the limestone quarries at São José de Itaboraí, Brazil. *In* M. Archer (ed.), *Possums and opossums: studies in evolution.* vol. 1, Sydney: Surrey Beatty & Sons, pp. 91–160.

Marshall, L. G., Case, J. A., & Woodburne, M. O. 1990. Phylogenetic relationships of the families of marsupials. *In* H. H. Genoways (ed.), *Current Mammology,* vol. 2. New York: Plenum Press, pp. 433–505.

Marshall, L. G., & Cifelli, R. L. 1990. Analysis of changing diversity patterns in Cenozoic land mammal age faunas, South America. *Palaeovertebrata* **19:** 160–210.

Marshall, L. G., & Kielan-Jaworowska, Z. 1992. Relationships of the dog-like marsupials, deltatheroidans and early tribosphenic mammals. *Lethaia* **25:** 361–74.

Marshall, L. G., Muizon, C. de, & Sigé, B. 1983a. *Peratherium altiplanense,* un notoungule du Cretacé superieur du Perou. *Palaeovertebrata Montpellier Mem. Extr.* **13:** 145–55.

Marshall, L. G., Muizon, C. de, & Sigé, B. 1983b. Late Cretaceous mammals (Marsupialia) from Bolivia. *Geobios* **16:** 739–45.

Marshall, L. G., Hoffstetter, R., & Pascual, R. 1983. Mammals and stratigraphy: geochronology of the continental mammal-bearing Tertiary of South America. *Pal. Montpellier Mem. Extr.* 1–93.

Marshall, L. G., & Muizon, C. de. 1984. Un nouveau marsupial didelphide (*Itaboraidelphis camposi* nov. gen., nov. sp.) du Paléocene Moyen (Itaboraien) de São José de Itaboraí (Brésil). *C. R. Acad. Sci. Paris* **299:** 1297–1300.

Marshall, L. G., & Muizon, C. de. 1988. The dawn of the Age of Mammals in South America. *Nat. Geog. Research* **4:** 23–55.

Marshall. L. G., & Muizon, C. de. 1992. Atlas photographique (MEB) des Metatheria et de quelques Eutheria du Paléocene inférieur de la formation Santa Lucia à Tiupampa (Bolivie). *Bull. Mus. Natl. Hist. nat., Paris 4e 14, section C* **1:** 63–91.

Maynard-Smith, J., Burian, R., Kauffman, S., Alberch, P., Campbell, J., Goodwin, B., Lande, R., Raup, D., & Wolpert, L. 1985. Developmental constraints and evolution. *Quart. Rev. Biol.* **60:** 265–87.

Mayr, E., & Ashlock, P. D. 1991. *Principles of Systematic Zoology.* New York: McGraw-Hill.

Mayr, E. 1981. Biological classification: toward a synthesis of opposing methodologies. *Science* **214:** 510–16.

McGowran, B. 1973. Rifting and drift of Australia and the migration of mammals. *Science* **180:** 759–61.

McKenna, M. C. 1973. Sweepstakes, filters, corridors, Noah's Arks, and beached Viking funeral ships in paleogeography. *In* D. H. Tarling & S. K. Runcorn *Implications of Continental Drift to the Earth Sciences.* Academic Press, London, pp. 291–304.

McKenna, M. C. 1980a. Eocene paleolatitude, climate, and mammals of Ellesmere Island. *Paleogeogr., Paleoclimatol., Palaeoecol.* **30:** 349–62.

McKenna, M. C. 1980b. Early history and biogeography of South America's extinct land mammals. *In* R. L. Ciohon & A. B. Chiarelli (eds.), *Evolutionary Biology of the New World Monkeys and Continental Drift.* New York: Plenum Press, pp. 43–77.

McKenna, M. C. 1987. Molecular and morphological analysis of high level mammalian interrelationships. *In* C. Patterson (ed.), *Molecules and Morphology in Evolution: Conflict or Compromise?* Cambridge: Cambridge University Press, pp. 53–93.

McManus, J. J. 1974. *Didelphis virginiana, Mammalian Species* **40:** 1–6.

Mellett, J. S. 1980. Function of the inflected mandibular angle in marsupials. *60th Annual Meet. Amer. Soc. Mammal. Abstract No.* 102.

Mickevich, M. F., & Weller, S. J. 1990. Evolutionary character analysis: tracing character change on a cladogram. *Cladistics* **6:** 137–70.

Moors, P. J. 1974. The foeto-maternal relationship and its significance in marsupial reproduction: a unifying hypothesis. *Aust. Mammal.* **1:** 263–7.

Morris, P., & Cobabe, E. 1991. Cuvier meets Watson and Crick: the utility of molecules as classical homologies. *Biol. J. Linn. Soc.* **44:** 307–24.

Mourier, T., Jaillard, E., Laubacher, G., Noblet, C., Pardo, A., Sigé, B., & Taquet, P. 1986. Découverte de restes dinosauriens et mammalien d'âge crétace supérieur à la base des couches rouges du synclinal de Bagua et paléogéographiques concernant la regression finicrétacée, *Bull. Soc. Geol. France* **8:** 171–5.

Muizon, C. de. 1991. La fauna de mammiferos de Tiupampa (Paleoceno inferior, Formation Santa Lucia), Bolivia. *In* R. Suarez-Soruco (ed.), *Fosiles y Facies de Bolivia.* vol. 1, *Vertebrados. Revista Tecnica de YPFB* **12:** 575–624.

Muizon, C. de, Marshall, L. G., & Sigé, B. 1984. The mammal fauna from the El Molino Formation (Late Cretaceous, Maestrichtian) at Tiupampa, southcentral Bolivia. *Bull. Mus. Nat. Hist. Paris* **Series 4, 6, section C(4):** 327–51.

Muizon, C. de, & Marshall, L. G. 1992. *Alcidedorbignya inopinata* (Mammalia: Pantodonta) from the early Paleocene of Bolivia: phylogenetic and paleobiogeographic implications. *J. Paleont.* **66:** 499–520.

Müller, F. 1967. Zum Vergleich der Ontogenesen von *Didelphis virginiana* und *Mesocricetus auratus. Rev. Suisse Zool.* **74:** 607–13.

Müller, F. 1968a. Die transitorischen Verschlusse in der postnatalen Entwicklung der Marsupialia. *Acta Anat.* **71:** 581–624.

Müller, F. 1968b. Zür Phylogenese des sekundaren Kiefergelenks. *Rev. Suisse Zool.* **75:** 373–414.

Müller, F. 1968c. Methodische Gesichtspunkte zum Stadium der Evolution der Säuger-Ontogenesetypen. *Rev. Suisse Zool.* **75:** 630–43.

Müller, F. 1969a. Verhaltnis von Korperentwicklung und Cerebralisation in Ontogenese und Phylogenese der Säuger. Versuch einer Ubersicht des Problems. *Verh. Naturf. Ges. Basel* **80:** 1–31.

Müller, F. 1969b. Zur fruhen Evolution der Säuger-Ontogenesetypen. Versuch einer Rekonstruction aufgrund der Ontogenese-Verhaltnisse bei den Marsupialia. *Acta Anat.* **74:** 297–404.

Müller, F. 1969c. Zur phylogenese des sekundaren Kiefergelenks: Zeugniswert diarthrognather Fossilien im Lichte neuer ontogenetischer Befunde. *Rev. Suisse Zool.* **76:** 710–15.

Müller, F. 1972–1973. Zur stammesgeschichtlichen Veranderung der Eutheria-Ontogenesen. Versuch einer Ubersicht aufgrund vergleichend morphologischer Studien an Marsupialia und Eutheria. *Rev. Suisse Zool.* **79.** Einfuhrung und 1. Teil: Zur Evolution der Geburtgestalt: Gestaltstadien der Eutheria. Fasc. **1,** no. **1,** 1972a, p. 1–97. 2. Teil: Ontogenesetypus und Cerebralisation. Fasc. **2,** no. **17,** 1972b. p. 501–566. 3. Teil: Zeitliche Aspekte in der Evolution der Ontogenesetypen. Fasc. **2,** no. **18,** 1972c. p. 567–611. 4. Spezieller Teil. Fasc. **4,** no. **65,** 1973. p. 1599–1685.

Munson, C. J. 1992. Postcranial descriptions of *Ilaria* and *Ngapakaldia* (Vombatiformes, Marsupialia) and the phylogeny of the vombatiforms based on postcranial morphology. *Univ. California Publ. Zool.* **125:** 1–99.

Murray, J. D., Donnellan, S., McKay, G. M., Rofe, R. H., Baverstock, P. R., Hayman, D. L., & Gelder,

M. 1990. The chromosomes of four genera of possums from the family Petauridae (Marsupialia: Diprotodonta). *Aust. J. Zool.* **38**: 33–9.

Murray, P. 1990a. Primitive marsupial tapirs (*Propalorchestes novaculacephalus* Murray and *P. ponticulus* sp. nov.) from the mid-Miocene of North Australia (Marsupialia: Palorchestidae). *The Beagle, Rec. North. Terr. Mus. Arts Sci.* **7**: 39–51.

Murray, P. 1990b. *Alkwertatherium webbi*, a new zygomaturine genus and species from the Late Miocene Alcoota local fauna, Northern Territory (Marsupialia: Diprotodontidae). *The Beagle, Rec. North. Terr. Mus. Arts Sci.* **7**: 53–80.

Murray, P., & Megirian, D. 1990. Further observations on the morphology of *Wakaleo vanderleueri* (Marsupialia: Thylacoleonidae) from the Mid-Miocene Camfield beds, Northern Territory. The Beagle, Rec. North. Terr. Mus. Arts Sci. **7**: 91–102.

Murray, P., Wells, R., & Plane, M. 1987. The cranium of the Miocene thylacoleonid marsupial *Wakaleo vanderleuri*: click go the shears – a fresh bite at thylacoleonid systematics. *In* M. Archer (ed.), *Possums and Opossums: Studies in Evolution*, vol. 2, Sydney: Surrey Beatty & Sons, pp. 433–66.

Neff, N. A. 1986. A rational basis for *a priori* character weighting. *Syst. Zool.* **35**: 110–23.

Nelson, G. J. 1971. Paraphyly and polyphyly: redefinitions. *Syst. Zool.* **20**: 471–2.

Nelson, G. J. 1992. Why, after all, must it? *Cladistics* **8**: 139–46.

Nelson, G. J., & Platnick, N. 1981. *Systematics and Biogeography – Cladistics and Vicariance.* New York: Columbia University Press.

Nelson, G. J., & Platnick, N. 1984. Systematics and evolution. *In* M. W. Ho & P. T. Saunders (eds.), *Beyond Neo Darwinism: An Introduction to the New Evolutionary Paradigm.* New York: Academic Press, pp. 143–58.

Nessov, L. A. 1985. [New mammals from the Cretaceous of Kizyl-Kum.] *Vestnik Leningradskogo Universiteta* **17**: 8–18.

Nessov, L. A., & Kielan-Jaworowska, Z. 1991. Evolution of the Cretaceous Asian therian mammals. *Contr. Paleontol. Mus., Univ. Oslo* **364**: 51–52.

Nicoll, M. E. 1987. Basal metabolic rates and energetics of reproduction in therian mammals: marsupials and placental compared. *Symp. Zool. Soc. London* **57**: 7–27.

Norris, C. A. 1993. Changes in the composition of the auditory bulla in southern Solomon Islands populations of the Grey cuscus, *Phalanger orientalis breviceps* (Marsupialia, Phalengeridae). *J. Zool. (London)* **107**: 93–106.

Novacek, M. J. 1977. Aspects of the problem of variation, origin, and evolution of the eutherian auditory bulla. *Mammal Rev.* **7**: 131–49.

Novacek, M. J., & Wyss, A. 1986. Origin and transformation of the mammalian stapes. *Contributions to Geology, University of Wyoming, Special Paper* **3**: 35–53.

Nowak, R. M., Paradiso, J. L. 1983. *Walker's Mammals of the World.* Baltimore: John Hopkins University Press.

Ogilby, J. D. 1892. Catalogue of Australian Mammalia, with introductory notes on general mammalogy. *Australian Museum, Sydney, Catalogue* **16**: 1–142.

O'Hara, R. J. 1988. Homage to Clio, or toward an historical philosophy for evolutionary biology. *Syst. Zool.* **37**: 142–55.

Ortiz Jaureguizar, E. & Pascual, R. 1989. South American land-mammal faunas during the Cretaceous-Tertiary transition: evolutionary biogeography. *Contrib. Simp. Cretacico Amer. Lat. Parte* **A**: 231–52.

Osgood, W. H. 1921. A monographic study of the American marsupial *Caenolestes. Field Mus. Nat. Hist., Zool. Ser.* **14**: 1–162.

Owen, R. 1839. Outlines of a classification of the Marsupialia. *Proc. Zool. Soc. London* **7**: 5–19.

Owen, R. 1866. *On the Anatomy of Vertebrates*, vol. II. London: Longmans, Green and Co.

Owen, R. 1868. *On the Anatomy of Vertebrates*, vol. III. London: Longmans, Green and Co.

Owen, R., 1840–1845, *Odontography*, vol I. London: Hippolyte Bailliere.

Packer, D. J., & Sarmiento, E. E. 1984. External and middle ear characteristics of primates, with reference to tarsier-anthropoid affinities. *Am. Mus. Nov.* **2787**: 1–23.

Padykula, H. A., & Taylor, J. M. 1977. Uniqueness of the bandicoot chorioallantoic placenta (Marsupialia: Peramelidae): cytological and evolutionary interpretations. *In* J. H. Calaby, & C. H. Tyndale-Biscoe *Reproduction and Evolution*, Sydney: Australian Academy of Science, pp. 303–11.

Parker, P. 1977. The evolutionary comparison of placental and marsupial patterns of reproduction. *In* B. Stonehouse & D. Gilmore (eds.), *The Biology of Marsupials.* London: Macmillan Press, pp. 273–86.

Parker, T. J. & Haswell, W. A. 1897. *A Text-book of Zoology.* vol. 2. New York: Macmillan and Co.

Pascual, R. 1980. Prepidolopidae, nueva familia de Marsupialia Didelphoidea del Eoceno Sudamericano. *Ameghiniana* **17**: 216–42.

Pascual, R. 1981. Adiciones al conocimiento do *Bonapartherium hinakusijum* (Marsupialia, Bonapartheriidae) del Eoceno temprano del Noroeste Argentino. *Anais II Congr. Latinoamer. Palcont.* **2**: 507–20.

Pascual, R. 1983. Nove dosos marsupiales Paleogenos de la Formacion Pozuelos (Grupo Pastos

Grandes) de la Puna, Salta, Argentina. *Ameghiniana* **20**: 265–80.

Pascual, R. 1986. Evolucion de los vertebrados Cenozoicos: sumario de los principales hitos. *Actas IV Congr. Arg. Paleont. Bioestr.* **2**: 209–18.

Pascual, R., & Bond, M. 1981. Epidolopinae *subfam. nov.* de los Polydolopidae (Marsupialia, Polydolopoidea). *Anais II Congr. Latinoamer. Paleont.* **2**: 479–88.

Pascual, R., & Bond, M. 1986. Evolucion de los marsupiales Cenozoicos de Argentina. *Actas IV Congr. Arg. Paleont. Bioestr.* **2**: 143–50,

Pascual, R., & Carlini, A. A. 1987. A new superfamily in the extensive radiation of South American Paleogene marsupials. *Fieldiana (Zoology), N. S.* **39**: 99–110.

Pascual, R., Carlini, A. A., & de Santis, L. J. M. 1986. Dentition and ways of life in Cenozoic South American rodent-like marsupials. Outstanding examples of convergence. *Mem. Mus. Nat. Hist. Nat., Ser. C (Science de la Terre)*, **53**: 217–26.

Pascual, R., & Ortiz Jaureguizar, E. 1990. Evolving climates and mammal faunas in Cenozoic South America. *J. Hum. Evol.* **19**: 23–60.

Patterson, B. D. 1952. Un nuevo y extraordinario marsupial Deseadiano. *Rev. Mus. Munic, Cien. Nat. Trad. de Mar del Plata* **1**: 39–44.

Patterson, B. D. 1956. Early Cretaceous mammals and the evolution of mammalian molar teeth. *Fieldiana, Geology* **13**: 1–105.

Patterson, B. D. 1958. Affinities of the Patagonian fossil mammal *Necrolestes*. Breviora: Mus. of Comp. Zool., Harvard **94**: 1–14.

Patterson, B. D., & Pascual, R. 1968. The fossil mammal fauna of South America. *Quart. Rev. Biol.* **43**: 409–51.

Patterson, C. 1981. Significance of fossils in determining evolutionary relationships. *Ann. Rev. Ecol. Syst.* **12**: 195–223.

Patterson, C. 1983. Aims and methods in biogeography. *In* R. W. Sims, J. H. Price, & P. E. s. Whalley (eds.), *Evolution, Time, and Space: the Emergence of the Biosphere*, New York: Academic Press, pp. 1–28.

Patterson, C. 1988. The impact of evolutionary theories on systematics. *In* D. L. Hawksworth (ed.), *Prospects in systematics*. Systematics Association Special Volume No. **36**. Oxford: Clarendon Press, pp. 59–91.

Paula Couto, C. de. 1952a. Novos elementos na fauna fossil de São José de Itaboraí. *Bol. Mus. Nacl. Geol.* **12**: 1–6.

Paula Couto, C. de. 1952b. Fossil mammals from the early Cenozoic in Brazil (Marsupialia: Polydolopidae and Borhyaenidae). *Amer. Mus. Novit.* **1559**: 1–26.

Paula Couto, C. de. 1952c. Fossil mammal from the Paleocene of Brazil. *C. R. Congress Geol. Internal, 19th Alger.* **15**: 101–06.

Paula Couto, C. de. 1959. Fossil mammals from the beginning of the Cenozoic in Brazil (Marsupialia: Didelphidae). *Amer. Mus. Novit.* **1567**: 1–26.

Paula Couto, C. de, 1961. Marsupialis fossiles do Paleoceno do Brasil. *Acad. Brasil. Cienc.* **33**: 321–33.

Paula Couto, C. de, 1962, Didelfideos fossiles del Paleoceno de Brasil. *Ciencias Zool.* **8**: 135–66.

Paula Couto, C. de. 1970. News on the fossil marsupials from the Riochican of Brasil. *An. Acad. Brasil. Cienc.* **42**: 19–34.

Platnick, N. 1977. Paraphyletic and polyphyletic groups. *Syst. Zool.* **26**: 195–200.

Pledge, N. S. 1987a. *Phascolarctos maris*, a new species of koala (Marsupialia: Phascolarctidae) from the Early Pliocene of South Australia. *In* M. Archer (ed.), *Possums and Opossums: Studies in Evolution*, vol. 1. Sydney: Surrey Beatty & Sons, pp. 327–30.

Pledge, N. S. 1987b. *Muramura williamsi*, a new genus and species of ?wynyardiid (Marsupialia: Vombatoidea) from the Middle Miocene Etadunna Formation of South Australia. *In* M. Archer (ed.), *Possums and Opossums: Studies in Evolution*, vol. 1. Sydney: Surrey Beatty & Sons, p. 393–400.

Pledge, N. S. 1987c. *Kuterintja ngama*, a new genus and species of vombatiform marsupial from the Medial Miocene Ngama local fauna of South Australia. *In* M. Archer (ed.), *Possums and Opossums: Studies in Evolution*, vol. 2. Sydney: Surrey Beatty & Sons, pp. 419–22.

Pond, C. M. 1977. The significance of lactation in the evolution of mammals. *Evolution* **31**: 177–99.

Presley, R. 1979. The primitive course of the internal carotid artery in mammals. *Acta Anat.* **103**: 238–44.

Presley, R. 1984. The tympanic cavity of Mesozoic mammals. *Third Symposium on Mesozoic Terrestrial Ecosystems, Tübingen*, Tübingen: Attempto Verlag, pp. 189–92.

Presley, R. 1993. Development and the phylogenetic features of the middle ear region. *In* F. S. Szalay, M. J. Novacek, & M. C. McKenna (eds.), *Mammalian phylogeny. Placentals*. New York: Springer-Verlag, pp. 21–29.

Pridmore, P. A. 1992. Trunk movements during locomotion in the marsupial *Monodelphis domestica* (Didelphidae). *J. Morph.* **211**: 137–46.

Prothero, D. R. 1983. The oldest mammalian petrosals from North America. *J. Paleontol.* **57**: 1040–6.

Rauscher, B. 1987. *Priscileo pitikantensis*, a new genus and species of thylacoleonid marsupial (Marsupialia: Thylacoleonidae) from the Miocene Etadunna Formation, South Australia. *In* M. Archer (ed.), *Possums and Opossums: Studies in Evolution.* vol. 2. Sydney: Surrey Beatty & Sons, pp. 423–32.

Raven, P. H., & Axelrod, D. I. 1975. History of the flora and fauna of Latin America. *Amer. Sci.* **63**: 420–9.

Reig, O. A. 1958. Notas para una actualization del conocimeinto de la fauna de la formacion Chapadmalal. II. Amphibia, Reptilia, Aves, Mammalia (Marsupialia, Didelphidae, Borhyaenidae). *Acta Geol. Lilloana* **2**: 255–83.

Reig, O. A. 1981. Teoria del origen y desarrollo de la fauna de mamiferos de America del Sur, *Monographiae Naturae, Mus. Munic. Cien. Natur. "Lorenzo Scaglia"* **1**: 1–162.

Reig, O. A., Kirsch, J. A. W., & Marshall, L. G. 1987. Systematic relationships of the living and neocenozoic American "opossum-like" marsupials (Suborder Didelphimorphia), with comments on the classification of these and of the Cretaceous and Paleogene New World and European metatherians. *In* M. Archer (ed.), *Possums and Opossums: Studies in Evolution*, vol. 1, Sydney: Surrey Beatty & Sons, pp. 1–89.

Reig, O. A., & Simpson, G. G. 1972. *Sparassocynus* (Marsupialia, Didelphidae), a peculiar mammal from the late Cenozoic of Argentina. J. Zool. London **167**: 511–39.

Renfree, M. B. 1993. Ontogeny, genetic control, and phylogeny of female reproduction in monotreme and therian mammals. *In* F. S. Szalay, M. J. Novacek, & M. C. McKenna (eds.), *Mammalian Phylogeny. Mesozoic Differentiation, Multituberculates, Monotremes, Early therians, and Marsupials*, New York: Springer-Verlag, pp. 4–20.

Rich, P. V. 1975. Antarctic dispersal routes, wandering continents, and the origin of Australia's non-passeriform avifauna. *Mem. Nat. Mus. Vic.* **36**: 63–126.

Rich, P. V., & Rich, T. H. V. 1983. The Central American dispersal route: biotic history and paleogeography. *In* D. H. Janzen (ed.), *Costa Rican Natural History*, Chicago: University of Chicago Press, pp. 12–34.

Rich, P. V., VanTets, G. F., & Knight, F. 1985. *Kadimakara: Extinct Vertebrates of Australia.* Victoria: Pioneer Design Studio.

Rich, T. H. V., Archer, R., & Tedford, R. H. 1978. *Raemeotherium yatkolai*, gen. et sp. nov., a primitive diprotodontid from the medial Miocene of South Australia. *Mem. Nat. Mus. Vict.* **39**: 85–91.

Rich, T. H. V., & Rich, P. V. 1987. New specimens of *Ngapakaldia* (Marsupialia: Diprotodontoidea) and taxonomic diversity in medial Miocene palorchestids. *In* M. Archer (ed.), *Possums and Opossums: Studies in Evolution*, vol. 2. Sydney: Surrey Beatty & Sons, pp. 467–76.

Richardson, K. C., Wooller, R. D., & Collins, B. G. 1986. Adaptations to a diet of nectar and pollen in the marsupial *Tarsipes rostratus* (Marsupialia: Tarsipedidae). *J. Zool. (London)* **208**: 285–97.

Ride, W. D. L. 1956. The affinities of *Burramys parvus* Broom a fossil phalangeroid marsupial, *Proc. Zool. Soc. London* **127**: 413–29.

Ride, W. D. L. 1962. On the evolution of Australian marsupials. *In* G. W. Leeper (ed.), *The Evolution of Living Organisms.* Melbourne: Melbourne University Press, pp. 281–306.

Ridley, M. 1986. *Evolution and Classification, the Reformation of Cladism*, London: Longman.

Rieppel, O. C. 1988. *Fundamentals of Comparative Biology.* Basel: Birkhauser Verlag.

Rigby, J. K., Jr., & Wolberg, D. L. 1987. The therian mammalian fauna (Campanian) of Quarry 1, fossil forest study area, San Juan Basin, New Mexico. *Geol. Soc. Amer. Special Paper* **209**: 51–79.

Riggs, E. S. 1934. A new marsupial saber-tooth from the Pliocene of Argentina and its relationships to other South American predaceous marsupials. *Trans. Am. Phil. Soc. n. s.* **24**: 1–32.

Rougier, G. W., Wible, J. R., & Hopson, J. A. 1992. Reconstruction of the cranial vessels in the Early Cretaceous mammal *Vincelestes neuquenianus*: implications for the evolution of the mammalian cranial vascular system. *J. Vert. Paleo.* **12**: 188–216.

Rowe, T. 1987. Definition and diagnosis in the phylogenetic system. *Syst. Zool.* **36**: 208–11.

Rowe, T. 1988. Definition, diagnosis and origin of Mammalia. *J. Vert. Paleont.* **8**: 241–6.

Rubin, C. T., & Lanyon, L. E. 1985. Regulation of bone mass by mechanical strain magnitude. *Calcif. Tissue Int.* **40**: 59–63.

Russel, L. S. 1984. Tertiary mammals of Saskatchewan. VII: Oligocene marsupials. *Life Sci. Contrib.* **139**: 1–31.

Russell, D. E. 1967. Le Paléocène Continental d'Amerique du Nord. *Mem. Mus. Nal. Hist. Nat.* (Ser C) **16**(2): 1–99.

Russell, D. E., & R.-J. Zhai. 1987. The Paleogene of Asia: mammals and stratigraphy. *Mem. Mus. Natl. Hist. Nat.* (Ser. C) **52**:

Russell, E. M. 1982. Patterns of parental care and parental investment in marsupials. *Biol. Rev.* **57**: 423–86.

Russell, E. M., & Renfree, M. B. 1989. Tarsipedidae. *In* D. W. Walton & G. R. Dyne (eds.), *Fauna of Australia.* vol. 1B, *Mammalia.* Canberra: Australian Government Publishing Service, pp. 771–82.

Sahni, A. 1972. The vertebrate fauna of the Judith River Formation, Montana. *Bull. Amer. Mus. Nat. Hist.* **147**: 323–412.

Sanson, G. D. 1978. The evolution and significance of mastication in the Macropodidae. *Aust. Mammal.* **2**: 23–8.

Sanson, G. D. 1980. The morphology and occlusion of molariform cheek teeth in some Macropodinae (Marsupialia, Macropodidae). *Aust. J. Zool.* **28**: 341–65.

Sarich, V. M. 1986. Transferrin, *Trans. Zool. Soc. London* **33**: 165–71.

Sarich, V. M. 1993. Some results of twenty-five years with the blood of mammals. *In* F. S. Szalay, M. J. Novacek, & M. C. McKenna (eds.), *Mammalian Phylogeny. Placentals.* New York: Springer-Verlag, pp. 103–14.

Sarmiento, E. E. 1983. The significance of the heel process in anthropoids. *Int. J. Primat.* **42**: 127–52.

Savage, D. E., & D. E. Russell. 1983. *Mammalian Paleofaunas of the World.* London: Addison-Wesley. Pub. Co.

Schaeffer, B. 1941. The morphological and functional evolution of the tarsus in amphibians and reptiles. *Bull. Amer. Mus. Nat. Hist.* **78**: 395–472.

Schaeffer, B., Hecht, M. K., & Eldredge, N. 1972. Phylogeny and paleontology. *In* T. Dobzhansky, M. K. Hecht, & W. C. Steere (eds.). *Evolutionary Biology*, vol. 6. New York: Appleton-Century-Crofts, pp. 31–46.

Segall, W. 1969. The middle ear region of *Dromiciops. Acta Anat.* **72**: 489–501.

Seligsohn, D. 1977. Analysis of species-specific molar adaptations in strepsirhine primates. *Contrib. Primatol.* **11**: 1–116.

Setoguchi, T. 1975. Paleontology and geology of the Badwater Creek area, central Wyoming. 11. Late Eocene marsupials. *Ann. Carnegie Mus.* **45**: 263–75.

Setoguchi, T. 1978. Paleontology and geology of the Badwater Creek area, central Wyoming. 16. The Cedar Ridge local fauna (Late Oligocene). *Bull. Carnegie Mus. Nat. Hist.* **9**: 1–61.

Sharman, G. B. 1970. Reproductive physiology of marsupials. *Science* **167**: 1221–8.

Sharman, G. B. 1982. Karyotypic similarities between *Dromiciops australis* (Microbiotheriidae, Marsupialia) and some Australian marsupials. *In* M. Archer (ed.). *Carnivorous Marsupials.* Sydney: Royal Zoological Society of New South Wales, pp. 711–14.

Shubin, N. H. & Alberch, P. 1986. A morphogenetic approach to the origin and basic organization of the tetrapod limb. *Evol. Biol.* **20**: 319–87.

Sigé, B. 1972a. Les Didelphoidea de Laguna Umayo Formation Vilquechico, Crétacé supérieur, Per-ou), et le peuplement marsupial d'Amérique du Sud. *C. R. Acad. Sci. Paris* **273(D)**: 2479–81.

Sigé, B. 1972b. La faunule de mammifères du Crétacé supérieur de Laguna Umayo (Andes peruviennes), *Bull. Mus. Nat. Hist. Nat.* **3**: 375–405.

Sigogneau-Russell, D. 1992. *Hypomylos phelizoni* nov. gen. nov. sp., une étape précoce de l'évolution de la molaire tribosphenique (Crétacé basal du Maroc). *Geobios* **25**: 389–93.

Sigogneau-Russell, D., Dashzeveg, D., & Russell, D. E. 1992. Further data on *Prokennalestes* (Mammalia, Eutheria *inc. sed.*) from the Early Cretaceous of Mongolia. *Zool. Scripta* **21**: 205–9.

Simons, E. L., & Bown, T. M. 1984. A new species of *Peratherium* (Didelphidae; Polyprotodonta): the first African marsupial. *J. Mammal.* **65**: 539–48.

Simpson, G. G. 1927. Mesozoic Mammalia. 8. Genera of Lance mammals other than multituberculates *Am. J. Sci.* (Ser. 5) **14**: 121–30.

Simpson, G. G. 1928. *A catalogue of the Mesozoic Mammalia in the Geological Department of the British Museum.* London: British Museum (Natural History).

Simpson, G. G. 1929,. American Mesozoic Mammalia. *Mem. Peabody Mus. Yale Univ.* **3**: 1–171.

Simpson, G. G. 1935a. Descriptions of the oldest known South American mammals, from the Rio Chico Formation. *Amer. Mus. Novit.* **793**: 1–25.

Simpson, G. G. 1935b. Note on the classification of recent and fossil opossums. *J. Mammal.* **16**: 134–7.

Simpson, G. G. 1944. *Tempo and Mode in Evolution.* New York: Columbia University Press.

Simpson, G. G. 1945. The principles of classification and a classification of mammals. *Bull. Amer. Mus. Nat. Hist.* **85**: 1–350.

Simpson, G. G. 1947. Holarctic mammalian faunas and continental relationships during the Cenozoic. *Bull. Geol. Soc. Amer.* **58**: 613–88.

Simpson, G. G. 1950. History of the fauna of Latin America. *Amer. Scientist* **38**: 361–89.

Simpson, G. G. 1953. *Evolution and Geography, an Essay on Historical Biogeography with Special Reference to Mammals.* Eugene, Oregon: Oregon State System of Higher Education.

Simpson, G. G. 1961a. *Principles of Animal Taxonomy* New York: Columbia University Press.

Simpson, G. G. 1961b. Historical zoogeography of Australian mammals. *Evolution* **15**: 431–46.

Simpson, G. G. 1965. *The Geography of Evolution: Collected Essays.* New York: Chilton Books.

Simpson, G. G. 1970a. The Argyrolagidae, extinct South American marsupials. *Bull. Mus. Comp. Zool.* **139**: 1–80.

Simpson, G. G. 1970b. Addition to knowledge of the

Argyrolagidae (Mammalia, Marsupialia) from the Late Cenozoic of Argentina. *Breviora Mus. Comp. Zool.* **361**: 1–9.

Simpson, G. G. 1970c. Addition to knowledge of *Groeberia* (Mammalia, Marsupialia) from the mid-Cenozoic of Argentina. *Breviora Mus. Comp. Zool.* **362**: 1–17.

Simpson, G. G. 1971. The evolution of marsupials in South America. *An. Acad. Brasil Cienc.* **43**(suplemento): 103–18.

Simpson, G. G. 1978. Early mammals in South America: fact, controversy, and mystery. *Proc. Amer. Phil. Soc.* **122**: 318–28.

Sinclair, W. J. 1906. Mammalia of the Santa Cruz beds: Marsupialia. *Rept. Princeton Univ. Exped. Patagonia* **4**: 333–460.

Slaughter, B. H. 1965. A new therian from the Lower Cretaceous (Albian) of Texas. *Postilla* **93**: 1–18.

Slaughter, B. H. 1968. Earliest known marsupials. *Science* **162**: 254–5.

Smith, A. P. 1980. *The Diet and Ecology Leadbeaters Possum and the Sugar Glider.* Unpublished Ph.D. Thesis. Melbourne: Monash University.

Smith, A. P., & Hume, I., eds. 1984. *Possums and Gliders.* Sydney: Surrey Beatty & Sons.

Smith, A. P., & Lee, A. K. 1984. The evolution of strategies for survival and reproduction in possums and gliders. *In* A. P. Smith & I. D. Hume (eds.), Sidney: Surrey Beatty & Sons, pp. 17–33.

Smith, J. 1612. *A Map of Virginia.* Oxford: J. Barnes.

Smith, J. M. B. 1972. Southern biogeography on the basis of continental drift: a review. *Aust. Mammal.* **1**: 213–29.

Smith, M. J. 1980. *The Marsupials of Australia.* Melbourne: Landsdowne Editions.

Sober, E. 1989. *Reconstructing the Past: Parsimony, Evolution, and Inference.* Cambridge: MIT Press.

Springer, M. S. 1987. Lower molars of *Litokoala* (Marsupials: Phascolarctidae) and their bearing on phascolarctid evolution. *In* M. Archer (ed.), *Possums and Opossums: Studies in Evolution,* vol. 1. Sydney: Surrey Beatty & Sons, pp. 319–25.

Springer, M. S. & Kirsch, J. A. W. 1989. Rates of single-copy DNA evolution in phalangeriform marsupials. *Mol. Biol. Evol.* **6**: 331–41.

Springer, M. S. & Kirsch, J. A. W. 1991. DNA hybridization, the compression effect, and the radiation of diprotodontian marsupials. *Syst. Zool.* **40**: 131–51.

Springer, M. S., Kirsch, J. A. W., Aplin, K., & Flannery, T. 1990. DNA hybridization, cladistics, and the phylogeny of phalangeriforme marsupials. *J. Mol. Evol.* **30**: 298–311.

Springer, M. S., & Woodburne, M. O. 1989. The distribution of some basicranial characters within the Marsupialia and a phylogeny of the Phalangeriformes. *J. Vert. Paleont.* **9**: 210–21.

Stein, B. R. 1981. Comparative limb myology of two opossums, *Didelphis* and *Chironectes. J. Morph.* **169**: 113–40.

Stephenson, N. G. 1967. Phylogenetic trends and speciation among wombats. *Rep. Aust. J. Zool.* **15**: 873–80.

Stirling, E. C. 1891a. Communications to the Society. *Proc. Zool. Soc. London* **1891**: 327–9.

Stirling, E. C. 1891b. Description of a new genus and species of marsupial. *"Notoryctes typhlops." Trans. Roy. Soc. South Aust.* **14**: 154–87.

Stirling, E. C. 1891c. Further notes on the habits and anatomy of *Notoryctes typhlops. Trans. Roy. Soc. South Aust.* **14**: 283–91.

Stirling, E. C. 1894. Supplemental note on the osteology of *Notoryctes typhlops. Trans. Roy. Soc. South Aust.* **18**: 1–2.

Stirton, R. A., Woodburne, M. O. & Plane, M. D. 1967. Tertiary Diprotodontidae from Australia and New Guinea, *Dept. Nat. Devel. Bur. Min. Resour., Geol. Geoph.* **85**: 1–160.

Stonehouse, B, & Gilmore, D. 1977. *The Biology of Marsupials.* New York: Macmillan Press, Ltd.

Storch, G. 1990. The Eocene mammalian fauna from Messel – a paleobiogeographical jigsaw puzzle. *In* G. Peters & R. Hutterer (eds.), *Vertebrates in the Tropics.* Bonn: Museum Alexander Koenig, pp. 23–32.

Strasser, E., & Dagosto, M. 1988. *The Primate Postcranial Skeleton.* San Diego: Academic Press.

Streilein, K. E. 1982. Behavior, ecology, and distribution of south American marsupials. *Spec. Pub. Pymatuning Lab. Ecol.* **6**: 231–50.

Stucky, R. K. 1984. Revision of the Wind River faunas, Early Eocene of central Wyoming. 5. Geology and biostratigraphy of the upper part of the Wind River Formation, Northeastern Wind River Basin. *Ann. Carnegie Mus.* **53**: 231–94.

Stucky, R. K. 1990. Evolution of land mammal diversity in North America during the Cenozoic. *Curr. Mammalogy* **2**: 375–432.

Stuessy, T. F. 1987. Explicit approaches for evolutionary classification. *Syst. Bot.* **12**: 251–62.

Sues, H.-D. 1986. Locomotion and body form in early therapsids (Dinocephalia, Gorgonopsia, and Therocephalia). *In* N. Hotton III et al. (eds.), *The Ecology and Biology of Mammal-like Reptiles* Washington D. C: Smithsonian Institution Press, pp. 61–70.

Sweet, G. 1904. Contributions to our knowledge of the anatomy of *Notoryctes typhlops,* Stirling. I & II. *Proc. Roy. Soc. Vic.* (n.s.) **17**: 76–111.

Sweet, G. 1906. Contributions to our knowledge of the anatomy of *Notoryctes typhlops,* Stirling. III. The eye. *Quart. J. Mic. Sci.* **50**: 547–71.

Sweet, G. 1907. Contributions to our knowledge of

the anatomy of *Notoryctes typhlops*, Stirling. IV & V. The skin, hair, and reproductive organs of *Notoryctes*. *Quart. J. Mic. Sci.* **51:** 325–44.

Szalay, F. S. 1975a. Phylogeny of primate higher taxa: The basicranial evidence. *In* W. P. Luckett & F. S. Szalay (eds.), *Phylogeny of the Primates, a Multidisciplinary Approach.* New York: Plenum Press, pp. 91–125.

Szalay, F. S. 1975b. Phylogeny, adaptations, and dispersal of the tarsiiform primates. *In* W. P. Luckett & F. S. Szalay (eds.), *Phylogeny of the Primates, a Multidisciplinary approach.* New York: Plenum Press, pp. 91–125.

Szalay, F. S. 1976. Systematics of the Omomyidae (Tarsiiformes, Primates) – taxonomy, phylogeny, and adaptations. *Bull. Amer. Mus. Nat. Hist.* **156:** 157–450.

Szalay, F. S. 1977a. Ancestors, descendants, sister groups, and testing of phylogenetic hypotheses. *Syst. Zool* **26:** 12–18.

Szalay, F. S. 1977b. Phylogenetic relationships and a classification of the eutherian Mammalia. *In* M. Hecht, P. C. Goody & B. M. Hecht (eds.), *Major Patterns in Vertebrate Evolution: Macroevolutionary Trends and Their Implications in Vertebrate Phylogeny.* New York: Plenum Press, pp. 315–74.

Szalay, F. S. 1981. Functional analysis and the practice of the phylogenetic method as reflected by some mammalian studies. *Amer. Zool.* **21:** 37–45.

Szalay, F. S. 1982. A new appraisal of marsupial phylogeny and classification. *In* M. Archer (ed.), *Carnivorous Marsupials.* Sydney: Royal Zoological Society of New South Wales, pp. 621–40.

Szalay, F. S. 1984. Arboreality: is it homologous in metatherian and eutherian mammals? *Evol. Biol.* **18:** 215–58.

Szalay, F. S. 1985. Rodent and lagomorph morphotype adaptations, origins, and relationships: some postcranial attributes analyzed. *In* W. P. Luckett & J. L. Hartenberger (eds.), *Evolutionary Relationships Among Rodents – a Multidisciplinary Analysis,* New York: Plenum Press, pp. 83–157.

Szalay, F. S. 1993a. Pedal evolution of mammals in the Mesozoic. *In* F. S. Szalay, M. J. Novacek, & M. C. McKenna (eds.), *Mesozoic Differentiation, Multituberculates, Monotremes, Early Therians, and Marsupials.* New York: Springer-Verlag, pp. 108–28.

Szalay, F. S. 1993b. Metatherian taxon phylogeny: evidence and interpretation from the cranioskeletal system. *In* F. S. Szalay, M. J. Novacek, & M. C. McKenna (eds.), *Mammalian Phylogeny. Mesozoic Differentiation, Multituberculates, Monotremes, Early Therians, and Marsupials.* New York: Springer-Verlag, pp. 216–41.

Szalay, F. S. 1993c. Species concepts: the tested, the untestable, and the redundant. *In* L. Martin & W. Kimbel (eds.), *Species, Species Concepts, and Primate Evolution.* New York: Plenum Press, pp. 21–41.

Szalay, F. S., & Bock, W. J. 1991. Evolutionary theory and systematics: relationships between process and patterns. *Z. Zool. Syst. Evolut.-forsch.* **29:** 1–39.

Szalay, F. S., and Costello, R. K. 1991. Evolution of permanent estrus displays in hominids. *J. Human Evol.* **20:** 439–464.

Szalay, F. S. & Dagosto, M. 1988. Evolution of hallucial grasping in the primates. *J. Human Evol.* **17:** 1–33.

Szalay, F. S., & Decker, R. L. 1974. Origins, evolution, and function of the tarsus in late Cretaceous eutherians and Paleocene primates. *In* F. A. Jenkins, Jr. (ed.), *Primate Locomotion,* New York: Academic Press, pp. 223–54.

Szalay, F. S., & Drawhorn, G. 1980. Evolution and diversification of the Archonta in an arboreal milieu. *In* W. P. Luckett (ed.), *Comparative Biology and Evolutionary Relationships of Tree Shrews.* New York: Plenum Press, pp. 133–69.

Szalay, F. S., & Langdon, J. H. 1986. The foot of *Oreopithecus*, an evolutionary assessment. *J. Human Evol.* **15:** 585–621.

Szalay, F. S., & Li, C.-K. 1986. Middle Paleocene euprimate from southern China and the distribution of the primates in the Paleogene. *J. Human Evol.,* **15:** 387–97.

Szalay, F. S., & McKenna, M. C. 1971. Beginning of the Age of Mammals in Asia: The Late Paleocene Gashato Fauna, Mongolia. *Bull. Amer. Mus. Nat. Hist.* **144:** 271–317.

Szalay, F. S., Novacek, M. J., & McKenna, M. C. (eds.). 1993a. *Mammalian Phylogeny. Mesozoic Differentiation, Multituberculates, Monotremes, Early Therians, and Marsupials.* New York: Springer-Verlag.

Szalay, F. S., Novacek, M. J., & McKenna, M. C. (eds.). 1993b. *Mammalian Phylogeny. Placentals.* New York: Springer-Verlag.

Szalay, F. S., Rosenberger, A. L., & Dagosto, M. 1987. Diagnosis and differentiation of the order Primates. *Yrbk. Phys. Anthropol.* **30:** 75–105.

Takahashi, F. 1974. Variaçao morfoligica de Incisivos em Didelfideos (Marsupialia – Didelphinae). *An. Acad. Brasil. Cienc.* **46:** 413–16.

Tandler, J. 1899. Zur vergleichenden Anatomie der Kopffarterien bein den Mammalia, *Denkschriften Akad Wissenschaften, Wien, Mathematisch-Naturwissenschaftliche Klasse* **67:** 677–784.

Tassy, P. 1988. The classification of Proboscidea: how many cladistic classifications? *Cladistics* **4:** 43–57.

Tate, G. H. H. 1947. Results of the Archbold Expeditions. 56. On the anatomy and classification of the Dasyuridae (Marsupialia). *Bull. Amer. Mus. Nat. Hist.* **88**: 97–156.

Tate, G. H. H. 1948. Results of the Archbold Expeditions. 60. Studies in the Peramelidae (Marsupialia), *Bull. Amer. Mus. Nat. Hist.* **92**: 313–46.

Tedford, R. H. 1966. A review of the macropodid genus *Sthenurus, Univ. Calif. Publ. Geol. Sci.,* **57**: 1–72.

Tedford, R. H. 1967. The fossil Macropodidae from Lake Menindee, New South Wales. *Univ. Calif. Publ. Geol. Sci.* **64**: 1–156.

Tedford, R. H. 1973. The diprotodonts of Lake Callabonna. *Aust. Nat. Hist.* 349–54.

Tedford, R. H. 1974. Marsupials and the new paleogeography. *Paleont. Miner.* **21**: 109–26.

Tedford, R. H. 1985. Late Miocene turn-over of the Australian mammal fauna. *South African J. Sci.* **81**: 262–3.

Tedford, R. H., Archer, M., Bartholmai, A., Plane, M., Pled., N. S., Rich, T., Rich, P., & Wells, R. T. 1977. The discovery of Miocene vertebrates, Lake Frome area, South Australia, B. M. R. *J. Aust. Geol. Geoph.* **2**: 53–7.

Tedford, R. H., Banks, M. R., Kem, N. R., McDougall, I., & Sutherland, F. L. 1975. Recognition of the oldest known fossil marsupials from Australia. Nature **255**: 141–2.

Tedford, R. H., & Woodburne, M. O. 1987. The Ilariidae, a new form of vombatiform marsupial from Miocene strata of South Australia and an evaluation of the homology of the molar cusps of the Diprotodontia. *In* M. Archer (ed.), *Possums and Opossums: Studies in Evolution,* vol. 2, Sydney: Surrey Beatty & Sons, pp. 401–18.

Temple-Smith, P. 1987. Sperm structure and marsupial phylogeny. *In* M. Archer (ed.), *Possums and Opossums: Studies in Evolution,* vol. 1. Sydney: Surrey Beatty & Sons, pp. 171–93.

Thomas, O. 1887. On the homologies and succession of the teeth in the Dasyuridae, with an attempt to trace the history of the evolution of mammalian teeth in general. *Phil. Trans. Roy. Soc. London* **B1887**: 443–62.

Thomas, O. 1888. Marsupialia. Catalogue of the Marsupialia and Monotremata in the collection of the British Museum (Natural History), London, 1888, pp. 1–401.

Thompson, S. D. 1987. Body size, duration of parental care, and the intrinsic rate of natural increase in eutherian and metatherian mammals. *Oecologia* **71**: 201–9.

Thomson, K. S. 1988. *Morphogenesis and Evolution.* New York: Oxford University Press.

Trofimov, B. A., & Szalay, F. S. 1993. New group of Asiatic marsupials (Order Asiadelphia) from the Late Cretaceous of Mongolia. *J. Vert. Paleont.* **13**(supplement 3): 60A.

Trouessart, R. L. 1879. Catalogue des mammifères vivants et fossiles . . . Insectivores. *Rev. Mag. Zool., Paris, ser.* **3** **7**: 219–85.

Trouessart, R. L. 1898. Catalogus Mammalian tam viventium quam fossilium. Berlin.

Turnbull, W. D., Rich, T. V. H., & Lundelius E. L., Jr. 1987a. The petaurids (Marsupialia: Petauridae) of the Early Pliocene Hamilton local fauna, Southwestern Victoria. *In* M. Archer (ed.), *Possums and Opossums: Studies in Evolution,* vol. 2, Sydney: Surrey Beatty & Sons, pp. 629–38.

Turnbull, W. D., Rich, T. H. V., & Lundelius E. L., Jr. 1987b. Pseudocheirids (Marsupialia: Pseudocheiridae) of the Early Pliocene Hamilton local fauna, Southwestern Australia. *In* M. Archer (ed.), *Possums and Opossums: Studies in Evolution,* vol. 2. Sydney: Surrey Beatty & Sons, pp. 689–713.

Turnbull, W. D., Rich, T. H. V., & Lundelius E. L., Jr. 1987c. Burramyids (Marsupialia: Burramyidae) of the Early Pliocene Hamilton local fauna, Southwestern Australia. *In* M. Archer (ed.), *Possums and Opossums: Studies in Evolution,* vol. 2. Sydney: Surrey Beatty & Sons, pp. 729–39.

Turnbull, W. D., & Segall, W. 1984. The ear region of the marsupial sabertooth, *Thylacosmilus:* influence of the sabertooth lifestyle upon it, and convergence with placental sabertooths. *J. Morph.* **181**: 239–70.

Tyndale-Biscoe, C. H. 1973. *Life of Marsupials.* New York: American Elsevier Publishing Co.

Tyndale-Biscoe, C. H., & Janssens, P. A. 1988. *The Developing Marsupial. Models for Biomedical Research.* Berlin: Springer-Verlag.

Tyndale-Biscoe, C. H., & Renfree, M. 1987. *Reproductive Physiology of Marsupials.* Cambridge: Cambridge University Press.

Van Valen, L. 1974. *Deltatheridium* and marsupials. *Nature* **248**: 165–6.

Van Valen, L. M. 1978. The beginning of the Age of Mammals. *Evol. Theory* **4**: 445–80.

Van Valen, L. M. 1988a. Faunas of a southern world. Nature **333**: 113.

Van Valen, L. M. 1988b. Paleocene dinosaurs or Cretaceous ungulates in South America? *Evolutionary Monographs* **10**: 1–79.

Van Valen, L. M. 1988c. Species, sets, and the derivative nature of philosophy. *Biol. Philos.* **3**: 49–66.

Van Valen, L. M. 1989. Metascience. *Evol. Theory* **9**: 99–103.

Veevers, J. J. 1991. Mid-Cretaceous tectonic climax, Late Cretaceous recovery, and Cainozoic relaxation in the Australian region. *Geol. Soc. Australia, Inc., Spec. Pub.* **18**: 1–14.

Villarroel, C., & Marshall, L. G. 1983. Two new late Tertiary marsupials (Hathlyacyninae and Sparassocyninae) from the Bolivian Altiplano. *J. Paleont.* **57**: 1061–6.

Villarroel, C., & Marshall, L. G. 1988. A new argyrolagoid (Mammalia: Marsupialia) from the Middle Miocene of Bolivia. *J. Paleont.* **62**: 463–7.

Warburton, F. E. 1967. The purposes of classifications. *Syst. Zool.* **16**: 241–5.

Waterhouse, G. R. 1938. *Catalogue of the Mammalia Preserved in the Museum of the Zoological Society.* 2nd ed., London: Richard and John E. Taylor.

Waters, B. T. 1967. *Osteology of Diprotodon.* M.A. Thesis, Berkeley: University of California.

Webster, D. B., & Webster, M. 1980. Morphological adaptations of the ear in the rodent family Heteromyidae. *Amer. Zool.* **20**: 247–54.

Wells, R. T. 1978. Field observations of the hairy-nosed wombat [*Lasiorhinus latifrons* (Owen)], *Aust. Wildlife Res.* **5**: 299–303.

Wells, R. T. 1989. Vombatidae. *In* D. W. Walton & G. R. Dyne (eds.), *Fauna of Australia.* vol. 1B, *Mammalia.* Canberra: Australian Government Publishing Service, pp. 755–67.

Wells, R. T., Horton, D. R., & Rogers, P. 1982. *Thylacoleo carnifex* Owen marsupial carnivore? *In* M. Archer (ed.), *Carnivorous Marsupials*, vol. 2. Sydney: Royal Zoological Society, New South Wales, pp. 573–86.

Wells, R. T., & Nichol, B., 1977. On the manus and pes of *Thylacoleo carnifex* Owen (Marsupialia). *Trans. Royal Soc. S. Australia* **101**: 139–46.

Westerman, M., & Edwards, D. 1991. The relationships of *Dromiciops australis* to other marsupials: data from DNA-DNA hybridisation studies. *Aust. J. Zool.* **39**: 123–30.

White, T. D. 1989. An analysis of epipubic bone function in mammals using scaling theory. *J. Theor. Biol.* **139**: 342–57.

Wible, J. R. 1984. *The Ontogeny and Phylogeny of the Mammalian Cranial Arterial Pattern.* Ph.D. Thesis, Durham: Duke University.

Wible, J. R. 1986. Transformations in the extracranial course of the internal carotid artery in mammalian phylogeny. *J. Vert. Paleont.* **6**: 313–25.

Wible, J. R. 1989. Differences in the petrosal bone among the major groups of Mesozoic and recent mammals. *J. Vert. Paleont.* **9**(supplement): 44A.

Wible, J. R. 1990. Petrosals of Late Cretaceous marsupials from North America, and a cladistic analysis of the petrosal in therian mammals. *J. Vert. Paleont.* **10**: 183–205.

Wible, J. R. 1991. Origin of Mammalia: the craniodental evidence reexamined. *J. Vert. Paleont.* **11**: 1–28.

Wiley, E. 0. 1981. *Phylogenetics: the Theory and Practice of Phylogenetic Systematics.* John New York: John Wiley and Sons.

Williams, C. A. 1986. An oceanwide view of Palaeogene plate tectonic events. *Paleogreogr. Paleoclimatol. Paleoecol.* **57**: 3–25

Wing, S. L., & Tiffney, B. H. 1987. The reciprocal interaction of angiosperm evolution and tetrapod herbivory. *Rev. Paleobot. Palynol.* **50**: 179–210.

Winge, H. 1893. Jordfunde og nulevende Pungdyr (Marsupialia) fra Lagoa Santa, Minas Geraes, Brasilien. *Med. Udsigt over Pungdyrenes Slaegtskab. E. Mus. Lundii* **11**: 1–149.

Winge, H. 1923. *Pattedyr-Slaegter. I. Monotremata, Marsupialia, Insectivora, Chiroptera, Edentata.* Copenhagen.

Winge, H. 1941. (written between 1887–1918; English translation of Winge, 1923). *The Interrelationships of the Mammalian Genera.* Copenhagen: C. A. Reitzels Forlag.

Wolff, R. G. 1987. Late Oligocene-Early Miocene didelphid marsupials from Florida. *J. Vert. Paleont.* **7**: 29A.

Wood, C. B., & Clemens, W. A. 1992. Enamel evidence for the phylogenetic position of the Stagodontidae (Marsupialia, Cretaceous), *J. Vert. Paleo.* **12**: 60A-61A.

Wood, H. E., II. 1924. The position of the Sparassodonts: with notes on the relationships and history of the Marsupialia. *Bull. Amer. Mus. Nat. Hist.* **51**: 77–101.

Wood-Jones, F. W. 1923–1925. *The Mammals of South Australia.* parts 1–3, Photolitho Reprint, 1968. Adelaide: A. B. James, Government Printer.

Wood-Jones, F. W. 1949. The study of a generalized marsupial (*Dasycercus cristicauda* Krefft). *Rep. Trans. Zool. Soc.* **26**: 409–501.

Woodburne, M. 0. 1984. *Wakiewakie lawsoni*, a new genus and species of Potoroinae (Marsupialia: Macropodidae) of medial Miocene age, South Australia. *J. Paleont.* **58**: 1062–73.

Woodburne, M. O. 1987. The Ektopodontidae, an unusual family of Neogene phalangeroid marsupials. *In* M. Archer (ed.), *Possums and Opossums: Studies in Evolution.* vol. 2. Sydney: Surrey Beatty & Sons, pp. 603–6.

Woodburne, M. O., & Clemens, W. A. 1986. *Revision of the Ectopodontidae (Mammalia; Marsupialia; Phalangeroidea).* Berkeley: University of California.

Woodburne, M.O., Pledge, N.S., & Archer, M. 1987b. The Miralinidae, a new family and two new species of phalangeroid marsupials from Miocene strata of South Australia. *In* M. Archer (ed.), *Possums and Opossums: Studies in Evolution*, vol. 2, Sydney: Surrey Beatty & Sons, pp. 581–602.

Woodburne, M. O., Tedford, R. H., & Archer, M. 1987b. New Miocene ringtail possums (Marsupialia: Pseudocheiridae) from South Australia. *In* M. Archer (ed.), *Possums and Opossums: Studies in Evolution*, vol. 2, Sydney: Surrey Beatty & Sons, pp. 639–79.

Woodburne, M. O., Tedford, R. H., Archer, M., & Pledge, M. S. 1987c. *Madakoala*, a new genus and two species of Miocene koalas (Marsupialia: Phascolarctidae) from South Australia, and a new species of *Perikoala*. *In* M. Archer (ed.), *Possums and Opossums: Studies in Evolution*, vol. 1. Sydney: Surrey Beatty & Sons, pp. 293–317.

Woodburne, M. O., & Zinsmeister, W. J. 1984a. Fossil land mammal from Antarctica. Science **218**: 284–286.

Woodburne, M. O., & Zinsmeister, W. J. 1984b. The first land mammal from Antarctica and its biogeographic implications. J. Paleont. **58**: 913–48.

Wyss, A. R., & de Queiroz, K. 1984. Phylogenetic methods and the early history of amniotes. *J. Vert. Paleo.* **4**: 604–8.

Zeller, U. 1989. Die Entwicklung und Morphologie des Schadels von *Ornithorhynchus anatianus* (Mammalia: Prototheria: Monotremata). *Abh. Senckenberg. Naturforsch. Ges.* **545**: 1–188.

Zinsmeister, W. J. 1982. Review of the Upper Cretaceous-Lower Tertiary sequence on Seymour Island, Antarctica. *J. Geol. Soc. London* **139**: 776–86.

# Index